# Livestock Management: Breeding and Raising Healthy Animals

# Livestock Management: Breeding and Raising Healthy Animals

Edited by Roger Greer

SYRAWOOD
PUBLISHING HOUSE
New York

Published by Syrawood Publishing House,
750 Third Avenue, 9th Floor,
New York, NY 10017, USA
www.syrawoodpublishinghouse.com

**Livestock Management: Breeding and Raising Healthy Animals**
Edited by Roger Greer



International Standard Book Number: 978-1-68286-584-2 (Hardback)



**Cataloging-in-Publication Data**

Livestock management : breeding and raising healthy animals / edited by Roger Greer.
p. cm.
Includes bibliographical references and index.
ISBN 978-1-68286-584-2
1. Livestock. 2. Animal breeding. 3. Animal health. I. Greer, Roger.
SF61 .L58 2018
636--dc23

Printed in China.

# TABLE OF CONTENTS

# PREFACE

The main aim of this book is to educate learners and enhance their research focus by presenting diverse topics covering this vast field. This is an advanced book which compiles significant studies by distinguished experts in the area of analysis. This book addresses successive solutions to the challenges arising in the area of application, along with it; the book provides scope for future developments.

Livestock refers to those domesticated animals which are raised for commercial purposes such as providing food or fiber or even serve as labor. These animals are mostly reared in an agricultural setting. Cows, chickens, goats, turkeys are the some of the most popularly raised livestock animals across the globe. This book aims to shed light on some of the most unexplored aspects of breeding livestock and their management. It will serve as a valuable source of reference for graduate and post-graduate students.

It was a great honour to edit this book, though there were challenges, as it involved a lot of communication and networking between me and the editorial team. However, the end result was this all-inclusive book covering diverse themes in the field.

Finally, it is important to acknowledge the efforts of the contributors for their excellent chapters, through which a wide variety of issues have been addressed. I would also like to thank my colleagues for their valuable feedback during the making of this book.

**Editor**

# 1

# Estimating Diagnostic Test Accuracies for *Brachyspira hyodysenteriae* Accounting for the Complexities of Population Structure in Food Animals

**Sonja Hartnack[1]*, Christina Nathues[2], Heiko Nathues[3,4], Elisabeth Grosse Beilage[5], Fraser Iain Lewis[1]**

**1** Section of Epidemiology, Vetsuisse Faculty, Zurich, Switzerland, **2** Veterinary Public Health Institute, Vetsuisse Faculty, Liebefeld, Switzerland, **3** Department of Production and Population Health, Royal Veterinary College London, London, United Kingdom, **4** Department of Clinical Veterinary Medicine, Vetsuisse Faculty, Zurich, Switzerland, **5** Field Station for Epidemiology, University of Veterinary Medicine Hannover, Foundation, Bakum, Germany

**Abstract**

For swine dysentery, which is caused by *Brachyspira hyodysenteriae* infection and is an economically important disease in intensive pig production systems worldwide, a perfect or error-free diagnostic test ("gold standard") is not available. In the absence of a gold standard, Bayesian latent class modelling is a well-established methodology for robust diagnostic test evaluation. In contrast to risk factor studies in food animals, where adjustment for within group correlations is both usual and required for good statistical practice, diagnostic test evaluation studies rarely take such clustering aspects into account, which can result in misleading results. The aim of the present study was to estimate test accuracies of a PCR originally designed for use as a confirmatory test, displaying a high diagnostic specificity, and cultural examination for *B. hyodysenteriae*. This estimation was conducted based on results of 239 samples from 103 herds originating from routine diagnostic sampling. Using Bayesian latent class modelling comprising of a hierarchical beta-binomial approach (which allowed prevalence across individual herds to vary as herd level random effect), robust estimates for the sensitivities of PCR and culture, as well as for the specificity of PCR, were obtained. The estimated diagnostic sensitivity of PCR (95% CI) and culture were 73.2% (62.3; 82.9) and 88.6% (74.9; 99.3), respectively. The estimated specificity of the PCR was 96.2% (90.9; 99.8). For test evaluation studies, a Bayesian latent class approach is well suited for addressing the considerable complexities of population structure in food animals.

**Editor:** Baochuan Lin, Naval Research Laboratory, United States of America

**Funding:** The authors have no support or funding to report. Data originate from routine sampling.

**Competing Interests:** The authors have declared that no competing interests exist.

* E-mail: sonja.hartnack@access.uzh.ch

## Introduction

### Currently used diagnostic tests for *Brachyspira hyodysenteriae*

Swine dysentery (SD) is caused by infection with the bacterium *Brachyspira hyodysenteriae* and imposes major economic losses on intensive pig production systems worldwide [1]. The affected animals, usually growing or finishing pigs, show non-specific clinical symptoms, such as diarrhoea, reduced growth and decreased food conversion which precludes a diagnosis based solely on pathognomonic signs. More specific signs, i.e. large amounts of mucus and often flecks of blood in the faeces, can become visible during the course of the disease, but this is not guaranteed. Finally, sub-clinical disease can occur. The accurate diagnosis of sub-clinical, as well as clinical disease, is a prerequisite for providing efficient treatment procedures, and to support prevention programmes in the field.

Animals in intensive pig production systems are reared in distinct population groups, i.e. herds or farms, and within each group may share multiple genetic and immunologic characteristics. Animals, therefore, from the same population group will likely be more similar to each other than to animals from other groups. In epidemiological terms this may result in over-dispersion, that is, the level of variation in disease prevalence (say) across multiple population groups (e.g. multiple herds or farms) may be far in excess of that allowed by commonly used statistical methods for estimating diagnostic accuracy. Failure to account for such population clustering may lead to highly unreliable results. Moreover, such adjustment for excess variation is common place in other types of epidemiological analyses, e.g. risk factor studies, but as yet rare in diagnostic test evaluation despite identical issues being present.

The development of methods for dealing with the complexity of population structures found in food animal production, such as clustered or hierarchical data, has been described as the single most important advancement [2] for animal health researchers. To ensure success, animal disease prevention programs, typically organized at a national level, need to consider the inherent complexity of the population structure, rather than simply treating the population as an assemblage of individual and unrelated or independent animals. With regard to data from diagnostic test studies, this also holds true for laboratories which receive samples for large-scale testing from national animal populations. The necessity to consider variation of test accuracies at farm level, where test errors might be clustered, has been stated by Donald and co-workers [3], [4]. The ultimate diagnosis of SD should be based on the isolation of the causative agent, but this procedure is extremely laborious, time-consuming and expensive, as the bacteria are fastidious and grow slowly. Another disadvantage of isolation is the fact that both confirmation and differentiation between *B. hyodysenteriae*, and other less- or non-pathogenic *Brachyspira* spp., are dependent upon colony morphology, pattern and intensity of haemolysis and other growth characteristics, which might be very similar among different subspecies [5]. Finally, isolation often fails to recover organisms when infected pigs have been submitted to antimicrobial therapy prior to sampling. This inevitably leads to a decreased diagnostic sensitivity of cultural isolation, and as such, cultural isolation cannot be assumed to be a perfect gold standard for diagnostic test evaluation studies.

Polymerase chain reaction (PCR) is increasingly used for diagnostic purposes, and produces more specific and sensitive results in the detection of *B. hyodysenteriae* when compared with other diagnostic methods [6], [7]. Others found PCR for *B. hyodysenteriae* being less sensitive than culture [8], underlining that none of these tests perform as a perfect gold standard, that is, completely error-free (display 100% diagnostic sensitivity and specificity). A number of false negative test results due to a less than perfect sensitivity, and thus falsely classified animals may contribute to the spread of a contagious agent. False positive test results may lead to welfare and ethical issues if healthy animals are subsequently culled. Reliable information about test accuracies, including information about how well a diagnostic test performs in a defined animal population, is crucial for efficient prevention and control programs.

Generally speaking, for diagnostic test evaluation, the difficulty of obtaining robust estimates for diagnostic sensitivity and specificity by comparing a new diagnostic test with an established standard test, which presumably is not a perfect gold standard, is due to the uncertainty associated with results from the standard test. If for example a highly sensitive PCR is compared with culture as a standard test, which might be highly specific but less sensitive, a number of samples might be correctly classified as being positive by the PCR, but – due to a lower sensitivity – be falsely classified as negative by culture. In this case, if the diagnostic specificity of the PCR is derived by determining the quotient of PCR negative test results divided by the number of culture negative results, then the diagnostic specificity of the PCR might be biased, being too low.

## Using latent class models to estimate test accuracies

Statistical methods have been developed to deal with situations of test evaluation where no gold standard test is readily available. Seminal work was done by Hui and Walter – [9] after whom the Hui-Walter paradigm was named (reviewed by [10] and [11] with regard to the underlying assumptions with regard to the characteristics of diagnostic test data). The term "latent" here refers to the fact that the dichotomized test results contain, only latently, statistical information about the parameters of interest i.e. sensitivities, specificities, and prevalences. These parameters are not directly observed but can be extracted from the data via an appropriate statistical model. There are many extensions to the original Hui-Walter paradigm. Most notably, models for the inclusion of conditional dependencies between diagnostic test accuracies have been developed [12–18]. Furthermore, latent class models are now recommended in the Standard Operating Procedures (SOP) of the OIE for validation and certification of diagnostic assays [19].

The key caveat in using latent class or variable models is that the parameter estimates obtained will be, to a greater or lesser extent, dependent upon the model used to extract the latent information from the observed data. Hence, particular attention must be paid when selecting an appropriate model formulation – one that best matches the type and structure of the study data. In respect of food animal production the complexity of the population structure from which study data is sampled is obviously of major importance. To date in veterinary medicine, diagnostic test accuracy studies rarely perform data analysis with regard to potential clustering, with only a few but notable exceptions [20–23].

The aim of the present study was to estimate test accuracies of a PCR and cultural testing for *B. hyodysenteriae* on 239 samples from 103 herds originating from routine diagnostic sampling at a laboratory in Bakum, Germany. Given the number of different population groups included in the study data the resulting accuracy estimates could then be considered as broadly applicable to the national swine population. The PCR has been designed to be used as a confirmatory test displaying a high diagnostic specificity.

# Material and Methods

## Samples

Faecal samples (n = 239) originating from 103 herds that had been submitted to the Field Station for Epidemiology in Bakum for the purpose of routine diagnostics due to diarrhoea in pigs were included in this study. These samples arrived in the laboratory during 2007 and 2011, and they were all examined for *B. hyodysenteriae* by both, PCR and cultural testing.

Upon arrival in the laboratory faecal samples were carefully homogenized and each time one aliquot of 200 mg was transferred to a microtube for storage at −20°C. From the remaining faeces a swab was obtained, transferred into a container with transport medium and subsequently shipped to the Institute of Bacteriology, University of Veterinary Medicine Hannover for cultural testing.

PCR was exactly performed as previously described [24]. This PCR had been designed to be used as a confirmatory test displaying a high diagnostic specificity in detecting *B. hyodysenteriae*, *B. pilosicoli* and/or *Lawsonia intracellularis*. For cultural isolation swabs were streaked on Columbia blood agar plates (Oxoid, Germany) and on TSA-plates (Trypticase Soy Agar) containing 0.1% yeast extract, 6 µg/ml vancomycin, 6.25 µg/ml colistin, 12.5 µg/µl rifampicin, 15.25 µg/ml spiramycin, 200 µg/ml spectinomycin, and 5% bovine blood. These were incubated anaerobically (AnaeroJar with AnaeroGen, Oxoid) at 42°C for six consecutive days. Suspected growth of *Brachyspira spp.* was confirmed and species were identified using nox-RFLP as described previously [25].

The raw data of the 239 faecal samples tested in parallel by PCR and culture are shown in table 1.

## Statistical analysis

A Bayesian latent class analysis was performed to obtain robust estimates for test accuracies of PCR and culture. It was assumed, as is standard practice, that sensitivities and specificities of the tests are constant across all animals. With the exception of culture, where the specificity was set to 1, uninformative ("flat") prior distributions on key model parameters were utilized (prior beta distribution (1,1) for the sensitivities of PCR and culture, for the specificity of PCR, and prior uniform distribution (−0.01,0.01) for conditional test dependence between the two sensitivities). A sensitivity analysis for three different beta priors was conducted utilising Betabuster (http://www.epi.ucdavis.edu/diagnostictests/betabuster.html) to define the two shape parameters. The priors

**Table 1.** Dichotomized diagnostic test results of 239 porcine faecal samples tested in parallel by culture and PCR for *Brachyspira hyodysenteriae.*

| | | Culture | |
|---|---|---|---|
| | | + | − |
| PCR | + | 52 | 18 |
| | − | 11 | 158 |

have been chosen as input information "to be 95% sure that the parameter of interest is greater than 0.3, 0.4 and 0.5 and the mode is at 0.6, 0.75 and 0.95" respectively. The model parameters were estimated using a hierarchical beta-binomial approach which allowed the prevalences for individual population groups (farms) to vary as a farm level random effect. In other words, we allow different farms to have potentially very different prevalences subject to the condition that the overall distribution of farms prevalences should follow a beta probability distribution (which is very flexible in terms of shape). Gamma distributed priors of (5, 0.01) were used to model the shape parameters of this beta distribution (describing the population level prevalence – i.e. the distribution of within farm prevalences across all farms). Three different gamma priors were tested in a sensitivity analysis.

The effect of conditional dependency between test results, specifically two-way covariance between test sensitivities, was assessed by formal model selection using DIC (deviance information criterion) as the goodness-of-fit criterion, with lower values indicating a better model fit [26]. All models were fitted using Markov chain Monte Carlo estimation through JAGS software (Just Another Gibbs Sampler) (http://mcmc-jags.sourceforge.net/) version 3.3.0. Good standard practice was followed doing MCMC running four chains independently for 500,000 iterations after a burn-in of 25,000 iterations, and thinning of (10), thus resulting 10,000 values to derive the posterior means. Code for the final model is presented in code S1. The output was analysed with the package coda [27] within the software R (http://www.r-project.org/) version 2.15.2. Gelman-Rubin statistics was used to assess mixing [28].

## Results

Two latent class models, one with and one without a covariance term have been tested to obtain estimates for the parameters of interest. Based on DIC as a decision criterion (without test sensitivity covariance: 453.6 and with tests sensitivity covariance: 454.3), the model without any covariance structure between tests was chosen due to the principle of parsimony. Additionally, the inclusion of the covariance yielded similar values for the parameters of interest. Posterior means and corresponding 95% credibility intervals for all parameters of interest are shown in table 2. Posterior estimates for the sensitivity of the PCR and culture, and the specificity, are shown in figures 1 and 2. Four chains were run from different starting points (figures S1–S3), and after burn-in converged to a common distribution. Additionally, convergence was also confirmed by the Gelman-Rubin statistics. Utilising an uninformative beta prior for the specificity of culture instead of fixing it –based on expert opinion equal to 1- led to virtually unchanged estimates for the sensitivity of culture and the specificity of PCR, but to an increased sensitivity of PCR of 83% with a wider 95% CI (67.2,98.7). The estimate of the specificity culture was 94.6% with a 95% CI (90.9,99.7). A sensitivity analysis of different beta priors is presented in table S1 indicating a minor influence of the different beta priors compared to non-informative priors. A sensitivity analysis for the gamma priors modelling the shape parameters suggested that the model would not converge (MCMC sampler failure) with wider priors e.g. (0.01,0.01) or (0.0001,0.0001), but did so with (0.5,0.0005) (results are present in table S2).

## Discussion/Conclusions

By applying a Bayesian latent class approach to data from large-scale routine laboratory testing, robust estimates for diagnostic test accuracies were obtained. The benefit of the random effects model becomes evident when un-informative priors for the sensitivities and specificities of PCR and culture are considered. In the case of a classical Hui-Walter model Bayesian inferences based on a single population with two tests will be imprecise e.g. a non-identifiable situation [10]. A reason for the sampler failing in the sensitivity analysis of the gamma priors is potentially explained by the variable number of samples tested per herd (a small number of herds had very few animals).

As – strictly speaking – diagnostic test accuracies are population dependent characteristics [29] (as opposed to some fixed intrinsic value related only to the diagnostic test being used), there is a need to evaluate diagnostics tests in the population of interest. The Field Station for Epidemiology in Bakum receives samples from a significant number of swine herds in North-West Germany, thus representing the population in the centre of the German pig industry. The obtained estimates are not necessarily directly applicable to another population e.g. a surveillance program aiming at identifying subclinical cases shedding presumably low levels of spirochetes or to a population consisting only of severe clinical cases of SD.

This approach allowed estimation of the specificity of the PCR without the need for a gold standard reference test, which is not available for *B. hyodysenteriae.* The reasonably good sensitivity and high specificity of the PCR compared to cultural testing, together with financial advantages makes the PCR a valuable diagnostic tool in which large numbers of samples can be tested for *B. hyodysenteriae.* From the perspective of a routine laboratory, the greatest advantage of the PCR is to enable the rapid implementation of therapeutic measures and preventive programmes in pig farms intended to control SD. Using standard isolation methods, the detection of *B. hyodysenteriae* usually requires 5–7 days. Conversely, PCR results can be provided in 2 days at the longest. Beside this advantage for PCR, this method is lacking the possibility to detect novel pathogenic spirochetes such as those recently identified in North America associated with SD [30]. In case of suspicion of e.g. weakly beta hemolytic *Brachyspira spp.* culture could be the method of choice.

The main novelty of the Bayesian latent class approach utilized here lies in its use of random effects to model the complex population structure of food animals, i.e. by allowing different farms to have very different prevalences (resulting from within farm clustering of disease). This is practically important, as without such flexibility the results of any analyses (i.e. estimates of sensitivity and specificities) are likely to be highly misleading.

In risk factor studies, accounting for clustering is considered good, indeed arguably essential, statistical practice. Diagnostic test evaluation studies, for tests applied at a regional or national level, have identical issues which should be accounted for given the complexities of population structure in food animal production.

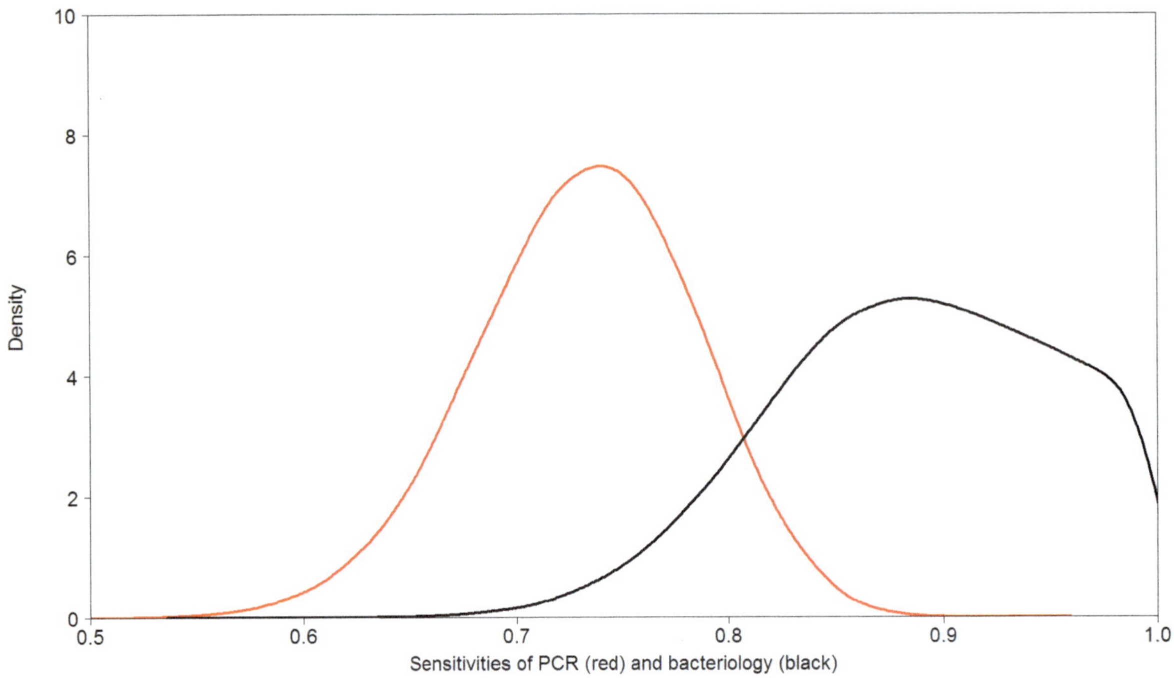


**Figure 1. Diagnostic sensitivities of bacteriology and PCR.** Diagnostic sensitivities culture and PCR for detection of *Brachyspira hyodysenteriae* infection in pigs, estimated by latent class analysis. Results are given in the form of posterior density distributions, and show that culture has a higher sensitivity than PCR (red =PCR, black = bacteriology).

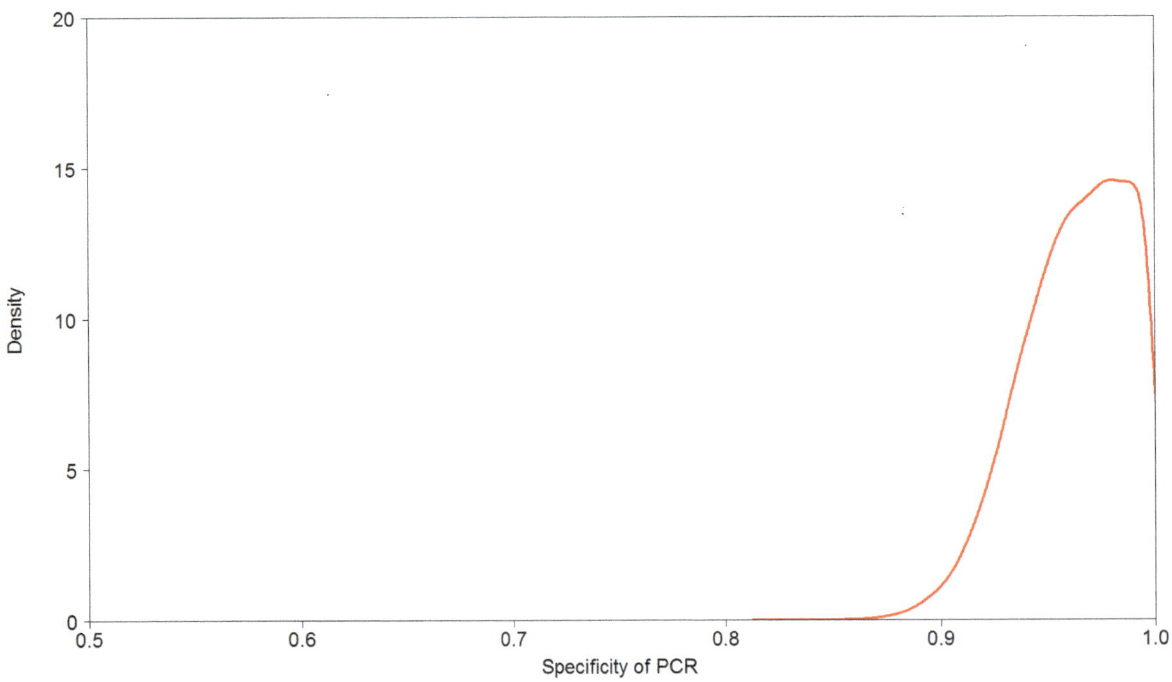


**Figure 2. Diagnostic specificity of PCR.** Diagnostic specificity of PCR for detection of *Brachyspira hyodysenteriae* infection in pigs, estimated by latent class analysis. Results are given in the form of a posterior density distribution.

**Table 2.** Estimates (posterior means) of diagnostic test accuracies for the diagnosis of *Brachyspira hyodysenteriae* by culture and PCR and their corresponding 95% credibility intervals resulting from a Bayesian latent class model.

| Culture | | PCR | |
|---|---|---|---|
| **Sensitivity** | **Specificity** | **Sensitivity** | **Specificity** |
| (95% CI) | | (95% CI) | (95% CI) |
| 88.6 | 1* | 73.2 | 96.2 |
| (74.9;99.3) | | (62.3;82.9) | (90.9;99.8) |

*Specificity of culture was fixed to 1.

Bayesian latent class approaches can be readily utilized for this purpose.

## Supporting Information

**Figure S1 Four simulated chains for the sensitivity of PCR.** To ensure numerical robustness of the estimated diagnostic sensitivity of PCR, multiple simulations were performed using a Bayesian latent class model. The posterior density estimates of the diagnostic sensitivities were compared across all four simulations to check that the results were similar. It can be seen that the curves (densities) are almost identical, therefore providing strong confidence in the results.

**Figure S2 Four simulated chains for the sensitivity of culture.** To ensure numerical robustness of the estimated diagnostic sensitivity of culture, multiple simulations were performed using a Bayesian latent class model. The posterior density estimates of the diagnostic sensitivities were compared across all four simulations to check that the results were similar. It can be seen that the curves (densities) are almost identical, therefore providing strong confidence in the results.

**Figure S3 Four simulated chains for the specificity of PCR.** To ensure numerical robustness of the estimated diagnostic specificity of PCR, multiple simulations were performed using a Bayesian latent class model. The posterior density estimates of the diagnostic sensitivities were compared across all four simulations to check that the results were similar. It can be seen that the curves (densities) are almost identical, therefore providing strong confidence in the results.

**Table S1 Sensitivity analysis for different beta priors for test accuracies.**

**Table S2 Sensitivity analysis with different priors for the gamma priors.**

**Code S1 Code for the final model.**

## Acknowledgments

We are thankful to Mechthild Busemann, Mechthild Sieve and Regina Tegeler from the Field Station for Epidemiology for technical assistance and assistance with microbiological procedures, respectively.

## Author Contributions

Conceived and designed the experiments: EGB HN CN. Analyzed the data: SH FL. Wrote the paper: SH FL CN HN.

## References


1. Alvarez-Ordonez A, Martinez-Lobo FJ, Arguello H, Carvajal A, Rubio P (2013) Swine Dysentery: Aetiology, Pathogenicity, Determinants of Transmission and the Fight against the Disease. Int J Environ Res Public Health 10, 1927–1947.
2. Dohoo IR (2008) Quantitative epidemiology: Progress and challenges. Prev Vet Med 86: 260–269.
3. Greiner M, Gardner IA (2000) Epidemiologic issues in the validation of veterinary diagnostic tests. Prev Vet Med 45: 3–22.
4. Donald AW, Gardner IA, Wiggins AD (1994) Cut-off points for aggregate herd testing in the presence of disease clustering and correlation of test errors. Prev Vet Med 19: 167–187.
5. Elder RO, Duhamel GE, Schafer RW, Mathiesen MR, Ramanathan M (1994) Rapid Detection of Serpulina hyodysenteriae in diagnostic specimens by PCR. J Clin Microbiol 32: 1497–1502.
6. La T, Phillips ND, Hampson DJ (2003) Development of a duplex PCR assay for detection of Brachyspira hyodysenteriae and Brachyspira pilosicoli in pig feces. J Clin Microbiol 41: 3372–3375.
7. Willems H, Reiner G (2010) A multiplex real-time PCR for the simultaneous detection and quantitation of Brachyspira hyodysenteriae, Brachyspira pilosicoli and Lawsonia intracellularis in pig faeces. Berl Munch Tierarztl Wochenschr 123: 205–209.
8. Rasback T, Fellstrom C, Gunnarsson A, Aspan A (2006) Comparison of culture and biochemical tests with PCR for detection of Brachyspira hyodysenteriae and Brachyspira pilosicoli. J Microbiol Methods 66: 347–353.
9. Hui SL, Walter SD (1980) Estimating the error rates of diagnostic tests. Biometrics 36: 167–171.
10. Johnson WO, Gastwirth JL, Pearson LM (2001) Screening without a "Gold Standard": The Hui-Walter paradigm revisited. Am J Epidemiol 153: 921–924.
11. Toft N, Jorgensen E, Hojsgaard S (2005) Diagnosing diagnostic tests: evaluating the assumptions underlying the estimation of sensitivity and specificity in the absence of a gold standard. Prev Vet Med 68: 19–33.
12. Vacek PM (1985) The effect of conditional dependence on the evaluation of diagnostic-tests. Biometrics 41: 959–968.
13. Gardner IA, Stryhn H, Lind P, Collins MT (2000) Conditional dependence between tests affects the diagnosis and surveillance of animal diseases. Prev Vet Med 45: 107–122.
14. Berkvens D, Speybroeck N, Praet N, Adel A, Lesaffre E (2006) Estimating disease prevalence in a Bayesian framework using probabilistic constraints. Epidemiology 17: 145–153.
15. Branscum AJ, Gardner IA, Johnson WO (2004) Bayesian modeling of animal- and herd-level prevalences. Prev Vet Med 66: 101–112.
16. Branscum AJ, Gardner IA, Johnson WO (2005) Estimation of diagnostic-test sensitivity and specificity through Bayesian modeling. Prev Vet Med 68: 145–163.
17. Enoe C, Georgiadis MP, Johnson WO (2000) Estimation of sensitivity and specificity of diagnostic tests and disease prevalence when the true disease state is unknown. Prev Vet Med 45: 61–81.
18. Georgiadis MP, Johnson WO, Gardner IA, Singh R (2003) Correlation-adjusted estimation of sensitivity and specificity of two diagnostic tests. J R Stat Soc Ser C Appl Stat 52: 63–76.
19. OIE (2009) Standard operating procedure (SOP) for validation and certification of diagnostic assays. Available: http://web.oie.int/VCDA/eng/en_fichier_SOP.pdf. Accessed 10. January 2014

20. Suess EA, Gardner IA, Johnson WO (2002) Hierarchical Bayesian model for prevalence inferences and determination of a country's status for an animal pathogen. Prev Vet Med 55: 155–171.
21. Hanson T, Johnson WO, Gardner IA (2003) Hierarchical models for estimating herd prevalence and test accuracy in the absence of a gold standard. Journal of Agricultural, Biological, and Environmental Statistics 8: 223–239.
22. Lewis FI, Brulisauer F, Cousens C, McKendrick IJ, Gunn GJ (2011) Diagnostic accuracy of PCR for Jaagsiekte sheep retrovirus using field data from 125 Scottish sheep flocks. Vet J 187: 104–108.
23. Lewis FI, Torgerson PR (2012) A tutorial in estimating the prevalence of disease in humans and animals in the absence of a gold standard diagnostic. Emerg Themes Epidemiol 9: 9.
24. Nathues H, Oliveira CJB, Wurm M, Grosse Beilage E, Givisiez PEN (2007) Simultaneous detection of Brachyspira hyodysenteriae, Brachyspira pilosicoli and Lawsonia intracellularis in porcine faeces and tissue samples by multiplex-PCR J Vet Med A Physiol Pathol Clin Med 54: 532–538.
25. Rohde J, Rothkamp A, Gerlach GF (2002) Differentiation of porcine Brachyspira species by a novel nox PCR-based restriction fragment length polymorphism analysis. J Clin Microbiol 40: 2598–2600.
26. Spiegelhalter DJ, Best NG, Carlin BR, van der Linde A (2002) Bayesian measures of model complexity and fit. J R Stat Soc Ser B Stat Methodol 64: 583–616.
27. Plummer M, Best N, Cowles KVK (2006) CODA: Convergence Diagnosis and Output Analysis for MCMC. R news 6: 7–11.
28. Gelman A, Rubin D (1992) Inference from Iterative Simulation using Multiple Sequences. Stat Sci 7: 457–511.
29. Sackett DL, Haynes RB (2002) Evidence base of clinical diagnosis - The architecture of diagnostic research. BMJ 324: 539–541.
30. Burrough ER, Strait EL, Kinyon JM, Bower LP, Madson DM, et al. (2012) Comparative virulence of clinical Brachyspira spp. isolates in inoculated pigs. J Vet Diagn Invest 24: 1927–1947.

2

# Consumption of Antimicrobials in Pigs, Veal Calves, and Broilers in The Netherlands: Quantitative Results of Nationwide Collection of Data in 2011

**Marian E. H. Bos[1]*, Femke J. Taverne[2], Ingeborg M. van Geijlswijk[2], Johan W. Mouton[3], Dik J. Mevius[4,5], Dick J. J. Heederik[1], on behalf of the Netherlands Veterinary Medicines Authority (SDa)**

**1** Division of Environmental Epidemiology, Institute for Risk Assessment Sciences, Utrecht University, Utrecht, The Netherlands, **2** Pharmacy, Faculty of Veterinary Medicine, Utrecht University, Utrecht, The Netherlands, **3** Department of Medical Microbiology, Radboud University Nijmegen Medical Centre, Nijmegen, The Netherlands, **4** Department of Infectious Diseases and Immunology, Faculty of Veterinary Medicine, Utrecht University, Utrecht, The Netherlands, **5** Central Veterinary Institute of Wageningen UR, Lelystad, The Netherlands

## Abstract

In 2011, Dutch animal production sectors started recording veterinary antimicrobial consumption. These data are used by the Netherlands Veterinary Medicines Authority to create transparency in and define benchmark indicators for veterinary consumption of antimicrobials. This paper presents the results of sector wide consumption of antimicrobials, in the form of prescriptions or deliveries, for all pig, veal calf, and broiler farms. Data were used to calculate animal defined daily dosages per year (ADDD/Y) per pig or veal calf farm. For broiler farms, number of animal treatment days per year was calculated. Furthermore, data were used to calculate the consumption of specific antimicrobial classes per administration route per pig or veal calf farm. The distribution of antimicrobial consumption per farm varied greatly within and between farm categories. All categories, except for rosé starter farms, showed a highly right skewed distribution with a long tail. Median ADDD/Y values varied from 1.2 ADDD/Y for rosé finisher farms to 83.2 ADDD/Y for rosé starter farms, with 28.6 ADDD/Y for white veal calf farms. Median consumption in pig farms was 9.3 ADDD/Y for production pig farms and 3.0 ADDD/Y for slaughter pig farms. Median consumption in broiler farms was 20.9 ATD/Y. Regarding specific antimicrobial classes, fluoroquinolones were mainly used on veal calf farms, but in low quantities: P75 range was 0 – 0.99 ADDD/Y, and 0 – 0.04 ADDD/Y in pig farms. The P75 range for 3$^{rd}$/4$^{th}$-generation cephalosporins was 0 – 0.07 ADDD/Y for veal calf farms, and 0 – 0.1 ADDD/Y for pig farms. The insights obtained from these results, and the full transparency obtained by monitoring antimicrobial consumption per farm, will help reduce antimicrobial consumption and endorse antimicrobial stewardship. The wide and skewed distribution in consumption has important practical and methodological implications for benchmarking, surveillance and future analysis of trends.

**Editor:** Jan Kluytmans, Amphia Ziekenhuis, The Netherlands

**Funding:** The work of the Netherlands Veterinary Medicines Authority is funded by the government and animal sectors. The study itself was not funded.

**Competing Interests:** The authors have declared that no competing interests exist.

* E-mail: m.e.h.bos@uu.nl

## Introduction

During the first decade of the 21$^{st}$ century, sales of antimicrobials for veterinary consumption strongly increased in the Netherlands. [1] This is partially attributed to the ban of antimicrobial growth promoters in the European Union, which was fully effective in 2006. However, factors like intensification of animal husbandry, increase of farm size, and changing criteria for animal feed quality, due to e.g. prion prevention and economic considerations (such as more expensive compounds), have been suggested to play a role as well. [2-4].

More recently, antimicrobial resistance has become a growing public health problem as a result of failure of empiric treatment. [5] Prevalence and spread of antimicrobial resistant commensal and pathogenic bacteria have increased as a result of the selection for resistance because of the widespread consumption of antimicrobials. Furthermore, the same antimicrobial anatomical therapeutic chemical (ATC) classes are used in both human and veterinary medicine. [6] In the Netherlands, consumption of antimicrobials in humans is among the lowest in Europe. [7] There is also a strict infection control policy in place in most of the Dutch hospitals, intended to minimize prevalence of antimicrobial resistant pathogens in Dutch health care facilities. Although this policy did result in limitation of MRSA prevalence, extended-spectrum $\beta$-lactamase (ESBL)-producing Enterobacteriaceae prevalence in the Netherlands is equivalent to most other EU countries participating in EARSS. [8] Outbreaks in hospitals and other health care facilities with antimicrobial resistant pathogens are immediately dealt with, often at high costs. [9–11] In recent years, the emergence in livestock of antimicrobial resistant bacteria, such as livestock-associated methicillin-resistant *Staphylococcus aureus* in 2004 [12,13] and the recent occurrence of ESBL-producing Enterobacteriaceae in the poultry production chain, led to new awareness on this issue. [14] It is still unclear to what extent antimicrobial resistant bacteria in livestock have an impact on the occurrence of antimicrobial resistant bacteria in humans, or how often the reverse is the case, but a recent study showed genetic similarities between resistant isolates found in chicken meat and humans. [15] Therefore, it is important to decrease prevalence of these bacteria in livestock, which can be achieved by, among others, decreasing antimicrobial usage in these populations. [16].

In 2010, the Dutch government demanded that veterinary consumption of antimicrobials should decrease to 50% in 2013, compared to 2009 (http://www.rijksoverheid.nl/documenten-en-publicaties/kamerstukken/2010/04/09/deskundigenberaad-rivm-en-reductie-antibioticumgebruik.html). In addition to this, the Health Council of the Netherlands advised to limit veterinary consumption of specified antimicrobials in order to preserve them for human medicine, as these are pivotal antimicrobials (e.g. fluoroquinolones and $3^{rd}/4^{th}$-generation cephalosporins). [17] The Health Council also advised to consider limitation of all antimicrobials with selection pressure potential for ESBLs. Thus, in addition to adaptation of the national veterinary guidelines for therapy, full transparency of consumption of antimicrobials on each individual farm was required in combination with a system of benchmarking. Information on antimicrobial consumption for individual farms can identify persistent high consumers, who should give priority to consumption reduction. An independent institution, the SDa (the Netherlands Veterinary Medicines Authority), was formed in 2010 with the main purpose of creating transparency in and setting benchmark indicators for consumption of antimicrobials in livestock production, based on the consumption data as presented in the present study. Moreover, the SDa monitors, analyses, and reports data on consumption of antimicrobials on a yearly basis, thus making trends in consumption patterns in the various sectors transparent.

National overall antimicrobials sales data have been reported in the European Union for several countries for several years through the European Surveillance of Veterinary Antimicrobial Consumption (ESVAC) project of the European Medicine Agency [18]. So far, the only country registering and reporting data on sector wide veterinary consumption of antimicrobials prescribed by veterinarians, is Denmark. [19] These data are collected since 2001 on a mandatory monthly basis in a central database, VetStat [20], and reported annually. [21] In these national reports, the calculation and reporting of antimicrobial drug consumption in animal daily doses (ADDkg) was done on a national level, comparable to the human method of defined daily dosage (DDD) calculations. However, in these reports, no insight is given into the shape and width of the distribution of farm consumption. Individual farm data are available and acted upon, when considered necessary, by the Danish Veterinary and Food Administration. [22].

This paper presents the first unique results of the complete consumption of antimicrobials as registered on individual farm level, for all pig, veal calf, and broiler farms in the Netherlands in 2011.

## Materials and Methods

### Data collection

Data were collected by the respective private animal sectors and sent to the SDa after encryption of identifiers, thus ensuring the privacy of the parties concerned. In brief, the data collection process can be described as follows: each time a veterinarian prescribes and supplies medicines, information on these deliveries is entered in Practice Management Systems (PMSs) and transferred to a central database. The majority of transfers takes place through VetCIS (www.vetcis.nl), a data hub system set up by a joint collaboration of the Royal Dutch Veterinary Association (KNMvD), the main veterinary drug wholesaler in the Netherlands AUV, and the association of the veterinary pharmaceutical industry in the Netherlands (FIDIN). An external audit has been performed on data entry by the veterinarians, showing that the maximal margin of error is 10–20%. Data in VetCIS is subsequently transferred to the private sector databases. This route accounts for approximately 70% of all transferred prescriptions. Part of the data is directly transferred from the PMS to the sector databases or entered by veterinarians through internet portals of the sector systems. Farmers and veterinarians have access to the central databases through internet portals to consult prescription data and consumption of veterinary medicines. The sector data systems are part of existing integral quality assurance systems, which aim to guarantee food safety for the consumer. As a part of these systems, each farm is visited at least once a year for inspection and certification by the private certifying authorities.

Data entered per medicine delivery includes: a unique farm identifier (UFI), a unique veterinarian identifier (UVI), EAN code (unique European Article Number), number of packages supplied, animal species, animal category, and delivery date. Linked to the EAN code the following data can be collected as well: REG NL number (Dutch authorisation number for veterinary pharmaceuticals), ATCvet code, administration route, product name, and content (including unit) of packaging. Product data are derived from the so-called Branche Code Table (BCT; provided by FIDIN). Besides these data, the $DD_{kg}$ (the Defined Dosage of medicine (g or ml) needed for the treatment of one kilogram of animal during one day) is derived from the veterinary medicine criterion (designated DG-standard) and registered in the databases. The DG-standard is a database defining standard doses for each animal species and route of administration for which an antimicrobial veterinary medicinal product is licensed in the Netherlands. In practice, DDkg is derived from the number of animal kg that, in accordance with the Specification of Product Characterization (SPC), can be treated with one ml, one gram or one piece (e.g. intramammary injector) of the specified veterinary medicinal product (VMP). It is combined with the treatment duration of this dose. For example: if the authorized dose of a VMP is 0.01 ml/kg, twice daily, this results in a treatable weight*days of (100 kg/ml*0.5 days) 50 kgdays per mL. In the case of a longacting VMP, e.g. 5 days treatment with one shot, dose again 0.01 ml/kg, this will result in (100 kg/ml*5 days) 500 kgdays per mL. The DG-standard is updated and controlled by the SDa and available on the website. The average number of animals present at the farm per year is collected on animal subcategory level, specified by age groups and farm types.

For the purpose of the first descriptive analyses, the SDa received a dataset per animal production sector containing a record for each farm with, besides the animal defined daily dosages per year (ADDD/Y; for explanation see further), UFI, UVI, data on the animal species and category, the average number of animals present on the farm during 2011 (per category and, where applicable, subcategory), empty barn period (where applicable; in days), and the period of registration of antimicrobial deliveries or prescriptions. Pig farms could be included twice in the dataset; the so-called "closed farms" breed and rear their own slaughter pigs. For the purpose of this study and benchmarking by the SDa, these farms were treated as two farm types: production and slaughter pig farms. All datasets were checked for consistency and in case of doubt, feedback was asked from the sector representatives. Where necessary, data were corrected. For veal calf farms, ADDD/Y were corrected for the periods when the barns were empty. In the pig husbandry sector, it was agreed that the - for most fattener pig husbandries consistent - factor of five days empty between two rounds of five months (3%), was not significant enough to correct for. In broilers, the empty period is already included in the variable "animal days". Coverage of the number of farms in these databases, related to Statistics Netherlands (www.cbs.nl), was 106% for pig farms, 110% for veal calf farms, and 122% for broiler farms. Discrepancies may be caused by differences in the definition of "farm" and timing of registration of the number of farms. On individual farms, the total of antimicrobial deliveries or prescriptions could be below 0 ADDD/Y. Reversed entries are made when unused antimicrobials are returned to the veterinarian or when a wrong entry was made.

For analysis of consumption of specific antimicrobial classes for each route of administration, the SDa also received prescription data for the veal calf and pig sectors. For broiler farms, these data were not yet completely available in sufficient detail and are not presented.

In these datasets, each veterinary drug delivery on a farm is recorded, containing, amongst others, identifiers, EAN code, number of packages, and delivery date. For calculation of ADDD/Y, prescription data was linked with the other datasets to add the number of animals present on the farm.

We will further use the term "antimicrobial consumption" as a synonym for "antimicrobial delivery", as we assume that each delivery is either consumed by animals, or returned to the veterinarian and a reversed entry is made for this delivery.

## Calculation of ADDD/Y and ATD/Y

The sectors calculated the consumption of antimicrobials per farm as animal defined daily dosages per year (ADDD/Y); apart from the denominator this is similar to the standard unit for consumption of antimicrobials in humans (DDD/1000 days). An ADDD/Y of 1 means that the average animal in the population was exposed to an antimicrobial for one day per year. This measure is similar to that proposed by ESVAC. [23].

To calculate ADDD/Y, two variables are needed. First, the total animal mass in kg that can be treated for one day with the amount of antimicrobials supplied to the farmer (treatable weight*days), which is derived from the DG-standard. Second, the mean total weight (kg) of animals present on the farm during 2011 (animal weight), which can be calculated according to standardised mean animal weights (determined per production type and animal category, see Table 1). By dividing these numbers, the number of animal daily dosages per year is obtained.

For broilers, ADDD/Y could not yet be determined, because these data did not include the variables needed to calculate treatable kilograms. Instead, animal treatment days per year (ATD/Y) were calculated. One difference between ADDD/Y and ATD/Y is that the number of treatment days is based on prescriptions by a veterinarian for a specific age category, independent of dosage or duration of effectiveness. Another difference is that the calculation of ATD/Y does not include a standardised mean weight for the animals. For calculation of ATD/Y, the numerator was the summation of the number of treatment days for all broilers present during the year (eg., when 100 animals were each treated for 5 days, the number of treatment days would be 100*5 = 500). The denominator was denoted by the number of animal days; i.e. the sum of the number of birds present per day for the year (eg., when during a whole year on average 100 animals are present, the number of animal days is 100*365 = 36500). By dividing these numbers and multiplying by 365, the number of ATD/Y was obtained: the number of days antimicrobials were administered to broilers on a farm per year (in our example: (500/36500)*365 = 5).

## Statistical analysis

Descriptive statistics were obtained using the MEANS and FREQ procedures in SAS® software version 9.2 (SAS Institute Inc., Cary, NC, USA). Statistical analyses were performed on the data by means of the REG and TTEST procedures. We applied univariate models with ADDD/Y as the dependent variable to test for the influence of farm characteristics (such as farm size) on antimicrobial use. R© open source software version 2.15.2 (R Foundation for Statistical Computing, Vienna, Austria) was used for creating (density) plots.

# Results

## Farm descriptives

The majority of the 2125 veal calf farms in the data consisted of white veal calf farms (n = 934, 44%), followed by rosé finisher farms (n = 671, 32%), rosé starter farms (n = 207, 10%), and farms on which both rosé starter and finisher calves were present, or a combination of white and rosé calves (n = 313, 15%) (figure 1). There were 14 combination farms in the data with no animals present; these have been excluded from the results. Registered farm sizes were diverse, ranging from one animal present in 2011 to over 3000 animals, with mean numbers of 606 (white veal), 285 (rosé starter), 210 (rosé finisher), and 437 (combination) animals per farm. There were 2528 production farms with sows and piglets in the data, and 5531 farms where slaughter pigs were kept. Two production farms had no animals in 2011; these have been excluded from further results. Again, registered pig farm sizes varied greatly, ranging from 2 sows to over 4000 for production farms, and from 2 slaughter pigs to 13000 for slaughter pig farms. Mean number of sows for production farms was 382, and for

**Table 1.** The applied standardised mean animal weight and treatment weight per animal category and age category (adapted from http://www.autoriteitdiergeneesmiddelen.nl/Userfiles/rapportage--sda-expertpanel-dataanalyse-2011-en-benchmarkindicatoren-2012.pdf).

| Animal species | Farm category | Age | Standardised mean weight |
|---|---|---|---|
| Veal calf | White | 0 - 222 days | 160 kg |
| | Rosé starter | 0 - 98 days | 77.5 kg |
| | Rosé finisher | 98 - 256 days | 232.5 kg |
| Pig | Production with sows and piglets | Combination of ages | 303.8 kg* |
| | Slaughter | 74 - 191 days | 70 kg |

*Combination of 1 sow (of 220 kg) + 5.5 piglets (of 12.5 kg each) + 0.14 gilts (of 107.5 kg each).

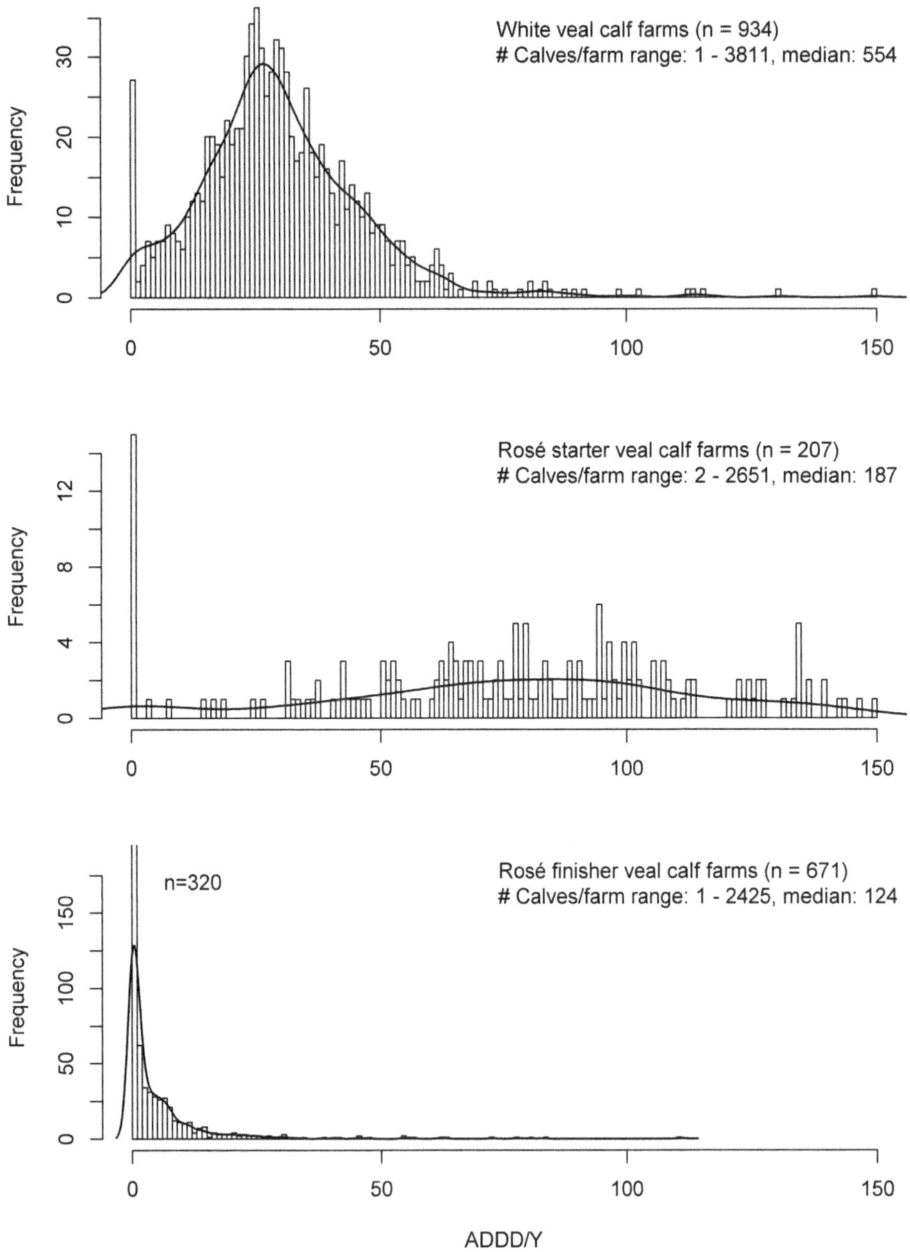


**Figure 1. Frequency distribution of ADDD/Y per veal calf farm ranging from 0 to 150.** The line demonstrates a density plot based on the histogram. Number of farms with ADDD/Y>150: 27 (white: 7, rosé start: 20) and number of farms with ADDD/Y<0: 4 (white: 3, rosé finisher: 1).

slaughter pig farms 1092 animals. The 732 broiler farms ranged from 620 to 434106 (mean: 67745) broilers.

## Antimicrobial consumption

The distributions of ADDD/Y and ATD/Y per farm are shown in Figures 1, 2, 3 for the various livestock categories. The distributions for white and rosé starter farms showed a large spread in ADDD/Y values. With exception of rosé starter farms, all distributions were highly skewed to the right and showed a long tail.

Table 2 shows the arithmetic and geometric means and standard deviations, and the median and 75$^{th}$ percentile (P75) for ADDD/Y for veal calf and pig farms, and ATD/Y for broiler farms. The highest arithmetic mean and median ADDD/Y were found in rosé starter farms (arithmetic mean/median ADDD/Y = 105.4/83.2, P75 = 110.0, 90$^{th}$ percentile (P90) = 149.5), and the lowest arithmetic mean and median ADDD/Y in rosé finisher farms (arithmetic mean/median ADDD/Y = 5.2/1.2, P75 = 6.0, P90 = 13.1). White veal calf farms had an intermediate consumption of antimicrobials (arithmetic mean/median ADDD/Y = 35.6/28.6, P75 = 38.9, P90 = 50.7). ADDD/Y values were not calculated for combination farms, because of their complex nature with regards to the categories of calves present.

In the pig sector, the arithmetic mean/median ADDD/Y were 16.9/9.3 (P75 = 20.8, P90 = 40.6) and 9.6/3.0 (P75 = 10.8, P90 = 22.5) for production and slaughter pig farms, respectively. Broiler farms had an arithmetic mean/median ADT/Y of 23.8/20.9 (P75 = 34.1, P90 = 50.0).

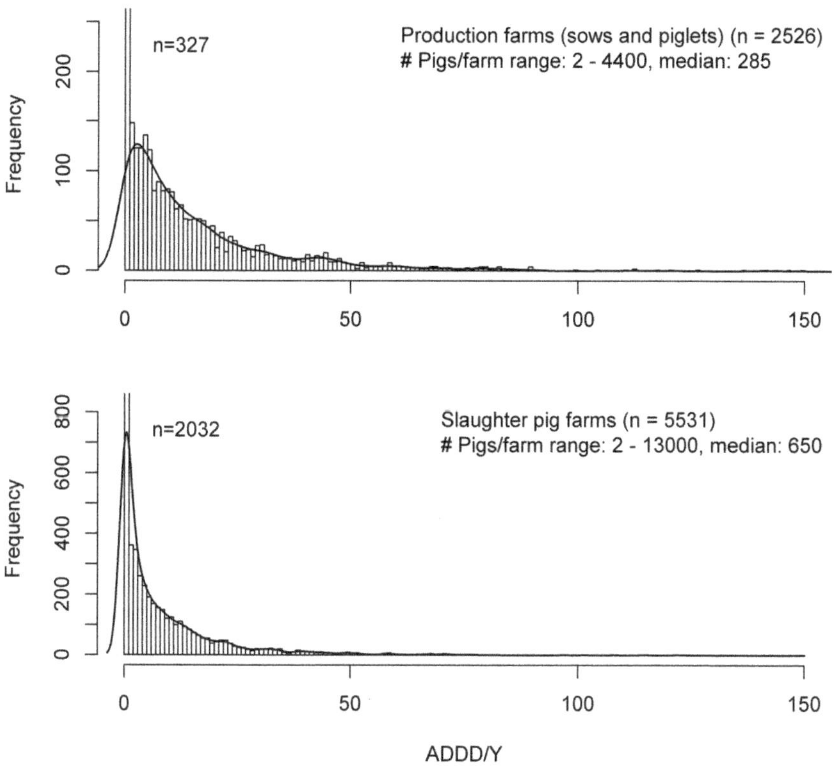


**Figure 2. Frequency distribution of ADDD/Y per pig farm ranging from 0 to 150.** The line demonstrates a density plot based on the histogram. Number of farms with ADDD/Y>150: 15 (sows: 3, finishers: 12) and number of farms with ADDD/Y<0: 1 (finisher: 1).

No univariate significant association was found between consumption of antimicrobials and the number of veal calves present as a continuous variable (p-values > 0.1), or as a categorical variable (smaller or larger than the median, p-values > 0.09). However, for the number of pigs present on the farm, a positive significant univariate association with consumption of antimicrobials was found (production farms: $\beta = 0.012$, 95% Confidence Interval (CI): 0.006 – 0.019, $p < 0.01$; slaughter pig farms: $\beta = 0.0017$, 95% CI: 0.0007 – 0.003, $p < 0.01$). A similar result was seen for production pig farms when analysing number of pigs as a categorical variable depending on the median. For the number of broilers present on the farm, a significant, albeit small, positive association with consumption of antimicrobials was found as well ($\beta = 0.00006$, 95% CI: 0.00004 – 0.00009, $p < 0.01$).

In figure 4 an overview is given of the mean consumption of antimicrobials per farm category in the pig and veal calf sectors, for specified ATCvet groups. The figure shows that tetracyclines were overall the most used antimicrobial group, followed by trimethoprim/sulfonamides and macrolides/lincosamides.

On 39% of 934 white veal calf farms, 3$^{rd}$/4$^{th}$-generation cephalosporins were consumed in 2011, with a P75 of 0.07 ADDD/Y. Fluoroquinolones were administered orally on 59% of the white veal calf farms (P75 = 0.32 ADDD/Y), and parenterally on 87% of these farms (P75 = 0.34 ADDD/Y). Rosé starter and finisher calf farms hardly used 3$^{rd}$/4$^{th}$-generation cephalosporins (13%/10% of 207 starter and 665 finisher farms, respectively, with P75 = 0 ADDD/Y). Fluoroquinolones were administered on most rosé starter farms in 2011 (40% of the farms orally: P75 =

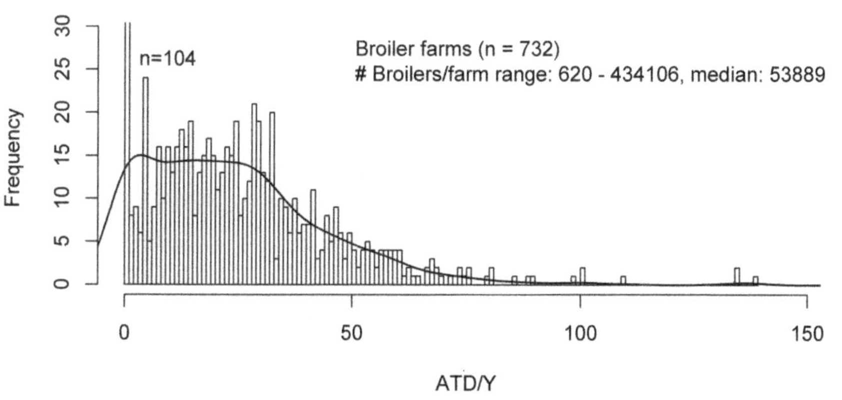


**Figure 3. Frequency distribution of ATD/Y per broiler farm ranging from 0 to 150.** The line demonstrates a density plot based on the histogram. Number of farms with ATD/Y>150: 0 and number of farms with ATD/Y<0: 0.

**Table 2.** Arithmetic mean and standard deviation, and geometric mean and standard deviation of ADDD/Y for veal calf and pig farms, and of ATD/Y for broiler farms.

| Animal species | Farm category | Arithmetic mean | Arithmetic standard deviation | Median | P75 | Geometric mean* | Geometric standard deviation* |
|---|---|---|---|---|---|---|---|
| Veal calf | All active farms** | 32.2 | 125.4 | 19.7 | 35.6 | 8.1 | 9.0 |
| | White | 35.6 | 111.5 | 28.6 | 38.9 | 23.2 | 3.1 |
| | Rosé starter | 105.4 | 159.1 | 83.2 | 110.0 | 52.5 | 6.5 |
| | Rosé finisher | 5.2 | 10.9 | 1.2 | 6.0 | 1.1 | 7.2 |
| Pig | Production with sows and piglets | 16.9 | 58.9 | 9.3 | 20.8 | 6.3 | 5.8 |
| | Slaughter | 9.6 | 48.0 | 3.0 | 10.8 | 1.8 | 8.4 |
| Poultry | Broiler | 23.8 | 20.6 | 20.9 | 34.1 | 9.9 | 7.5 |

*In order to calculate the natural log, ADDD/Y ≤ 0 was set at 0.1.
**Farms where no animals were present were excluded, 2111 farms are included in this category.

0.50 ADDD/Y, and 75% parenterally: P75 = 0.99 ADDD/Y), but hardly on rosé finisher farms (2% of the farms orally: P75 = 0 ADDD/Y, and 24% parenterally: P75 = 0 ADDD/Y).

Consumption of these antimicrobials was low on pig farms. On 22% (P75 = 0 ADDD/Y) and 4% (P75 = 0 ADDD/Y) of 2494 production and 5441 slaughter pig farms 3rd/4th-generation cephalosporins were administered, respectively. Fluoroquinolones were administered orally on 4% (P75 = 0 ADDD/Y) and 0.1% (P75 = 0 ADDD/Y) of the production and slaughter pig farms, respectively. Parenterally, these antimicrobials were administrated on 36% (P75 = 0.04 ADDD/Y) and 8% (P75 = 0 ADDD/Y) of the pig production and slaughter pig farms.

## Discussion

The unique data presented in this study represent the first time that consumption of antimicrobials, in the form of prescriptions and/or deliveries on the farm, is available on this level for each pig, veal calf, and broiler farm in the Netherlands. The data show that large differences exist between farms within the different categories regarding consumption of antimicrobials. Consumption also greatly varies between farm categories. In most categories analysed here, the majority of farms have low consumption of antimicrobials. However, a minority of farms have a very high ADDD/Y. The insights obtained from these data aid the SDa with their aim to define benchmark criteria for antimicrobial

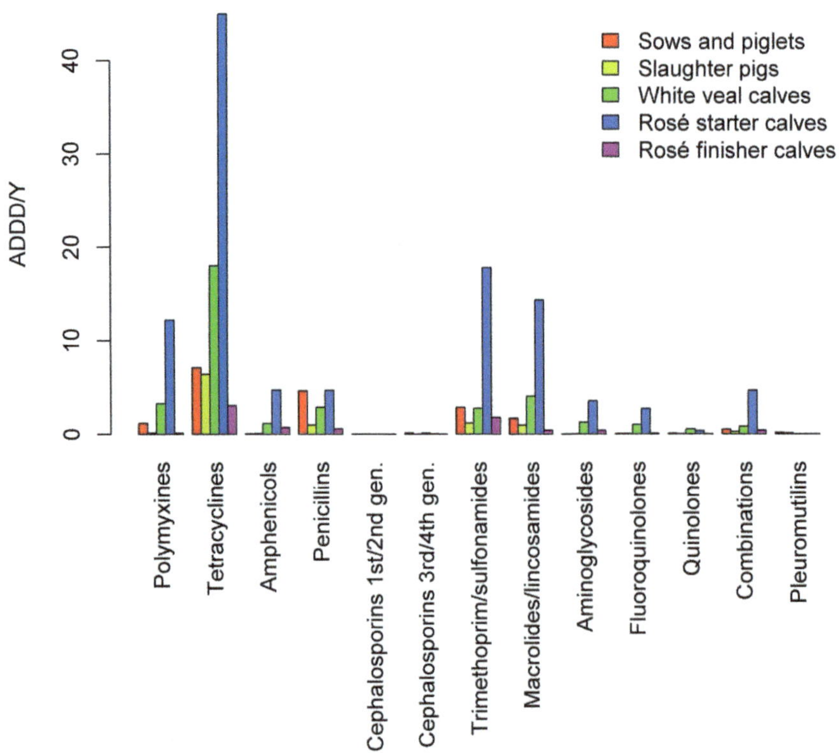


**Figure 4. Mean ADDD/Y per farm per animal sector, given for 13 ATCvet classes.**

consumption, which should ultimately lead to reduction in consumption.

ADDD/Y is a measure for the number of days on a farm antimicrobials have been administered. It is important to realize that this does not mean that each and every animal has received that calculated ADDD/Y, but that it in fact is a measure per animal position available on the farm. ADDD/Y is also based on standard doses as determined by the SDa, which in general is the mean dose according to the Specification of Product Characterization (SPC) of the specific veterinary medicinal product. In reality, under- or overdosing may occur frequently. [24] Other factors influencing the calculated ADDD/Y include in the numerator long acting products and in the denominator the selection of the average weight (such as the combination of sows and piglets). For long acting products a treatment duration factor is added to the ADDD calculation (for instance a single shot of a product acting for 5 days results in 5 times the number of ADDD compared to a single shot of a direct acting drug).

On mixed farms, such as combination veal calf farms (with both rosé and white veal calves present) and "closed" pig farms (with both sows and piglets and slaughter pigs), it is possible that antimicrobials have not been registered for the correct animal category. This may lead to misclassification and a biased ADDD/Y per farm. However, this effect is expected to be small over a farm category. In the current analyses, closed pig farms have been included as two farms, and combination veal farms have been ignored. In future analyses, these farms will be considered as distinct farm categories.

Clear differences exist in antimicrobial consumption between animal sectors and farm categories, but these are difficult to compare due to differences in farm management. The veal calf sector has a higher consumption, which might be a consequence of the way calves are collected. Veal calf farms receive calves from dairy farms from the Netherlands (approximately 50%) and countries throughout Europe (predominantly Germany). These calves are a product of the dairy industry. [25] Sober calves arrive once or twice a year (usually an all-in, all-out system is applied), after transport often creating an optimal atmosphere for pathogen transmission. Frequently, calves receive antimicrobial treatments in the first few weeks after arrival, which means that farms with two production cycles in a year, likely will have a higher ADDD/Y then farms with only one starting phase of a production cycle in that particular year. [24–26].

Dutch pig farms have more closed production systems than veal calf farms, or have animal supply from limited sources. In broilers, infection control is well possible by hygienic measures between production rounds. However, broiler production in the Netherlands is not integrated in production chains that include breeders, hatcheries and broiler farms. As a result, infection control is also not optimally integrated along these production chains. Currently, best practices to control infectious diseases and the health status of animals is a priority in the Netherlands, both within animal production sectors and within the veterinary society.

Comparisons between animal sectors or farm categories should be made with care because of the influence of animal weight on the calculated ADDD/Y. Calculation of ADDD/Y is based on a standardised average weight for animals. For adult animals, e.g. sows, this weight will be steady over the course of a year, but for fattener animals, e.g. white veal calves and slaughter pigs, this weight will strongly increase during the production cycle. Recent studies showed that the majority of antimicrobial use in pigs is administered before they are 10 weeks of age, i.e. at lower weight, which is comparable to treatments in veal calves. [26,27] For the category of white veal calves and slaughter pigs, depending on the exact timing of the treatment over- but more likely underestimation of the ADDD/Y can occur. Use of actual weights during treatment could improve the ADDD/Y estimate but requires a more detailed and accurate registration. [24].

To date, Denmark is the only other country publishing antimicrobial consumption figures annually, based on prescriptions by veterinarians, for all farms in an animal sector. [21,28] In the DANMAP reports, antimicrobial drug consumption is presented as animal daily doses (ADDkg), related to total biomass-year-at-risk, kg-meat-produced, or number of animals produced. ADDkg is given for each age-group and species, e.g. piglets, weaners, fattening pigs or sows, but no mean farm ADD is reported in the reports. [19] Denmark does, however, calculate farm level ADDs for their "yellow card" system, which is designed to control veterinary antimicrobial consumption. [22] In this system, pig farms are given a "yellow card" when they consume more than twice the average consumption. Farms with a "yellow card" have to implement antimicrobial restrictive measures.

The European Medicines Agency is currently considering which technical units of measurements and indicators should be used for future collection of and reporting on national veterinary antimicrobial drug consumption, as part of the European Surveillance of Veterinary Antimicrobial Consumption project. [23] Although this will provide valuable information on national level, neither the proposed ESVAC indicators for the near future nor the DANMAP indicators as reported up till now, consider the practical and methodological implications of the wide distribution of antimicrobial consumption on farms, as shown in this paper. Our farm level data, reporting on all individual farms nationwide, create a breakthrough in analytical possibilities and benchmarking options, as we clearly demonstrate that normal distribution statistics are not appropriate for describing or analysing antimicrobial consumption, and that differences in consumption of antimicrobial agents amount to one, sometimes two orders of magnitude between farms.

In previous years, veterinary consumption of antimicrobials in the Netherlands was reported for a selection of farms (based on stratified sampling) and in terms of total national sales data. [1,29] Bondt *et al.* used national sales data to compare veterinary antimicrobial exposure between Denmark and the Netherlands, demonstrating that data on consumption per animal species is a necessity to be able to adequately compare countries. [30] When comparing our data (table 2) with those of the MARAN-report on 2011, we demonstrate that outcomes based on a sample of farms may give a biased estimate of antimicrobial consumption. [31] The MARAN-report shows consistently lower results when comparing to the arithmetic means in table 2. The mean arithmetic consumption in daily doses per animal year (DD/AY) reported by MARAN is 13 (95% CI: 10 – 16) for pig production farms, 8 (95% CI: 5 – 11) for slaughter pig farms, 25 (95% CI: 23 – 26) for veal calf farms, and 16 (95% CI: 12 – 21) for broiler farms. However, when we compare the MARAN results with our calculated geometric means, our results are much lower. This shows that the long right tail in the distributions strongly influences the average consumption in a farm category, and emphasizes the need for reporting distributions of consumption on farm level. The use of a disproportional stratified random sampling design for the MARAN-report probably also affected the outcomes, where the weighted results of a selection of farms are extrapolated to the whole sector.

So far, the reason for the large and right skewed between-farm spread in consumption of antimicrobials is unknown. In veal calf farms the number of production cycles started in 2011, may have contributed to the skewedness of the distribution, as younger calves are more susceptible to infections and therefore will receive more treatments with antimicrobials. This may lead to a relatively higher consumption in farms with more production cycles in a year. A similar wide between-farm spread has been shown by other studies. [2,32] For slaughter pig farms, farm system (farrow-to-finish or specialized slaughter pig farms) and number of pigs were shown to influence consumption of antimicrobials, and for sow farms this was farm system (specialized sow farm), number of sows and regional population density. [2] We analysed the influence of farm size on consumption of antimicrobials, as earlier reports on Dutch sentinel farms found an association in pig production (but not in poultry). [3] In our data, this association was found to be significant for the pig and broiler farms, but in univariate analyses only, therefore not accounting for the influence of other factors. The influence of farm size needs to be studied in more detail.

For 2012, the SDa have defined benchmark action criteria for antimicrobial consumption. Because of the wide distributions, the initial aim is to limit the tail of the distribution, and benchmarks focus at the 75$^{th}$ percentile of the consumption in 2011. Farms that exceed these benchmarks are obliged to promptly undertake measures to decrease their consumption of antimicrobials. Moreover, in general median consumption in 2011 minus 20% was defined as the upper limit for the target benchmark criterion for appropriate antimicrobial consumption. This target will be re-evaluated in 2015. The benchmark criterion for action will annually be re-evaluated, because it is anticipated that removing the long tail in the population will have a major impact on the average consumption in a sector. When all farms above the action benchmark criterion reduce their antimicrobial consumption to the upper limit of the target benchmark criterion, the overall decrease of antimicrobial consumption in each of the farm categories will be substantial. Early 2013, data for 2012 have become available that will allow the evaluation of the achieved reduction and persistent high users. Data on broiler farms will be provided similar to the other animal sectors, facilitating calculation of ADDD/Y for broiler farms instead of ATD/Y. In addition, antimicrobial prescription behaviour will also be monitored on the level of veterinarians.

In conclusion, the analysis of unique data provided to the SDa by the animal sectors in the Netherlands, shows that consumption of antimicrobials varies strongly between animal production sectors and farm categories, and also within farm categories. The wide and skewed distribution in consumption has important methodological implications for benchmarking, surveillance, and future analyses of trends over the years within farms, within animal sectors, and between animal sectors. The full transparency obtained on antimicrobial consumption per farm, as shown in the data collected for the SDa, enables targeted measures to reduce and improve the quality (in terms of very restricted antimicrobial use of specified groups like fluoroquinolones and cephalosporins) of antimicrobial consumption, and serves as a tool for both farmers and veterinarians. These will include measures to improve health status and control of infectious diseases on those farms where consumption of antimicrobials is (consistently) highest. This should result in optimal reduction of antibiotic consumption combined with improved health control. In the following years, the SDa will report on within-(sub)sector trends, expanding to other animal sectors, as well as on within-farm trends and trends on prescriptions by veterinarians.

## Acknowledgments

We gratefully acknowledge the Product Board for Livestock and Meat, the Product Board for Poultry and Eggs, Stichting Kwaliteitsgarantie Vleeskalversector, Co-More, De Groene Belangenbehartiger BV, and De Gezondheidsdienst voor Dieren for giving permission to use the data they provided.

## Author Contributions

Conceived and designed the experiments: IvG JM DM DH. Analyzed the data: MB FT IvG JM DM DH. Wrote the paper: MB FT IvG JM DM DH.

## References

1. Bondt N, Puister-Jansen LF, Ge L, van der Veen HB, Bergevoet RHM, et al. (2012) Trends in veterinary antibiotic use in the Netherlands 2004–2012 (LEI 12-109). LEI of Wageningen UR, the Hague, the Netherlands. Available: http://www.wageningenur.nl/upload_mm/8/7/f/e4deb048-6a0c-401e-9620-fab655287fbc_Trends%20in%20use%202004-2012.pdf. Accessed 15 March 2013.
2. Van der Fels-Klerx HJ, Puister-Jansen LF, van Asselt ED, Burgers SL (2011) Farm factors associated with the use of antibiotics in pig production. J Anim Science 89: 1922–1929.
3. VANTURES, the Veterinary Antibiotic Usage and Resistance Surveillance Working Group (2009) Monitoring of antimicrobial resistance and antibiotic usage in animals in the Netherlands in 2006/2007 (MARAN-2007). (Mevius DJ, Wit B, Van Pelt W, eds) CVI of Wageningen UR, Lelystad, the Netherlands. Available: http://edepot.wur.nl/148647. Accessed 26 March 2013.
4. Sapkota AR, Lefferts LY, McKenzie S, Walker P (2007) What do we feed to food-production animals? A review of animal feed ingredients and their potential impacts on human health. Environ Health Perspect 115: 663–670.
5. World Health Organisation (2012) WHO fact sheet on antimicrobial resistance (194). Available: http://www.who.int/mediacentre/factsheets/fs194/en/index.html. Accessed 15 March 2013.
6. Moulin G, Cavalié P, Pellanne I, Chevance A, Laval A, et al. (2008) A comparison of antimicrobial usage in human and veterinary medicine in France from 1999 to 2005. J Antimicrob Chemother 62: 617–625.
7. European Centre for Disease Prevention and Control (2012) Summary of the latest data on antibiotic consumption in the European Union. Available: http://www.ecdc.europa.eu/en/eaad/documents/eaad-2011-summary-antimicrobial-consumption-data.pdf. Accessed: 15 March 2013.
8. European Centre for Disease Prevention and Control (2012) Antimicrobial resistance surveillance in Europe 2011. Annual report of the European Antimicrobial Resistance Surveillance Network (EARS-Net). ECDC. Stockholm, Sweden. Available: http://www.ecdc.europa.eu/en/publications/Publications/Forms/ECDC_DispForm.aspx?ID = 998. Accessed: 15 March 2013.
9. Schop WA, Trompenaars M, Tjon-a-Tsien A, Lavooij P, Richardus JH (2012) De GGD en bestrijding van resistente bacteriën: de Klebsiella Oxa-48-uitbraak in het Maasstad Ziekenhuis in Rotterdam in 2011. Infectieziekten Bulletin 23: 77–79.
10. Van der Zee A, Hendriks WD, Roorda L, Ossewaarde JM, Buitenwerf J (2013) Review of a major epidemic of methicillin resistant Staphylococcus aureus: The costs of screening and consequences of outbreak management. Am J Infect Control 41: 204–209.
11. Van Rijen MM, Kluytmans JA (2009) Costs and benefits of the MRSA Search and Destroy policy in a Dutch hospital. Eur J Clin Microbiol Infect Dis 28: 1245–1252.
12. Armand-Lefevre L, Ruimy R, Andremont A (2005) Clonal Comparison of Staphylococcus aureus Isolates from Healthy Pig Farmers, Human Controls, and Pigs. Emerg Infect Dis 11: 711–714.
13. Voss A, Loeffen F, Bakker J, Klaassen C, Wulf M (2005) Methicillin-resistant Staphylococcus aureus in Pig Farming. Emerg Infect Dis 11: 1965–1966.
14. Dierikx C, van der Goot J, Fabri T, van Essen-Zandbergen A, Smith H, et al. (2013) Extended-spectrum-beta-lactamase- and AmpC-beta-lactamase-producing Escherichia coli in Dutch broilers and broiler farmers. J antimicrob chemother 68: 60–67.
15. Kluytmans JA, Overdevest IT, Willemsen I, Kluytmans-van den Bergh MF, van der Zwaluw K, et al. (2013) Extended-spectrum beta-lactamase-producing Escherichia coli from retail chicken meat and humans: comparison of strains, plasmids, resistance genes, and virulence factors. Clin Infect Dis 56: 478–487.
16. Liebana E, Carattoli A, Coque TM, Hasman H, Magiorakos AP, et al. (2013) Public health risks of enterobacterial isolates producing extended-spectrum β-lactamases or AmpC β-lactamases in food and food-producing animals: An EU perspective of epidemiology, analytical methods, risk factors, and control options. Clinical Infect Dis 56: 1030–1037.

17. Health Council of the Netherlands (2011) Antibiotics in food animal production and resistant bacteria in humans (2011/16). Health Council of the Netherlands, the Hague, the Netherlands. Available: http://www.gezondheidsraad.nl/sites/default/files/201116E_Antibiotica_in_food_animal.pdf. Accessed: 15 March 2013.
18. European Medicines Agency (2012) Sales of veterinary antimicrobial agents in 19 EU/EEA countries in 2010 (EMA/88728/2012). Available: http://www.ema.europa.eu/docs/en_GB/document_library/Report/2012/10/WC500133532.pdf. Accessed: 15 March 2013.
19. Jensen VF, Jacobsen E, Bager F (2004) Veterinary antimicrobial-usage statistics based on standardized measures of dosage. Prev Vet Med 64: 201–215.
20. Jensen VF, Jakobsen L, Emborg H-D, Seyfarth AM, Hammerum AM (2006) Correlation between apramycin and gentamicin use in pigs and an increasing reservoir of gentamicin-resistant Escherichia coli. J Antimicrob Chemother 58: 101–107.
21. The Danish Integrated Antimicrobial Resistance Monitoring and Research Programme (2012) DANMAP 2011 - Use of antimicrobial agents and occurrence of antimicrobial resistance in bacteria from food animals, food and humans in Denmark (ISSN 1600-2032). Available: http://www.danmap.org/Downloads/~/media/Projekt%20sites/Danmap/DANMAP%20reports/Danmap_2011.ashx. Accessed: 15 March 2013.
22. Alban L, Dahl J, Andreasen M, Petersen JV, Sandberg M (2013) Possible impact of the "yellow card" antimicrobial scheme on meat inspection lesions in Danish finisher pigs. Prev Vet Med 108: 334–341.
23. European Surveillance of Veterinary Antimicrobial Consumption (2012) European Medicines Agency. ESVAC reflection paper on collecting data on consumption of antimicrobial agents per animal species, on technical units of measurement and indicators for reporting consumption of antimicrobial agents in animals (draft) (EMA/286416/2012-CONSULTATION). Available: http://www.ema.europa.eu/docs/en_GB/document_library/Scientific_guideline/2012/12/WC500136456.pdf. Accessed: 15 March 2013.
24. Pardon B, Catry B, Dewulf J, Persoons D, Hostens M, et al. (2012) Prospective study on quantitative and qualitative antimicrobial and anti-inflammatory drug use in white veal calves. Journal of Antimicrobial Chemotherapy doi:10.1093/jac/dkr570.
25. Bos MEH, Graveland H, Portengen L, Wagenaar JA, Heederik DJJ (2012) Livestock-associated MRSA prevalence in veal calf production is associated with farm hygiene, use of antimicrobials, and age of the calves. Prev Vet Med 105: 155–159.
26. Graveland H, Wagenaar JA, Verstappen KM, Oosting-van Schothorst I, Heederik DJ, et al. (2012) Dynamics of MRSA carriage in veal calves: a longitudinal field study. Prev Vet Med 107: 180–186.
27. Callens B, Persoons D, Maes D, Laanen M, Postma M, et al. (2012) Prophylactic and metaphylactic antimicrobial use in Belgian fattening pig herds. Prev Vet Med 106: 53–62.
28. Jensen VF, Enøe C, Wachmann H, Nielsen EO (2010) Antimicrobial use in Danish pig herds with and without postweaning multisystemic wasting syndrome. Prev Vet Med 95: 239–247.
29. Central Veterinary Institute of Wageningen University and Research Centre (CVI) (2012) Monitoring of antimicrobial resistance and antibiotic usage in animals in the Netherlands in 2010/2011 (MARAN-2012) (Mevius DJ, Koene MGJ, Wit B, van Pelt W, Bondt N, eds) CVI of Wageningen UR, Lelystad, the Netherlands. Available: http://edepot.wur.nl/215472. Accessed 15 March 2013.
30. Bondt N, Frokjaer Jensen V, Puister-Jansen LF, Van Geijlswijk IM (2013) Comparing antimicrobial exposure based on sales data. Prev Vet Med 108: 10–20.
31. Bondt N, Puister-Jansen LF, Ge L, van der Veen HB, Bergevoet RHM, et al. (2012) Trends in veterinary antibiotic use in the Netherlands 2005-2011 (LEI 12-057). LEI of Wageningen UR, the Hague, the Netherlands. Available: http://www.lei.dlo.nl/wever.internet/MARAN/report%20Veterinary%20antibiotic%20usage.pdf. Accessed: 15 March 2013.
32. Timmerman T, Dewulf J, Catry B, Feyen B, Opsomer G, et al. (2006) Quantification and evaluation of antimicrobial drug use in group treatments for fattening pigs in Belgium. Prev Vet Med 74: 251–263.

# Phylogenetic Analysis Reveals Common Antimicrobial Resistant *Campylobacter coli* Population in Antimicrobial-Free (ABF) and Commercial Swine Systems

**Macarena P. Quintana-Hayashi, Siddhartha Thakur***

Department of Population Health and Pathobiology, College of Veterinary Medicine, North Carolina State University, Raleigh, North Carolina, United States of America

**Abstract**

The objective of this study was to compare the population biology of antimicrobial resistant (AR) *Campylobacter coli* isolated from swine reared in the conventional and antimicrobial-free (ABF) swine production systems at farm, slaughter and environment. A total of 200 *C. coli* isolates selected from fecal, environmental, and carcass samples of ABF ($n = 100$) and conventional ($n = 100$) swine production systems were typed by multilocus sequence typing (MLST). Sequence data from seven housekeeping genes was analyzed for the identification of allelic profiles, sequence types (STs) and clonal complex determination. Phylogenetic trees were generated to establish the relationships between the genotyped isolates. A total of 51 STs were detected including two novel alleles (*glnA* 424 and *glyA* 464) and 14 novel STs reported for the first time. The majority of the *C. coli* isolates belonged to ST-854 (ABF: 31, conventional: 17), and were grouped in clonal complex ST-828 (ABF: 68%, conventional: 66%). The mean genetic diversity (*H*) for the ABF (0.3963+/−0.0806) and conventional (0.4655+/−0.0714) systems were similar. The index of association ($I_A^S$) for the ABF ($I_A^S = 0.1513$) and conventional ($I_A^S = 0.0991$) *C. coli* populations were close to linkage equilibrium, indicative of a freely recombining population. Identical STs were detected between the pigs and their environment both at farm and slaughter. A minimum spanning tree revealed the close clustering of *C. coli* STs that originated from swine and carcass with those from the environment. In conclusion, our study reveals a genotypic diverse *C. coli* population that shares a common ancestry in the conventional and ABF swine production systems. This could potentially explain the high prevalence of antimicrobial resistant *C. coli* in the ABF system in the absence of antimicrobial selection pressure.

**Editor:** Maria-Jose Figueras, Unidad de Microbiología, Facultad de Medicina Universidad Rovira, Spain

**Funding:** The funders had no role in study design, data collection and analysis, decision to publish, or preparation of the manuscript. This study was funded by the USDA-National Research Initiative grant, USDA-National Integrated Food Safety Initiative grant and the College of Veterinary Medicine at NC State University grant.

**Competing Interests:** The authors have declared that no competing interests exist.

* E-mail: sthakur@ncsu.edu

## Introduction

*Campylobacter* is one of the leading causes of foodborne diarrheal illness in the U.S. and a significant public health concern worldwide [1]. The majority of human cases are attributed to *C. jejuni*, although human infections due to *C. coli* are estimated to be underreported [2]. Food animals like poultry, cattle and swine are common hosts and important reservoirs of *Campylobacter* species. Swine are known to be a main animal reservoir of *C. coli* [3],while *C. jejuni* is prevalent in poultry [4] and cattle [5].

The use of antimicrobials in food animals for disease treatment, prophylaxis, and to enhance animal growth [6,7] has been under scrutiny due to associations with the development of resistant bacteria to important antimicrobials in human medicine [8]. Antimicrobial resistant (AR) *Campylobacter* have been shown to be shed by swine, poultry, and cattle reared in conventional production systems that use antimicrobials for treatment and growth promotion [9–11]. However, studies have demonstrated that AR *Campylobacter* strains are also prevalent in farm animals from antimicrobial-free (ABF) and organic production systems [12,13]. Concerns of *C. coli* as a zoonotic pathogen have been highlighted due to the presence of AR strains in higher frequency than *C. jejuni* [14]. *C. coli* from swine have exhibited a diversity of multidrug-resistant (MDR) profiles that persist throughout the production stages at farm and slaughter, even in the absence of antimicrobial selection pressure [11]. The prevalence of *C. coli* strains exhibiting higher frequency of resistance to antimicrobials than *C. jejuni* further highlights the importance of conducting molecular epidemiologic studies of this zoonotic pathogen.

Multilocus sequence typing (MLST) has been a valuable technique for the determination of epidemiological associations between human *Campylobacter* cases and sources of infection [15], as well as for the characterization of bacterial population structures [16]. MLST schemes have been developed for *C. jejuni* [16] and *C. coli* [17] among various bacterial species, indexing variation in seven housekeeping genes. This method has several advantages, among which are the generation of portable data, and the availability of a centralized internet database [18]. Previous work performed by our laboratory have used MLST for molecular epidemiologic studies of *C. coli* in swine [3,19,20], and for the genotyping of human and retail meat isolates [21].

We recently reported the high prevalence of MDR *C. coli* from pigs reared in conventional and ABF production systems at farm, slaughter, and the environment. To better understand the persistence of MDR *C. coli* strains in swine production, we proposed to genotypically characterize the *C. coli* population in the ABF and conventional swine production systems at farm, slaughter and environment using MLST. An important objective was to determine whether the same *C. coli* population existed in the two production systems, and whether the presence or absence of antimicrobial use selected for a particular population subset. Our aim was to determine (i) the genetic variability or similarity of the ABF and conventional *C. coli* populations, (ii) whether the phenotypic diversity of *C. coli* observed at farm, slaughter, and environment is replicated at the genotypic level, and most importantly (iii) the role played by the environment as a potential reservoir of MDR *C. coli* for pigs at farm and slaughter in the two production systems.

## Materials and Methods

### Origin of *C. coli* Isolates

The *C. coli* isolates were recovered as part of a two year longitudinal study involving pigs reared in two distinct production systems in North Carolina. Pigs reared under the conventional systems received antimicrobials for therapeutic and growth promotion purposes, while in ABF systems there was absence of antimicrobial use. The details regarding antimicrobials administered to conventional pigs, sample collection, *C. coli* prevalence in pigs at farm, slaughter, and their environment, and phenotypic and molecular characterization of isolates were previously reported [11]. Briefly, a total of thirty conventional and eight ABF farms were selected for sampling. Fecal samples from cohorts of 35 pigs and their environment (feed, water, soil, drag swabs, lagoon, and truck swabs) were collected from sows post-farrowing and piglets (7-to-10 days of age), twice at the nursery stage (4 and 7 weeks of age), and twice again at the finishing (16 and 26 weeks of age) stages. Sampling of carcasses from the same cohort of pigs (post-evisceration and post-chill) and mesenteric lymph nodes (MLN) was completed, while the lairage and truck swab samples were collected as part of the slaughter environment.

Overall, the *Campylobacter* prevalence on farm was 69.3% in the pigs (ABF: 72.9%, conventional: 66.6%) and 26.3% in their environment (ABF: 20.5%, conventional: 29.9%). The *Campylobacter* prevalence at slaughter was 23% in carcass (ABF: 27.6%, conventional: 19.8%) and 16.1% at the slaughter plant environment (ABF: 14.8%, conventional: 17.8%). *C. coli* was the predominant *Campylobacter* species identified (99.7%) out of the 2,908 that were isolated. The rest of the isolates (0.3%) corresponded to *C. jejuni.*

### Multilocus Sequence Typing

A total of 200 *C. coli* isolates (Table S1) were selected for MLST representing swine and environmental isolates at farm (farrowing, nursery, and finishing stages), and slaughter (carcass and environment) of the ABF ($n = 100$) and conventional ($n = 100$) production systems. Isolate selection was representative of the total number of isolates and antimicrobial susceptibility phenotypes of different swine cohorts at farm, slaughter, and environment. Bacterial DNA was purified with a DNeasy Blood & Tissue kit (Qiagen, Valencia, CA) and quantitated (NanoDrop 2000c; Thermo Scientific Inc., Waltham, MA). DNA purity was assessed based on the $A_{260}/A_{280}$ ratio. Seven housekeeping genes (*aspA, glnA, gltA, glyA, pgm, tkt, uncA*) between 402–507 bp were PCR amplified as previously described for *C. coli* MLST typing [17]. PCR products were enzymatically purified (ExoSAP-IT; Affymetrix Inc., Santa Clara, CA) prior to sequencing. Sequencing reactions consisted of 1 µl BigDye (Applied Biosystems, Carlsbad, CA), 1 µl of cleaned PCR product (6–7 ng/µl), 1 µl 3.2 pmol sequencing primer, 2 µl of 5× buffer, and $dH_2O$ to a final volume of 10 µl. Cycle sequencing of cleaned reactions (CleanSEQ; Beckman Coulter, Brea, CA) was performed in a ABI 3730×l DNA Analyzer (Applied Biosystems), and sequencing parameters consisted of 96°C for 1 min, 25 cycles of 96°C for 10 s, 50°C for 5 s, and 60°C for 4 min. Final hold was at 4°C.

### Genetic Diversity and Phylogenetic Analysis

Sequence data from the ABF and conventional *C. coli* isolates was analyzed and queried against the *Campylobacter* database at http://pubmlst.org/campylobacter/for the determination of allelic profiles. New allele sequences and sequence types (STs) were submitted at http://pubmlst.org and incorporated in the *Campylobacter* database. The genetic diversity at individual loci, mean genetic diversity (*H*) and standardized index of association ($I^s_A$) between loci were calculated with LIAN program version 3.5 [22]. Linkage equilibrium, characterized by statistical independence of alleles at all loci and indicative of a freely recombining population, was determined by an $I^s_A$ equal to zero. Related STs corresponding to *C. coli* isolates from ABF and conventional production systems were grouped and predicted founders were identified using eBURST v3 algorithm [23]. Groups of STs were defined under a conservative approach by sharing alleles at ≥6 of the 7 loci. The ratio of non-synonymous to synonymous substitutions per site ($d_N/d_S$) were computed in S.T.A.R.T.2 software [24] using the Nei-Gojobori method. A radial neighbor-joining tree was elaborated with the concatenated allele sequences in MEGA software version 5.0 [25] for the unique STs in the ABF and conventional production systems. BioNumerics version 6.6 (Applied Maths Inc., Austin, TX) was utilized in the construction of a minimum spanning tree with the complete ABF and conventional *C. coli* ST data.

## Results

### Multilocus Sequence Typing (MLST) of *C. coli* Isolates

A total of 51 STs were observed, with unique STs (detected only in specific production systems) in ABF ($n = 13$) and conventional ($n = 32$) *C. coli* isolates. Two novel *C. coli* alleles (*glnA* 424 and *glyA* 464) and 14 novel STs (ABF, $n = 2$; conventional, $n = 12$) were identified and reported for the first time. These new STs were identified at farm, slaughter, and environment of ABF and conventional production systems. Common STs were detected between ABF and conventional *C. coli* isolates at farm, slaughter and/or environment as described in Table 1.

Clonal complex ST-828 was the single lineage identified and was represented by 68% of the ABF and 66% of conventional *C. coli* isolates. ST-854 was the most prevalent ST in both production systems (ABF: 31%, conventional: 17%). STs belonging to ABF *C. coli* isolates were assigned by eBURST to one main group formed by 58 isolates and 11 STs, with ST-828 as the predicted founder of the group. Conventional *C. coli* STs were assigned to two groups. The primary founder for the first group was ST-1096, and the group was constituted by 75 isolates and 22 STs. Alternatively, a presumed founder was not identified for the second group due to the limited number of STs ($n = 2$). Eight ABF and 14 conventional STs were considered as singletons.

### Genetic Diversity of *C. coli* from ABF and Conventional Systems

Genetic diversity at individual loci for the ABF and conventional *C. coli* allelic profiles is shown in Table 2. We detected a

**Table 1.** Common STs between ABF and conventional *C. coli* isolates.

| | ST-828 [a] | | ST-854 | | ST-1068 | | ST-1096 | | ST-1185 | | ST-1186 | |
|---|---|---|---|---|---|---|---|---|---|---|---|---|
| Source | ABF (n=3) | Conv (n=4) | ABF (n=31) | Conv (n=17) | ABF (n=12) | Conv (n=6) | ABF (n=1) | Conv (n=3) | ABF (n=5) | Conv (n=3) | ABF (n=17) | Conv (n=1) |
| Pigs | 3 (100) | 0 | 14 (45.2) | 3 (17.6) | 3 (25) | 3 (50) | 0 | 2 (66.7) | 3 (60) | 3 (100) | 6 (35.3) | 0 |
| Farm env [b] | 0 | 2 (50) | 14 (45.2) | 10 (58.8) | 5 (41.7) | 3 (50) | 0 | 1 (33.3) | 0 | 0 | 6 (35.3) | 0 |
| Carcass | 0 | 0 | 3 (9.7) | 1 (5.9) | 3 (25) | 0 | 0 | 0 | 1 (20) | 0 | 4 (23.5) | 0 |
| Slaughter env | 0 | 2 (50) | 0 | 3 (17.6) | 1 (8.3) | 0 | 1 (100) | 0 | 1 | 0 | 1 (5.9) | 1 (100) |

[a]No. of isolates (%).
[b]Environment.

mean genetic diversity (*H*) of 0.3963+/−0.0806 and 0.4655+/−0.0714 among *C. coli* isolates from the ABF and conventional production systems, respectively. A greater difference in genetic diversity was observed in the *pgm* (ABF: 0.0396, conventional: 0.3448), *tkt* (ABF: 0.6135, conventional: 0.7042), and *uncA* (ABF: 0.5228, conventional: 0.6121) genes. Allelic diversity was higher in the *glyA* locus of both systems with a greater number of alleles (ABF: 6, conventional: 7). The index of association ($I_A^S$) close to zero indicated that the *C. coli* populations were in proximity to linkage equilibrium (ABF: 0.1513, conventional: 0.0991). The non-synonymous to synonymous substitution ratio ($d_N/d_S$) was determined to be less than one at each locus as shown in Table 2. Testing for selection suggested that nucleotide changes observed did not result in amino acid changes.

## Distribution of *C. coli* STs and AR Profiles at Farm, Slaughter and Environment

Clonal complex ST-828 consisted of AR and pansusceptible *C. coli* strains from the ABF and conventional production systems. At farm, predominant ST-854 was observed in *C. coli* isolates from five ABF cohorts (A2, A3, A4, A5, and A8) and their environment (Table S1). Within an ABF cohort, ST-854 was found at farm, in swine and environmental isolates, and at slaughter in carcass only. The majority of ABF isolates with ST-854 were associated with MDR profiles (AZI-ERY-TEL-CLI, $n = 7$; and AZI-ERY-TET, $n = 2$), TET resistance ($n = 16$). Conventional isolates assigned to ST-854 were also distributed at farm ($n = 10$) and slaughter ($n = 3$) however, the majority were observed in the respective environments. Conventional *C. coli* isolates with ST-854 predominantly exhibited the MDR phenotype CIP-TET-NAL ($n = 10$).

ST-1186 was mainly distributed in two ABF cohorts (A6 and A7) at farm, from farrowing to finishing, and at slaughter. MDR profiles belonging to ST-1186 were AZI-ERY-TET-TEL ($n = 14$), and AZI-ERY-TET ($n = 3$). Resistance to seven antimicrobials (AZI-CIP-ERY-TET-NAL-TEL-CLI) was observed in ABF *C. coli* isolates ($n = 3$) with ST-5377, which was able to persist from farm to slaughter.

Slaughter carcasses presented unique STs ($n = 5$) absent at the farm and slaughter plant environment. The majority of these STs ($n = 4$) were from *C.coli* strains exhibiting resistance to TET, AZI-ERY-TET-TEL-CLI, AZI-ERY-TET and AZI, ERY, TET, TEL. *C. coli* isolates from ABF lairage presented a greater diversity of STs ($n = 8$) compared to conventional lairage isolates ($n = 3$). Furthermore, half of the ABF lairage STs ($n = 4$) exhibited MDR profiles (ST-1068, AZI-ERY-CLI; ST-1096, AZI-ERY-TEL-CLI; ST-1186, AZI-ERY-TET-TEL; ST-5377, AZI-CIP-ERY-TET-NAL-TEL-CLI). Lairage and truck isolates from the slaughter plant environment of both production systems presented unique STs that were not observed previously at farm (ABF: 4; conventional: 4).

**Table 2.** Genetic diversity of *C. coli* from ABF and conventional production systems.

| | ABF | | | Conventional | | |
|---|---|---|---|---|---|---|
| Locus | Genetic diversity | Alleles (*n*) | $d_N/d_S$ [a] | Genetic diversity | Alleles (*n*) | $d_N/d_S$ |
| *aspA* | 0.1836 | 3 | 0.1026 | 0.2014 | 3 | 0.1026 |
| *glnA* | 0.5065 | 3 | 0.2747 | 0.5800 | 5 | 0.0913 |
| *gltA* | 0.3578 | 2 | – | 0.2818 | 4 | 0.1121 |
| *glyA* | 0.5505 | 6 | 0.0388 | 0.5341 | 7 | 0.0317 |
| *pgm* | 0.0396 | 2 | 0 | 0.3448 | 4 | 0 |
| *tkt* | 0.6135 | 5 | 0.0780 | 0.7042 | 6 | 0.0620 |
| *uncA* | 0.5228 | 3 | 0 | 0.6121 | 4 | 0 |
| *H* [b] | 0.3963+/−0.0806 | | | 0.4655+/−0.0714 | | |
| $I_A^S$ | 0.1513 | | | 0.0991 | | |

[a]Non-synonymous to synonymous ratio.
[b]Mean genetic diversity.
$I_A^S$ Standardized index of association.

### Phylogenetic Analysis of ABF and Conventional *C. coli* STs

The evolutionary relationships of the unique ABF ($n = 13$) and conventional ($n = 32$) *C. coli* STs, exclusive to either production system, are represented in a radial neighbor-joining tree (Figure 1). These unique STs were observed in ABF and conventional *C. coli* isolates at farm, slaughter and environment after repeated sampling. The phylogenetic tree displays the close clustering of these ABF and conventional STs with overlap suggesting genetically similar populations. As the tree branch lengths are equivalent to the evolutionary distances, divergence of some of the conventional and ABF *C. coli* STs from the main cluster is observed. This is suggestive of rapidly evolving *C. coli* STs at farm and slaughter. Reference is made to the distant cluster formed by ST-5770, ST-1827, ST-1109, ST-5772, which included only conventional MDR *C. coli* strains. When queried against the MLST database, these *C. coli* STs revealed shared allele's (*aspA* 33, *glnA* 153, *glnA* 39, and *gltA* 30) with *C. jejuni* isolates from U.S. cattle and humans. Other outlier STs were observed in ABF (ST-5766, ST-1450, and ST-5764) and conventional (ST-5248, ST-1097, and ST-4938) *C. coli* isolates. The minimum spanning tree (Figure 2) created with all the ABF and conventional *C. coli* ST data revealed the close clustering of swine and carcass STs with those of the environment at farm and slaughter, the majority of which exhibited one allele difference. As shown in this figure, several STs (ST-854, ST-1068, ST-1096, ST-1185, ST-1186) shared isolates from farm, slaughter, and environment of ABF and conventional production systems.

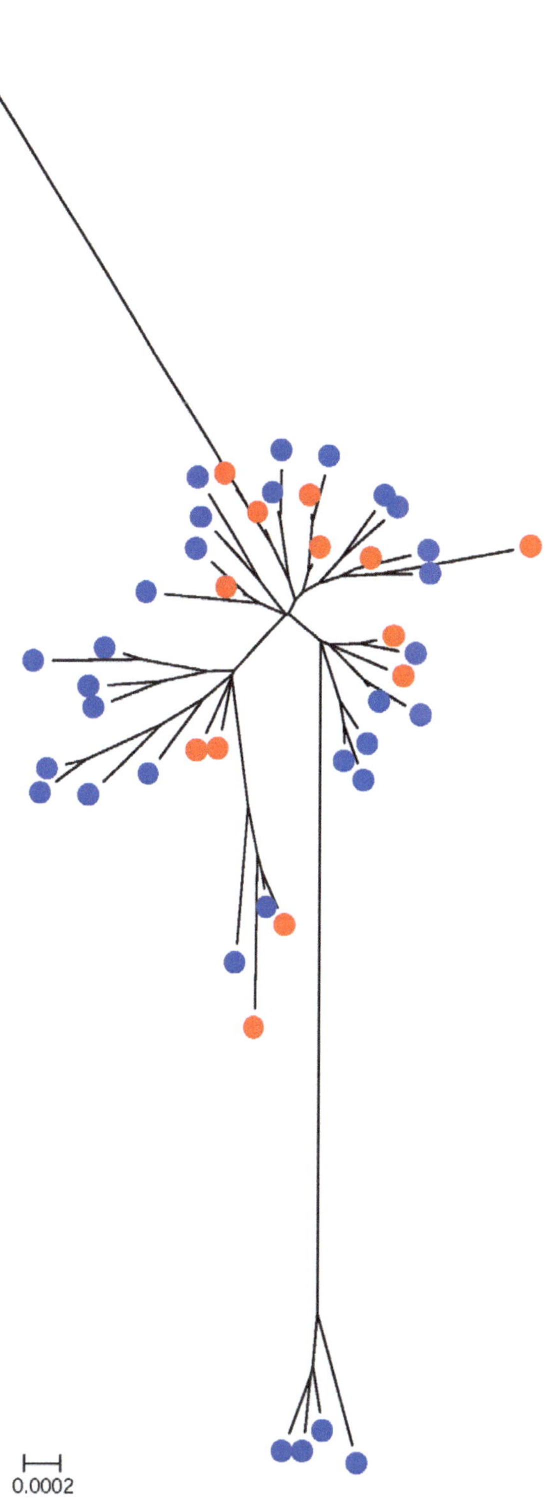


**Figure 1. Radial neighbor-joining tree of the unique ABF and conventional *C. coli* STs.** ABF and conventional STs are represented by red and blue circles, respectively.

## Discussion

There has been extensive debate regarding associations of antimicrobial use in food animals and the development of antimicrobial resistant bacterial strains that can be transferred to humans [7]. In the U.S swine industry antimicrobials are commonly used for therapeutic and non-therapeutic purposes like growth promotion [11,26]. There has been tremendous pressure on the food animal industry to withdraw the use of sub-therapeutic doses of antimicrobials for growth promotion purposes. A subsequent outcome due to increasing trends in AR bacterial pathogens has been the growth of the organic and ABF food animal industry which prohibits the use of antimicrobials for growth promotion. We recently reported the high prevalence of AR *C. coli* isolated from ABF pigs at farm and slaughter that were never exposed to antimicrobials [11].

Consequently, we aimed to determine the population structure of *C. coli* isolated from ABF reared pigs, and compared it to the *C. coli* population recovered from conventional reared pigs using MLST. The phenotypic diversity of AR *C. coli* from ABF and conventional systems was confirmed at the genotypic level. The linkage equilibrium values of the two *C. coli* swine populations were suggestive of a recombining and weakly clonal population. Based on $I_A^s$ estimates, a lower degree of clonality was observed when compared to results from a previous study (ABF: 0.279, conventional: 0.535) in the same geographic region [3], albeit from different swine farms, no environmental isolates studied, and almost eight years of difference with the present study. The weakly clonal population of *C. coli* contrasts with with highly clonal bacterial populations like those of *Escherichia coli* and *Salmonella* [27]. Recombination has proven to be higher in *C. coli* when compared to *C. jejuni* with a more efficient exchange of alleles [28,29]. The observation of more novel STs compared to alleles also suggests that recombination events are more prone to occur than mutations. The majority of these novel STs belonged to MDR *C. coli* strains at farm, slaughter, and environment. Recombination could explain the genetic diversity observed in AR *C. coli* from ABF and conventional swine systems, which can rapidly occur even in the absence of selective pressure [30].

Clonal complex ST-828 was the single lineage identified in ABF and conventional *C. coli* isolates. This genotypic group seemed to

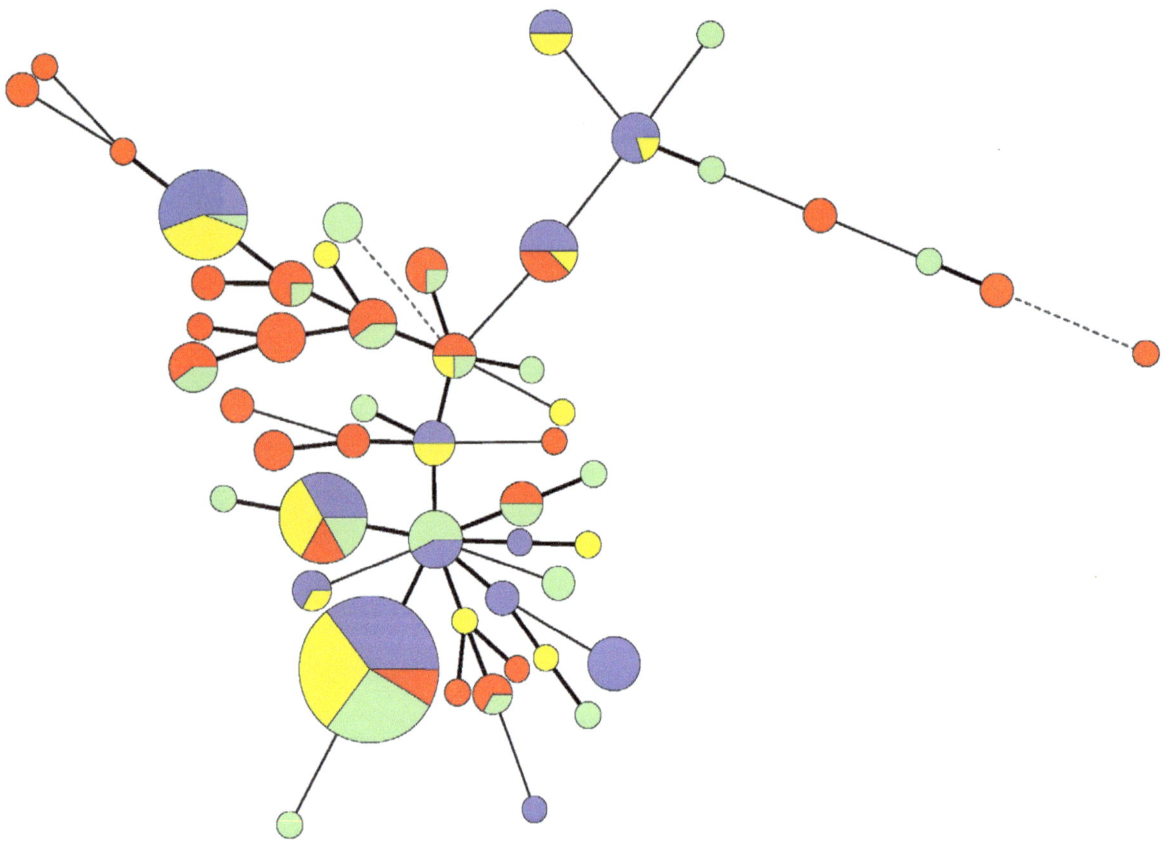

**Figure 2. Minimum spanning tree (MST) of *C. coli* isolates from ABF and conventional production systems.** Each ST is represented by a node. The size of each node is proportional to the number of strains that comprise that ST. Pie charts with differential colors represent *C. coli* sources: ABF pigs and carcass (purple), ABF environment at farm and slaughter (yellow), conventional pigs and carcass (red), and conventional environment at farm and slaughter (green). Allele differences are represented by thick bold branch lines (single locus variants), thin continuous lines (double locus variants), and dashed lines (three allele differences). Unique STs of the pigs, carcass, and environment at farm and slaughter from ABF and conventional systems were distinguished by single colored nodes.

be very well adapted in these two swine productions systems as previously reported [19], with a similar distribution at farm, slaughter, and environment. In the U.S, swine is the main source representing this lineage followed by chickens, cattle and sporadic human cases [19,31]. Within this lineage is ST-854, a prevalent ST in ABF and conventional *C. coli* isolates, which in the U.S has been exclusively reported in pigs [3,31]. Niche adaptation of clonal complexes has also been described in *C. jejuni* isolated from humans, poultry, cattle, and sheep [32]. In our study, ST-1096 (member of the ST-828 complex) was identified as the predicted founder for *C. coli* from conventional production systems. This ST is present in pigs, chickens, human stools, and environmental waters (http://pubmlst.org/campylobacter). A high frequency of *C. coli* ST-1068 from ABF swine farms was observed, even though cattle have been reported as the major source of this ST [31]. In the present study, half of the ABF swine farms sampled raised cattle in their fields, representing a potential for the exchange of *C. coli* between cattle and pigs.

Identical STs were observed in *C. coli* isolates from ABF and conventional farm, slaughter, and environment. This was also true between different farms within a production system, suggesting these STs were stable enough to persist and disseminate throughout the different production stages at farm. Pigs from the same cohort also shared the same STs at farm. Novel STs were well adapted to either one of the production systems, while others like ST-854 were detected in both. Sow and piglet STs were diverse enough that genotypic similarities between them were not observed in the majority of the ABF and conventional isolates. This is consistent with previous work where sows and piglets from the same barn did not share the same genotypes by pulse field gel electrophoresis [33]. As previously described [10], associations between ST and AR profile could not be established in our study. *C. coli* isolates with the same ST could exhibit resistance to multiple AR profiles and vice versa. An exception was the ABF MDR pattern resistant to seven antimicrobials AZI-CIP-ERY-TET-NAL-TEL-CLI, which was exclusively observed in isolates with ST- 5377.

We identified diverse STs at the ABF and conventional slaughter plants. In particular, the ABF lairage displayed more STs from AR *C. coli* strains than the conventional lairage. These STs were unique and not observed at the farm level, suggesting the lairage is a potential place for the transmission of AR *C. coli* between pigs from different ABF farms. Unique STs absent at farm or at the slaughter plant environment were also observed in *C. coli* isolates from ABF and conventional swine carcasses at post-evisceration, post-chill, and in MLN samples. It was interesting to note that these diverse *C. coli* STs exhibited varied AR profiles (AZI-ERY-TEL-CLI, AZI-ERY-TET, and AZI-ERY-TET-TEL). Furthermore, some of these STs were present within the same cohort of pigs at different slaughter stages including post-chill, signaling cross-contamination of carcasses with MDR *C. coli* at the slaughter plant level. Genotypes specific to the slaughter plant demonstrate that pigs that do not carry MDR *C. coli* at farm can acquire these AR strains during processing and present them at the final end product. Cross-contamination has also been associated with the slaughtering process of chickens [34]. A study comparing *Campylobacter* genotypes from a chicken flock before and after slaughter revealed a greater diversity at this stage, with 48% of the carcass STs undetected prior to slaughter [35].

As demonstrated by our results, the swine environment plays a key role in the persistence of MDR *C. coli* strains at farm and slaughter despite distinct management practices. This is specifically important when we consider the presence of AR resistant *C. coli* strains in ABF pigs which are raised in the absence of antimicrobial exposure. AR *C. coli* strains were able to persist and disseminate in the absence of antimicrobial selection pressure, likely explained by a common population structure between *C. coli* from conventional and ABF systems. There was a clear overlap of swine and environmental STs, although we cannot ascertain the direction in which *C. coli* was transmitted. Additionally, the diversity of STs present in the environment at farm and slaughter, including unique STs, suggest the environment plays a role in increasing the genetic diversity, as has been previously reported [36].

In conclusion, phylogenetic analysis revealed a genotypically diverse *C. coli* population with the presence of *C. coli* isolates sharing a common ancestry in both production systems.

## Acknowledgments

The authors would like to thank the swine producer's and swine processing facilities in North Carolina for sample collection. This publication made use of the *Campylobacter* Multi Locus Sequence Typing website (http://pubmlst.org/campylobacter/) [18].

## Author Contributions

Conceived and designed the experiments: ST. Performed the experiments: MQH. Analyzed the data: MQH. Contributed reagents/materials/analysis tools: MQH ST. Wrote the paper: MQH ST.

## References


1. Mead PS, Slutsker L, Dietz V, McCaig LF, Bresee JS, et al. (1999) Food-related illness and death in the United States. Emerg Infect Dis 5: 607–625.
2. Gürtler M, Alter T, Kasimir S, Fehlhaber K (2005) The importance of *Campylobacter coli* in human campylobacteriosis: prevalence and genetic characterization. Epidemiol Infect 133: 1081–1087.
3. Thakur S, Gebreyes WA (2005) *Campylobacter coli* in swine production: antimicrobial resistance mechanisms and molecular epidemiology. J Clin Microbiol 43: 5705–5714.
4. Hermans D, Pasmans F, Messens W, Martel A, Van Immerseel F, et al. (2012) Poultry as a host for the zoonotic pathogen *Campylobacter jejuni*. Vector Borne Zoonotic Dis 12: 89–98.
5. Grove-White DH, Leatherbarrow AJ, Cripps PJ, Diggle PJ, French NP (2010) Temporal and farm-management-associated variation in the faecal-pat prevalence of *Campylobacter jejuni* in ruminants. Epidemiol Infect 138: 549–558.
6. McEwen SA, Fedorka-Cray PJ (2002) Antimicrobial use and resistance in animals. Clin Infect Dis 34 Suppl 3: S93–S106.
7. Dibner JJ, Richards JD (2005) Antibiotic growth promoters in agriculture: history and mode of action. Poult Sci 84: 634–643.
8. Castanon JI (2007) History of the use of antibiotic as growth promoters in European poultry feeds. Poult Sci 86: 2466–2471.
9. Châtre P, Haenni M, Meunier D, Botrel MA, Calavas D, et al. (2010) Prevalence and antimicrobial resistance of *Campylobacter jejuni* and *Campylobacter coli* isolated from cattle between 2002 and 2006 in France. J Food Prot 73: 825–831.
10. Gu W, Siletzky RM, Wright S, Islam M, Kathariou S (2009) Antimicrobial susceptibility profiles and strain type diversity of *Campylobacter jejuni* isolates from turkeys in eastern North Carolina. Appl Environ Microbiol 75: 474–482.
11. Quintana-Hayashi MP, Thakur S (2012) Longitudinal study of the persistence of antimicrobial-resistant *Campylobacter* strains in distinct swine production systems on farms, at slaughter, and in the environment. Appl Environ Microbiol 78: 2698–2705.
12. Rollo SN, Norby B, Bartlett PC, Scott HM, Wilson DL, et al. (2010) Prevalence and patterns of antimicrobial resistance in *Campylobacter* spp. isolated from pigs reared under antimicrobial-free and conventional production methods in eight states in the Midwestern United States. J Am Vet Med Assoc 236: 201–210.
13. Tadesse DA, Bahnson PB, Funk JA, Thakur S, Morrow WE, et al. (2011) Prevalence and antimicrobial resistance profile of *Campylobacter* spp. isolated from conventional and antimicrobial-free swine production systems from different U.S. regions. Foodborne Pathog Dis 8: 367–374.
14. Luangtongkum T, Jeon B, Han J, Plummer P, Logue CM, et al. (2009) Antibiotic resistance in *Campylobacter*: emergence, transmission and persistence. Future Microbiol 4: 189–200.
15. Sheppard SK, Dallas JF, Strachan NJ, MacRae M, McCarthy ND, et al. (2009) *Campylobacter* genotyping to determine the source of human infection. Clin Infect Dis 48: 1072–1078.
16. Dingle KE, Colles FM, Wareing DR, Ure R, Fox AJ, et al. (2001) Multilocus sequence typing system for *Campylobacter jejuni*. J Clin Microbiol 39: 14–23.
17. Dingle KE, Colles FM, Falush D, Maiden MC (2005) Sequence typing and comparison of population biology of *Campylobacter coli* and *Campylobacter jejuni*. J Clin Microbiol 43: 340–347.
18. Jolley KA, Maiden MC (2010) BIGSdb: Scalable analysis of bacterial genome variation at the population level. BMC Bioinformatics 11: 595.
19. Thakur S, Morrow WE, Funk JA, Bahnson PB, Gebreyes WA (2006) Molecular epidemiologic investigation of *Campylobacter coli* in swine production systems, using multilocus sequence typing. Appl Environ Microbiol 72: 5666–5669.
20. Thakur S, Gebreyes WA (2010) Phenotypic and genotypic heterogeneity of *Campylobacter coli* within individual pigs at farm and slaughter in the U.S. Zoonoses Public Health 57 Suppl 1: 100–106.
21. Thakur S, White DG, McDermott PF, Zhao S, Kroft B, et al. (2009) Genotyping of *Campylobacter coli* isolated from humans and retail meats using multilocus sequence typing and pulsed-field gel electrophoresis. J Appl Microbiol 106: 1722–1733.
22. Haubold B, Hudson RR (2000) LIAN 3.0: detecting linkage disequilibrium in multilocus data. Linkage Analysis. Bioinformatics 16: 847–848.
23. Feil EJ, Li BC, Aanensen DM, Hanage WP, Spratt BG (2004) eBURST: inferring patterns of evolutionary descent among clusters of related bacterial genotypes from multilocus sequence typing data. J Bacteriol 186: 1518–1530.
24. Jolley KA, Feil EJ, Chan MS, Maiden MC (2001) Sequence type analysis and recombinational tests (START). Bioinformatics 17: 1230–1231.
25. Tamura K, Peterson D, Peterson N, Stecher G, Nei M, et al. (2011) MEGA5: molecular evolutionary genetics analysis using maximum likelihood, evolutionary distance, and maximum parsimony methods. Mol Biol Evol 28: 2731–2739.
26. Engberg J, Aarestrup FM, Taylor DE, Gerner-Smidt P, Nachamkin I (2001) Quinolone and macrolide resistance in *Campylobacter jejuni* and *C. coli*: resistance mechanisms and trends in human isolates. Emerg Infect Dis 7: 24–34.
27. Smith JM, Smith NH, O'Rourke M, Spratt BG (1993) How clonal are bacteria? Proc Natl Acad Sci U S A 90: 4384–4388.
28. Gripp E, Hlahla D, Didelot X, Kops F, Maurischat S, et al. (2011) Closely related *Campylobacter jejuni* strains from different sources reveal a generalist rather than a specialist lifestyle. BMC Genomics 12: 584.
29. Litrup E, Torpdahl M, Nielsen EM (2007) Multilocus sequence typing performed on *Campylobacter coli* isolates from humans, broilers, pigs and cattle originating in Denmark. J Appl Microbiol 103: 210–218.
30. de Boer P, Wagenaar JA, Achterberg RP, van Putten JP, Schouls LM, et al. (2002) Generation of *Campylobacter jejuni* genetic diversity in vivo. Mol Microbiol 44: 351–359.
31. Miller WG, Englen MD, Kathariou S, Wesley IV, Wang G, et al. (2006) Identification of host-associated alleles by multilocus sequence typing of *Campylobacter coli* strains from food animals. Microbiology 152: 245–255.
32. Dingle KE, Colles FM, Ure R, Wagenaar JA, Duim B, et al. (2002) Molecular characterization of *Campylobacter jejuni* clones: a basis for epidemiologic investigation. Emerg Infect Dis 8: 949–955.
33. Hume ME, Droleskey RE, Sheffield CL, Harvey RB (2002) *Campylobacter coli* pulsed field gel electrophoresis genotypic diversity among sows and piglets in a farrowing barn. Curr Microbiol 45: 128–132.
34. Johnsen G, Kruse H, Hofshagen M (2006) Genotyping of *Campylobacter jejuni* from broiler carcasses and slaughterhouse environment by amplified fragment length polymorphism. Poult Sci 85: 2278–2284.
35. Colles FM, McCarthy ND, Sheppard SK, Layton R, Maiden MC (2010) Comparison of *Campylobacter* populations isolated from a free-range broiler flock before and after slaughter. Int J Food Microbiol 137: 259–264.
36. Lévesque S, St-Pierre K, Frost E, Arbeit RD, Michaud S (2012) Use of amplified-fragment length polymorphism to study the ecology of *Campylobacter jejuni* in environmental water and to predict multilocus sequence typing clonal complexes. Appl Environ Microbiol 78: 2470–2473.

# 4

# Cross-Sectional Study on Prevalence and Molecular Characteristics of Plasmid Mediated ESBL/AmpC-Producing *Escherichia coli* Isolated from Veal Calves at Slaughter

**Joost Hordijk[1,2]*, Jaap A. Wagenaar[1,2], Arie Kant[2], Alieda van Essen-Zandbergen[2], Cindy Dierikx[2], Kees Veldman[2], Ben Wit[3], Dik Mevius[1,2]**

**1** Department of Infectious Diseases and Immunology, Faculty of Veterinary Medicine, Utrecht University, Utrecht, The Netherlands, **2** Central Veterinary Institute, part of Wageningen UR, Lelystad, The Netherlands, **3** Netherlands Food and Consumer Product Safety Authority, Utrecht, The Netherlands

## Abstract

***Objectives:*** The presence of ESBL/AmpC-producing *E. coli* in cattle has been reported previously, however information on veal calves is limited. This study describes the prevalence and molecular characteristics of *E. coli* with non-wild type susceptibility to cefotaxime in veal calves at slaughter.

***Methods:*** Faecal samples from 100 herds, 10 individual animals per herd, were screened for *E. coli* with non-wild type susceptibility for cefotaxime. Molecular characterization of ESBL/AmpC genes and plasmids was performed on one isolate per herd by microarray, PCR and sequence analysis.

***Results:*** 66% of the herds were positive for *E. coli* with non-wild type susceptibility for cefotaxime. Within-herd prevalence varied from zero to 90%. 83% of *E. coli* producing ESBL/AmpC carried $bla_{CTX-M}$ genes, of which $bla_{CTX-M-1}$, $bla_{CTX-M-14}$ and $bla_{CTX-M-15}$ were most prevalent. The dominant plasmids were IncI1 and IncF-type plasmids.

***Conclusions:*** A relatively high prevalence of various $bla_{CTX-M}$ producing *E. coli* was found in veal calves at slaughter. The genes were mainly located on IncI1 and IncF plasmids.

**Editor:** Qijing Zhang, Iowa State University, United States of America

**Funding:** This study was funded by the Dutch Ministry of Economic Affairs, Agriculture and Innovation (project 3201949) and the Product Boards for Livestock and Meat (project 08.30.002). The funders had no role in study design, data collection and analysis, decision to publish, or preparation of the manuscript.

**Competing Interests:** The authors have declared that no competing interests exist.

* E-mail: J.Hordijk@uu.nl

## Introduction

The development of resistance to extended spectrum cephalosporins (ESC) has evolved rapidly world-wide in both clinical settings as well as in the community. Resistance to ESC is mainly caused by various extended spectrum ß-lactamases (ESBL) or AmpC ß-lactamases (AmpC) that are often found in *Enterobacteriaceae* and hydrolyse the ß-lactam ring of these antibiotics [1]. These enzymes are encoded by genes that are frequently located on mobile genetic elements (plasmids) [2], which have the ability to transfer horizontally within and between different bacterial species. Novel ESBL/AmpC genes or gene variants are reported on a regular basis (http://www.lahey.org/studies) and are classified based on either their functional characteristics [3] or primary structure [4].

Food-producing animals have been suggested as the primary reservoir of zoonotic foodborne pathogens, including antimicrobial resistant bacteria [5]. Faecal carriage of ESBL/AmpC-producing *E. coli* in cattle has been reported previously [6,7,8,9,10]. In addition, a recent study showed that in the period from 2005 to 2010, resistance to ESC in veal calves was mainly caused by plasmid mediated beta-lactamases, in contrast to the years before 2005 (Hordijk *et al*, submitted for publication). Furthermore, beef, chicken and pig meat products have also been found to be positive for ESBL/AmpC-producing bacteria [11], which may constitute a transmission route to humans. However, studies are difficult to compare, because isolation methods and sample sizes vary greatly. When cattle is studied, the type of cattle included is not always specified. For instance dairy farming is very different from veal calves by means of housing, life expectation and exposure to antibiotics. The majority of veal farms maintain an all-in, all-out system, while dairy farms generally maintain a closed production system. Almost each individual animal on a veal calf farm originates from another dairy farm. Furthermore, compared to dairy cattle, veal calves are more frequently exposed to antimicrobials [12]. These factors may influence the prevalence and genetic characteristics of antimicrobial resistance genes in isolates from these animals. Furthermore, antimicrobial susceptibility data from national surveillance studies have shown that isolates from veal calves are often multi resistant [12].

Therefore, the aim of this study was to determine the prevalence of *E. coli* with reduced susceptibility to ESC in veal calf herds. Moreover the ESBL/AmpC genes and plasmids on which they are located were identified to determine their molecular characteristics in relation to those found in other food-producing animals and humans.

## Materials and Methods

### Sampling design and isolation of Escherichia coli

From January to December 2011, fecal samples from 10 individual veal calves were taken from 100 slaughter batches (1000 samples in total). The samples were taken as part of the Dutch national control program on antimicrobial resistance by the Netherlands Food and Consumer Product Safety Authority (NVWA). All slaughter batches originated from different herds. To ensure anonymity, the origin of samples was aggregated at province level. The sampled slaughter batches were equally distributed over the five slaughter houses in the Netherlands and the sampling took place equally distributed over the year. All fecal samples were individually screened for ESC reduced susceptible *E. coli* using selective enrichment broth. Fecal swabs were taken from the colon of the carcass after evisceration and subsequently transferred to the laboratory and placed in a tube with 1 ml Luria-Bertani broth (Beckton Dickinson) supplemented with 1 mg/L cefotaxime (Sigma-Aldrich, Germany) (LB+). LB+ tubes were incubated aerobically at 37°C for 20±2 hours. Subsequently, the overnight LB+ culture was inoculated on a MacConkey agar plate (Becton Dickinson), which was supplemented with 1 mg/L cefotaxime (MC+). All MC+ plates were incubated aerobically at 37°C for 20±2 hours. From each MC+ plate showing growth, one colony typical for *E. coli* was selected, subcultured on blood agar, and suspended the next day in 1 ml of buffered peptone water supplemented with 30% glycerol and stored at −80°C, pending analysis. All selected *E. coli* were inoculated in 5 ml Gersbach medium and incubated at 37°C for 20±2 hours for *E. coli* confirmation. One droplet of Kovac's Indole reagent (Merck, Germany) was added to the Gersbach culture to check for tryptophan reduction following the requirements for *E. coli* identification stated by the European Food Safety Authority (EFSA) [13]. All indole-positive isolates were considered *E. coli.* From all isolates showing growth on MacConkey, one randomly picked isolate was selected per herd for antimicrobial susceptibility testing and molecular characterization.

### Antimicrobial susceptibility testing

All selected isolates (n = 66) were tested for susceptibility to antimicrobials by broth micro-dilution according to ISO standard 20776-1:2006 using microtitre trays with a custom made dehydrated panel of antibiotics (Sensititre, Trek Diagnostic Systems, Basingstoke, UK). The following antibiotics were included: ampicillin, cefotaxime, ceftazidime, tetracycline, sulfamethoxazole, trimethoprim, ciprofloxacin, nalidixic acid, chloramphenicol, florfenicol, gentamicin, kanamycin, streptomycin and colistin. All results were interpreted using cut-off values defined for epidemiological purposes as recommended by the European Committee on Antimicrobial Susceptibility Testing (EUCAST; http://mic.eucast.org/Eucast2/), apart from sulfamethoxazole, for which the clinical break point defined by the Clinical and Laboratory Standards Institute (CLSI) was used [14]. Values above these cut-offs and break points were interpreted as resistant (sulfamethoxazole) or non-wild type (all other antimicrobials) [15].

### Characterization of ESBL/AmpC genes and plasmids

All selected isolates were screened for ESBL/AmpC genes using the tube based amr-ve-05 microarray (Alere, Tilburg, the Netherlands) [16]. Beta-lactamase gene families identified by this microarray were subsequently characterized by PCR and sequence analysis as described previously [17]. For sequence analysis the following additional primers were used: TEM-Fseq: 5′-GCCAACTTACTTCTGACAACG, CMY-F-838: 5′-TGGCGTATTGGCGATATGTA and CMY-R-857: 5′-TACATATCGCCAATACGCCA. Primers and conditions used to identify and sequence the modified aminoglycoside resistance gene *aac(6′)Ib-cr* have been described previously [18]. Plasmids were isolated using a modified miniprep method as follows. One colony of interest was inoculated in 3 ml LB broth and incubated overnight. Subsequently 1.5 ml of the culture was transferred and spinned down at 14000 rpm for 5 minutes. The pellet was suspended in 60 μl TEG buffer (25 mM Tris/Hcl; 10 mM EDTA; 50 mM glucose). Then, 120 μl NaOH/SDS (0.2 M/1%) was added and the whole sample was placed on ice for 5 minutes. Subsequently, 90 μl NaOH (3 M) was added and the sample was placed on ice for 5 minutes. The suspension was spinned down at 14000 rpm for 5 minutes. The supernatant was transferred and 270 μl LiCl (5 M) was added. After 10 minutes, the suspension was spinned down at 14000 rpm for 5 minutes. The supernatant was transferred and 1 ml of EtOH (96%) was added. The suspension was spinned down at 14000 rpm for 10 minutes and the pellet was resuspended in 200 μl EtOH (70%). The suspension was spinned down at 13000 rpm for 10 minutes and the pellet was resuspended in 20 μl $H_2O$. Plasmids were transformed into Electromax DH10B cells by electroporation (Invitrogen, USA) by mixing 2 μl of plasmid DNA suspension with 20 μl competent cells. The cells were electroporated under the following conditions: 1.25 kV, 200 ohm, 25 μFar. Transformants were subsequently plated on LB agar plates supplemented with 1 mg/L cefotaxime in order to selectively isolate transformants with ESBLs carrying plasmids. PCR-based replicon typing (PBRT) was conducted on the transformants to identify the replicon type of the plasmid inside the transformant [19]. Plasmid MLST (pMLST) [20] and replicon sequence typing (RST) were used to further characterize IncI1 and IncF plasmids, respectively [21]. Plasmids that were negative in the PBRT analysis were designated "not typable" (nt). The size of these plasmids was determined by PFGE with S1-nuclease digestion as described previously [22]. Because this study focused on plasmid mediated genes, another ESBL/AmpC suspected isolate from the same herd was included as a replacement if the micro array results were negative. For isolates in which no plasmid mediated ESBL/AmpC gene was detected, the promoter region of the chromosomal *ampC* gene was sequenced using primers that have been described previously [23]. Specific *ampC*-types were designated as described by Mulvey *et al* [24]. In addition a combination disk test was performed as described previously on all isolates that were negative in the array [17].

### Ethics Statement

Samples were taken from the colon of the veal calves after slaughter as part of the Dutch national control program on antimicrobial resistance as required by the zoonosis directive of the European Parliament (2003/99/EC). Since the sampling of calves was not performed on live animals, no approval from the ethical committee was required. In line with 'Regulation (EC) No 854/2004 of the European Parliament and of the Council, article 4 section 8', concerning specific rules for the organisation of official controls on products of animal origin intended for human consumption, the NVWA is the designated authority in the Netherlands to take samples for surveillance purposes. The NVWA granted permission to analyse the samples as described in this paper.

## Results

### Prevalence of non-wild type cefotaxime susceptible E. coli

From the 100 herds that were screened for the presence of *E. coli* with non-wild type susceptibility at slaughter, 66% were found positive for *E. coli* with non-wild type susceptibility to cefotaxime. The within-herd prevalence of *E. coli* with non-wild type susceptibility varied greatly, ranging from zero to 90%. However, in the majority of the positive farms, less than five out of ten fecal samples harbored *E. coli* with non-wild type cefotaxime susceptibility. From all positive farms 90% was located in the eastern and southern provinces. This is also where veal farming is concentrated. In addition, 86% of the ESBL/AmpC-producing *E. coli* was multi-drug resistant, showing non-wild type susceptibility to three or more antimicrobial classes, including ß-lactams (Tables 1 and 2).

Of all selected isolates with a non-wild type susceptibility to cefotaxime, 83% carried genes belonging to the $bla_{CTX-M}$ gene family (Table 1+2). The predominant genes within this $bla_{CTX-M}$ group, were $bla_{CTX-M-1}$ (45.5%), $bla_{CTX-M-14}$ (16.4%), and $bla_{CTX-M-15}$ (21.8%). One of the isolates harbouring $bla_{CTX-M-1}$ also harboured $bla_{TEM-52c}$. The remaining $bla_{CTX-M}$ variants were $bla_{CTX-M-2/97}$ (5.5%), $bla_{CTX-M-3}$ (3.6%) and $bla_{CTX-M-32}$ (7.2%). From 11 isolates (17%) that did not harbour $bla_{CTX-M}$, three harboured $bla_{TEM-52c}$ (4.5%), and one $bla_{CMY-2}$ (1.5%). In 6 isolates (9%), only mutations in the promoter region of the chromosomal *ampC* were detected. Finally, in one isolate (1.5%) neither plasmid mediated ESBL/AmpC genes nor mutations in the promoter region of the chromosomal *ampC* were detected. This isolate only harboured the ß-lactamases $bla_{TEM-1c}$ and $bla_{OXA-1}$.

The predominant plasmid types among all isolates harbouring ESBL/AmpC genes were IncI1 and IncF. Both IncI1 and IncF plasmids were present in 26% of the isolates depicted in Tables 1 and 2. The highest diversity of plasmids was observed among $bla_{CTX-M-1}$ positive isolates, which was also the most abundant ESBL variant. In 9 isolates, $bla_{CTX-M-1}$ was located on IncI1 with six different IncI1 pMLST types (Table 1). From these six different pMLST types, four types had not yet been reported in the pMLST database (http://www.pubmlst.org/plasmid, last accessed: 20 March 2013). Furthermore, in 7 isolates $bla_{CTX-M-1}$ was located on IncF plasmids with four different IncF RST profiles. From these four different profiles, two had not yet been reported in the pMLST database. The remaining $bla_{CTX-M-1}$ genes were located on IncB/O (three isolates), IncN (three isolates) and IncK (one isolate). In two isolates the plasmid was not typable. The second largest group of ESBL genes, $bla_{CTX-M-15}$, was predominantly carried on IncF plasmids, of which four different RST types were found among seven isolates (Table 1). From these seven isolates, five also harboured $bla_{OXA-1}$ and *aac(6')Ib-cr*. The remaining $bla_{CTX-M-15}$ genes were located on IncI1 (two isolates), IncHI2 (one isolate) and two on not typable plasmids. The $bla_{CTX-M-14}$ genes were mainly carried on IncK plasmids (Table 2), but also on IncI1, of which all three plasmids had the same pMLST sequence type (ST80), and one on IncF (F2; A-; B-)). From the isolates harbouring $bla_{CTX-M-2/97}$ or $bla_{CTX-M-3}$, all genes were located on different plasmids (Table 1+2). All plasmids carrying $bla_{CTX-M-32}$, were not typable. Furthermore, four isolates harboured $bla_{TEM-52c}$, of which three were carried on IncI1 plasmids with two different pMLST types (ST10 and ST36), both belonging to the same clonal complex (CC5) (Table 2). From the fourth $bla_{TEM-52c}$, which was found in an isolate also harbouring $bla_{CTX-M-1}$, transformation of the plasmid failed, so no replicon type was established. In the single isolate harbouring an *ampC* gene, $bla_{CMY-2}$ was located on an IncK plasmid (Table 2). Finally, one transformant harbouring $bla_{CTX-M-2}$ showed two replicon types, namely IncP and IncHI2 (Table 2).

## Discussion

This study showed that 66% of the slaughter batches of veal calves were positive for fecal carriage *E. coli* with a non-wild type susceptibility to cefotaxime. In 68% of the positive herds, five or less animals out of ten were found carrier of these isolates. Since only one isolate per herd was included in the molecular analysis, no conclusions can be drawn towards clonal diversity of ESBL/AmpC genes and plasmids in herds with multiple positive animals. Further in depth studies should be performed to assess whether a clonal spread within herds or a high diversity of ESBL/AmpC genes and/or plasmids exists within these herds. Factors that may influence the prevalence and spread of these resistance determinants within herds, including the use of antimicrobials and farm management, should be taken into account to analyze possible differences between farms.

Studies have been performed on fecal carriage of ESBL/AmpC-producing *E. coli* in cattle at slaughter in several countries, and the prevalence varied greatly. In Poland no ESBL/AmpC-producing *E. coli* were observed [25], in contrast to France (5.8%) [10], Switzerland (16%) [26], Hong Kong (3.1%) [27] and Japan (31.3%) [28]. However, while all studies were performed on individual animals, comparing these prevalence data should be performed with care since not all studies used the same selection methods (*e.g.* the use of enrichment broth and/or different selective media), which may have resulted in different screening sensitivities.

The vast majority of plasmid mediated ESBL/AmpC genes in the present study belonged to the $bla_{CTX-M}$ gene family, of which $bla_{CTX-M-1}$ was most abundant, followed by $bla_{CTX-M-15}$ and $bla_{CTX-M-14}$ (Table 1+2). These findings are in line with the observed trend in the last five years of a retrospective study performed on faecal samples from veal calves collected at farms from 1997 to 2010 [29]. Studies from other countries, that focused on ESBLs in general or $bla_{CTX-M}$ genes in specific, confirm the relatively high abundance of $bla_{CTX-M-1}$, $bla_{CTX-M-14}$ and/or $bla_{CTX-M-15}$ genes in either healthy or sick cattle [6,8,10,30]. In contrast, in the United States, $bla_{CMY-2}$ is the predominant gene found in cattle and $bla_{TEM}$ ESBL genes were found at a low level [31,32]. Furthermore, a study from Hong Kong reported that $bla_{CTX-M-13}$ was the only ESBL observed in cattle at slaughter [27] and $bla_{CTX-M-2}$ was reported to be predominant in cattle in Japan [33]. This indicates that there are geographical differences in the prevalence of resistance genes.

This study showed that ESBL/AmpC genes found in veal calves were predominantly located on IncI1 and IncF plasmids. These plasmids belong to the most commonly reported plasmid families in *Enterobacteriaceae* [34]. Interestingly, a relatively high proportion of IncI1 and IncF plasmid subtypes (mainly associated with $bla_{CTX-M-1}$) had not yet reported in the pMLST database (Table 1). IncI1 sequence types (ST) in combination with $bla_{CTX-M-1}$ have been reported either commonly (CC3, ST7) or occasionally (ST58) [34,35,36]. To our knowledge, the combination of $bla_{CTX-M-1}$ with IncF type plasmids has not been reported previously. The IncF RST-type found in this study (F2;A-;B-) has frequently been associated with $bla_{CTX-M-15}$ [34,37]. Furthermore, we have also

**Table 1.** *E. coli* carrying $bla_{CTX-M}$ group 1 genes with their corresponding plasmids and resistance profile.

| | | | Plasmid | | |
|---|---|---|---|---|---|
| ESBL | Strain | Other ß-lactamase | Replicon | incI1 pMLST/ incF RST [1, 2] | Non-wild type susceptibility[3] |
| CTX-M-01 | OT-ESBL-0199 | TEM-1a, OXA-1 | inc nt | (97 kb)[5] | Amp-Ctx-Caz-Tet-Chl-Str-Kan |
| | OT-ESBL-0294 | | inc nt | (97 kb)[5] | Amp-Ctx-Caz-Smx-Tmp-Gen |
| | OT-ESBL-0285 | | incB/O | | Amp-Ctx-Caz-Tet-Smx-Tmp-Str-Kan |
| | OT-ESBL-0547 | | incB/O | | Amp-Ctx-Caz-Tet-Smx-Gen-Str-Kan |
| | OT-ESBL-0589 | | incB/O | | Amp-Ctx-Caz-Tet-Smx-Tmp-Cip-Nal-Chl-Str |
| | OT-ESBL-0054 | | incF | F2; A-; B- | Amp-Ctx-Caz-Tet-Smx-Ffn-Chl-Gen-Str |
| | OT-ESBL-0198 | | incF | F2; A-; B- | Amp-Ctx-Caz-Tet- Gen-Str |
| | OT-ESBL-0591 | | incF | F2; A-; B- | Amp-Ctx-Caz-Tet-Ffn-Chl-Str |
| | OT-ESBL-0600 | | incF | F17*; A-; B- | Amp-Ctx-Caz-Tet-Gen |
| | OT-ESBL-0382 | | incF | F13#; A-; B20# | Amp-Ctx-Caz-Tet-Smx-Str-Kan |
| | OT-ESBL-0519 | | incF | F35#; A-; B- | Amp-Ctx-Caz-Tet |
| | OT-ESBL-0565 | | incF | F35#; A-; B_ | Amp-Ctx-Caz-Tet |
| | OT-ESBL-0328 | | incI1 | ST3; CC 3 | Amp-Ctx-Caz-Smx-tmp |
| | OT-ESBL-0359 | | incI1 | ST3; CC 3 | Amp-Ctx-Caz-Tet-Smx-Tmp-Str |
| | OT-ESBL-0406 | | incI1 | ST58; CC 58 | Amp-Ctx-Caz-Tet-Smx-Str-Kan |
| | OT-ESBL-0018 | | incI1 | ST58; CC 58 | Amp-Ctx-Caz-Tet-Smx-Tmp-Str |
| | OT-ESBL-0062 | | incI1 | new; 1 2 8# 3 3 | Amp-Ctx-Caz-Tet-Cip-Nal-Chl-Str |
| | OT-ESBL-0441 | | incI1 | new; 1 5* 17 1 7 | Amp-Ctx-Caz-Tet-Str |
| | OT-ESBL-0450 | | incI1 | new; 1 5̂16 1 7 | Amp-Ctx-Caz-Tet-Smx-Cip-Nal-Chl-Str |
| | OT-ESBL-0477 | | incI1 | new; 1 9# 17 1 7 | Amp-Ctx-Caz |
| | OT-ESBL-0546 | | incK | | Amp-Ctx-Caz-Tet-Smx-Cip-Nal-Chl-Str-Col |
| | OT-ESBL-0262 | OXA-1 | incN | | Amp-Ctx-Caz-Tet-Smx-Tmp-cip-Nal-Ffn-Chl-Str-Kan |
| | OT-ESBL-0361 | TEM-1b | incN | | Amp-Ctx-Caz-Tet-Smx-Ffn-Chl-Str-Kan |
| | OT-ESBL-0434 | | incN | | Amp-Ctx-Caz |
| CTX-M-01 | OT-ESBL-0414 | | incI1 | ST58; CC 58 | Amp-Ctx-Caz-Tet-Smx-Tmp-Str-Kan |
| + TEM-52c | | | unknown | (145 kb)[5] | |
| CTX-M-03 | OT-ESBL-0437 | TEM-1b | incB/O | | Amp-Ctx-Caz-Tet-Smx-Tmp-Cip-Nal-Gen-Str-Kan |
| | OT-ESBL-0567 | | incF | F35#; A-; B- | Amp-Ctx-Caz-Tet-Smx-Tmp-Cip-Nal-Str-Kan |
| CTX-M-15 | OT-ESBL-0327 | | inc nt | (110 kb)[5] | Amp-Ctx-Caz |
| | OT-ESBL-0502 | qnrS[4] (other plasmid) | inc nt | (40 kb)[5] | Amp-Ctx-Caz-Cip |
| | OT-ESBL-0028 | | incF | F2; A-; B- | Amp-Ctx-Caz-Tet-Smx-Tmp-Cip-Nal-Str |
| | OT-ESBL-0031 | OXA-1, (aac6′-Ib-cr)[4] | incF | F31; A4; B1 | Amp-Ctx-Caz-Tet-Cip-Nal-Kan |
| | OT-ESBL-0156 | OXA-1, (aac6′-Ib-cr)[4] | incF | F31; A4; B1 | Amp-Ctx-Caz-Tet-Smx-Tmp-Cip-Nal-Gen-Str-Kan |
| | OT-ESBL-0221 | OXA-1, (aac6′-Ib-cr)[4] | incF | F31; A4; B1 | Amp-Ctx-Caz-Tet-Smx-Tmp-Cip-Nal-Gen-Str-Kan |
| | OT-ESBL-0161 | TEM-1b, OXA-1, (aac6′-Ib-cr)[4] | incF | F31; A4; B1 | Amp-Ctx-Caz-Tet-Smx-Tmp-Cip-Nal-Gen-Str-Kan |
| | OT-ESBL-0563 | OXA-1, (aac6′-Ib-cr)[4] | incF | F31; A-; B1 | Amp-Ctx-Caz-Tet-Smx-Tmp-cip-Nal-Ffn-Chl-Gen-Str-Kan |
| | OT-ESBL-0534 | TEM-1b | IncF | F46; A-; B20 | Amp-Ctx-Caz-Tet-Smx-Tmp-cip-Nal-Ffn-Chl-Gen-Str-Col |
| | OT-ESBL 0256 | | IncI1 | ST31; CC 31 | Amp-Ctx-Caz-Tet-Smx-Tmp-cip-Nal-Ffn-Chl-Gen-Str-Kan |
| | OT-ESBL-0549 | | incI1 | ST31; CC 31 | Amp-Ctx-Caz-Cip-Nal |
| | OT-ESBL-0443 | | incHI2 | | Amp-Ctx-Caz-Tet-Smx-Tmp-Cip-Nal-Ffn-Chl-Gen-Kan |
| CTX-M-32 | OT-ESBL-0163 | | inc nt | (40 kb)[5] | Amp-Ctx-Caz-Tet-Smx-Tmp-Cip-Nal-Chl-Str-Kan |
| | OT-ESBL-0173 | | inc nt | (40 kb)[5] | Amp-Ctx-Caz-Tet-Smx-Tmp-Cip-Nal-Chl-Str-Kan |
| | OT-ESBL-0449 | | inc nt | (40 kb)[5] | Amp-Ctx-Caz-Tet-Smx-Tmp-Cip-Nal-Ffn-Chl-Str-Kan |
| | OT-ESBL-0473 | | inc nt | (40 kb)[5] | Amp-Ctx-Caz-Tet-Smx-Tmp-Cip-Nal-Chl-Str-Kan |

[1]The following deviations were found in IncI1 pMLST types: ESBL-0062: trbA8#: nucleotide 371 T>C; ESBL-0441: ardA5*: 59 G>A, 137 A>T, 271 C>G, 331 C>T; ESBL-0450: ardA5̂: 199 G>A, 271 C>G, 331 C>T; ESBL-0477: ardA9#: 199 G>A/C

[2]The following deviations were found in IncF RST types: ESBL-0301: FIB-26#: 251 G>T; ESBL-0382: FII-13#: 77 A>G and FIB-20#: 194 G>A, 203 T>C; ESBL-0519/0565/0567: FII-35#: 22 A>T, 38+39>CG insertion.

[3]Amp = ampicillin, Ctx = cefotaxime, Caz = ceftazidime, Nal = nalidixic acid, Cip = ciprofloxacin, Ffn = florfenicol, Chl = chloramphenicol, Tet = tetracycline, Smx =

sulfamethoxazole, Tmp = trimethoprim, Str = streptomycin, Kan = kanamycin, Gen = gentamicin), Col = colistin.
[4]*qnrS* and *aac(6')lb-cr* are not ß-lactamase genes, but cause reduced susceptibility to quinolones. *aac(6')lb-cr* also causes reduced susceptibility to aminoglycosides.
[5]Plasmid size in kilo bases (kb) of non-typeble plasmids.

found $bla_{CTX-M-1}$ in combination with IncB/O plasmids. This has also been reported in isolates from humans [38] and horses [11]. The $bla_{CTX-M-1}$ associated with IncN was previously observed in human isolates [38,39], poultry [40], and in both porcine isolates and their farm workers, indicating transmission may occur between food-producing animals and their care takers [41].

In this study, $bla_{CTX-M-15}$ was mainly located on IncF plasmids, a combination that has been reported frequently [34,39,42,43]. In addition, the combination of $bla_{CTX-M-15}$ with $bla_{OXA-1}$ and *aac6'-Ib-cr* (the latter causing reduced susceptibility to aminoglycosides and fluoroquinolones), has been reported previously [44]. This combination was observed in the predominant IncF type found in this study carrying $bla_{CTX-M-15}$ (F31;A4;B1). Furthermore, both the IncF RST types F2;A-B- and F31;A4;B1 and the IncI1 pMLST type ST31 that carry $bla_{CTX-M-15}$ that were reported in this study have also been isolated from cattle in France [37]. The $bla_{CTX-M-14}$ genes in our study were mainly located on IncK plasmids. This has also been reported in calves in the UK [45], as well as in isolates from humans and turkey [46,47]. None of the plasmids harbouring $bla_{CTX-M-32}$ were typable by PBRT, making it difficult to draw any conclusions towards genetic characteristics. For these isolates, we determined the size of the ESBL harbouring plasmid. The combination of $bla_{TEM-52}$ with IncI1 (ST36) has been reported in cattle in France [48], but both ST36 and ST10 in combination with $bla_{TEM-52}$ have also been reported in humans and poultry isolates [38] and are the most commonly found types [34]. This shows that various combinations of ESBL/AmpC genes and plasmids are widely distributed in isolates of both animal and human origin. The IncHI2/P multi replicon plasmid harboring $bla_{CTX-M-2}$ was not confirmed as a multi replicon by Southern blot hybridization. However, similar IncHI2/P plasmids harboring $bla_{CTX-M-2}$ were reported previously and have been confirmed as multi replicons [49]. Additional experiments such as PFGE, RFLP or sequence analysis of the whole plasmid are required to determine whether the plasmids with similar replicon types, pMLST-type or RST-type are clonally related. Based on the data presented in this study we cannot determine whether the overlapping genes and plasmids reside in *E. coli* that are host specific (animal or human) or that they proliferate well in both human and animal hosts.

**Table 2.** *E. coli* carrying ESBL/AmpC genes other than $bla_{CTX-M}$ group 1 with their corresponding plasmids and resistance profile.

| ESBL | Strain | Other ß-lactamase | Plasmid: Replicon | Plasmid: incI1 pMLST/incF RST | Non-wild type susceptibility[1] |
|---|---|---|---|---|---|
| CTX-M-02 | OT-ESBL-0301 | TEM-1b | incF | F1; A6; B26# | Amp-Ctx-Caz-Tet-Smx-Tmp-cip-Nal-Ffn-Chl-Gen-Str-Kan |
| | OT-ESBL-0310 | | incHI1 | | Amp-Ctx-Caz-Tet-Smx-Tmp-Ffn-Chl-Gen-Kan |
| | OT-ESBL-0514 | | incP, incHI2 | | Amp-Ctx-Caz-Tet-Smx-Tmp-Ffn-Chl-Str-Kan |
| CTX-M-14 | OT-ESBL-0403 | | incF | F2; A-; B- | Amp-Ctx-Caz-Tet-Smx-Tmp-cip-Nal-Ffn-Chl-Gen-Str-Kan-Col |
| | OT-ESBL-0021 | TEM-1b | incl1 | ST80 | Amp-Ctx-Caz-Tet-Smx-Tmp-Cip-Nal-Ffn-Chl-Str |
| | OT-ESBL-0336 | | incl1 | ST80 | Amp-Ctx-Caz-Tet-Smx-Cip-Nal-Ffn-Chl-Str-Kan |
| | OT-ESBL-0337 | | incl1 | ST80 | Amp-Ctx-Caz-Tet-Smx-Cip-Nal-Ffn-Chl-Str-Kan |
| | OT-ESBL-0058 | | incK | | Amp-Ctx-Caz-Tet-Smx-tmp-Str |
| | OT-ESBL-0291 | | incK | | Amp-Ctx-Caz-Tet-Smx-Tmp-Str |
| | OT-ESBL-0380 | TEM-1b | incK | | Amp-Ctx-Caz-Tet-Smx-Tmp-cip-Nal-Ffn-Chl-Gen-Str-Kan |
| | OT-ESBL-0405 | | incK | | Amp-Ctx-Caz-Tet-Gen |
| | OT-ESBL-0590 | | incK | | Amp-Ctx-Caz-Tet-Smx-Tmp-Cip-Nal-Chl-Gen-Str |
| TEM-52c | OT-ESBL-0192 | | incl1 | ST36; CC 5 | Amp-Ctx-Caz |
| | OT-ESBL-0364 | | incl1 | ST36; CC 5 | Amp-Ctx-Caz-Smx-tmp-Cip-Nal-Chl-Str |
| | OT-ESBL-0392 | TEM-1b | incl1 | ST10; CC 5 | Amp-Ctx-Caz-Tet-Smx-Tmp-Chl-Str |
| CMY-2 | OT-ESBL-0357 | | incK | | Amp-Ctx-Caz-Tet-Str |
| ampC-type-3 | OT-ESBL-0281 | | - | | Amp-Ctx-Caz-Tet-Smx-Str-Kan |
| | OT-ESBL-0386 | TEM-1b | - | | Amp-Ctx-Caz-Tet-Smx-Ffn-Chl-Str |
| | OT-ESBL-0453 | | - | | Amp-Ctx-Caz |
| | OT-ESBL-0599 | TEM-1a | - | | Amp-Ctx-Caz-Tet-Smx-Tmp-Str-Kan |
| | OT-ESBL-0601 | TEM-1b | - | | Amp-Ctx-Caz-Tet-Smx-Ffn-Chl-Str |
| *ampC*-type-11var | OT-ESBL-0543 | TEM-1a | - | | Amp-Ctx-Caz-Tet-Smx-Tmp-Str |
| Unknown | OT-ESBL-0261 | TEM-1c, *OXA*-1, ampc WT | - | | Amp-Ctx-Caz-Tet-Smx-Tmp-cip-Nal-Ffn-Chl-Str-Col |

[1]Amp = ampicillin, Ctx = cefotaxime, Caz = ceftazidime, Nal = nalidixic acid, Cip = ciprofloxacin, Ffn = florfenicol, Chl = chloramphenicol, Tet = tetracycline, Smx = sulfamethoxazole, Tmp = trimethoprim, Str = streptomycin, Kan = kanamycin, Gen = gentamicin), Col = colistin.

In six herds, only promoter mutations of the chromosomal *ampC* gene were detected (Table 2) and no isolates harbouring plasmid mediated ESBL/AmpC genes were found within the same herd. Since this study focused on the dissemination of plasmid mediated ESBL/AmpC genes, this prevalence is likely to be an underestimate of the actual number of herds that are positive for chromosomal *ampC* genes with promoter variations that lead to non-wild type susceptibility to cefotaxime. Furthermore, we based the absence of plasmid mediated ESBL/AmpC genes on array data. Novel genes or ESBL/AmpC variants not present on the array will have been missed. The phenotypic combination disk test was performed on all isolates negative in the array in order to reduce the chance of missing novel genes.

The fact that international trafficking of live calves from many different dairy farms to Dutch veal calf farms happens on a large scale may contribute to the high dissemination observed in this study. This may also explain the high diversity in gene/plasmid combinations compared to Dutch poultry, in which $bla_{CTX\text{-}M\text{-}1}$ is highly prevalent and commonly associated with IncI1 [36].

## Conclusion

A relatively high percentage of slaughter batches (66%) were found positive for *E. coli* with a non-wild type susceptibility to cefotaxime. The within-herd prevalence varied greatly from zero to 90% positive. In the majority of herds positive for ESBL/AmpC-producing *E. coli*, the within-farm prevalence was below 50%. Furthermore, plasmid mediated resistance to cefotaxime was predominantly caused by enzymes encoded by the $bla_{CTX\text{-}M}$ gene family. Many gene-plasmid combinations found in this study have also been found in cattle in other countries, indicating that there is a non-local dissemination of resistance determinants. However, since the number of ESBL/AmpC-producing *E. coli* was not quantified, no conclusions can be drawn about the actual risk of presence.

## Acknowledgments

We are very thankful to the Netherlands Food and Consumer Product Safety Authority (NVWA) for collecting the samples and providing laboratory assistance. We thank the research committee of the Task Force MRGA/MRSA for their constructive support for this project.

## Author Contributions

Conceived and designed the experiments: DM BW AEZ CD KV. Performed the experiments: AEZ AK KV. Analyzed the data: JH AEZ AK. Contributed reagents/materials/analysis tools: JH AK AEZ BW. Wrote the paper: JH JW DM CD AK KV BW.

## References

1. Bush K, Jacoby GA (2010) Updated functional classification of beta-lactamases. Antimicrob Agents Chemother 54: 969–976.
2. Carattoli A (2009) Resistance plasmid families in Enterobacteriaceae. Antimicrob Agents Chemother 53: 2227–2238.
3. Bush K, Jacoby GA, Medeiros AA (1995) A functional classification scheme for beta-lactamases and its correlation with molecular structure. Antimicrob Agents Chemother 39: 1211–1233.
4. Ambler RP (1980) The structure of beta-lactamases. Philos Trans R Soc Lond B Biol Sci 289: 321–331.
5. Carattoli A (2008) Animal reservoirs for extended spectrum beta-lactamase producers. Clin Microbiol Infect 14 Suppl 1: 117–123.
6. Geser N, Stephan R, Hachler H (2012) Occurrence and characteristics of extended-spectrum beta-lactamase (ESBL) producing Enterobacteriaceae in food producing animals, minced meat and raw milk. BMC Vet Res 8: 21.
7. Guerra B, Junker E, Schroeter A, Malorny B, Lehmann S, et al. (2003) Phenotypic and genotypic characterization of antimicrobial resistance in German *Escherichia coli* isolates from cattle, swine and poultry. J Antimicrob Chemother 52: 489–492.
8. Horton RA, Randall LP, Snary EL, Cockrem H, Lotz S, et al. (2011) Fecal carriage and shedding density of CTX-M extended-spectrum beta-lactamase-producing *Escherichia coli* in cattle, chickens, and pigs: implications for environmental contamination and food production. Appl Environ Microbiol 77: 3715–3719.
9. Liebana E, Batchelor M, Hopkins KL, Clifton-Hadley FA, Teale CJ, et al. (2006) Longitudinal farm study of extended-spectrum beta-lactamase-mediated resistance. J Clin Microbiol 44: 1630–1634.
10. Madec JY, Lazizzera C, Chatre P, Meunier D, Martin S, et al. (2008) Prevalence of fecal carriage of acquired expanded-spectrum cephalosporin resistance in Enterobacteriaceae strains from cattle in France. J Clin Microbiol 46: 1566–1567.
11. Rodriguez I, Barownick W, Helmuth R, Mendoza MC, Rodicio MR, et al. (2009) Extended-spectrum beta-lactamases and AmpC beta-lactamases in ceftiofur-resistant *Salmonella enterica* isolates from food and livestock obtained in Germany during 2003–07. J Antimicrob Chemother 64: 301–309.
12. MARAN (2012) Monitoring of Antimicrobial Resistance and Antibiotic Usage in Animals in The Netherlands. Central Veterinary Institute (CVI) of Wageningen UR; http://www.wageningenur.nl/nl/Expertises-Dienstverlening/Onderzoeksinstituten/central-veterinary-institute/Publicaties-CVI/MARAN-Rapporten.htm.
13. EFSA (2007) Report from the Task Force on Zoonoses Data Collection including guidance for harmonized monitoring and reporting of antimicrobial resistance in commensal *Escherichia coli* and *Enterococcus spp.* from food animals. EFSA Journal 96: 1–44
14. CLSI (2010) Performance Standards for Antimicrobial Susceptibility Testing; 20th Informational Suppplement: Supplemental Table 2A-S1., M100-S20; Wayne, PA USA: CLSI.
15. Schwarz S, Silley P, Simjee S, Woodford N, van Duijkeren E, et al. (2010) Editorial: assessing the antimicrobial susceptibility of bacteria obtained from animals. J Antimicrob Chemother 65: 601–604.
16. Batchelor M, Hopkins KL, Liebana E, Slickers P, Ehricht R, et al. (2008) Development of a miniaturised microarray-based assay for the rapid identification of antimicrobial resistance genes in Gram-negative bacteria. Int J Antimicrob Agents 31: 440–451.
17. Dierikx CM, van Duijkeren E, Schoormans AH, van Essen-Zandbergen A, Veldman K, et al. (2012) Occurrence and characteristics of extended-spectrum-beta-lactamase- and AmpC-producing clinical isolates derived from companion animals and horses. J Antimicrob Chemother 67: 1368–1374.
18. Park CH, Robicsek A, Jacoby GA, Sahm D, Hooper DC (2006) Prevalence in the United States of *aac(6')-Ib-cr* encoding a ciprofloxacin-modifying enzyme. Antimicrob Agents Chemother 50: 3953–3955.
19. Carattoli A, Bertini A, Villa L, Falbo V, Hopkins KL, et al. (2005) Identification of plasmids by PCR-based replicon typing. J Microbiol Methods 63: 219–228.
20. Garcia-Fernandez A, Chiaretto G, Bertini A, Villa L, Fortini D, et al. (2008) Multilocus sequence typing of IncI1 plasmids carrying extended-spectrum beta-lactamases in *Escherichia coli* and *Salmonella* of human and animal origin. J Antimicrob Chemother 61: 1229–1233.
21. Villa L, Garcia-Fernandez A, Fortini D, Carattoli A (2010) Replicon sequence typing of IncF plasmids carrying virulence and resistance determinants. J Antimicrob Chemother 65: 2518–2529.
22. Barton BM, Harding GP, Zuccarelli AJ (1995) A general method for detecting and sizing large plasmids. Anal Biochem 226: 235–240.
23. Caroff N, Espaze E, Berard I, Richet H, Reynaud A (1999) Mutations in the *ampC* promoter of *Escherichia coli* isolates resistant to oxyiminocephalosporins without extended spectrum beta-lactamase production. FEMS Microbiol Lett 173: 459–465.
24. Mulvey MR, Bryce E, Boyd DA, Ofner-Agostini M, Land AM, et al. (2005) Molecular characterization of cefoxitin-resistant *Escherichia coli* from Canadian hospitals. Antimicrob Agents Chemother 49: 358–365.
25. Wasyl D, Hasman H, Cavaco LM, Aarestrup FM (2011) Prevalence and Characterization of Cephalosporin Resistance in Nonpathogenic *Escherichia coli* from Food-Producing Animals Slaughtered in Poland. Microb Drug Resist 18: 79–82.
26. Geser N, Stephan R, Kuhnert P, Zbinden R, Kaeppeli U, et al. (2011) Fecal carriage of extended-spectrum beta-lactamase-producing Enterobacteriaceae in swine and cattle at slaughter in Switzerland. J Food Prot 74: 446–449.
27. Duan RS, Sit TH, Wong SS, Wong RC, Chow KH, et al. (2006) *Escherichia coli* producing CTX-M beta-lactamases in food animals in Hong Kong. Microb Drug Resist 12: 145–148.
28. Hiroi M, Yamazaki F, Harada T, Takahashi N, Iida N, et al. (2012) Prevalence of extended-spectrum beta-lactamase-producing *Escherichia coli* and *Klebsiella pneumoniae* in food-producing animals. J Vet Med Sci 74: 189–195.
29. Hordijk J, Wagenaar JA, van de Giessen AW, Dierikx C, van Essen-Zandbergen A, et al. (2013) Increasing prevalence and diversity of ESBL/AmpC-type beta-lactamase genes in *Escherichia coli* isolated from veal calves from 1997 to 2010. J Antimicrob Chemother 68: (in press).
30. Zheng H, Zeng Z, Chen S, Liu Y, Yao Q, et al. (2012) Prevalence and characterisation of CTX-M beta-lactamases amongst *Escherichia coli* isolates from healthy food animals in China. Int J Antimicrob Agents 39: 305–310.

31. Donaldson SC, Straley BA, Hegde NV, Sawant AA, DebRoy C, et al. (2006) Molecular epidemiology of ceftiofur-resistant *Escherichia coli* isolates from dairy calves. Appl Environ Microbiol 72: 3940–3948.
32. Frye JG, Fedorka-Cray PJ, Jackson CR, Rose M (2008) Analysis of Salmonella enterica with reduced susceptibility to the third-generation cephalosporin ceftriaxone isolated from U.S. cattle during 2000-2004. Microb Drug Resist 14: 251–258.
33. Shiraki Y, Shibata N, Doi Y, Arakawa Y (2004) *Escherichia coli* producing CTX-M-2 beta-lactamase in cattle, Japan. Emerg Infect Dis 10: 69–75.
34. Carattoli A (2011) Plasmids in Gram negatives: molecular typing of resistance plasmids. Int J Med Microbiol 301: 654–658.
35. Dierikx C, van der Goot J, Fabri T, van Essen-Zandbergen A, Smith H, et al. (2013) Extended-spectrum-beta-lactamase- and AmpC-beta-lactamase-producing *Escherichia coli* in Dutch broilers and broiler farmers. J Antimicrob Chemother 68: 60–7.
36. Leverstein-van Hall MA, Dierikx CM, Cohen Stuart J, Voets GM, van den Munckhof MP, et al. (2011) Dutch patients, retail chicken meat and poultry share the same ESBL genes, plasmids and strains. Clin Microbiol Infect 17: 873–880.
37. Madec JY, Poirel L, Saras E, Gourguechon A, Girlich D, et al. (2012) Non-ST131 *Escherichia coli* from cattle harbouring human-like *bla*(CTX-M-15)-carrying plasmids. J Antimicrob Chemother 67: 578–581.
38. Leverstein-van Hall MA, Dierikx CM, Cohen Stuart J, Voets GM, van den Munckhof MP, et al. (2011) Dutch patients, retail chicken meat and poultry share the same ESBL genes, plasmids and strains. Clin Microbiol Infect 17: 873–880.
39. Novais A, Canton R, Moreira R, Peixe L, Baquero F, et al. (2007) Emergence and dissemination of Enterobacteriaceae isolates producing CTX-M-1-like enzymes in Spain are associated with IncFII (CTX-M-15) and broad-host-range (CTX-M-1, -3, and -32) plasmids. Antimicrob Agents Chemother 51: 796–799.
40. Bortolaia V, Larsen J, Damborg P, Guardabassi L (2011) Potential Pathogenicity and Host Range of Extended-Spectrum beta-Lactamase-Producing *Escherichia coli* Isolates from Healthy Poultry. Appl Environ Microbiol 77: 5830–5833.
41. Moodley A, Guardabassi L (2009) Transmission of IncN plasmids carrying blaCTX-M-1 between commensal *Escherichia coli* in pigs and farm workers. Antimicrob Agents Chemother 53: 1709–1711.
42. Coque TM, Novais A, Carattoli A, Poirel L, Pitout J, et al. (2008) Dissemination of clonally related *Escherichia coli* strains expressing extended-spectrum beta-lactamase CTX-M-15. Emerg Infect Dis 14: 195–200.
43. Hopkins KL, Liebana E, Villa L, Batchelor M, Threlfall EJ, et al. (2006) Replicon typing of plasmids carrying CTX-M or CMY beta-lactamases circulating among *Salmonella* and *Escherichia coli* isolates. Antimicrob Agents Chemother 50: 3203–3206.
44. Woodford N, Carattoli A, Karisik E, Underwood A, Ellington MJ, et al. (2009) Complete nucleotide sequences of plasmids pEK204, pEK499, and pEK516, encoding CTX-M enzymes in three major *Escherichia coli* lineages from the United Kingdom, all belonging to the international O25:H4-ST131 clone. Antimicrob Agents Chemother 53: 4472–4482.
45. Teale CJ, Barker L, Foster AP, Liebana E, Batchelor M, et al. (2005) Extended-spectrum beta-lactamase detected in *E. coli* recovered from calves in Wales. Vet Rec 156: 186–187.
46. Stokes MO, Cottell JL, Piddock LJ, Wu G, Wootton M, et al. (2012) Detection and characterization of pCT-like plasmid vectors for *bla*CTX-M-14 in *Escherichia coli* isolates from humans, turkeys and cattle in England and Wales. J Antimicrob Chemother 67: 1639–1644.
47. Valverde A, Canton R, Garcillan-Barcia MP, Novais A, Galan JC, et al. (2009) Spread of *bla*(CTX-M-14) is driven mainly by IncK plasmids disseminated among *Escherichia coli* phylogroups A, B1, and D in Spain. Antimicrob Agents Chemother 53: 5204–5212.
48. Haenni M, Saras E, Metayer V, Doublet B, Cloeckaert A, et al. (2012) Spread of the *bla*TEM-52 gene is mainly ensured by IncI1/ST36 plasmids in *Escherichia coli* isolated from cattle in France. J Antimicrob Chemother 67: 2774–6.
49. Dierikx C, van Essen-Zandbergen A, Veldman K, Smith H, Mevius D (2010) Increased detection of extended spectrum beta-lactamase producing *Salmonella enterica* and *Escherichia coli* isolates from poultry. Vet Microbiol 145: 273–8.

# 5

# ESBL-Producing *Enterobacteriaceae*: Occurrence, Risk Factors for Fecal Carriage and Strain Traits in the Swiss Slaughter Cattle Population Younger than 2 Years Sampled at Abattoir Level

**Martin Reist[1], Nadine Geser[2], Herbert Hächler[2], Sara Schärrer[1], Roger Stephan[2]***

1 Veterinary Public Health Institute, Vetsuisse Faculty University of Bern, Bern, Switzerland, 2 Institute for Food Safety and Hygiene, Vetsuisse Faculty University of Zurich, Zurich, Switzerland

## Abstract

During the past decade extended-spectrum beta-lactamase (ESBL) producing *Enterobacteriaceae* have become a matter of great concern in human and veterinary medicine. In this cross-sectional study fecal swabs of a geographically representative number of Swiss cattle at slaughterhouse level were sampled i) to determine the occurrence of ESBL producing *Enterobacteriaceae* in the Swiss slaughter cattle population younger than 2 years, and ii) to assess risk factors for shedding ESBL producing *Enterobacteriaceae*. In total, 48 (8.4%; 95% C.I. 6.3–11.1%) independent ESBL producing *Enterobacteriaceae* were detected among the 571 tested animals. Species identification revealed 46 *E. coli* strains, one *Enterobacter cloacae* and one *Citrobacter youngae*. In view of beta-lactam antibiotics, all 48 isolates were resistant to ampicillin, cephalothin and cefpodoxime. Forty-five (93.8%) isolates were resistant cefuroxime; one (2.1%) isolate to cefoxitin, 28 (58.3%) isolates to cefotaxime, 2 (4.2%) isolates to ceftazidime, and 2 (4.2%) isolates to cefepime. Risk factors for shedding ESBL producing *Enterobacteriaceae* were (i) age (OR 0.19 and 0.12 in age category 181 d to 1y and 1y to 2 y compared to ≤180 d), (ii) primary production type, meaning dairy compared to beef on farm of origin (OR 5.95), and (iii) more than 1 compared to less than 1 animal movement per d per 100 animals on farm of origin (OR 2.37).

**Editor:** Odir A. Dellagostin, Federal University of Pelotas, Brazil

**Funding:** This study was funded by the Swiss Federal Veterinary Office (FVO). The funders had no role in study design, data collection and analysis, decision to publish, or preparation of the manuscript.

**Competing Interests:** The authors have declared that no competing interests exist.

* E-mail: stephanr@fsafety.uzh.ch

## Introduction

Antimicrobial resistance in bacteria has emerged as a problem in both human and veterinary medicine. One of the currently most important resistance mechanisms in *Enterbacteriaceae*, which reduces the efficacy even of modern expanded-spectrum cephalosporins (except cephamycins and carbapenems) and monobactams is based on plasmid-mediated production of enzymes that inactivate these compounds by hydrolyzing their beta-lactam ring. Such resistance is encoded by an increasing number of different point-mutational variants, called extended spectrum beta-lactamases (ESBL), of classical broad-spectrum beta-lactamases (BSBL): most are derivates of TEM and SHV beta-lactamase families, whereas other groups, such as CTX-M, OXA, PER and VEB beta-lactamases have been described more recently [1]. The phenotypical difference between BSBLs and ESBLs is that the latter efficiently hydrolyze $3^{rd}$- and $4^{th}$-generation cephalosporins, in addition to penicillins and $1^{st}$ and $2^{nd}$ generation cephalosporins as the BSBLs are capable of. ESBLs are inhibited by clavulanic acid, sulbactam and tazobactam [2], a feature that is used (i) as a criterion for classification of β-lactamases and (ii) for diagnostic ESBL detection purposes. As a matter of growing concern, resistance caused by ESBLs is often associated with resistance to other classes of antibiotics like fluoroquinolones, aminoglycosides and trimethoprim-sulfamethoxazole [3,4].

Since the first description of ESBL producing *Enterobacteriaceae* isolated from hospitalized humans [5], many nosocomial outbreaks have been reported. However, since a few years, there is an increase in the detection of ESBL producing strains in the community [6,7]. More recently, some reports have alerted about the dissemination of ESBL producing *E. coli* in healthy cattle in several countries in Europe and USA [8,9,10] or in cattle derived food products like meat and raw milk [11,12,13]. Therefore, the impact of healthy farm animals as a reservoir for an input of ESBL producing *E. coli* in the food processing chain has to be assessed.

The aim of the present study was to assess the prevalence of ESBL producing *Enterobacteriaceae* in the Swiss cattle population younger than two years, and to assess risk factors for shedding ESBL producing *Enterobacteriaceae* based on a slaughterhouse monitoring approach that is aiming at achieving a geographically representative sample. Whilst for monitoring purposes a geographically representative set is easily achieved for samples taken on a farm level, where farms to be sampled can be determined in advance by applying a stratified randomization scheme, this is typically more difficult for samples taken at the slaughterhouse.

**Table 1.** Description of risk factors studied in univariate and multivariate logistic regression models.

| Risk factor | Category | Number of animals | Proportion of positive samples |
|---|---|---|---|
| Age Class | ≤180 d | 248 | 15.7% |
| | 181 d–1 y | 148 | 3.4% |
| | >1 y | 175 | 2.3% |
| Production Type | Meat | 244 | 2.5% |
| | Dairy | 327 | 12.8% |
| Number of cattle on farm | ≤30 | 173 | 6.4% |
| | 31–60 | 211 | 8.1% |
| | >60 | 187 | 10.7% |
| Animal movements to farm per day | ≤0.5 | 428 | 6.3% |
| | >0.5 | 143 | 14.7% |
| Animal movements to farm per day per 100 animals | ≤1/d/100 animals | 461 | 6.9% |
| | >1/d/100 animals | 110 | 14.6% |
| Number of animals dying on farm per 100 days | 0 | 235 | 10.2% |
| | ≤1/100 d | 185 | 4.3% |
| | >1/100 d | 151 | 10.6% |
| Number of animals dying on farm per 100 days per 100 animals | 0 | 235 | 10.2% |
| | ≤2/100 d | 200 | 4.0% |
| | >2/100 d | 136 | 11.8% |

## Materials and Methods

### Sampling

Representative samples for the cattle population younger than 2 years were taken at the slaughterhouse level. A minimum required sample size of 385 randomly selected animals was calculated with the assumptions of an infinite population size, a prevalence of 50%, a desired confidence level of 95.0% and an absolute error of 5% (Win Episcope 2.0 software http://www.clive.ed.ac.uk/winepiscope). The samples were randomly taken at the five biggest cattle abattoirs (A: Zurich; B: Hinwil; C: St. Gallen; D: Oensingen; E: Estavayer-le-Lac), where over 75% of Swiss cattle of the targeted age group are slaughtered. Because animals originating from farms located south of the Alps are typically underrepresented in the slaughter population of these 5 biggest abattoirs, samples from the biggest abattoir in the canton of Ticino (F: Bellinzona) and samples from several smaller abattoirs from the canton of Valais were additionally collected to guarantee a geographically representative distribution. From all the abattoirs we obtained the permission to use these animal parts.

Only one sample was taken per animal holding of origin. The number of samples in the sampling frame collected from each slaughterhouse was proportional to the number of cattle slaughtered at each establishment per year. Based on these data, a random sampling plan was conceived.

Faecal swabs were collected from November 2010 to September 2011 from 571 healthy animals younger than 2 years at slaughter. These animals had no contact to older cattle populations on the transport to the slaughterhouse, in the slaughterhouse or during the slaughtering procedure. Therefore, contamination by animals of older age classes after having left the farm of origin can be excluded. The samples were taken through opening the large intestine with sterile scissors after evisceration. Unique animal identification numbers registered in the Swiss central animal movement database (AMD) were recorded.

### Microbiological analysis

Each swab was incubated for 24 hours at 37°C in 10 ml EE Broth (BD, Franklin Lakes, USA) for enrichment. One loop of the enriched faecal samples were inoculated onto Brilliance ESBL agar (Oxoid, Hampshire, UK) and incubated at 37°C for 24 hours under aerobic conditions. All grown colonies were selected and sub-cultured onto Triple Sugar Iron (TSI) agar (BD, Franklin Lakes, USA) at 37°C for 24 hours. By the oxidase test and the assessment of lactose fermentation, non-fermenters were discarded, and oxidase-negative colonies were subjected to identification by API ID 32 E (bioMérieux, Marcy l'Etoile, France).

### Antimicrobial susceptibility testing and ESBL detection

All isolated strains were subjected to susceptibility testing for 18 antimicrobial agents by the disc diffusion method and evaluated according to CLSI criteria [14]. Strains exhibiting intermediate resistance were classified as susceptible. The antibiotics tested were: ampicillin (AM, 10 µg), amoxicillin/clavulanic acid (AMC, 20/10 µg), cephalothin (CF, 30 µg), cefuroxime (CXM, 30 µg), cefoxitin (FOX, 30 µg), cefpodoxime (CPD, 10 µg), cefotaxime (CTX, 30 µg), ceftazidime (CAZ, 30 µg), cefepime (FEP, 30 µg), ciprofloxacin (CIP, 5 µg), nalidixic acid (NA, 30 µg), gentamicin (GM, 10 µg), streptomycin (S, 10 µg), trimethoprim-sulfamethoxazole (SXT, 1.25/23.75 µg), chloramphenicol (C, 30 µg) and tetracycline (TE, 30 µg). The amoxicillin/clavulanic acid disc was placed between the cefpodoxime and the ceftazidime discs, and the synergy effect was documented. The strains, which showed a synergy effect, were then confirmed as ESBL producers on Muller-Hinton agar plates using E-Test-ESBL strips containing cefotaxime, cefepime or ceftazidime alone and in combination with clavulanic acid (bioMérieux, Marcy l'Etoile, France).

### Risk Factor Analysis

Risk factors were calculated based on data obtained from the Swiss AMD. The individual animal identification number of the

**Table 2.** Prevalences of animals with ESBL positive isolates.

| | No. positives | No. negatives | prevalence | 95% confidence interval * |
|---|---|---|---|---|
| Total | 48 | 523 | 8.41% | 6.32–11.07% |
| ≤180 d | 39 | 209 | 15.73% | 11.55–21.00% |
| 181 d–1 y | 5 | 143 | 3.38% | 1.25–8.12% |
| >1 y | 4 | 171 | 2.29% | 0.73–6.12% |
| Meat | 6 | 238 | 2.46% | 1.00–5.53% |
| Dairy | 42 | 285 | 12.84% | 9.51–17.08% |
| ≤1 mv./d/100 animals | 32 | 429 | 6.94% | 4.87–9.76% |
| >1 mv./d/100 animals | 16 | 94 | 14.55% | 8.80–22.85% |

*95% confidence interval with Yates' continuity correction.

slaughtered and sampled animals served as a unique identifier to determine the farm of origin, date of birth and to access all incoming movement data corresponding to the farm of origin and the length of stay of each individual animal of the sample. Risk factors derived from AMD data comprised age, main production type on the farm of origin (dairy vs. beef and fattening), farm size, number of animal movements and number of animals found death or euthanized on farm. The full list of studied risk factors and their categorization is given in Table 1.

## Statistical Analysis

Statistical analyses were performed in R version 2.13 for Mac OS. The significance level was set at p≤0.05. ESBL prevalence and their Yates' continuity corrected 95% confidence intervals were calculated applying exact binomial tests. Logistic regression models were applied for risk factor analyses. First, a univariate model was calculated for each risk factor given in Table 1. Risk factors with a p-value <0.25 were retained for multivariate analyses. To avoid colinearity, the correlation structure of these retained risk factors was assessed. Of each pair of correlating risk factors, only the one showing the more significant association with the dependent variable was retained for the multivariate model. Multivariate models were fitted by backward elimination procedures. According to Hosmer and Lemeshow [15] confounders were eliminated if they did not importantly change the estimates of the significant predictors. All two ways interactions were tested.

## Results

The sampling regimen resulted in a geographically representative sample population that was uniformly spread over the entire territory of Switzerland. Farms of origin of cattle shedding ESBL-producing *Enterobacteriaceae* are as well uniformly spread over the densely populated Swiss midlands as well as over the canton of Ticino in the south east of Switzerland. There were no cattle shedding ESBL-producing *Enterobacteriaceae* found originating from the canton of Valais in the south west of Switzerland.

ESBL-producing *Enterobacteriaceae* were detected in 48 (8.4%; 95% C.I. 6.3–11.1%) of the 571 tested animals. ESBL prevalence in different age classes, different production types and different animal movement activities in the farm of origin are summarized in Table 2. Results of the final multivariate logistic regression models applied for risk factor analysis are shown in Tables 3 and 4. Animals from within the age class ≤180 days were at a significantly higher risk of shedding ESBL-producing *Enterobacteriaceae* than animals from age classes 180 days to 1 year or 1 year to 2 years. Animals originating from farms with primary production type "dairy" were at a 5.95 times greater risk of shedding ESBL-producing *Enterobacteriaceae* than animals originating from farms with primary production type "beef". Finally, animals originating from farms with more than one animal movement per day per 100 animals were at a 2.37 times higher risk of shedding ESBL-producing *Enterobacteriaceae* than animals originating form farms with less than one animal movement per day100 animals.

In total, 48 isolates from 48 different animals showing the ESBL phenotype were further characterized. Species identification revealed 46 *E. coli*, one *Enterobacter cloacae* and one *Citrobacter youngae* (Figure 1). In view of β-lactam antibiotics, all 48 isolates were resistant to ampicillin (AM), the 1$^{st}$-generation cephalosporin, cephalothin (CF) and the 3$^{rd}$-generation cephalosporin, cefpodoxime (CPD) (Figure 1). Forty-five (93.8%) isolates were resistant to the 2$^{nd}$-generation cephalosporin cefuroxime (CXM); one (2.1%) isolate to cefoxitin (FOX), 28 (58.3%) isolates to cefotaxime (CTX), 2 (4.2%) isolates to ceftazidime (CAZ), and 2 (4.2%) isolates to the 4$^{th}$-generation cephalosporin cefepime (FEP). Seventeen strains (35.4%) showed resistance to amoxicillin-clavulanic acid (AMC). Besides β-lactam resistance, susceptibility

**Table 3.** Farm level risk factors for ESBL shedding.

| Risk factor | | OR | 95% Conf. Int. | | p-value |
|---|---|---|---|---|---|
| Prod. type | meat | 1.00 | | | |
| | dairy | 5.95 | 2.48 - | 14.30 | <0.001 |
| Movements | ≤1/d/100 anim. | 1.00 | | | |
| | >1/d/100 anim. | 2.37 | 1.23 - | 4.57 | 0.010 |

Multivariate logistic regression model.

**Table 4.** Animal level risk factors for ESBL shedding.

| Risk factor | | OR | 95% Conf. Int. | | p-value |
|---|---|---|---|---|---|
| Age | ≤180 d | 1.00 | | | |
| | 181 d–1 y | 0.19 | 0.07 - | 0.49 | <0.001 |
| | >1 y | 0.12 | 0.04 - | 0.36 | <0.001 |

Logistic regression model.

| | | Species | | | Resistances | | | | | | | | | | | | | | | |
|---|---|---|---|---|---|---|---|---|---|---|---|---|---|---|---|---|---|---|---|---|
| | | | | | β-lactame antibiotics | | | | | | | | | other antibiotics | | | | | | |
| slaughterhouse | sample No. | Escherichia coli | Citrobacter yongae | Enterobacter cloacae | AM | AMC | CF | CXM | FOX | CPD | CTX | CAZ | FEP | GM | S | SXT | Te | NA | CIP | C |
| A | 46 | ■ | | | ■ | | ■ | ■ | | ■ | | | | ■ | ■ | ■ | ■ | | | ■ |
| | 68 | ■ | | | ■ | | ■ | ■ | | ■ | ■ | | | | | | | ■ | | |
| | 104 | ■ | | | ■ | | ■ | ■ | | ■ | | | | ■ | ■ | ■ | ■ | | | ■ |
| | 112 | ■ | | | ■ | | ■ | ■ | | ■ | ■ | | | ■ | ■ | ■ | ■ | | | ■ |
| | 115 | ■ | | | ■ | | ■ | ■ | | ■ | | | | | | ■ | | ■ | ■ | ■ |
| | 128 | ■ | | | ■ | | ■ | ■ | | ■ | | | ■ | | ■ | ■ | ■ | | | |
| | 136 | ■ | | | ■ | | ■ | ■ | | ■ | ■ | | | | | | ■ | ■ | ■ | |
| | 142 | ■ | | | ■ | | ■ | ■ | | ■ | | | | ■ | ■ | ■ | ■ | ■ | ■ | ■ |
| | 142b | ■ | | | ■ | | ■ | ■ | | ■ | ■ | | | ■ | ■ | ■ | ■ | | | ■ |
| | 142g | | ■ | | ■ | | ■ | ■ | | ■ | ■ | | | ■ | ■ | ■ | ■ | | | ■ |
| B | 240 | ■ | | | ■ | ■ | ■ | ■ | | ■ | | | | ■ | | | | | | |
| | 248 | ■ | | | ■ | ■ | ■ | ■ | | ■ | ■ | | | ■ | ■ | | ■ | | | |
| | 250 | ■ | | | ■ | ■ | ■ | ■ | | ■ | | | | ■ | ■ | ■ | ■ | ■ | ■ | ■ |
| | 329 | ■ | | | ■ | | ■ | ■ | | ■ | | | | | | | | | | |
| | 340 | | | ■ | ■ | ■ | ■ | ■ | ■ | ■ | | | | | | ■ | ■ | | | ■ |
| | 466 | ■ | | | ■ | | ■ | ■ | | ■ | | | | ■ | | ■ | | | | |
| | 467 | ■ | | | ■ | | ■ | ■ | | ■ | | | | | | | | | | |
| | 469 | ■ | | | ■ | | ■ | ■ | | ■ | | | | ■ | ■ | ■ | ■ | ■ | ■ | ■ |
| C | 422 | ■ | | | ■ | | ■ | ■ | | ■ | | | | | | | ■ | | | |
| | 522 | ■ | | | ■ | | ■ | ■ | | ■ | ■ | | | ■ | ■ | ■ | ■ | | | |
| | 531 | ■ | | | ■ | | ■ | ■ | | ■ | ■ | | | | ■ | ■ | | | | |
| | 534 | ■ | | | ■ | | ■ | ■ | | ■ | ■ | | | | | ■ | ■ | | | ■ |
| | 542 | ■ | | | ■ | | ■ | ■ | | ■ | ■ | | | | | ■ | ■ | ■ | ■ | |
| D | 643 | ■ | | | ■ | | ■ | | | ■ | | | | | | | | | | |
| | 644 | ■ | | | ■ | | ■ | | | ■ | | | | | | | | | | |
| | 745 | ■ | | | ■ | | ■ | ■ | | ■ | ■ | | | | ■ | ■ | ■ | ■ | ■ | ■ |
| | 747 | ■ | | | ■ | | ■ | ■ | | ■ | ■ | | | | | ■ | ■ | | | |
| E | 769 | ■ | | | ■ | | ■ | ■ | | ■ | | | | ■ | ■ | ■ | ■ | | | ■ |
| | 772 | ■ | | | ■ | | ■ | ■ | | ■ | | | | | ■ | ■ | ■ | | | |
| | 776 | ■ | | | ■ | ■ | ■ | ■ | | ■ | ■ | | | ■ | ■ | ■ | ■ | ■ | ■ | |
| | 780 | ■ | | | ■ | | ■ | ■ | | ■ | ■ | | | | ■ | ■ | ■ | ■ | ■ | ■ |
| | 794 | ■ | | | ■ | ■ | ■ | | | ■ | ■ | | | | ■ | ■ | ■ | ■ | ■ | ■ |
| | 847 | ■ | | | ■ | ■ | ■ | ■ | | ■ | ■ | ■ | ■ | ■ | | | ■ | ■ | ■ | ■ |
| | 850 | ■ | | | ■ | ■ | ■ | ■ | | ■ | ■ | | | ■ | ■ | | ■ | ■ | ■ | |
| | 851 | ■ | | | ■ | ■ | ■ | ■ | | ■ | ■ | | | ■ | ■ | ■ | ■ | | | ■ |
| | 856 | ■ | | | ■ | ■ | ■ | ■ | | ■ | ■ | ■ | | ■ | ■ | ■ | ■ | ■ | ■ | |
| | 858 | ■ | | | ■ | ■ | ■ | ■ | | ■ | ■ | | | ■ | ■ | ■ | ■ | ■ | ■ | ■ |
| | 866 | ■ | | | ■ | ■ | ■ | ■ | | ■ | ■ | | | ■ | | | ■ | ■ | ■ | |
| | 867 | ■ | | | ■ | ■ | ■ | ■ | | ■ | ■ | | | ■ | | | ■ | ■ | ■ | |
| | 868 | ■ | | | ■ | ■ | ■ | ■ | | ■ | ■ | | | ■ | | | ■ | ■ | ■ | |
| | 878 | ■ | | | ■ | ■ | ■ | ■ | | ■ | ■ | | | ■ | ■ | ■ | ■ | ■ | ■ | |
| | 890 | ■ | | | ■ | ■ | ■ | ■ | | ■ | | | | ■ | | ■ | ■ | | | ■ |
| | 921 | ■ | | | ■ | ■ | ■ | ■ | | ■ | ■ | | | ■ | ■ | ■ | ■ | ■ | ■ | |
| F | 1103 | ■ | | | ■ | | ■ | ■ | | ■ | ■ | | | ■ | ■ | ■ | ■ | | | ■ |
| | 1112 | ■ | | | ■ | | ■ | ■ | | ■ | ■ | | | | ■ | | ■ | | | ■ |
| | 1119 | ■ | | | ■ | | ■ | ■ | | ■ | ■ | | | | | | | | | ■ |
| | 1127 | ■ | | | ■ | | ■ | ■ | | ■ | | | | | | | | | | |
| | 1129 | ■ | | | ■ | | ■ | ■ | | ■ | | | | | ■ | | | ■ | ■ | |

**Figure 1. Characteristics and antimicrobial susceptibility profiles of ESBL-producing *Enterobacteriaceae* isolated from cattle younger than 2 years.** Symbols: black square, positive result or resistant to a specific antimicrobial agent; white square, negative result or susceptible to a specific antimicrobial. Abbreviations: AM, ampicillin (resistant ≤13 mm); AMC, amoxicillin-clavulanic acid (resistant ≤13 mm); CF, cephalothin (resistant ≤14 mm); CXM, cefuroxime (resistant ≤14 mm); FOX, cefoxitin (resistant ≤14 mm); CPD, cefpodoxime (resistant ≤17 mm); CTX, cefotaxime (resistant ≤14 mm); CAZ, ceftazidime (resistant ≤14 mm); FEP, cefepime (resistant ≤14 mm); GM, gentamicin (resistant ≤10 mm); S, streptomycin (resistant ≤11 mm); SXT, trimethoprim-sulfamethoxazole (resistant ≤10 mm); TE, tetracycline (resistant ≤11 mm); NA, nalidixic acid (resistant ≤13 mm); CIP, ciprofloxacin (resistant ≤15 mm); C, chloramphenicol (resistant ≤12 mm). Discrimination between "susceptible" and "resistant" was strictly according to CLSI interpretive criteria. It should be noted, however, that for clinical and therapeutic purposes, ESBL producers should generally be reported resistant to cephalosporins of all 4 generations and monobactams.

to other classes of antibiotics was tested. Hereby, 26 (54.2%) isolates were resistant to gentamicin (GM), 27 (56.25%) to streptomycin (S), 21 (43.8%) to nalidixic acid (NA), 20 (41.7%) to ciprofloxacin (CIP), 36 (75.0%) to tetracycline (TE), 22 (45.8%) to chloramphenicol (C), and 30 (62.5%) to trimethoprim-sulfamethoxazole (SXT).

## Discussion

In this cross-sectional study, fecal samples of a geographically representative cattle subpopulation for Switzerland were sampled at slaughterhouse level to determine the occurrence of ESBL producing *Enterobacteriaceae* in the Swiss cattle population younger than 2 years, and to assess risk factors for shedding such organisms.

To assess the population prevalence of a pathogen or indicator organism on a national level, a geographically representative set of samples needs to be selected. Whilst this is easily achieved for samples taken on a farm level, where farms to be sampled can be determined in advance by applying a stratified randomization scheme, this is typically more difficult for samples taken at the slaughterhouse. This study demonstrated that the goal of a geographically representative survey could be achieved with a carefully planned sampling scheme involving the major cattle slaughterhouses and additional small abattoirs in areas disconnected from the main animal traffic.

To our knowledge, this is the first study on risk factors for shedding ESBL producing *Enterobacteriaceae* working with risk factors derived from animal movement data stored in a national AMD. Performing risk factor analysis with such data offered several advantages compared to risk factors derived form surveys, especially when the samples were collected at the abattoir and not on the farm. Getting information on animal movements and mortalities from 571 farms of origin of slaughtered animals via telephone or email surveys would be very time consuming. Furthermore, such information would be very imprecise compared to data from the Swiss AMD whose content is very accurate and complete.

The overall prevalence of cattle hosting ESBL producing *Enterobacteriaceae* found in this study (8.4%; 95% C.I. 6.3–11.1%) is slightly lower than found in previous, smaller scaled Swiss studies that reported prevalences of 13.7% [16] and 17.1% [10]. Recent studies from other countries reported a very low (0.2%) prevalence in cattle in Korea [17] or even zero prevalence in cattle in Tunisia [18]. Studies from European countries reported 35.4% (22.2–50.5% C.I. 95%) in North West England and North Wales [19] and 4.8% in France [20].

One of the significant risk factors for shedding ESBL producing *Enterobacteriaceae* obtained in the final multivariate model was age, with cattle over 6 months being at a considerably lower risk than calves (younger than 6 months) This is in accordance with a previous study of Geser et al. [16] who reported a overall prevalence of 13.7% in cattle as opposed to a prevalence of 25.3% among calves.

Animals from dairy farms were at a 5.95 times higher risk for hosting ESBL producing *Enterobacteriaceae* than animals from beef or fattening farms. This might be explained by differences between production types with respect to farming practices and antimicrobial compounds applied. In Switzerland, sales of cephalosporins have increased over the past years, especially the sales of 3$^{rd}$ and 4$^{th}$ generation cephalosporins for intramammary application [21]. Snow et al. [19] reported that in North West England and North Wales farms that had used 3$^{rd}$ or 4$^{th}$ generation cephalosporins in livestock during the previous 12 months were nearly 4 times more likely to host ESBL *E. coli*. On Swiss dairy farms, calves not in consideration for breeding are either fattened on their farm of birth or they are sold to fattening farms at a very young age. Those fattened on the dairy farms are primarily fed with milk. For economic reasons, milk that cannot be put on the market because of elevated cell counts or because of recent antimicrobial treatments is often fed to calves. Moreover, ESBL hosting *Enterobacteriaceae* present on dairy farms might be transmitted to calves by the fecal-oral route. Calves later to be fattened on fattening farms leave their dairy farms very few weeks after birth and are thus at a lower risk for acquiring ESBL producing *Enterobacteriaceae* than animals staying on the dairy farms until slaughter. In comparison to dairy farms, beta-lactam antibiotics, especially 3$^{rd}$ and 4$^{th}$ generation cephalosporins, do not represent the predominantly used antimicrobial compounds on beef and fattening farms. The fact that among 40 animals originating from the canton of Valais no ESBL producers were found, might also be explained by the farming type, as meat production is predominant over milk production in this area.

The number of animal movements per farm per day per 100 animals is a factor related to introduction of new stock, and it was associated with presence of ESBL *Enterobacteriaceae*. Animals from farms with a high number of animal incoming movements in relation to total farm size were at a 2.37 times higher risk of hosting ESBL producing *Enterobacteriaceae* than animals from farms with a lower level of animal traffic. This is in agreement with Snow et al. [19] who reported that operating a closed farm policy reduced the risk of the farm having ESBL *E. coli* compared to farms that were open and did not quarantine new cattle.

The confirmation of relatively high rates of ESBL producers in cattle and the high diversity among the isolates are worrisome and indicate an established reservoir, especially in dairy farms. A prudent use of antibiotics, especially of 3$^{rd}$ and 4$^{th}$ generation cephalosporins, restrictions in feeding milk of treated cows to calves and management improvements to facilitate the operation of closed herd policies could represent modes of action towards reduction of ESBL prevalence in cattle. Experimental studies would be needed to assess the effectiveness of such measures. Moreover, this study showed that with a carefully planned sampling scheme involving the major cattle slaughterhouses and – in addition – small abattoirs in areas disconnected from the main animal traffic, geographically representative surveys can be achieved by taking random samples at the abattoir.

## Acknowledgments

The authors thank the staff of the slaughterhouses for facilitating access to their premises and for the assistance during the collection of data. Moreover, we would like to thank Dr. Franceso Albanese and Dr. Margot Chastonnay for their help in the collection of the samples in the cantons of Ticino and Valais.

## Author Contributions

Conceived and designed the experiments: MR RS. Performed the experiments: NG HH RS. Analyzed the data: MR SS RS. Wrote the paper: MR SS RS.

## References

1. Coque TM, Baquero F, Cantòn R (2008) Increasing prevalence of ESBL-producing Enterobacteriaceae in Europe. Eurosurveill 13:19044.
2. Bradford PA (2001) Extended-spectrum beta-lactamases in the 21st century: characterization, epidemiology, and detection of this important resistance threat. Clin Microbiol Rev 14:933–951.
3. Cantòn R, Coque TM (2006) The CTX-M β-lactamase pandemic. Curr Opin Microbiol 9:466–475.
4. Gniadkowski M (2001) Evolution and epidemiology of extended-spectrum beta-lactamases (ESBLs) and ESBL-producing microorganisms. Clin Microbiol Infect 7:597–608.
5. Knothe H, Shah P, Krcmery V, Antal M, Mitsuhashi S (1983) Transferable resistance to cefotaxime, cefoxitin, cefamamdole and defuroxime in clinical isolates of *Klebsiella pneumoniae* and *Serratia marcescens*. Infect 11:317–317.
6. Mesa RJ, Blanc V, Blanch AR, Cortés P, González JJ, et al. (2006) Extended-spectrum b-lactamase-producing *Enterobacteriaceae* in different environments (humans, food, animal farms and sewage). J Antimicrob Chemother 58:211–215.
7. Geser N, Stephan R, Korczak BM, Beutin L, Hächler H (2012) Molecular identification of extended-spectrum-β-lactamase genes from *Enterobacteriaceae* isolated from healthy human carriers in Switzerland. Antimicrob Agents Chemother 56:1609–1612.
8. Meunier D, Jouy E, Lazizzera C, Kobisch M, Madec JY (2006) CTX-M-1- and CTX-M-15-type beta-lactamases in clinical *Escherichia coli* isolates recovered from food-producing animals in France. Int J Antimicrob Agents 28:402–407.
9. Wittum TE, Mollenkopf DF, Daniels JB, Parkinson AE, Mathews JL, et al. (2010) CTX-M-type expended-spectrum beta-lactamases present in *Escherichia coli* from the feces of cattle in Ohio, United States. Foodborne Path Dis 7:1575–1579.
10. Geser N, Stephan R, Kuhnert P, Zbinden R, Kaeppeli U, et al. (2011) Fecal carriage of extended-spectrum ß-lactamase-producing *Enterobacteriaceae* in swine and cattle at slaughter in Switzerland. J Food Prot 74:446–449.
11. Jensen LB, Hasman H, Agersø Y, Emborg HD, Aarestrup FM (2006) First description of an oxyimino-cephalosporin-resistant, ESBL-carrying *Escherichia coli* isolated from meat sold in Denmark. J Antimicrob Chemother 57:793–794.
12. Hammad AM, Ahmed AM, Ishida Y, Shimamoto T (2008) First characterization and emergence of SHV-60 in raw milk of a healthy cow in Japan. J Vet Med Sci 70:1269–1272.
13. Doi Y, Paterson DL, Egea P, Pascual A, López-Cerero L, et al. (2010) Extended-spectrum and CMY-type beta-lactamase-producing *Escherichia coli* in clinical samples and retail meat from Pittsburgh, USA and Seville, Spain. Clin Microbiol Infect 16:33–38.
14. CLSI (2008) Clinical and Laboratory Standards Institute: Performance Standards for Antimicrobial Susceptibility Testing; Eighteenth Informational Supplement. CLSI Document M100–S18, Wayne, PA.
15. Hosmer DW, Lemeshow S (1989) Applied logistic regression. Wiley and Sons, New York, NY, USA.
16. Geser N, Stephan R, Hächler H (2012) Occurrence and characteristics of extended-spectrum β-lactamase (ESBL) producing *Enterobacteriaceae* in food producing animals, minced meat and raw milk. BMC Vet Res 8:21.
17. Tamang MD, Nam HM, Kim SR, Chae MH, Jang GC, et al. (2013) Prevalence and molecular characterization of CTX-M β-lactamase-producing *Escherichia coli* isolated from healthy swine and cattle. Foodborne Path Dis 10:13–20.
18. Ben Sallem R, Ben Slama K, Sáenz Y, Rojo-Bezares B, Estepa V, et al. (2012) Prevalence and characterization of extended-spectrum beta-lactamase (ESBL)- and CMY-2-producing *Escherichia coli* Isolates from healthy food-producing animals in Tunisia. Foodborne Path Dis 9:1137–1142.
19. Snow LC, Warner RG, Cheney T, Wearing H, Stokes M, et al. (2012) Risk factors associated with extended spectrum beta-lactame *Escherichia coli* (CTX-M) on dairy farms in North West England and North Wales. Prevent Vet Med 106: 225–234.
20. Hartmann A, Locatelli A, Amoureux L, Depret G, Jolivet C, et al. (2012) Occurrence of CTX-M producing *Escherichia coli* in soils, cattle, and farm environment in France (Burgundy Region). Front Microbiol 3:83.
21. Anonymous ARCH-VET (2011) Swiss annual report on usage of antimicrobials in veterinary medicine and on monitoring antimicrobial resistance in food animals. Available: http://www.swissmedic.ch/marktueberwachung/00147/00644/index.html?lang=de. Accessed 2013 May 13.

# Factors Associated with Pleurisy in Pigs: A Case-Control Analysis of Slaughter Pig Data for England and Wales

**Henrike C. Jäger[1¤], Trevelyan J. McKinley[1], James L. N. Wood[1], Gareth P. Pearce[1], Susanna Williamson[2], Benjamin Strugnell[3], Stanley Done[3], Henrike Habernoll[4], Andreas Palzer[5], Alexander W. Tucker[1]***

**1** Department of Veterinary Medicine, University of Cambridge, Cambridge, United Kingdom, **2** Veterinary Laboratories Agency (AHVLA), Bury St Edmunds, Suffolk, United Kingdom, **3** Veterinary Laboratories Agency (AHVLA), West House, Thirsk, North Yorkshire, United Kingdom, **4** BQP Ltd, Stradbroke Business Centre, Stradbroke, Suffolk, United Kingdom, **5** Ludwig-Maximilians-Universität München, Clinic for Swine, Sonnenstrasse, Oberschleissheim, Germany

## Abstract

A case-control investigation was undertaken to determine management and health related factors associated with pleurisy in slaughter pigs in England and Wales.

***Methods:*** The British Pig Executive Pig Health Scheme database of abattoir pathology was used to identify 121 case (>10% prevalence of pleurisy on 3 or more assessment dates in the preceding 24 months) and 121 control units (≤5% prevalence of pleurisy on 3 or more assessment dates in the preceding 24 months). Farm data were collected by postal questionnaire. Data from respondents (70 cases and 51 controls) were analysed using simple logistic regression models with Bonferroni corrections. Limited multivariate analyses were also performed to check the robustness of the overall conclusions.

***Results and Conclusions:*** Management factors associated with increased odds of pleurisy included no all-in all-out pig flow (OR 9.3, 95% confidence interval [CI]: 3.3–29), rearing of pigs with an age difference of >1 month in the same airspace (OR 6.5 [2.8–17]) and repeated mixing (OR 2.2 [1.4–3.8]) or moving (OR 2.2 [1.5–3.4]) of pigs during the rearing phase. Those associated with decreased odds of pleurisy included filling wean-to-finish or grower-to-finish systems with piglets from ≤3 sources (OR 0.18 [0.07–0.41]) compared to farrow-to-finish systems, cleaning and disinfecting of grower (ORs 0.28 [0.13–0.61] and 0.29 [0.13–0.61]) and finisher (ORs 0.24 [0.11–0.51] and 0.2 [0.09–0.44]) accommodation between groups, and extended down time of grower and finisher accommodation (OR 0.84 [0.75–0.93] and 0.86 [0.77–0.94] respectively for each additional day of downtime). This study demonstrated the value of national-level abattoir pathology data collection systems for case control analyses and generated guidance for on-farm interventions to help reduce the prevalence of pleurisy in slaughter pigs.

**Editor:** Todd Davis, Centers for Disease Control and Prevention, United States of America

**Funding:** The authors thank the British Pig Executive (BPEX - a part of the UK's Agricultural and Horticultural Development Board) for funding this project and providing the BPHS data base. The funders had no role in study design, data collection and analysis, decision to publish, or preparation of the manuscript.

**Competing Interests:** Dr H. Habernoll worked, until June 2011, as a veterinarian for BQP Ltd - a commercial pig production company. She contributed time (paid by BQP) to the design and testing of a questionnaire, but was not involved in the analysis of data.

* E-mail: awt1000@cam.ac.uk

¤ Current address: Garth Partnership, Straight Lane, Beeford, East Yorkshire, United Kingdom

## Introduction

Pleurisy is defined as inflammation of the pleural membranes, the serosal surfaces of the lung and chest cavity that facilitates smooth inflation of the lung. It is a particular problem in the pig industry [1] and is evident at necropsy or slaughter as fibrinous or fibrous adhesions between the lung lobes (visceral pleurisy) and/or the lungs and chest wall (parietal pleurisy). Interest in the economic and welfare impacts of pleurisy has increased since the high prevalence of this condition in finisher pigs has become apparent [1]. The economic impacts require further investigation, but chronic pleurisy is associated with increased time to slaughter [2]. It also causes problems in abattoirs because carcases require trimming causing extra labour, slower production line speeds, and result in increased waste. Respiratory disease is known to have significant negative impacts on indicators of pig welfare [3].

Pleurisy is a common finding in slaughter pigs in the UK, as evidenced by data from the systematic abattoir pathology recording under the British Pig Executive's (BPEX) Pig Health Scheme (BPHS); data provided to us from 14 abattoirs showed that of 15,237 slaughter consignments between July 2005 and October 2008, 80% were affected by pleurisy. Within these consignments, at the individual pig level 12.5% of 641,763 pigs were affected. Studies in other countries have found similar and even increasing pleurisy prevalence over the last 20 years (Table 1). Pleurisy is a multifactorial syndrome that can be caused by a number of different infections and which is predisposed to by a range of different management factors.

Previous studies of management factors associated with pleurisy in pigs have identified some common management factors, as well as some regional differences. The most important risk factors found in previous studies were related to transmission of infections at herd or pig level such as pig density in neighbourhood [4,5], poor biosecurity [5], increased herd size [6] or number of pigs per pen [7], lack of complete all-in/all-out practice [4,8], and mixing of pigs

**Table 1.** Pleurisy prevalence, presented as percentage of individual affected pigs, in EU countries.

| Country | Period | Prevalence |
|---|---|---|
| Belgium | 2000 | 16% [5] |
| | 2009 | 20.8% [7] |
| Denmark | 1987 | 14 [9] |
| | 1998 | 24% [29] |
| | 2000 | 25% [4] |
| Netherlands | 1990 | 12% [14] |
| | 2004 | 22.5% [14] |
| Norway | 1991 | 41% [12] |
| Spain | 2009 | 26.8% [8] |
| UK | 1988 | 16% [1] |

in the finishing stage [4]. But whereas Maes (2001) detected a higher prevalence of pleurisy in slaughter pigs in January/February in Belgium, with more severe lesions in March/April, in the Netherlands Elbers (1992) found highest prevalence in June/August.

The presence of antibodies to *Actinobacillus pleuropneumoniae* (APP) is associated with pleurisy either alone [6,7,9,10] or in combination with Porcine Reproductive and Respiratory Syndromevirus (PRRSV) [8]. Also *Mycoplasma hyopneumoniae* (M. hyo) [7,11], *Mycoplasma hyorhinis* [12] and Swine Influenza virus (SIV) [6] have been shown to be associated with higher frequency of pleurisy. More recently PCV2 has also been suggested to be associated with increased levels of pleurisy [13], and in addition porcine atrophic rhinitis (PAR) has been associated with pleurisy in Denmark [6,9].

Understanding the health associated factors and clinical signs in live pigs with pleurisy would permit more effective and timely targeting of control measures, since often the disease is only apparent at slaughter. However, work in this area has been limited—coughing and lethargy are considered to be indicative, but not specific for pleurisy, but attempts to identify pigs suffering from pleurisy pre-mortem based on pyrexia and dyspnoea have not been successful [14].

The present analysis focused on management and health-related associative factors for pleurisy and took into account the three main types of slaughter pig production systems relevant in the European Union (farrow-to-finish, wean-to-finish, grow-to-finish). Most previous studies looked at only one [5] or two types of production systems [8,9]. A case-control analysis was conducted, using retrospective abattoir pathology data collected at national level within the BPHS over the previous two years. Due to the ubiquity of pleurisy in the UK, pig units were defined as cases or controls based on consistently high or low pleurisy prevalence at unit level. One goal was to demonstrate the value of a nation-wide abattoir pathology database in identifying these consistent case and control units since it provided objective data representing around 80% of the farm assurance accredited English and Welsh production base. Herd specific information on management practices and health observations were gathered by a postal questionnaire from units that met the criteria for case or control.

## Materials and Methods

### Selection of target units based on pre-existing abattoir pathology data

The British Pig Executive (BPEX), representing English and Welsh levy paying pig producers, launched the BPHS abattoir pathology monitoring scheme database in 2005 [15]. BPHS is considered a comprehensive representation of the slaughter pig population in England and Wales since it captures data from approximately 75% of all commercial slaughter herds (1036 of a total 1400 herds, based on 2010 data) [16]. For a given consignment of slaughter pigs, each containing from 10 to >200 pigs, assessments are recorded from every second pig on the slaughter-line up to a maximum sample size of 50 pigs per consignment. The scheme operates at the 14 largest pig abattoirs in England and Wales using 37 specialist veterinarian assessors to collect on-line pathology data on 1 to 4 assessment days per month depending on the size of the abattoir. Assessment days rotate ensuring each day of the week is represented allowing every herd to be assessed at least once a quarter. Standardisation of assessment data between abattoirs and assessors is monitored by the scheme and includes regular training and rotation of assessors [15,16].

Criteria for case and control definitions were developed from this pre-existing database, taking into account the distribution of the data, and aiming to avoid data collected from small sample populations or from producers that recorded highly variable pleurisy prevalence over time. The database was used to identify all producers that had 50 slaughter pigs assessed on at least three occasions in the 24 months prior to October 2008 (778 (56%) producers of a total of approximately 1400 commercial herds) (Table 2). Fifty nine percent of consignments assessed for these producers had at least a 5% prevalence of pleurisy during the 24 month period but the prevalence was highly variable on some units. As such it was felt important to define a case-control measure based on *consistency* of prevalence of pleurisy over time, in order to attempt to separate units with endemic pleurisy problems from those that exhibited more transient occurrences. *Cases* were defined as those that had >10% of pleurisy-affected pigs in each of the three most recent consignments in the 24 month period prior to October 2008, and *controls* were those that had ≤5% of pleurisy-affected pigs in each of the three most recent consignments in that same period. Selection of these cut-offs was based on examining the distribution of the full dataset while attempting to balance study power and maximum discrimination of case and control groups. Indicative sample size calculations were done on the basis of a single factor analysis and indicated that data would be needed from 105 case units and 105 control units to detect statistical significance ($p<0.05$) of a risk factor found in 20% of the control units that had an odds ratio of 2.5, with a desired study power of 80%.

### Questionnaire to collect farm-level information

Herd health and management data were gathered by a closed-question postal questionnaire sent to 242 units (121 cases, 121 controls) followed up by telephone liaison with the farm manager and the appropriate private veterinarian. Respondents were not informed of their case/control categorisation in order to minimise selection bias. A pilot questionnaire was validated at three units before dispatch. The questions were composed to ensure clarity for producers and sufficient detail for statistical analysis. An outline of investigated variable factors is presented in Table 3.

### Processing and statistical analysis of data

Data were stored and manipulated in Microsoft Access and Excel (Microsoft 2007). All statistical analyses were conducted in the R statistical language (R Core Development Team 2008).

The questionnaire was stratified into a series of categories, representing different characteristics of a unit. These were: general farm information (including production type), mortality and productivity, health status, herd environment and herd management. To explore the data in a systematic manner we stratified the

**Table 2.** The number (%) of herds at each level of the sampling strategy.

| Herds (cases and controls) | Number (%) |
|---|---|
| Commercial slaughter-pig holdings in England and Wales | 1400 (100%) [16] |
| Herds sampled by BPHS scheme (data for 2010) | 1036 (74% of 1400) [16] |
| Herds with 50 pigs sampled by BPHS on at least 3 occasions prior to October 2008 | 778 (56% of 1400) |
| Number of eligible cases | 121(16% of 778) |
| Number of eligible controls | 306 (39% of 778) |
| Total number of eligible herds | 427 herds (55% of 778; 31% of 1400) |
| Number of dispatched questionnaires | 242 (121 cases, 121 controls) |
| Number of completed questionnaires | 121 (50% of 242; 16% of 778; 9% of 1400) |
| | 51 cases (7% of 778) |
| | 70 controls (9% of 778) |
| Number of herds included in univariable model | 121 |
| Number of herds included in multivariable model | 121 |

The number (%) of herds at each level of the sampling strategy, including the number of eligible case and control herds, as a proportion of the total number of commercial slaughter-pig herds in England and Wales.

**Table 3.** Outline of variables included in a questionnaire addressed to pig farms.

| Variable | Levels (if applicable) |
|---|---|
| Production unit type (and number of sources where applicable) | Farrow-finish/wean-finish/grow-finish |
| All-in/All-out pig flow | By unit/room/pen |
| Number of finisher places | value |
| Distance to next pig unit (km) | value |
| Experience of senior stockman (years) | value |
| Ongoing training of stockmen | Yes/No |
| Accommodation systems (for weaning −30 kg, and 30 kg - slaughter) | Fully slatted/part slatted/straw yards/assisted ventilation |
| Number of times pigs moved after weaning | value |
| Number of times pigs mixed after weaning | value |
| Is airspace shared by pigs of >1 month age gap? | Yes/no |
| Maximum number of pigs in shared airspace | value |
| Feeding regime (for 7–30 kg, for 30–50 kg, and for 50 kg - slaughter) | Meal/pellets/wet feed |
| | Home-mixed/purchased compound/by-product |
| | Ad libitum/restrict fed |
| Medication: number at group level | Product/duration/in feed or water/reason |
| Medication: individual treatments: | Number in past week/reason |
| Farmer observations of disease (main effect: none, few, many; where an age effect requested this is 7–30 kg & >30 kg; data requested for 2008 & 2007) | Scours (by age)/sneezing (by age)/coughing (by age)/dyspnoea (by age)/meningitis/wasting (by age)/sudden deaths (by age)/porcine dermatitis and nephropathy syndrome (PDNS)/other |
| Farmer or herd vet knowledge of specific disease status (believed present, confirmed by vet, believed absent, not known) | porcine reproductive and respiratory syndrome (PRRS))/A. pleuropneumoniae (APP)/Glasser's Disease/enzootic pneumonia (EP)/post-weaning multisystemic wasting syndrome (PMWS) |
| Vaccination of finisher pigs | Absence of any vaccination/EP (one or 2 dose regime)/Porcine circovirus type 2 (PCV2)/PRRS/Glasser's Disease/Ileitis/Other |
| Post-weaning mortality | Values for 2008, 2007, 2006 |
| Mortality recording system type | Computer/other |
| Vet health plan in place on unit | Yes/No |

Outline of variables included in a questionnaire addressed to pig farms defined as case (pleurisy prevalence consistently >10%) or control (pleurisy prevalence consistently <5%) to seek relationships between pleurisy and production unit type, key indicators of general management, and health observations.

variables into two main groups: those that corresponded to farm management characteristics (for which the influence is possibly independent of the disease status), and disease associated factors (those factors that were directly dependent on the disease status of the farm).

It was necessary to re-categorise some of the categorical variables to ensure that there were >5 observations in any level of the factor and also to aid interpretation. Variables having large numbers of missing values (>60) were removed at the outset, as were those categorical variables that had <5 samples in a group and could not be easily re-categorised. Within each group of variables (e.g. management characteristics and disease associated characteristics) the data were screened by applying a simple logistic regression model to each variable in turn, using a chi-squared likelihood ratio test (LRT) [17], and correcting for multiple comparisons using Bonferroni step-down procedures. The extent and distribution of missing values precluded the development of a comprehensive multivariable regression model. However, it was possible to produce a limited multivariable model examining relationships between pleurisy and some of the more important management related factors obtained from the univariate analyses (see results sections for further discussion). Variable selection was conducted using forwards stepwise selection routines and Akaike Information Criterion (AIC) (using the MASS package in R [18]), including only those variables where $p = 0.05$ or less in the Bonferroni corrected LRT results. Collinearity between variables was assessed by examining the standard errors. As such, in addition to the univariable results we also present some further discussion regarding associations between some of the explanatory variables based on the constrained multivariable models. As a result of the aforementioned limitations, we did not explore interaction effects in this instance. Goodness-of-fit was assessed using the le Cessie-van Houwelingen normal test statistic for the unweighted sum of squared errors [19,20], as implemented in the "Design" package in R [21]. Discriminatory power was assessed using the Area Under the Receiver Operating Characteristic Curve (AUC), using the "verification" package [22]. Each observation with a standardised Pearson residual of >2 was removed from the final model in turn to check for undue influence due to outliers.

## Results

### Recruitment of respondent farms

Overall there were 126 respondent farms from the original 242 targeted: 51 cases, 70 controls, with 2 questionnaires unusable due to incorrect herd identification. Three had ceased business. Hence the overall usable response rate was 50%. The mean, minimum and maximum pleurisy prevalences across case producers were 29.5%, 12% and 76.7%. Across control producers the mean pleurisy prevalence was 1.6%, ranging from a minimum of 0% to a maximum of 3.3%.

### Management factors

The univariable results for management related risk factor analysis are shown in Table 4. Absence of all-in/all-out (AIAO) pig herd management was an important factor associated with increased pleurisy (OR 9.3) compared to complete AIAO. All-in/all-out by room was similar to no all-in/all-out practice (OR 0.96). Keeping pigs of more than one month age difference in the same airspace was associated with increased pleurisy prevalence (OR 6.5). In addition there was an association between moving and mixing of pigs on farms and higher levels of pleurisy (OR 2.2 and 2.2 per move/mix respectively). Partial slatted flooring for weaners was a strongly associated factor (OR 21.4), but had a very wide confidence interval (3.7–400).

Factors associated with reduced prevalence of pleurisy included wean-to-finish and grow-to-finish production systems compared to farrow-to-finish systems (OR 0.10 and 0.45 respectively), cleaning and disinfection on finishing batches (ORs 0.24 and 0.20 for cleaning and disinfecting respectively), and on grower batches (ORs 0.28 and 0.29 respectively). Also associated was purchasing feed for growers as compared to home-mixing of feed (OR 0.22). Farrow-to-finish production was associated with higher levels of pleurisy than multisite operations that sourced pigs from other breeding units. However, the protective effect became less strong (and statistically insignificant) when these grow-outs sourced from >3 units (ORs 0.18 for ≤3 sources compared to 0.69 for >3 sources). Finally, longer periods of downtime between grower and finisher batches were associated with reduced pleurisy prevalence (ORs 0.84 and 0.86 for each additional day of downtime respectively).

Due to the stratified nature of some of the variables (e.g. grow-to-finish units do not have weaner accommodation), and the within-unit heterogeneity (particularly with regards to some of the accommodation types), it was difficult to design a sensible multivariable model that included all of the variables, such that there were sufficient samples to produce reasonable statistical power. Instead, we restricted attention to some of the more important variables identified in Table 4. Since we needed complete data in order to use stepwise selection, we excluded variables that had more than 5 missing values (leaving 10/15 variables). Then we excluded all batches that had any missing values across these 10 remaining variables (leaving 110 batches). We then fitted a forward stepwise selection model and report the results in Table 5.

Interestingly, the strongest variable from the univariable analysis (herd management) was the first to be added, and remained in the model until the final step, where it seems that the combination of cleaning between batches (growers), air-space shared by multiple age groups, and number of moves rendered herd management unnecessary to remain in the model. There was a strong association between shared air and herd management (only 2/30 herds with shared air = true practiced AIAO, compared to 57/80 herds with shared air = false), and also between the number of moves and herd management (median of 1 move for AIAO systems and 3 moves for non-AIAO systems). The association with cleaning between batches and herd management was less pronounced. This final model showed no statistically significant lack-of-fit ($p = 0.15$) and showed a relatively good discriminatory power (AUC = 0.83). Overall, three observations had an absolute standardised Pearson residual of >2 and <2.5, and three more of >2.5. Removing these in turn made negligible difference to the parameter estimates.

### Disease associated factors

Case units had an increased post-weaning mortality, dyspnoea (both<30 kg and >30 kg in weight), coughing (>30 kg) and increased odds of farmer declared positive status for APP. Also, increased frequency of group medication was associated with pleurisy (Table 6).

The median post-weaning mortality rate between 2006 and 2008 (Figure 1) was consistently higher in case units (by 3.3%) (2006: case = 7.7%, control = 5%; 2007: case = 7.7%, control = 4%; 2008: case = 6%, control = 4%. All figures are median values).

## Discussion

The BPHS database, which represents approximately 74% of slaughter pig production in England and Wales, proved suitable

**Table 4.** Analysis of management related factors related to pleurisy in slaughter pigs.

| Variable | Adj. LRT p-value | n | Type | Levels | OR | Lower 95% CI | Upper 95% CI |
|---|---|---|---|---|---|---|---|
| Herd management | 0.00 | 117 | - | AIAO | - | - | - |
| | | | - | By room | 0.96 | 0.05 | 7.2 |
| | | | - | Mixed | 8.2 | 3.0 | 24 |
| | | | - | None | 9.3 | 3.3 | 29 |
| Shared air | 0.00 | 121 | - | False | - | - | - |
| | | | - | True | 6.5 | 2.8 | 17 |
| Number moves (per move) | 0.00 | 119 | - | - | 2.2 | 1.5 | 3.4 |
| Production type | 0.00 | 121 | - | Farrow-to-finish | - | - | - |
| | | | - | Wean-to-finish | 0.10 | 0.03 | 0.28 |
| | | | - | Grow-to-finish | 0.45 | 0.18 | 1.1 |
| Disinfect between batches | 0.00 | 121 | Finisher | False | - | - | - |
| | | | | True | 0.20 | 0.09 | 0.44 |
| Downtime (per add. day) | 0.00 | 81 | Grower | - | 0.84 | 0.75 | 0.93 |
| Partial slatted | 0.01 | 80 | Weaner | False | - | - | - |
| | | | | True | 21 | 3.7 | 400 |
| Number source units | 0.01 | 116 | - | 0 | - | - | - |
| | | | - | <=3 | 0.18 | 0.07 | 0.41 |
| | | | - | >3 | 0.69 | 0.13 | 4.0 |
| Clean between batches | 0.01 | 121 | Finisher | False | - | - | - |
| | | | | True | 0.24 | 0.11 | 0.51 |
| Downtime (per add. day) | 0.01 | 83 | Finisher | - | 0.86 | 0.77 | 0.94 |
| Feed origin | 0.02 | 104 | Grower | Homemix | - | - | - |
| | | | | Purchased | 0.22 | 0.09 | 0.52 |
| Number mixes (per mix) | 0.03 | 120 | - | - | 2.2 | 1.4 | 3.8 |
| Disinfect between batches | 0.04 | 121 | Grower | False | - | - | - |
| | | | | True | 0.29 | 0.13 | 0.61 |
| Clean between batches | 0.04 | 121 | Grower | False | - | - | - |
| | | | | True | 0.28 | 0.13 | 0.61 |

Results of independent logistic regression models fitted to each management variable in turn, showing odds ratios (OR) and 95% confidence intervals for the variables shown to be statistically significant at the 5% level from univariable logistic regression models using likelihood ratio tests (LRT) with Bonferroni adjustments. Continuous and discrete variables are shown with a dash in the "Levels" column, with the OR corresponding to the OR per unit increase; for the categorical variables the OR is relative to the referent level, which is always shown first.

**Table 5.** Results from a constrained multiple regression model.

| Variable | Type | Level | OR | Lower 95% CI | Upper 95% CI |
|---|---|---|---|---|---|
| Clean between batches | Grower | False | - | - | - |
| | | True | 0.33 | 0.11 | 0.89 |
| Number of moves (per move) | - | - | 2.3 | 1.5 | 3.8 |
| Shared air | - | False | - | - | - |
| | - | True | 4.0 | 1.4 | 12 |

Results from a constrained multiple regression model fitted to ten variables across 110 batches to further investigate the relationship between management factors and pleurisy in slaughter pigs. Continuous (or discrete) variables are shown with a dash in the "Levels" column, with the OR corresponding to the OR per unit increase; for the categorical variables the OR is relative to the referent level, which is always shown first.

for the purpose of identifying case and control units. However, many units within it had a large variation in pleurisy prevalence over the 24 month period studied. Because of this we imposed a strict definition of *consistency* in pleurisy levels over time in our case/control definitions. Hartley (1988) made the same observation regarding pleurisy variability and concluded that this was due to disease dynamics and variation in susceptibility of disease influenced by the environment and management. This may also be impacted by differences from batch to batch in sourcing and mixing of pigs that comprise a batch on entry to a given wean- or grow-to-finish system such that the same unit could have a history of highly variable pleurisy prevalence over time. Chance variation in the infections introduced with different pig batches could be important. The case/control definitions used here provided a metric for distinguishing between *consistently* higher or lower risk units, and must be interpreted as such.

Within responding units there were varying degrees of missing data. This was partly to do with unforeseen heterogeneity in management practices. For example, many units used multiple

**Table 6.** Analysis of health related factors related to pleurisy in slaughter pigs.

| Variable | Adj. LRT p-value | n | Levels | OR | Lower 95% CI | Upper 95% CI |
|---|---|---|---|---|---|---|
| Mortality 2007 | 0.00 | 117 | - | 1.5 | 1.3 | 1.9 |
| APP( farmer or vet declared) | 0.00 | 92 | Absent | - | - | - |
| | | | Present | 8.8 | 3.4 | 25 |
| Mortality 2008 | 0.00 | 114 | - | 1.3 | 1.1 | 1.6 |
| Mortality 2006 | 0.00 | 111 | - | 1.3 | 1.1 | 1.5 |
| Dyspnoea (>30 kg) 2007 | 0.00 | 121 | Absent | - | - | - |
| | | | Present | 4.8 | 2.2 | 11 |
| Dyspnoea (>30 kg) 2008 | 0.01 | 121 | Absent | - | - | - |
| | | | Present | 4.1 | 1.9 | 9.0 |
| Cough (>30 kg) 2007 | 0.03 | 121 | Absent | - | - | - |
| | | | Present | 4.4 | 1.8 | 12 |
| Number of group medications | 0.04 | 117 | 0 | - | - | - |
| | | | 1–2 | 3.6 | 1.5 | 10 |
| | | | >=3 | 9.6 | 2.7 | 40 |
| Cough (>30 kg) 2008 | 0.05 | 121 | Absent | - | - | - |
| | | | Present | 4.0 | 1.7 | 10.4 |
| Dyspnoea (<30 kg) 2007 | 0.05 | 80 | Absent | - | - | - |
| | | | Present | 4.9 | 1.9 | 14 |

Results of independent logistic regression models fitted to each disease associated variable in turn, showing odds ratios (OR) and 95% confidence intervals for the variables shown to be statistically significant at the 5% level from likelihood ratio tests (p-value) with Bonferroni adjustments. Continuous (or discrete) variables are shown with a dash in the "Levels" column, with the OR corresponding to the OR per unit increase; for the categorical variables the OR is relative to the referent level, which is always shown first.

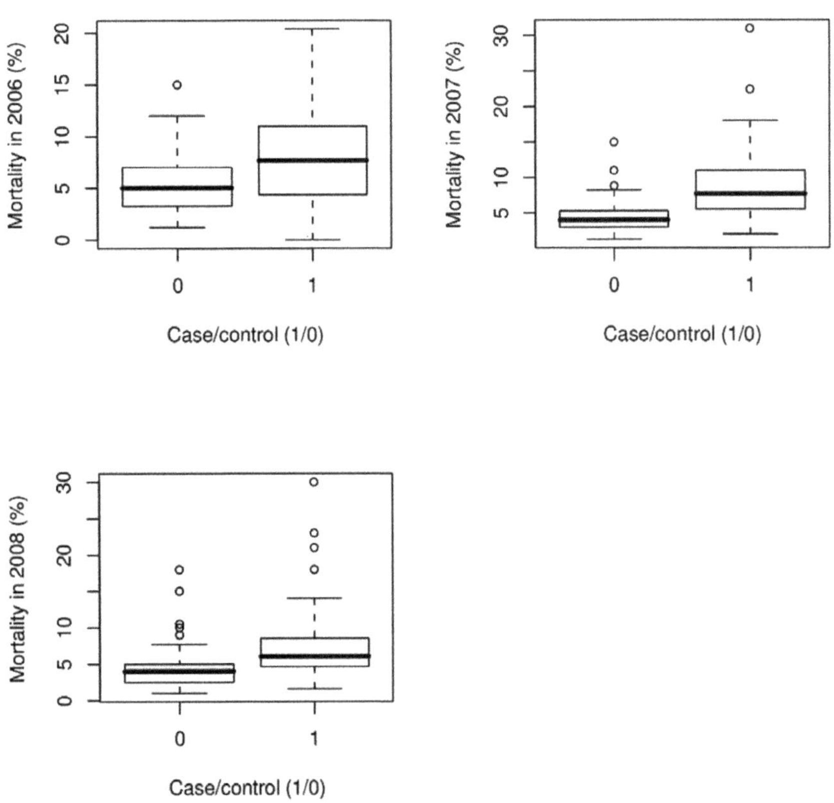


**Figure 1. Post-weaning mortality distributions, shown as percentages, for pig farms categorised as pleurisy affected (case) or less affected (control) for 2006–2008.**

accommodation types, sometimes for different age groups. These relationships were not clear before the study, but meant that it was difficult to stratify these variables in a sensible manner without incorporating missing information (e.g. stratifying accommodation by age group meant that grow-to-finish units would have missing values for weaning-age variables). Furthermore, there was also a tendency for respondents not to complete all questions. These limitations emphasise the importance of designing data capture questionnaires in a way that maximises the collection of relevant data but minimises the potential for missing data.

Since the definition of cases and controls was determined before recruitment, and the classification was unknown to the respondent, this should reduce the impact of selection bias. Nonetheless, more control farms replied than cases (59% and 41% respectively). We were unable to identify any systematic bias in terms of explanatory variables since we had no data from non-responders. However, the differing response rates suggest that there may be a relationship between producers' 'attitudes' to communication about this on-farm health issue and the prevalence of pleurisy. Similar future studies should take account of these differing response rates and factor in the need for follow-up phone calls to responders. Finally, the analysis only included units that had 50 pigs assessed (i.e. 100 or more pigs submitted) on each of 3 successive occasions and, although this means that the results might not extrapolate to small-scale producers, it nevertheless provides information about farm management and health characteristics that are associated with consistently high or low levels of pleurisy in larger, more economically significant, units.

We used a series of univariable logistic regression models using a conservative Bonferroni step-down multiple adjustment procedure [23]. One limitation of this approach is that it is difficult to assess the impact of confounding and effect interactions. As such the individual factors obtained from the univariable analyses that were associated with increased or decreased odds of pleurisy must be viewed in terms of providing information about potential foci for control and intervention that could be tested, and are discussed in the context of other studies and/or prior knowledge. Due to the stratified nature of some of the variables, and the degree-of-missing data, it was only possible to fit a multivariable model to a subset of the data to explore limited associations. However, caution must be used in the interpretation of these results, due to the limited scope of the variables included in the analysis. Nonetheless they further highlight the importance of the variables that were also identified in the univariable analysis.

The results of univariable analysis indicated that failure to implement strict AIAO (by unit or building) was strongly associated with increased pleurisy and this was in line with previous studies [4]. In contrast, the final multivariable model contained cleaning between batches (growers), air-space shared by multiple age-groups, and number of moves but not AIAO. Interestingly AIAO remained in the multivariable analysis until the final step of the procedure before dropping out. Cleaning between batches and avoidance of sharing airspace by pigs of different ages, factors that are both present in the final multivariable model, are important contributory elements of effective AIAO management. Not practising AIAO potentially allows diseases to circulate because susceptible pigs are continuously introduced and older pigs can pass on infections to the younger generation [2]. The univariable analysis findings that repeated mixing, moving, the co-existence of pigs of >1 month age difference in the same air space, and failures in cleaning or disinfection were also factors associated with increased pleurisy reinforced the biological relevance of this observation since these are key practical components of an AIAO management system.

Conversely, implementing AIAO by room, as opposed to by building or unit, was associated with increased pleurisy in the univariable analysis. It seems that there is sometimes confusion about the definition of AIAO – a management system that segregates pigs of a defined age span (e.g. 3 weeks) in an airspace that is separate from groups of other aged pigs throughout their life. A key part of AIAO is that the segregated airspace or accommodation is fully emptied before repopulation occurs. AIAO can break disease cycles, but only if the entire population is included in the process. Our data suggested that AIAO by room cannot be regarded as effective AIAO. In most cases, although the situation varies from farm to farm, a room is not separated enough from other pigs to allow calling the process of emptying a room 'all-out' or filling a room 'all-in'.

The odds of pleurisy increased each time pigs were mixed (univariable analysis) or moved (univariable and multivariable models). Moving and mixing are stressors for pigs which may impact on immunity [24], and are opportunities for pathogens such as APP to spread to susceptible pigs [25]. Although identifying the role of specific infections in causing pleurisy was not a central aim of the current work, vet or farmer-declared presence of clinical APP on the farm was associated with higher levels of pleurisy. APP status might have been determined by clinical or serological status. Vaccination against APP might have impacted on the serological status, or masked clinical disease, but vaccination against this organism is very uncommon in England and Wales. The role of APP in pleurisy is supported by several serological studies [6,7,8,9,10].

A number of previously undescribed protective factors were identified in this analysis. Firstly, cleaning and disinfection of grower and finisher accommodation between batches was identified in the univariable model, with cleaning of grower pens remaining in the final multivariable model. Secondly, increased "down time" between batches for finisher and grower accommodation was identified in the univariable model. These are issues that have previously been identified as important associative factors relating to enteric disease [26] but less so in the context of respiratory disease. Nevertheless, cleaning might be expected to contribute to respiratory health through reduced levels of dust, environmental bacteria and fungal spores. Resting buildings allows complete drying after disinfection and would be expected to optimise killing of important respiratory pathogens. This has been demonstrated in pig transport trailers for PRRSV [27] but studies of total aerobic bacterial counts were unable to show an effect of down time (Amass 2007). This is nevertheless an important area for future investigation since the presence of organic matter can significantly affect environmental survival of respiratory pathogens such as APP (Gottschalk 2006).

Compared to farrow-to-finish (FF) operations, grow-to-finish (GF) but especially wean-to-finish (WF) systems showed lower levels of pleurisy (GF OR = 0.45; WF OR = 0.1) according to the univariable analysis. The continuous presence of breeding and growing pigs on FF units may be responsible for continuous circulation of infections. Strict AIAO production, at building level, on FF units in the UK is extremely unlikely to occur and pigs must progress through what is often a closely located set of buildings. On the other hand, WF and GF units are more suited to strict AIAO, in spite of the fact that their population usually involves the mixing of pigs from different breeding sources. The observed additional protective effect of WF units over GF units is worthy of further investigation. Of potential importance might be the residual colostrally derived passive immunity at mixing during population in WF units. Population (and mixing of sources) on GF units takes place after the decline of passive immunity with, potentially, a consequential increase in the effective population of susceptible pigs. Also, or alternatively, if infections causing pleurisy spread soon after mixing on AIAO WF units, pigs have a longer period until slaughter during which lesions may resolve.

Another apparently protective factor identified in the univariable analysis was sourcing of piglets to WF or GF sites from ≤3 units in comparison to the single sourcing associated with farrow-finish (no external sources). This association was weaker when a batch was sourced from >3 breeding units. The protective effect over FF may be in part a proxy for the management conditions of WF and GF farms, although the reduced protective effect when more than 3 sources are taken is consistent with the notion that an increase in the likelihood of introduction of disease occurs when sourcing piglets from higher numbers of different units. The use of purchased grower feed versus home mixed feed was found to be associated with lower prevalence of pleurisy (OR = 0.2) but the absence of associations relating to feed at the finisher or weaner stages suggests that this finding may be an artefact, or may be correlated to other factors such as production type (home mixing is more common on FF units in the UK) but this could not be ascertained in the current project.

Regarding associations between pleurisy prevalence and disease related factors, the univariable study differentiated clinical signs by age group (< and >30 kg) and year (2007 and 2008). Similar to previous studies where observable respiratory disease in late finishing was associated with the presence of pleurisy [8], the present study found dyspnoea and coughing in pigs >30 kg were associated with pleurisy in 2007 and 2008. In 2007 dyspnoea in pigs <30 kg could also be related to increased pleurisy in slaughter pigs, but this effect was not observed in 2008. However, these clinical observations are not specific for pleurisy and may indicate other, co-existent, respiratory diseases. Previous research has indicated a link between pleurisy prevalence and prevalence of pneumonia [28], but more recent work suggests this relationship may not be straightforward since lesions of pneumonia were negatively associated with pleurisy lesions [5,10]. Much opportunity remains to understand how pleurisy relates to pneumonia in pigs and how it might be detected ante mortem.

Increased mortality was consistently and strongly associated with the units being defined as cases in each of the 3 years for which data was requested. This basis of this association is worthy of further investigation because, on one hand, it is another indication that pleurisy is a disease of generally lower health status units and, on the other, an indication of the economic consequences of pleurisy on units where it is a consistent problem. As a proxy for the overall health of a unit, increased numbers of group level medication periods in the post-weaning period were associated with units with consistent pleurisy. While this observation would be consistent with a tendency for pleurisy to occur on units of generally lower health status and with higher consequent production costs, it is probable that some of these additional medications would have been a direct consequence of pleurisy.

In conclusion, this study identified management and health related factors associated with pleurisy based on a questionnaire across 121 respondent units producing slaughter pigs and a national abattoir pathology surveillance database – demonstrating the value of this national disease surveillance system. The identified factors were mostly related to transmission of infectious diseases and the analyses highlighted the importance of AIAO but also a group of management factors associated with it. In addition, farrow-finish management systems were shown to be particularly at risk of consistent pleurisy, in part likely due to the difficulty in implementing strict AIAO in these systems in the UK. Since implementation of complete AIAO management, for example at the building or unit level, has significant cost implications a better understanding of the relative importance of specific management factors that contribute to AIAO and which can be implemented in any production system, is of value to the industry.

## Acknowledgments

The authors thank Dr Barbara Wieland and Dr Pablo Alarcon Lopez, both of the Royal Veterinary College, UK, for their assistance with design and review of the postal questionnaire and the sharing of relevant data.

## Author Contributions

Conceived and designed the experiments: HCJ TJM JLNW GPP SW BS SD HH AP AWT. Performed the experiments: HCJ TJM AWT. Analyzed the data: HCJ TJM JLNW AWT. Contributed reagents/materials/analysis tools: GPP SW BS SD HH AP. Wrote the paper: HCJ TJM JLNW AWT. Obtained BPHS dataset: HCJ TJM. Designed questionnaire: HCJ TJM JLNW GPP SW BS SD HH AP AWT. Tested questionnaire: HCJ HH.

## References


1. Rubies X, Kielstein P, Costa L, Riera P, Artigas C, et al. (1999) Prevalence of Haemophilus parasuis serovars isolated in Spain from 1993 to 1997. Vet Microbiol 66: 245–248.
2. Sorensen V, Jorsal SE, Mousing J (2006) Diseases of the respiratory tract. In: Straw B, Zimmermann J, D'Allaire S, Taylor D, eds. Diseases of Swine. 9 ed. Oxford: Blackwell Scientific. pp 149–177.
3. Schimmel D, Kielstein P, Hass R (1985) [Serological typing of Haemophilus parasuis]. Arch Exp Veterinarmed 39: 944–947.
4. Cleveland-Nielsen A, Nielsen EO, Ersboll AK (2002) Chronic pleuritis in Danish slaughter pig herds. Prev Vet Med 55: 121–135.
5. Maes DG, Deluyker H, Verdonck M, F C, Miry C, et al. (2001) Non-infectious factors associated with macroscopic and microscopic lung lesions in slaughter pigs from farrow-to-finish herds. Vet Record 148: 41–46.
6. Mousing J, Lybye H, Barfod K, Meyling A, Rønsholt L, et al. (1990) Chronic pleuritis in pigs for slaughter: an epidemiological study of infcetious and rearing system-related risk factors. Prev Vet Med 9: 107–119.
7. Meyns T, Van Steelant J, Rolly E, Dewulf J, Haesebrouck F, et al. (2011) A cross-sectional study of risk factors associated with pulmonary lesions in pigs at slaughter. Vet J 187: 388–392.
8. Fraile L, Alegre A, Lopez-Jimenez R, Nofrarias M, Segales J (2010) Risk factors associated with pleuritis and cranio-ventral pulmonary consolidation in slaughter-aged pigs. Vet J 184(3): 326–33.
9. Enøe C, Mousing J, Schirmer AL, Willeberg P (2002) Infectious and rearing-system related risk fcators for chronic pleuritis in slaughter pigs. Prev Vet Med 54: 337–349.
10. Wiegand M, Kielstein P, Pohle D, Rassbach A (1997) [Examination of primary SPF swine after experimental infection with Haemophilus parasuis. Clinical symptoms, changes in hematological parameters and in the parameters of the cerebrospinal fluid]. Tierarztl Prax 25: 226–232.
11. Sørensen V, Jorsal SE, Mousing J (2006) Diseases of the Respiratory System. In: Straw BE, Zimmermann JJ, D'Allaire S, Taylor DJ, eds. Diseases of Swine: Blackwell Publishing. pp 149–177.
12. Falk K, Hoie S, Lium BM (1991) An abattoir survey of pneumonia and pleuritis in slaughter weight swine from 9 selected herds. II. Enzootic pneumonia of pigs: microbiological findings and their relationship to pathomorphology. Acta Vet Scand 32: 67–77.
13. Kielstein P, Rapp-Gabrielson VJ (1992) Designation of 15 serovars of Haemophilus parasuis on the basis of immunodiffusion using heat-stable antigen extracts. J Clin Microbiol 30: 862–865.
14. Augustijn M, Stockhofe-Zurwieden N, Nielen M, Jirawattanapong P, Cruijsen A, et al. (2008) The etiology of chronic pleuritis is pigs: a clinical, pathological and serological study. Proceedings of the 20th Congress of the International Pig Veterinary Society, Durban. OR.05.12.
15. Anon (2008) The British Pig Health Scheme 2003–2008.: British Pig executive.
16. Sanchez-Vazquez MJ, Strachan WD, Armstrong D, Nielen M, Gunn GJ (2011) The British pig health schemes: integrated systems for large-scale pig abattoir lesion monitoring. Vet Rec 169: 413.
17. Hosmer D, Lemeshow S Applied Logistic Regression: John Wiley and Sons.
18. Venables WN, Ripley BD (2002) Modern Applied Statistics with S. New York: Springer.
19. le Cessie S, JC vH (1991) Biometrics 47: 1267–1282.
20. Hosmer D, Hosmer T, Lemeshow S, le Cessie S (1997) A comparison of goodness-of-fit tests for the logistic regression model. Stat in Med 16: 965–980.
21. Harrell F Design Package in R Package. 2.3-0 ed.

22. Program N-RA (2010) verification: Forecast. Verification utilities. R package. 1.31 ed.
23. Martin J, Gunther H, Janetschke P, Schonherr W, Kielstein P (1977) [Contribution to the experimental hemophilus infection (Haemophilus para-haemolyticus, Haemophilus parasuis) in SPF piglets. 2. Comparative pathology and histology]. Arch Exp Veterinarmed 31: 347–357.
24. Salak-Johnson JL, McGlone JJ (2007) Making sense of apparently conflicting data: stress and immunity in swine and cattle. J Anim Sci 85: E81–88.
25. Gottschalk M, Taylor DJ (2006) Actinobacillus pleuropneumoniae. In: Straw B, Zimmermann JJ, D'Allaire S, Taylor DJ, eds. Diseases Of Swine 9 ed: Blackwell Publishing. pp 563–576.
26. Janetschke P, Kielstein P, Schonherr W, Martin J, Gunther H (1977) [Contribution to the experimental haemophilus infection (haemophilus parahemolyticus, haemophilus parasuis) in specific pathogen-free piglets. 1. microbiology, experimental arrangement, results]. Arch Exp Veterinarmed 31: 129–137.
27. Zimmermann J, Benfield DA, Murthaugh MP, Osorio F, Stevenson GW, et al. (2006) Porcine Reproductive and Respiratory Syndrome Virus (Porcine Arterivirus). In: Straw B, Zimmermann JJ, D'Allaire S, Taylor DJ, eds. Diseases Of Swine 9 ed: Blackwell Publishing. pp 387–417.
28. Elbers ARW, Tielen MJM, Snijders JMA, Cromwijk WAJ, Hunneman WA (1992) Epidemiological studies on lesions in finishing pigs in the Netherlands. I. Prevalence, seasonality and interrelationship. Prev Vet Med 14: 217–231.
29. Christensen G, Enoe C (1999) The prevalence of pneumonia, pleuritis, pericarditis, and liver spots in Danish slaughter pigs in 1998, including comparison with 1994. Danish Veterinary Journal 82: 1006–1015.

# 7

# A Comparison of Non-Typhoidal *Salmonella* from Humans and Food Animals Using Pulsed-Field Gel Electrophoresis and Antimicrobial Susceptibility Patterns

**Carol H. Sandt[1]*, Paula J. Fedorka-Cray[2], Deepanker Tewari[4], Stephen Ostroff[3], Kevin Joyce[5], Nkuchia M. M'ikanatha[3]**

**1** Bureau of Laboratories, Pennsylvania Department of Health, Exton, Pennsylvania, United States of America, **2** United States Department of Agriculture, Agricultural Research Service, Athens, Georgia, United States of America, **3** Division of Infectious Disease Epidemiology, Pennsylvania Department of Health, Harrisburg, Pennsylvania, United States of America, **4** Pennsylvania Veterinary Laboratory, Pennsylvania Department of Agriculture, Harrisburg, Pennsylvania, United States of America, **5** National Antimicrobial Resistance Monitoring System, Centers for Disease Control and Prevention, Atlanta, Georgia, United States of America

## Abstract

Salmonellosis is one of the most important foodborne diseases affecting humans. To characterize the relationship between *Salmonella* causing human infections and their food animal reservoirs, we compared pulsed-field gel electrophoresis (PFGE) and antimicrobial susceptibility patterns of non-typhoidal *Salmonella* isolated from ill humans in Pennsylvania and from food animals before retail. Human clinical isolates were received from 2005 through 2011 during routine public health operations in Pennsylvania. Isolates from cattle, chickens, swine and turkeys were recovered during the same period from federally inspected slaughter and processing facilities in the northeastern United States. We found that subtyping *Salmonella* isolates by PFGE revealed differences in antimicrobial susceptibility patterns and, for human *Salmonella*, differences in sources and invasiveness that were not evident from serotyping alone. Sixteen of the 20 most common human *Salmonella* PFGE patterns were identified in *Salmonella* recovered from food animals. The most common human *Salmonella* PFGE pattern, Enteritidis pattern JEGX01.0004 (JEGX01.0003ARS), was associated with more cases of invasive salmonellosis than all other patterns. In food animals, this pattern was almost exclusively (99%) found in *Salmonella* recovered from chickens and was present in poultry meat in every year of the study. Enteritidis pattern JEGX01.0004 (JEGX01.0003ARS) was associated with susceptibility to all antimicrobial agents tested in 94.7% of human and 97.2% of food animal *Salmonella* isolates. In contrast, multidrug resistance (resistance to three or more classes of antimicrobial agents) was observed in five PFGE patterns. Typhimurium patterns JPXX01.0003 (JPXX01.0003 ARS) and JPXX01.0018 (JPXX01.0002 ARS), considered together, were associated with resistance to five or more classes of antimicrobial agents: ampicillin, chloramphenicol, streptomycin, sulfonamides and tetracycline (ACSSuT), in 92% of human and 80% of food animal *Salmonella* isolates. The information from our study can assist in source attribution, outbreak investigations, and tailoring of interventions to maximize their impact on prevention.

**Editor:** Herman Tse, The University of Hong Kong, Hong Kong

**Funding:** This study was supported in part by the Pennsylvania Department of Health through Centers for Disease Control and Prevention grant (ELC-04040) for National Antimicrobial Resistance Monitoring. The funders had no role in study design, data collection and analysis, decision to publish, or preparation of the manuscript.

**Competing Interests:** The authors have declared that no competing interests exist.

* E-mail: csandt@pa.gov

## Introduction

In the United States, non-typhoidal *Salmonella enterica* subsp. *enterica* cause an estimated one million episodes of salmonellosis each year [1] and are the leading cause of hospitalization and death from foodborne illness. The resulting annual economic burden, based on the costs of medical treatment, lost productivity and premature death, is estimated to be in the range of $3.3-4.4 billion [2,3].

PulseNet is the national molecular surveillance network for foodborne infections and includes in its network the laboratories of state, territorial, and local public health departments, federal food regulatory agencies, veterinary agencies, and agricultural agencies. PulseNet was established by the Centers for Disease Control and Prevention (CDC) and the Association of Public Health Laboratories in 1996 to reduce the time needed to detect, investigate, and control multistate outbreaks caused by foodborne bacterial pathogens. PulseNet laboratories subtype these pathogens using pulsed-field gel electrophoresis (PFGE) and upload the PFGE patterns to a centralized database at CDC [4,5].

The National Antimicrobial Resistance Monitoring System (NARMS) is a national public health surveillance system that tracks antimicrobial resistance in foodborne bacteria. The NARMS program was established in 1996 as a partnership between the U.S. Food and Drug Administration (FDA), CDC, and the U.S. Department of Agriculture (USDA) and is described on the FDA website [6]. The animal arm of NARMS resides in the USDA-Agricultural Research Service (ARS) laboratory in Athens, GA. In addition to monitoring antimicrobial susceptibility, NARMS partners collaborate on epidemiologic and microbiologic research studies. NARMS also examines foodborne bacteria for genetic relatedness using PFGE, and PFGE patterns are entered into USDA's VetNet database [7]. The food animal arm of NARMS is described on the USDA web site [8]. PulseNet and VetNet work synergistically to provide information that is important for public health. The PFGE protocols are highly standardized protocols developed by PulseNet to facilitate inter-laboratory comparisons [9].

As part of communicable disease control reporting requirements in Pennsylvania, clinical laboratories routinely submit *Salmonella* isolates to the Pennsylvania Department of Heath Bureau of Laboratories (BOL). At BOL, Salmonella isolates are biochemically identified, serotyped and subtyped by PFGE. USDA's Food Safety and Inspection Service (FSIS) samples food animals for *Salmonella* during slaughter and processing. *Salmonella* isolates from food animals are tested for susceptibility to antimicrobial agents and then subtyped via PFGE by the USDA-ARS Laboratory in Athens, Georgia.

Our objective was to compare clinical isolates of non-typhoidal *Salmonella* recovered from humans (human *Salmonella*) received as part of routine surveillance at the BOL in Pennsylvania from 2005 through 2011 with *Salmonella* isolates recovered during the same period from food animals (food animal *Salmonella*) at slaughter and processing facilities in the northeastern United States. The most common PFGE patterns observed in human *Salmonella* served as the reference set and included associated invasiveness and antimicrobial susceptibility profiles. Our hypothesis was that subtyping *Salmonella* isolates by PFGE could reveal differences within serotypes in terms of antimicrobial susceptibility patterns and, for human *Salmonella*, differences in food animal sources and invasiveness that were not evident from serotyping alone.

## Materials and Methods

### Sample Sources and Processing

Human and food animal non-typhoidal *Salmonella* isolates received between January 1, 2005 and December 31, 2011, were included in the study. At BOL, human *Salmonella* isolates received as part of the state's routine operations were grown, identified, and serotyped by the Bacteriology Section, using standard procedures [10,11]. Food animal *Salmonella* isolates were recovered from carcass rinsates (chickens), carcass swabs (turkeys, cattle, and swine), and ground products (chicken, turkey, and beef) during slaughter and processing at federally inspected facilities in the northeast (Pennsylvania, Maine, Vermont, New Hampshire, New York, Maryland, Connecticut, Rhode Island, Massachusetts, Delaware, New Jersey, Indiana, Ohio, Michigan and Washington, DC) as previously described [12].

PFGE testing of human *Salmonella* isolates was conducted by the BOL according to the CDC-standardized procedure used by all PulseNet-certified laboratories [9]. Gel images were analyzed using BioNumerics software Version 6.6 (Applied Maths, Saint-Martens-Latem, Belgium). All non-typhoidal human *Salmonella* isolates received at the BOL were evaluated by PFGE, except when the same serotype of *Salmonella* was recovered more than once from a patient within a six-month period. In this case, only the first isolate received was tested. PFGE testing of food animal *Salmonella* isolates was done by the animal arm of the NARMS, located in Athens, GA, as previously described [7].

### PFGE Pattern Names

PFGE fingerprints of human *Salmonella* were maintained in a local Pennsylvania database and submitted to CDC's PulseNet national *Salmonella* database where they were assigned pattern names [4,5]. The USDA maintains a similar database called USDA-VetNet for PFGE fingerprints of *Salmonella* isolated from food animals [7]. Food animal *Salmonella* pattern names were assigned by USDA-VetNet as previously described [7]. Isolates in the VetNet database were compared to the PulseNet database to capture matching PulseNet pattern names.

### Common Human *Salmonella* Patterns and Shared Common Patterns

The 20 most frequently identified PFGE patterns among human *Salmonella* isolates and with at least two isolates in each year of the study were designated as the most common human *Salmonella* patterns. These 20 patterns included patterns that occurred both sporadically ($n = 4{,}471$) and linked to known outbreaks ($n = 251$). The USDA-VetNet database was then searched for matching patterns in *Salmonella* isolates recovered from food animals during slaughter and processing in northeastern U.S. facilities. The most common human patterns that were also identified in food animal *Salmonella* and are defined here as "shared common patterns."

### Antimicrobial Susceptibility Testing

Susceptibility to the following classes of antimicrobial agents was tested using the Sensititre semi-automated broth microdilution antimicrobial susceptibility system (Trek Diagnostic Systems Inc., Cleveland, Ohio), with minimum inhibitory concentrations evaluated according to Clinical Laboratory Standards Institute (CLSI) guidelines [13]: (antimicrobial agents in parentheses; resistance breakpoints in brackets): aminoglycosides (amikacin [≥64 μg/mL], gentamicin [≥16 μg/mL], kanamycin [≥64 μg/mL] and streptomycin [≥64 μg/mL]), ß-lactam/ß-lactamase inhibitor combinations (amoxicillin/clavulanic acid [≥32/16 μg/mL]), cephems (cefoxitin [≥32 μg/mL], ceftiofur [≥8 μg/mL] and ceftriaxone [≥4 μg/mL]), penicillins (ampicillin [≥32 μg/mL]), quinolones (ciprofloxacin [≥4 μg/mL]and nalidixic acid [≥32 μg/mL]), folate pathway inhibitors (sulfisoxazole [≥512 μg/mL], trimethoprim/sulfamethoxazole [≥4/76 μg/mL]), phenicols (chloramphenicol [≥32 μg/mL]), and tetracyclines (tetracycline [≥16 μg/mL]). For antimicrobial agents without CLSI approved standards, NARMS interpretive criteria as established by the NARMS working group were used and quality control strains were as previously described [8]. Multidrug resistance was defined as resistance to three or more classes of antimicrobial agents. Of the 4,235 human *Salmonella* isolates with common shared patterns, 467 (11%) were tested for susceptibility to antimicrobial agents. Limited resources precluded testing all of the isolates. The isolates chosen for susceptibility testing included 267 randomly selected isolates originating from Pennsylvania that were tested by the human arm of NARMS as part of its routine surveillance program [14]. The Pennsylvania Department of Agriculture tested an additional 200 human *Salmonella* isolates sampled from the 16 shared common patterns. NARMS tested all 275 of the food animal *Salmonella* isolates having shared common patterns for antimicrobial susceptibility as previously described [12].

### Correlation of Invasiveness with PFGE Patterns

Isolation of *Salmonella* from human blood was used as an indicator of invasive disease [15]. The statistical association between each PFGE pattern and invasiveness was tested via a 2×2 contingency table and evaluated on the basis of the conditional maximum likelihood estimate of Odds Ratio (OR) and the mid-p test (two-tailed p value) [16]. PFGE patterns with $OR \geq 1$ and $p \leq 0.05$ were interpreted as associated with increased invasiveness; patterns with OR≤1 and p≤0.05 were interpreted as associated with decreased invasiveness.

### Correlation of Antimicrobial Resistance with Invasiveness

A two-tailed p value obtained with a Fisher's exact test (GraphPad QuickCalcs, GraphPad Software, Inc., San Diego, CA) was used to evaluate the relationship between antimicrobial resistance and invasiveness.

Regarding data for *Salmonella* recovered from animals: ACUC approval was not needed as the isolates were obtained as part of the National Antimicrobial Resistance Monitoring System (NARMS) from the USDA Food Safety and Inspection Service (FSIS) as part of their Salmonella PR/HACCP verification testing program (http://www.ars.usda.gov/Main/docs.htm?docid=6750&page=2). Accessed 2013 September 19.

## Results

### PFGE Patterns Found both in Human and Food Animal *Salmonella* Isolates

A total of 11,967 human *Salmonella* and 2,187 food animal *Salmonella* isolates were submitted for laboratory testing during the study period (2005–2011). The food animal *Salmonella* isolates were recovered from a total of 65,655 animals tested for *Salmonella* during slaughter and processing at federally inspected facilities in the northeastern U.S., and represented an overall yield of 3.3% (Table 1). The 65,655 animals tested for *Salmonella* included 42,368 (64.5%) cattle, 10,412 (15.9%) swine, 9,661(14.7%) chickens and 3,214 (4.9%) turkeys. *Salmonella* was isolated from a total of 2,187 samples (Table 1). Of the 2,187 *Salmonella*-positive samples, 1,194 (54.6%) were recovered from chickens, 472 (21.6%) from cattle, 282 (12.9%) from turkeys and 239 (10.9%) from swine. A total of 2,083 of the 2,187 food animal *Salmonella* isolates were available to VetNet for PFGE, antimicrobial susceptibility testing and comparison with human *Salmonella* isolates.

### Common Human *Salmonella* Patterns

The 20 most common human *Salmonella* XbaI patterns are shown in Fig. 1 and described in Table 2. These patterns represent 39% of the human *Salmonella* isolates (4,722/11,967) from Pennsylvania described in this paper. Six serotypes–Berta, Enteritidis, Heidelberg, Newport, Thompson, and Typhimurium (including variant 5-)–and one antigenic formula (I 4,[5],12:i:-) were represented among these 20 patterns. Each XbaI pattern is shown with all associated BlnI patterns that occurred at least three times in the dataset. One to seven different BlnI patterns were associated with a particular XbaI pattern.

Sixteen (80%) of the 20 most common human *Salmonella* patterns shown in Fig. 1 were also found among the 2,083 food animal *Salmonella* isolates (Table 2). A total of 4,235 (35%) of the 11,967 human *Salmonella* isolates shared these 16 PFGE patterns with 275 (13%) of the 2,083 food animal *Salmonella* isolates. All six serotypes and the antigenic formula identified in the 20 most common human *Salmonella* patterns were also represented among the shared common patterns.

The most frequently observed shared common pattern from human *Salmonella* isolates was Enteritidis pattern JEGX01.0004 (JEGX01.0003ARS), which accounted for 16% of all human *Salmonella*. Among food animal *Salmonella*, this pattern was the second most common pattern, accounting for 5% of all food animal *Salmonella* isolates. Serotype Kentucky pattern JGPX01.0027 (JGPX01.0220 ARS) was the most common PFGE pattern in food animal *Salmonella*, occurring in 13% of food animal

**Table 1.** Food Animal Sources of *Salmonella* Recovered at Federally Inspected Slaughter and Processing Facilities in the Northeastern U.S.[1] with Human Isolates Shown for Reference.

| Year | No. Chickens Tested | No. (%) Chickens Positive for *Salmonella* | No. Cattle Tested | No. (%) Cattle Positive for *Salmonella* | No. Turkeys Tested | No. (%) Turkeys Positive for Salmonella | No. Swine Tested | No. (%) Swine Positive for *Salmonella* | Total No. Samples Inspected | Total No. (%) Positive for *Salmonella* | No. Human Salmonella Received | No. 20 Most Common Human *Salmonella* |
|---|---|---|---|---|---|---|---|---|---|---|---|---|
| 2005 | 1714 | 315 (18.4) | 9,754 | 102 (1.0) | 175 | 39 (22.3) | 2,191 | 61 (2.8) | 13,834 | 517 (3.7) | 1,537 | 549 |
| 2006 | 1,947 | 271 (13.9) | 8,449 | 117 (1.4) | 730 | 74 (10.1) | 2,346 | 78 (3.3) | 13,472 | 540 (4.0) | 1,686 | 749 |
| 2007 | 1,684 | 172 (10.2) | 6,539 | 76 (1.2) | 709 | 49 (6.9) | 1,643 | 27 (1.6) | 10,575 | 324 (3.1) | 1,829 | 706 |
| 2008 | 1,002 | 103 (10.3) | 6,241 | 90 (1.4) | 276 | 43 (15.6) | 1,342 | 26 (1.9) | 8,861 | 262 (3.0) | 1,680 | 645 |
| 2009 | 989 | 95 (9.6) | 4,005 | 19 (0.5) | 376 | 29 (7.7) | 1,099 | 13 (1.2) | 6,469 | 156 (2.4) | 1,706 | 662 |
| 2010 | 1,254 | 146 (11.6) | 3,099 | 29 (0.9) | 573 | 38 (6.6) | 1,137 | 24 (2.1) | 6,063 | 237 (3.9) | 1,749 | 716 |
| 2011 | 1,071 | 92 (8.6) | 4,281 | 39 (0.9) | 375 | 10 (2.7) | 654 | 10 (1.5) | 6,381 | 151 (2.4) | 1,778 | 695 |
| TOTAL | 9,661 | 1,194 (12.4) | 42,368 | 472 (1.1) | 3,214 | 282 (8.8) | 10,412 | 239 (2.3) | 65,655 | 2,187 (3.3) | 11,967 | 4722 |

[1]Pennsylvania, Maine, Vermont, New Hampshire, New York, Maryland, Connecticut, Rhode Island, Massachusetts, Delaware, New Jersey, Indiana, Ohio, Michigan and Washington, DC.

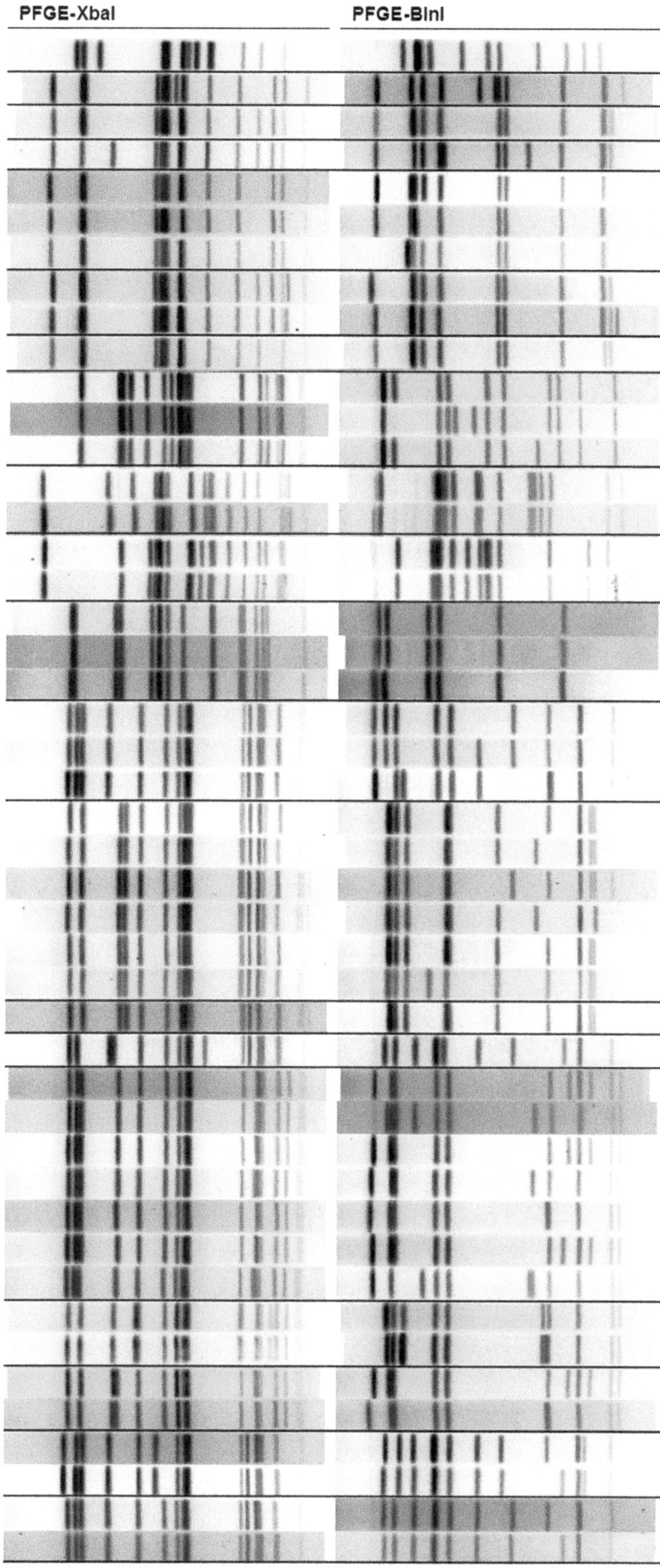


| Serotype | PFGE-XbaI-pattern | PFGE-BlnI-pattern |
|---|---|---|
| Berta | JAXX01.0001 | JAXA26.0004 |
| Enteritidis | JEGX01.0002 | JEGA26.0010 |
| Enteritidis | JEGX01.0004 | JEGA26.0002 |
| Enteritidis | JEGX01.0005 | JEGA26.0004 |
| Enteritidis | JEGX01.0009 | JEGA26.0016 |
| Enteritidis | JEGX01.0009 | JEGA26.0017 |
| Enteritidis | JEGX01.0009 | JEGA26.0036 |
| Enteritidis | JEGX01.0021 | JEGA26.0002 |
| Enteritidis | JEGX01.0021 | JEGA26.0005 |
| Enteritidis | JEGX01.0034 | JEGA26.0002 |
| Heidelberg | JF6X01.0022 | JF6A26.0001 |
| Heidelberg | JF6X01.0022 | JF6A26.0002 |
| Heidelberg | JF6X01.0022 | JF6A26.0013 |
| Newport | JJPX01.0011 | JJPA26.0044 |
| Newport | JJPX01.0011 | JJPA26.0194 |
| Newport | JJPX01.0061 | JJPA26.0021 |
| Newport | JJPX01.0061 | JJPA26.0062 |
| Thompson | JP6X01.0001 | JP6A26.0012 |
| Thompson | JP6X01.0001 | JP6A26.0024 |
| Thompson | JP6X01.0001 | JP6A26.0026 |
| Typhimurium | JPXX01.0001 | JPXA26.0001 |
| Typhimurium | JPXX01.0001 | JPXA26.0063 |
| Typhimurium | JPXX01.0001 | JPXA26.0084 |
| Typhimurium | JPXX01.0003 | JPXA26.0003 |
| Typhimurium | JPXX01.0003 | JPXA26.0007 |
| Typhimurium | JPXX01.0003 | JPXA26.0042 |
| Typhimurium | JPXX01.0003 | JPXA26.0106 |
| Typhimurium | JPXX01.0003 | JPXA26.0248 |
| Typhimurium. | JPXX01.0003 | JPXA26.0488 |
| Typhimurium. | JPXX01.0018 | JPXA26.0156 |
| I 4,[5],12:i:- | JPXX01.0026 | JPXA26.0033 |
| Typhimurium | JPXX01.0146 | JPXA26.0021 |
| Typhimurium | JPXX01.0146 | JPXA26.0172 |
| Typhimurium | JPXX01.0146 | JPXA26.0174 |
| Typhimurium | JPXX01.0146 | JPXA26.0291 |
| Typhimurium | JPXX01.0146 | JPXA26.0292 |
| Typhimurium | JPXX01.0146 | JPXA26.0297 |
| Typhimurium | JPXX01.0146 | JPXA26.0313 |
| Typhimurium | JPXX01.0302 | JPXA26.0183 |
| Typhimurium | JPXX01.0302 | JPXA26.0234 |
| Typhimurium | JPXX01.0604 | JPXA26.0174 |
| Typhimurium | JPXX01.0604 | JPXA26.0292 |
| I 4,[5],12:i:- | JPXX01.0621 | JPXA26.0055 |
| I 4,[5],12:i:- | JPXX01.0621 | JPXA26.0057 |
| I 4,[5],12:i:- | JPXX01.1212 | JPXA26.0108 |
| I 4,[5],12:i:- | JPXX01.1212 | JPXA26.0181 |

**Figure 1. DNA fingerprints of the 20 most common human *Salmonella* patterns.** PFGE was conducted as described in Materials and Methods, with restriction endonucleases XbaI and BlnI. BlnI patterns represented at least three times in the dataset are shown.

*Salmonella* isolates (data not shown). This pattern was rarely seen in BOL human *Salmonella* isolates (0.25%).

Of the 275 food animal *Salmonella* isolates with shared common patterns, 238 (87%) were recovered from chicken, 16 (6%) from cattle, 11 (4%) from turkey and 10 (4%) from swine (Table 2). Sixty seven percent (n = 186) of these 275 food animal *Salmonella* isolates belonged to serotype Enteritidis, and all but two isolates were recovered from chicken. The most common Enteritidis pattern in *Salmonella* from chickens was JEGX01.0004 (JEGX01.0003ARS) (n = 105/184; 57%). Chicken was also the most common source of Heidelberg pattern JF6X01.0022 (JF6X01.0015 ARS), with 25 of 30 isolates (83%) recovered from chicken (Table 2).

In contrast to the strong association of S. Enteritidis and Heidelberg patterns with chicken, Typhimurium pattern JPXX01.0003 (JPXX01.0003 ARS) (n = 18) was observed in *Salmonella* recovered in comparable numbers from cattle (n = 6), chicken (n = 5), turkey (n = 4) and swine (n = 3). The second most common food animal Typhimurium pattern, JPXX01.0604 (JPXX01.0079 ARS) (n = 10) was primarily recovered from chicken (n = 7). The two isolates of pattern JPXX01.0018 (JPXX01.0002 ARS) were recovered from cattle and swine, respectively, and Typhimurium pattern JPXX01.0146 (JPXX01.0081 ARS) (n = 5) was recovered primarily from swine (n = 4). Berta pattern JAXX01.0001 (JAXX01.0003 ARS) (n = 3) was found exclusively in turkey.

All 16 shared common patterns were found in human *Salmonella* during each of the seven years of the study. Three of the 16 shared common patterns, Enteritidis patterns JEGX01.0004 (JEGX01.0003 ARS), JEGX01.0005 (JEGX01.0002 ARS) and JEGX01.0034 (JEGX01.0005 ARS), were also found in food animal *Salmonella* during each of the seven years of the study. Similarly, Heidelberg pattern JF6X01.0022 (JF6X01.0015 ARS) was recovered from food animals in each of five years, and Typhimurium pattern JPXX01.0003 (JPXX01.0003 ARS) in each of four years (data not shown).

## Unshared PFGE Patterns

A total of 261 different patterns were identified among the 11,967 human *Salmonella* isolates in Pennsylvania, and 794 were identified among the 2,083 food animal *Salmonella* isolates (data not shown). Many of the identified human patterns were not shared by food animal *Salmonella* isolates, and many of the identified food animal *Salmonella* patterns were not shared by human *Salmonella* isolates. Of the 2,083 food animal *Salmonella* isolates, only 1,034 (50%) shared PFGE patterns with the human *Salmonella* patterns that had been assigned pattern names. Of the 11,967 human *Salmonella* isolates, only 5,266 (44%) had named patterns found among the 2,083 food animal *Salmonella* isolates recovered from northeastern slaughter and processing facilities.

Of the 20 most common human *Salmonella* patterns (Fig. 1 and Table 2), two Enteritidis patterns (JEGX01.0002 and JEGX01.0009) and two Typhimurium/I 4,[5],12:i:- patterns (JPXX01.0026 and JPXX01.1212) were not observed among food animal *Salmonella* isolates. Pattern JEGX01.0002 was found to be associated with travel. Travel histories were available for 200 persons associated with *Salmonella* pattern JEGX01.0002; of these, 164 (82%) reported traveling outside the U.S., notably to the Dominican Republic (n = 71) and Mexico (n = 53).

## Association of Invasiveness with PFGE Patterns

Isolation from blood was used as an indicator of invasiveness. A total of 646 *Salmonella* isolates from humans were recovered from blood, representing 5.4% of the 11,967 human *Salmonella* isolates tested (Table 2). Four patterns were found to be significantly associated with increased frequency of isolation from blood: Enteritidis JEGX01.0004 (JEGX01.0003 ARS) (OR = 1.30; p = 0.01), Enteritidis JEGX01.0005 (JEGX01.0002 ARS) (OR = 1.60; p = 0.014), Enteritidis JEGX01.0009 (JEGX01.0022 ARS) (OR = 2.06; p = 0.02) and Heidelberg JF6X01.002 (JF6X01.0015 ARS) (OR = 2.10; p = 0.01). Five patterns were found to be significantly associated with decreased frequency of isolation from blood: Enteritidis JEGX01.0034 (JEGX01.0005 ARS) (OR = 0.10; p = 0.00), Newport patterns JJPX01.0011 (JJPX01.0204 ARS) (OR = 0.00; p = 0.05) and JJPX01.0061 (JJPX01.0069 ARS) (OR = 0.00; p = 0.02), and Typhimurium patterns JPXX01.0146 (JPXX01.0081 ARS) (OR = 0.06; p = 0.00) and JPXX01.0302 (JPXX01.0106 ARS) (OR = 0.36; p = 0.05).

## Antimicrobial Susceptibility Associated with Shared Common Patterns

Of the 4,235 human *Salmonella* isolates with shared common patterns, 467 (11%) were tested for susceptibility to antimicrobial agents (Table 3). Of these 467, a total of 367 (79%) were pansusceptible, 100 (21%) were resistant to at least one class of antimicrobial agent, 75 (16%) were MDR, and 62 (13%) were resistant to five or more classes of antimicrobial agents. Of the 275 food animal *Salmonella* isolates, 226 (82%) were pansusceptible, 49 (18%) were resistant to at least one class of antimicrobial agents, 40 (14%) were MDR, and 16 (6%) were resistant to five or more classes of antimicrobial agents.

Multidrug resistance (resistance to ≥3 classes of antimicrobial agents) was associated with eight of the 16 shared common patterns in human *Salmonella* (Table 3). Five of these patterns were also associated with multidrug resistance in food animal *Salmonella*: Berta pattern JAXX01.0001 (JAXX01.0003 ARS), Heidelberg pattern JF6X01.0022 (JF6X01.0015 ARS), Typhimurium patterns JPXX01.0003 (JPXX01.0003 ARS) and JPXX01.0018 (JPXX01.0002 ARS), and I 4,[5],12:i:- pattern JPXX01.0621 (TERX01.0001 ARS). The incidence of multidrug resistance in both human and food animal *Salmonella* isolates exceeded 90% for Typhimurium patterns JPXX01.0003 (JPXX01.0003 ARS) and JPXX01.0018 (JPXX01.0002 ARS) (Table 3), patterns that differ by two bands (Fig. 1). When these two patterns were considered together, a total of 58 (89%) of the 65 human and 15 (75%) of the 20 food animal Typhimurium isolates were resistant to the following five antimicrobial agents: ampicillin, chloramphenicol, streptomycin, sulfisoxazole, and tetracycline (the ACSSuT phenotype and resistance to five classes of antimicrobial agents). Two of the human *Salmonella* isolates exhibited resistance to amoxicillin/clavulanic acid and ceftiofur in addition to the ACSSuT phenotype. The food animal *Salmonella* isolates exhibiting the ACSSuT phenotype were recovered from cattle (n = 5), chickens (n = 4), swine (n = 3) and turkeys (n = 3). An additional isolate recovered from a turkey exhibited the ACSSuT phenotype plus resistance to gentamicin, and three additional isolates (two from cattle; one from swine) lacked only the streptomycin resistance of this phenotype (ACSuT). One human isolate and three food animal isolates with patterns JPXX01.0003 (JPXX01.0003 ARS)

**Table 2.** Pennsylvania's Most Common Human *Salmonella* PFGE Patterns Also Found in *Salmonella* Recovered from Food Animals Processed in the Northeastern U.S.

| Serotype | PulseNet XbaI Pattern[1] | No. (% of All Human *Salmonella*)[2] | No. Linked to Outbreaks | No. from Blood | % from Blood within Pattern[3] | Association of PFGE Pattern with Blood[3] Odds Ratio (95% CI); p | Corresponding VetNet Pattern[4] | No. (%)[5] of Food Animal *Salmonella* | No. from Cattle | No. from Chicken | No. from Swine | No. from Turkey |
|---|---|---|---|---|---|---|---|---|---|---|---|---|
| Berta | JAXX01.0001 | 49 (0.41) | 0 | 5 | 10.20 | 2.00 (0.70-4.75); 0.17 | JAXX01.0003 ARS | 3 (0.14) | 0 | 0 | 0 | 3 |
| Enteritidis | JEGX01.0002 | 235 (1.96) | 3 | 14 | 5.96 | 1.11 (0.62-1.87); 0.68 | JEGX01.0017 ARS | 0 | 0 | 0 | 0 | 0 |
| Enteritidis | JEGX01.0004 | 1,968 (16.45) | 79 | 130 | 6.61 | 1.30 (1.06-1.58); 0.01 | JEGX01.0003 ARS | 106 (5.09) | 1 | 105 | 0 | 0 |
| Enteritidis | JEGX01.0005 | 426 (3.56) | 12 | 35 | 8.22 | 1.60 (1.11-2.26); 0.01 | JEGX01.0002 ARS | 65 (3.12) | 1 | 64 | 0 | 0 |
| Enteritidis | JEGX01.0009 | 144 (1.20) | 1 | 15 | 10.42 | 2.06 (1.16-3.46); 0.02 | JEGX01.0022 ARS | 0 | 0 | 0 | 0 | 0 |
| Enteritidis | JEGX01.0021 | 331 (2.77) | 24 | 21 | 6.34 | 1.19 (0.74-1.84); 0.43 | JEGX01.0010 ARS; JEGX01.0052 ARS[6] | 1 (0.05) | 0 | 1 | 0 | 0 |
| Enteritidis | JEGX01.0034 | 176 (1.47) | 11 | 1 | 0.57 | 0.10 (0.00-0.49); 0.00 | JEGX01.0005 ARS | 14 (0.67) | 0 | 14 | 0 | 0 |
| Heidelberg | JF6X01.0022 | 151 (1.26) | 18 | 16 | 10.60 | 2.10 (1.21-3.48); 0.01 | JF6X01.0015 ARS | 30 (1.44) | 3 | 25 | 0 | 2 |
| Newport | JJPX01.0011 | 54 (0.45) | 0 | 0 | 0 | 0.00 (0.00-1.00); 0.05 | JJPX01.0204 ARS | 1 (0.05) | 0 | 1 | 0 | 0 |
| Newport | JJPX01.0061 | 72 (0.60) | 28 | 0 | 0 | 0.00 (0.00-0.74); 0.02 | JJPX01.0069 ARS | 1 (0.05) | 0 | 0 | 1 | 0 |
| Thompson | JP6X01.0001 | 54 (0.45) | 0 | 3 | 5.56 | 1.03 (0.25-2.95); 0.90 | JP6X01.0024 ARS | 1 (0.05) | 0 | 1 | 0 | 0 |
| Typhimurium | JPXX01.0001 | 53 (0.44) | 2 | 3 | 5.66 | 1.05 (0.26-3.02); 0.87 | JPXX01.0021 ARS | 1 (0.05) | 0 | 1 | 0 | 0 |
| Typhimurium | JPXX01.0003 | 113 (0.94) | 2 | 9 | 7.96 | 1.52 (0.72-2.91); 0.24 | JPXX01.0003 ARS | 18 (0.86) | 6 | 5 | 3 | 4 |
| Typhimurium | JPXX01.0018 | 81 (0.68) | 0 | 2 | 2.47 | 0.44 (0.07-1.51); 0.24 | JPXX01.0002 ARS | 2 (0.10) | 1 | 0 | 1 | 0 |
| Typhimurium | JPXX01.0146 | 278 (2.32) | 16 | 1 | 0.36 | 0.06 (0.00-0.31); 0.00 | JPXX01.0081 ARS | 5 (0.24) | 0 | 1 | 4 | 0 |
| Typhimurium | JPXX01.0302 | 149 (1.25) | 46 | 3 | 2.01 | 0.36 (0.09-0.99); 0.05 | JPXX01.0106 ARS | 2 (0.10) | 1 | 1 | 0 | 0 |
| Typhimurium | JPXX01.0604 | 61 (0.51) | 3 | 1 | 1.64 | 0.29 (0.01-1.49); 0.18 | JPXX01.0079 ARS | 10 (0.48) | 1 | 7 | 0 | 2 |
| Typhimurium; I 4,[5],12:i:- | JPXX01.0026 | 61 (0.51) | 4 | 1 | 1.64 | 0.29 (0.01-1.49); 0.18 | JPXX01.0308 ARS; TERX01.0010 ARS | 0 | 0 | 0 | 0 | 0 |
| I 4,[5],12:i:- | JPXX01.0621 | 219 (1.83) | 2 | 8 | 3.65 | 0.66 (0.30-1.28); 0.25 | TERX01.0001 ARS | 15 (0.72) | 2 | 12 | 1 | 0 |
| Typhimurium; I 4,[5],12:i:- | JPXX01.1212 | 47 (0.39) | 0 | 3 | 6.38 | 1.20 (0.29-3.45); 0.72 | JPXX01.0304 ARS; TERX01.0008 ARS | 0 | 0 | 0 | 0 | 0 |
| TOTAL | | 4,722 (39.46) | 251 | 271 | NA | | TOTAL | 275 (13.20) | 16 | 238 | 10 | 11 |

[1]Pattern names include a three-letter code that indicates serotype (e.g., JEG for serotype Enteritidis in the pattern names JEGX01.0004 and JEGA26.0002), followed by a three-character code for. restriction enzyme (e.g., X01 for XbaI in JEGX01.0004; A26 for BlnI in JEGA26.0002), and a four-digit number for the unique strain designation (e.g., 0004 in JEGX01.0004) (Gerner-Smidt et al., 2006).
[2]Percentage of total number of human *Salmonella* isolates (11,967).
[3]Number of human *Salmonella* isolates recovered from blood for a particular pattern expressed as a percentage of the total number from all sources for that PFGE pattern; e.g., for Berta XbaI. pattern JAXX01.0001, $(5/49) \times 100 = 10.20\%$. Statistical method for determining Odds Ratio is described in Materials and Methods.
[4]VetNet's pattern nomenclature conforms to PulseNet's nomenclature, with three exceptions: (1) PulseNet uses a single three-letter code (JPX) to designate both serotype Typhimurium and antigenic formula I 4,[5],12:i:- whereas VetNet uses two different codes, JPX for Typhimurium and TER for I 4,[5],12:i:-; (2) the four-digit strain designation codes are different in the two systems; and (3) "ARS" (for Agricultural Research Service of the USDA) is appended to VetNet pattern names.
[5]Percentage of the total number of animal *Salmonella* isolates (2,083).
[6]VetNet identified two different patterns within PulseNet pattern JEGX01.0021.

**Table 3.** Antimicrobial resistance associated with the 16 most common human *Salmonella* PFGE patterns shared by food animal *Salmonella*.

| Human *Salmonella* | | | | | | | | | Food Animal *Salmonella* | | | | | | | |
|---|---|---|---|---|---|---|---|---|---|---|---|---|---|---|---|---|
| Serotype[1] | PulseNet Xbal Pattern | No. Tested | No. (%) Pansusceptible[7] | No. (%)[5] Resistant to 1-2 Antimicrobial Classes[7] | No. (%)[5] MDR[2] | No. (%)[5] Resistant to ≥5 Antimicrobial Classes[7] | No. Resist-ant to Nal[3] | No. Resist-ant to Axo[3] | Corresponding VetNet-Xbal-pattern | No. Tested | No. (%) Pansusceptible[7] | No. (%)[6] Resistant to 1-2 Antimicrobial Classes[7] | No. (%)[6] MDR[2] | No. (%)[6] Resistant to ≥5 Antimicrobial Classes[7] | No. Resist-ant to Nal[3] | No. Resist-ant to Axo[3] |
| Berta | JAXX01.0001 | 16 | 9 (56.3) | 2 (12.5) | 5 (31.3) | 1 (6.3) | 1 | 2 | JAXX01.0003 ARS | 3 | 0 (0.0) | 0 (0.0) | 3 (100.0) | 0 (0.0) | 0 | 1 |
| Enteritidis | JEGX01.0004 | 114 | 108 (94.7) | 5 (4.4) | 1 (0.9) | 0 (0.0) | 1 | 0 | JEGX01.0003 ARS | 106 | 103 (97.2) | 3 (2.8) | 0 (0.0) | 0 (0.0) | 0 | 0 |
| Enteritidis | JEGX01.0005 | 25 | 25 (100.0) | 0 (0.0) | 0 (0.0) | 0 (0.0) | 0 | 0 | JEGX01.0002 ARS | 65 | 65 (100.0) | 0 (0.0) | 0 (0.0) | 0 (0.0) | 0 | 0 |
| Enteritidis | JEGX01.0021 | 33 | 27 (81.8) | 6 (7.3) | 0 (0.0) | 0 (0.0) | 0 | 0 | JEGX01.0010 ARS JEGX01.0052 ARS[4] | 1 | 0 (0.0) | 1 (100.0) | 0 (0.0) | 0 (0.0) | 0 | 0 |
| Enteritidis | JEGX01.0034 | 22 | 21 (95.5) | 1 (4.5) | 0 (0.0) | 0 (0.0) | 0 | 0 | JEGX01.0005 ARS | 14 | 12 (85.7) | 2 (14.3) | 0 (0.0) | 0 (0.0) | 0 | 0 |
| Heidelberg | JF6X01.0022 | 14 | 13 (92.9) | 0 (0.0) | 1 (7.1) | 0 (00.) | 0 | 1 | JF6X01.0015 ARS | 30 | 18 (60.0) | 2 (6.7) | 10 (33.3) | 0 (0.0) | 0 | 10 |
| Newport | JJPX01.0011 | 9 | 9 (100.0) | 0 (0.0) | 0 (0.0) | 0 (0.0) | 0 | 0 | JJPX01.0204 ARS | 1 | 1 (100.0) | 0 (0.0) | 0 (0.0) | 0 (0.0) | 0 | 0 |
| Newport | JJPX01.0061 | 13 | 13 (100.0) | 0 (0.0) | 0 (0.0) | 0 (00.) | 0 | 0 | JJPX01.0069 ARS | 1 | 1 (100.0) | 0 (0.0) | 0 (0.0) | 0 (0.0) | 0 | 0 |
| Thompson | JP6X01.0001 | 8 | 8 (100) | 0 (0.0) | 0 (0.0) | 0 (0.0) | 0 | 0 | JP6X01.0024 ARS | 1 | 1 (100.0) | 0 (0.0) | 0 (0.0) | 0 (0.0) | 0 | 0 |
| Typhimurium | JPXX01.0001 | 13 | 10 (76.9) | 1 (7.7) | 2 (15.4) | 1 (7.7) | 0 | 1 | JPXX01.0021 ARS | 1 | 1 (100.0) | 0 (0.0) | 0 (0.0) | 0 (0.0) | 0 | 0 |
| Typhimurium | JPXX01.0003 | 38 | 0 (0.0) | 1 (2.6) | 37 (97.4) | 37 (97.4) | 0 | 0 | JPXX01.0003 ARS | 18 | 0 (0.0) | 1 (5.6) | 17 (94.4) | 14 (78.0) | 0 | 0 |
| Typhimurium | JPXX01.0018 | 27 | 0 (0.0) | 2 (7.4) | 25 (92.6) | 23 (85.2) | 0 | 0 | JPXX01.0002 ARS | 2 | 0 (0.0) | 0 (0.0) | 2 (100.0) | 2 (100.0) | 0 | 0 |
| Typhimurium | JPXX01.0146 | 24 | 23 (95.8) | 1 (4.2) | 0 (0.0) | 0 (0.0) | 0 | 0 | JPXX01.0081 ARS | 5 | 5 (100.0) | 0 (0.0) | 0 (0.0) | 0 (0.0) | 0 | 0 |
| Typhimurium | JPXX01.0302 | 22 | 20 (90.9) | 1 (4.5) | 1 (4.5) | 0 (0.0) | 1 | 0 | JPXX01.0106 ARS | 2 | 2 (100.0) | 0 (0.0) | 0 (0.0) | 0 (0.0) | 0 | 0 |
| Typhimurium | JPXX01.0604 | 13 | 13 (100.0) | 0 (0.0) | 0 (0.0) | 0 (0.0) | 0 | 0 | JPXX01.0079 ARS | 10 | 10 (100.0) | 0 (0.0) | 0 (0.0) | 0 (0.0) | 0 | 0 |
| I 4,[5],12:i:- | JPXX01.0621 | 76 | 68 (89.5) | 5 (6.6) | 3 (3.9) | 0 (0.0) | 0 | 1 | TERX01.0001 ARS | 15 | 7 (46.7) | 0 (0.0) | 8 (53.3) | 0 (0.0) | 0 | 7 |
| TOTAL | | 467 | 367 (78.6) | 25 (5.4) | 75 (16.1) | 62 (13.3) | 3 | 5 | TOTAL | 275 | 226 (82.2) | 9 (3.3) | 40 (14.5) | 16 (5.8) | 0 | 18 |

[1]Typhimurium includes var. O5-.
[2]Resistant to at least three classes of antimicrobial agents.
[3]Abbreviations: nalidixic acid (Nal); ceftriaxone (Axo).
[4]VetNet identified two different patterns within PulseNet pattern JEGX01.0021.
[5]Based on 467 isolates tested for antimicrobial susceptibility.
[6]Based on 275 isolates tested for antimicrobial susceptibility.
[7]Antimicrobial classes (antimicrobial agents in parentheses; resistance breakpoints in brackets): aminoglycosides (amikacin [≥64 μg/mL], gentamicin [≥16 μg/mL], kanamycin [≥64 μg/mL] and streptomycin [≥64 μg/mL]), ß-lactam/ß-lactamase inhibitor combinations (amoxicillin/clavulanic acid [≥32/16 μg/mL]), cephems (cefoxitin [≥32 μg/mL], ceftiofur [≥8 μg/mL] and ceftriaxone [≥4 μg/mL]), penicillins (ampicillin [≥32 μg/mL]), quinolones (ciprofloxacin [≥4 μg/mL]and nalidixic acid [≥32 μg/mL]), folate pathway inhibitors (sulfisoxazole [≥512 μg/mL], trimethoprim/sulfamethoxazole [≥4/76 μg/mL]), phenicols (chloramphenicol [≥32 μg/mL]), and tetracyclines (tetracycline [≥16 μg/mL]).

and JPXX01.0018 (JPXX01.0002 ARS) were resistant only to streptomycin and sulfisoxazole. Collectively Typhimurium patterns JPXX01.0003 (JPXX01.0003 ARS) and JPXX01.0018 (JPXX01.0002 ARS) were associated with resistance to five or more classes of antimicrobial agents in 60 (92%) of the 65 human *Salmonella* and 16 (80%) of the 20 food animal *Salmonella.*

Of the 15 food animal *Salmonella* isolates exhibiting I 4,[5],12:i:- pattern JPXX01.0621 (TERX01.0001 ARS), seven were resistant to amoxicillin/clavulanic acid, ampicillin, cefoxitin, ceftiofur and ceftriaxone. Of the 76 susceptibility-tested human *Salmonella* isolates exhibiting patternJPXX01.0621 (TERX01.0001 ARS), only one exhibited the exact resistance pattern. A second human *Salmonella* isolate exhibited a closely related resistance profile: the isolate was intermediate instead of resistant to ceftriaxone. The resistance patterns of the additional MDR food animal *Salmonella* isolates and human *Salmonella* isolate with pattern JPXX01.0621 (TERX01.0001 ARS) did not match.

Although we did not find resistance to the fluoroquinolone ciprofloxacin in any human or food animal *Salmonella* isolate, three human *Salmonella* isolates, Berta pattern JAXX01.0001, Enteritidis pattern JEGX01.0004 (JEGX01.0003 ARS) and Typhimurium pattern JPXX01.0302 (JPXX01.0106 ARS), were resistant to nalidixic acid (Nal), an indicator of decreasing susceptibility to quinolones [14].

### Invasiveness and Antimicrobial Resistance to Antimicrobial Agents

Of the 467 human *Salmonella* isolates tested for antimicrobial susceptibility, 367 were pansusceptible, and 100 were resistant to one or more antimicrobial agents (Table 3). Twelve (3.2%) of the 367 pansusceptible isolates and 10 (10%) of the 100 isolates resistant to one or more antimicrobial agents were recovered from blood. This difference was found to be significant ($p = 0.0129$). Included among the 10 resistant blood isolates were four Typhimurium pattern JPXX01.0003 (JPXX01.0003 ARS) isolates, two Enteritidis pattern JEGX01.0021 (JEGX01.0010 ARS and JEGX01.0052 ARS), one each of Typhimurium patterns JPXX01.0018 (JPXX01.0002 ARS) and JPXX01.0146 (JPXX01.0081 ARS), one Berta pattern JAXX01.0001 (JAXX01.0003 ARS ), and one nalidixic acid-resistant Enteritidis pattern JEGX01.0004 (JEGX01.0003 ARS).

## Discussion

We compared clinical isolates of *Salmonella* recovered from ill humans received as part of routine operations in Pennsylvania during 2005-2011 with *Salmonella* isolates collected during the same period from food animals at federally inspected slaughter and processing facilities in the northeastern U.S. We also correlated PFGE patterns of *Salmonella* from humans with animal sources, invasiveness, and antimicrobial susceptibility profiles. Our analysis was based on the most common human *Salmonella* PFGE patterns identified during this period, regardless of serotype. This approach differs from many studies because of its focus on PFGE patterns rather than serotypes and on PFGE patterns that are frequently observed with sporadic disease, rather than solely on patterns linked to outbreaks. However, it is important to note that outbreak isolates were a portion of the overall number of isolates in this study. This is the first study to compare the PFGE patterns of human *Salmonella* isolates recovered during routine surveillance with food animal *Salmonella* isolates recovered at federally inspected slaughter and processing facilities. We found that subtyping *Salmonella* isolates by PFGE revealed differences within serotypes in terms of antimicrobial susceptibility patterns and, for human *Salmonella*, differences in food animal sources and invasiveness that were not evident from serotyping alone.

We found that most (16) but not all of the 20 most common PFGE patterns identified in human *Salmonella* in Pennsylvania were also found in *Salmonella* recovered from food animals in the northeast. The fact that some patterns within a particular serotype were observed in human *Salmonella* but not food animal *Salmonella* and vice versa demonstrates the utility of subtyping isolates within a serotype. Others have previously demonstrated PFGE pattern diversity within a serotype [17-20], but such diversity is a critical observation that is often overlooked. PFGE pattern diversity may explain why morbidity and mortality vary within a serotype and could be useful in assessing the effectiveness of control measures.

Further, the absence of some of the most common human *Salmonella* PFGE patterns from food animal *Salmonella* isolates suggests that sources other than food animals play a role in clinical illness: e.g., fresh produce, nuts and other foods [21,22], suggesting that there is a vast biodiversity associated with food production that is often overlooked.

Although chickens represented only 14.7% of the 65,655 food animals tested, they represented 54.6% of the 2,187 food animals that were positive for *Salmonella* and were the source of 87% of the isolates with shared common patterns. Chicken consumption has been recognized as one of the risk factors for developing salmonellosis caused by serotypes Enteritidis [23-25] and Heidelberg [26], and our PFGE results support these associations. Enteritidis pattern JEGX01.0004 (JEGX01.0003 ARS) was the most common human *Salmonella* pattern in Pennsylvania, accounting for 1,968 (16%) of the 11,967 human Salmonella isolates and 130 (20%) of the 646 invasive salmonellosis cases. The presence of *Salmonella* Enteritidis pattern JEGX01.0004 (JEGX01.0003) in chickens during every year of this study suggests the potential for a significant impact on public health and requires further investigation to determine what attributes sustain the persistence of this strain and what measures could reduce its incidence. This further supports our assertion that consideration of strain diversity (i.e., PFGE pattern diversity) is a critical factor for developing control measures.

The frequent occurrence of human salmonellosis associated with pattern JEGX01.0004 (JEGX01.0003 ARS) makes case investigations difficult unless there is a strong epidemiological connection linking the cases. Lacking such a connection, salmonellosis cases associated with this pattern are generally given less scrutiny by epidemiologists, in spite of the fact that the number of cases associated with this pattern in Pennsylvania ($n = 1{,}968$) is much greater than the number of identified outbreak-associated isolates for all patterns combined ($n = 251$) (Table 2). S. Enteritidis pattern JEGX01.0004 (JEGX01.0003 ARS) was associated with a national outbreak involving approximately 1,939 illnesses [27] and recalls of more than 500 million shell eggs [28]. New regulations governing egg safety in the U.S. [29] may help to reduce the incidence of salmonellosis associated with pattern JEGX01.0004 (JEGX01.0003 ARS). However, since our study has demonstrated seven years of this pattern's high level of persistence in poultry meat, it is reasonable to conclude that *Salmonella* Enteritidis pattern JEGX01.0004 (JEGX01.0003 ARS) is persistent and endemic in poultry. Efforts focused on reducing *Salmonella* in broiler flocks have met with very encouraging results in Denmark [30,31] and the United Kingdom [32]. Data collected in the coming years will provide critical information regarding the potential reduction of *Salmonella* Enteritidis pattern JEGX01.0004 (JEGX01.0003 ARS) in the United States.

Eleven of the 16 shared PFGE patterns were associated with susceptibility to all antimicrobial agents tested or resistance to fewer than three classes of antimicrobial agents in food animal *Salmonella* (Table 3). This is encouraging from an antimicrobial resistance perspective, indicating that not all PFGE subtypes and serotypes acquire the same degree or type of resistance. Multidrug resistance was associated with Berta pattern JAXX01.0001 (JAXX01.0003 ARS), Heidelberg pattern JF6X01.0022 (JF6X01.0015 ARS), Typhimurium patterns JPXX01.0003 (JPXX01.0003 ARS) and JPXX01.0018 (JPXX01.0002 ARS), and I 4,[5],12:i:- pattern JPXX01.0621 (TERX01.0001 ARS) in both human and food animal *Salmonella*. The high degree of correspondence between PFGE patterns and antimicrobial resistance profiles for Typhimurium patterns JPXX01.0003 (JPXX01.0003 ARS) and JPXX01.0018 (JPXX01.0002 ARS) in human and food animal *Salmonella* isolates contributes to the existing body of evidence linking contaminated food and human salmonellosis. Supporting this correlation is the fact that ACSSuT-associated *Salmonella* pattern JPXX01.0003 (JPXX01.0003 ARS) was linked to a 2003 outbreak of *Salmonella*-contaminated ground beef [33]. The broad representation of MDR patterns JPXX01.0003 (JPXX01.0003 ARS) and JPXX01.0018 (JPXX01.0002 ARS) in all four types of food animals contrasts sharply with the strong association of most other patterns with chicken (Table 2), although this may be confounded by differences in the number of animals of each type that were tested (Table 1). Monitoring the incidence of these patterns and their associated resistance phenotypes could aid in assessing the effect of efforts to reduce the use of antimicrobial agents for growth promotion and disease prevention in food animals on public health.

Differences in the percentage of blood isolates representing each pattern suggest that there may be differences in invasiveness among *Salmonella* having different patterns, even for patterns within a serotype. Further work is needed to determine why certain patterns were associated with increased or decreased invasiveness.

In our sample of 467 human *Salmonella* isolates that were tested for susceptibility to antimicrobial agents, we did not observe a direct correlation between resistance and invasiveness within specific PFGE patterns; however, we did find that the percentage of antimicrobial-resistant isolates that were recovered from blood (10%) was approximately three times the percentage found to be pansusceptible (3.2%; $p = 0.0129$). Further work is needed to confirm this correlation.

For some of the four unshared common patterns, sources other than food animals slaughtered or processed in northeastern U.S. facilities are likely. For example, we found that S. Enteritidis pattern JEGX01.0002 was associated with travel outside the U.S., especially to the Dominican Republic and Mexico. Another common human *Salmonella* pattern that was not noted in food animal *Salmonella* isolates, Enteritidis pattern JEGX01.0009, has been associated with imported food: *Salmonella* with this pattern has been recovered from shrimp imported from Bangladesh and pompano from Taiwan (PulseNet national PulseNet *Salmonella* database was searched on 2012-06-11 and included data uploaded between 2008-06-12 and 2012-06-11).

In some cases, the absence of common human *Salmonella* PFGE patterns in food animal *Salmonella* may have resulted from testing an insufficient number of animals. From 2005 through 2011, the number of food animals tested for *Salmonella* in northeastern slaughter and processing facilities declined by more than half–from 13,834 in 2005 to 6,381 in 2011. Although Typhimurium pattern JPXX01.1212 and I 4,[5],12:i:- pattern JPXX01.0026 were not observed in our sample of food animal *Salmonella*, they have repeatedly been observed in *Salmonella* recovered from chicken (PulseNet data; searches of the PulseNet national *Salmonella* database were conducted on 2012-06-11 and included data uploaded between 2008-06-12 and 2012-06-11). The absence of these patterns in *Salmonella* recovered from food animals in the northeast may simply reflect the limited amount of testing during the study period. Alternatively, the food responsible for illness may have come from outside the northeastern U.S., as some of the patterns have been observed in food animals from other areas in the U.S. (data not shown).

### Limitations

This study has several limitations. The first is the absence of data for testing food animal *Salmonella* isolates with a second-enzyme (BlnI). Testing human *Salmonella* isolates with BlnI has helped to discriminate within most PFGE patterns produced with XbaI. Beginning in 2011, VetNet specimens have been tested with BlnI, so this information will be available for future studies. A second limitation is the reduced number of samples collected by FSIS for *Salmonella* testing at federally inspected slaughter and processing plants in the northeast in recent years. A third limitation is the fact that only 11% of the human *Salmonella* isolates with the 16 common shared PFGE patterns were tested for susceptibility to antimicrobial agents. Finally, we recognize that not all food slaughtered and processed in a particular geographic region is consumed in the same region and that different regions of the U.S. vary in their poultry production concentrations.

## Conclusions

This work shows that 16 of the 20 most common PFGE patterns found in human *Salmonella* in Pennsylvania were also found in *Salmonella* recovered from food animals sampled at federally inspected slaughter and processing facilities in the northeastern U.S. Multidrug resistance was correlated with five PFGE patterns shared by food animal and human *Salmonella*: Berta pattern JAXX01.0001 (JAXX01.0003 ARS), Heidelberg pattern JF6X01.0022 (JF6X01.0015 ARS), Typhimurium patterns JPXX01.0003 (JPXX01.0003 ARS) and JPXX01.0018 (JPXX01.0002 ARS), and I 4,[5],12:i:- pattern JPXX01.0621 (TERX01.0001 ARS). The most common human pattern, *S.* Enteritidis patterns JEGX01.0004 (JEGX01.0003 ARS), accounting for 16% of all human *Salmonella*, was also common in *Salmonella* recovered from food animals, especially chicken. Our findings suggest an association between the most prevalent forms of human salmonellosis and contaminated meat and poultry. They also show a persistence of Enteritidis pattern JEGX01.0004 (JEGX01.0003 ARS), the association of differences in invasiveness with different PFGE patterns, and the tendency for some but not all PFGE subtypes within a serotype to acquire multidrug resistance. Comparing the PFGE fingerprint patterns and antimicrobial susceptibility profiles of *Salmonella* from humans with those from food animals inspected in slaughter and processing facilities located in a specific geographic region provides information that may assist in source attribution and outbreak investigations.

## Acknowledgments

We acknowledge important contributions from the following partners: Todd A. Lauze, U.S. Department of Agriculture, Food Safety and Inspection Service, Eastern Laboratory, Athens, Georgia, USA, for providing the food animal sampling data in Table 1; Kirsten Waller, Pennsylvania Department of Health Bureau of Epidemiology, USA, for providing the outbreak data for Table 2 and assistance with statistical analysis; FDA for PFGE testing of *Salmonella* whose patterns matched those of some common human PFGE patterns not shared by food animal *Salmonella*; CDC PulseNet for assigning PFGE pattern names; BOL

Bacteriology Section for identifying and serotyping human *Salmonella*; BOL Molecular Microbiology Section for PFGE testing.

## Author Contributions

Conceived and designed the experiments: CHS PJF-C. Performed the experiments: CHS DT KJ. Analyzed the data: CHS PJF-C SO NMM. Wrote the paper: CHS PJF-C DT KJ SO NMM.

## References


1. Scallan E, Hoekstra RM, Angulo FJ, Tauxe RV, Widdowson MA, et al. (2011) Foodborne illness acquired in the United States–major pathogens. Emerg Infect Dis 17: 7–15.
2. Hoffmann M, Luo Y, Lafon PC, Timme R, Allard MW, et al. (2013) Genome sequences of Salmonella enterica serovar Heidelberg isolates isolated in the United States from a multistate outbreak of human Salmonella infections. Genome announcements 1.
3. Scharff RL (2012) Economic Burden from Health Losses Due to Foodborne Illness in the United States. Journal of Food Protection 75: 123–131.
4. Swaminathan B, Barrett TJ, Hunter SB, Tauxe RV (2001) PulseNet: the molecular subtyping network for foodborne bacterial disease surveillance, United States. Emerg Infect Dis 7: 382–389.
5. Gerner-Smidt P, Hise K, Kincaid J, Hunter S, Rolando S, et al. (2006) PulseNet USA: a five-year update. Foodborne Pathog Dis 3: 9–19.
6. National Antimicrobial Resistance Monitoring System-FDA http://www.fda.gov/AnimalVeterinary/SafetyHealth/AntimicrobialResistance/NationalAntimicrobialResistanceMonitoringSystem/default.htm.Food and Drug Administration. Accessed 2013 September 19.
7. Jackson CR, Fedorka-Cray PJ, Wineland N, Tankson JD, Barrett JB, et al. (2007) Introduction to United States Department of Agriculture VetNet: status of Salmonella and Campylobacter databases from 2004 through 2005. Foodborne Pathog Dis 4: 241–248.
8. National Antimicrobial Resistance Monitoring System-USDA http://www.ars.usda.gov/Main/docs.htm?docid = 6750.United States Department of Agriculture. Accessed 2013 September 19.
9. Ribot EM, Fair MA, Gautom R, Cameron DN, Hunter SB, et al. (2006) Standardization of pulsed-field gel electrophoresis protocols for the subtyping of Escherichia coli O157:H7, Salmonella, and Shigella for PulseNet. Foodborne Pathog Dis 3: 59–67.
10. Brenner F, McWhorter-Murlin A (1998) Identification and serotyping of Salmonella and an update of the Kauffman-White scheme. Atlanta: Centers for Disease Control and Prevention.
11. Farmer III J (1995) Enterobacteriaceae: Introduction and Identification In: Manual of Clinical Microbiology 6 th ed./editor in chief, P. Murray editors: Baron E, Pfaller M, Tenover F, Yolken R American Society for Microbiology Washington DC.
12. USDA. National Antimicrobial Resistance Monitoring System – Enteric Bacteria AAN (2012) 2010 NARMS Animal Arm Annual Report. Athens, GA: U.S. Department of Agriculture, Agricultural Research Service.
13. Clinical Laboratory Standards Institute W, Pennsylvania, (2010) Performance Standards for Antimicrobial Susceptibility Testing; Twentieth Informational Supplement. CLSI-document M100-S20. Clinical Laboratory Standards Institute.
14. Centers for Disease Control and Prevention (2012) National Antimicrobial Resistance Monitoring System for Enteric Bacteria (NARMS): Human Isolates Final Report, 2010. Atlanta: Centers for Disease Control.
15. Vugia DJ, Samuel M, Farley MM, Marcus R, Shiferaw B, et al. (2004) Invasive Salmonella infections in the United States, FoodNet, 1996–1999: incidence, serotype distribution, and outcome. Clin Infect Dis 38 Suppl 3: S149–156.
16. Dean AG, Sullivan KM, Soe MM (2013) OpenEpi: Open Source Epidemiologic Statistics for Public Health, Version 3.01. http://www.openepi.com/v37/Menu/OE_Menu.htm. updated 2013/04/06, accessed 2013/08/22. Accessed 2013 September 19.
17. Adaska JM, Silva AJ, Berge ACB, Sischo WM (2006) Genetic and Phenotypic Variability among Salmonella enterica Serovar Typhimurium Isolates from California Dairy Cattle and Humans. Applied and Environmental Microbiology 72: 6632–6637.
18. Woo YK, Lee SH (2006) Genetic diversity of multi-resistant Salmonella enterica serotype Typhimurium isolates from animals and humans. J Microbiol 44: 106–112.
19. Soyer Y, Alcaine SD, Schoonmaker-Bopp DJ, Root TP, Warnick LD, et al. (2010) Pulsed-field gel electrophoresis diversity of human and bovine clinical Salmonella isolates. Foodborne Pathog Dis 7: 707–717.
20. Zou M, Keelara S, Thakur S (2012) Molecular characterization of Salmonella enterica serotype Enteritidis isolates from humans by antimicrobial resistance, virulence genes, and pulsed-field gel electrophoresis. Foodborne Pathog Dis 9: 232–238.
21. Centers for Disease Control and Prevention (2013) Surveillance for Foodborne Disease Outbreaks – United States, 2009–2010. Morbidity and Mortality Weekly Report 63: 41–56.
22. Painter JA HR, Ayers T, Tauxe RV, Braden CR, Angulo FJ, et al (2013) Attribution of foodborne illnesses, hospitalizations, and deaths to food commodities by using outbreak data, United States, 1998–2008. Emerg Infect Dis 19.
23. Altekruse SF, Bauer N, Chanlongbutra A, DeSagun R, Naugle A, et al. (2006) Salmonella enteritidis in broiler chickens, United States, 2000–2005. Emerg Infect Dis 12: 1848–1852.
24. Kimura AC, Reddy V, Marcus R, Cieslak PR, Mohle-Boetani JC, et al. (2004) Chicken consumption is a newly identified risk factor for sporadic Salmonella enterica serotype Enteritidis infections in the United States: a case-control study in FoodNet sites. Clin Infect Dis 38 Suppl 3: S244–252.
25. Marcus R, Varma JK, Medus C, Boothe EJ, Anderson BJ, et al. (2007) Re-assessment of risk factors for sporadic Salmonella serotype Enteritidis infections: a case-control study in five FoodNet Sites, 2002–2003. Epidemiol Infect 135: 84–92.
26. Currie A, MacDougall L, Aramini J, Gaulin C, Ahmed R, et al. (2005) Frozen chicken nuggets and strips and eggs are leading risk factors for Salmonella Heidelberg infections in Canada. Epidemiology and infection 133: 809–816.
27. Centers for Disease Control and Prevention (2010) Investigation update: multistate outbreak of human Salmonella Enteritidis infections associated with shell eggs Atlanta, Georgia. US Department of Health and Human Services, CDC.
28. Allard MW, Luo Y, Strain E, Pettengill J, Timme R, et al. (2013) On the Evolutionary History, Population Genetics and Diversity among Isolates of Salmonella Enteritidis PFGE Pattern JEGX01. 0004. PloS one 8: e55254.
29. Food and Drug Administration US (2009) Prevention of Salmonella Enteritidis in shell eggs during production, storage, and transportation; final rule. Fed Regist. 33030–33101.
30. Anonymous (2012) Annual Report on Zoonoses in Denmark 2011. National Food Institute, Technical University of Denmark.
31. Wegener HC (2010) Danish initiatives to improve the safety of meat products. Meat Science 84: 276–283.
32. O'Brien SJ (2012) The "Decline and Fall" of Nontyphoidal Salmonella in the United Kingdom. Clin Infect Dis cis967 first published online November 19, 2012 doi:101093/cid/cis967.
33. Dechet AM, Scallan E, Gensheimer K, Hoekstra R, Gunderman-King J, et al. (2006) Outbreak of multidrug-resistant Salmonella enterica serotype Typhimurium Definitive Type 104 infection linked to commercial ground beef, northeastern United States, 2003–2004. Clin Infect Dis 42: 747–752.

# Effectiveness of Simulated Interventions in Reducing the Estimated Prevalence of *Salmonella* in UK Pig Herds

**Alexander D. C. Berriman[1], Damian Clancy[2], Helen E. Clough[1], Derek Armstrong[3], Robert M. Christley[1]***

1 Institute of Infection and Global Health, Department of Epidemiology and Population Health, University of Liverpool, Neston, Cheshire, United Kingdom, 2 Department of Mathematical Sciences, University of Liverpool, Liverpool, United Kingdom, 3 BPEX, Stoneleigh Park, Warwickshire, United Kingdom

**Abstract**

*Salmonella* spp are a major foodborne zoonotic cause of human illness. Consumption of pork products is believed to be a major source of human salmonellosis and *Salmonella* control throughout the food-chain is recommended. A number of on-farm interventions have been proposed, and some have been implemented in order to try to achieve *Salmonella* control. In this study we utilize previously developed models describing *Salmonella* dynamics to investigate the potential effects of a range of these on-farm interventions. As the models indicated that the number of bacteria shed in the faeces of an infectious animal was a key factor, interventions applied within a high-shedding scenario were also analysed. From simulation of the model, the probability of infection after *Salmonella* exposure was found to be a key driver of *Salmonella* transmission. The model also highlighted that minimising physiological stress can have a large effect but only when shedding levels are not excessive. When shedding was high, weekly cleaning and disinfection was not effective in *Salmonella* control. However it is possible that cleaning may have an effect if conducted more often. Furthermore, separating infectious animals, shedding bacteria at a high rate, from the rest of the population was found to be able to minimise the spread of *Salmonella*.

**Editor:** Maurice J. Chacron, McGill University, Canada

**Funding:** This work was funded by the Biotechnology and Biological Sciences Research Council (BBSRC) and BPEX, as part of a Collaborative Award in Science and Engineering (CASE) studentship (http://www.bbsrc.ac.uk/funding/studentships/case.aspx). The funders had no role in study design, data collection and analysis, decision to publish, or preparation of the manuscript.

**Competing Interests:** The authors have declared that no competing interests exist.

* E-mail: robc@liverpool.ac.uk

## Introduction

*Salmonella* species are a major cause of zoonotic disease and *Salmonella* spp can be found in many products intended for human consumption, for example eggs, poultry and pork. Consequently, *Salmonella* control at the point of production (i.e. on-farm) is considered important. Although there are more than 2,500 different *Salmonella* serovars [1], *S.* Typhimurium continues to be the most commonly isolated serovar in pigs in the United Kingdom (UK), which has remained the case for a number of years [2,3]. An abattoir study in 2003 showed that 23.4% ($CI_{95}$: 19.9–27.3%) of pigs were *Salmonella* positive [4]. Although *Salmonella* prevalence differed between regions ($\approx$19%–30% [4]), an average prevalence of 23.4% could be representative for the UK. Whilst *Salmonella* infection on-farm might be high, procedures in place in the abattoir (such as scalding, singeing and polishing) can considerably reduce *Salmonella* prevalence, resulting in a low carcase prevalence. However, the bacteria can still survive during these processes [5–7] and the resulting carcase prevalence can vary depending on abattoir [2]. Consequently, pork and pork products are considered to be a principal source of human food-borne infections [8]. In the UK 10,071 confirmed cases of human salmonellosis were reported in 2009 [9]; however, the true number of cases is unknown. It is unclear how many cases are directly a result of pork and pork products; in Denmark, domestic pork was estimated to have caused between 3.7%–11.2% of human salmonellosis cases in 2011 [10]. In order to minimise the risk to humans, efforts can be made to reduce the prevalence of infection within commercial herds. Mathematical models can be used to investigate mechanisms that drive *Salmonella* transmission, and can be used to help inform decision making. Within the pig's life cycle, it is the finishing stage which poses the biggest risk to public health. As such, analysing interventions imposed during this stage is of particular interest.

In Berriman et al. [11], models of *Salmonella* transmission around varying structures of British pig grower-finisher farms have been developed and analysed. These models account for the varying flooring types used within the UK, whereby some are 'solid floored units' and some are 'slatted floored units.' The *Salmonella* dynamics on each type of unit are potentially quite different. For the slatted unit, the basic reproduction number $R_0$ was calculated, being defined to be the average number of secondary infections produced when one infected individual is introduced into a naïve host population (for example, [12]). By computing the value of $R_0$, key drivers of *Salmonella* transmission could be investigated. Note that although $R_0$ is a useful tool and is widely used in epidemiology, its calculation is insufficient in order to determine whether there is a food risk, since this is dependent on the time frame of infection. This is highlighted through the development of a simple deterministic model in [11]; with base parameter values (Table 1) $R_0$ was found to be 0.8204, below the threshold value of 1, but nevertheless infection was found to persist over the time frame relevant to the finishing stage of production. It is possible that the presence of carrying animals keeps the infection

**Table 1.** Definitions of the parameters used in the model.

| Definition (units) | Parameter | Estimate | Reference |
|---|---|---|---|
| Number of pigs per pen | $N$ | 25 | [41] |
| Number of pens on either side of a corridor | *PensPerSide* | 20 | [41] |
| Infection rate | $\beta$ | Assume $1.67\times10^{-3}$ | - |
| The rate at which a pig ceases to remain infectious (day$^{-1}$) | $\gamma$ | 1/26 = 0.03846 | [14,19] |
| The rate at which a carrier becomes re-infectious (day$^{-1}$) | $\delta$ | 1/108 = 0.00926 | - |
| The rate at which a pig ceases to carry the bacteria (day$^{-1}$) | $\varepsilon$ | 1/60 = 0.01667 | [14] |
| Loss of immunity (day$^{-1}$) | $\nu$ | 0.5 | - |
| Shedding rate (cfu day$^{-1}$) | $\lambda$ | $2.25\times10^{4}$ | [42,43] |
| Proportion of cfu present ingested (day$^{-1}$): Slatted | $\kappa$ | $4.23\times10^{-4}$ | [43,44] |
| Solid | | $3.17\times10^{-5}$ | |
| Bacteria death rate (day$^{-1}$) | $l$ | 1/84 = 0.01190 | [45] |
| Probability of infection from bacterial consumption | $p$ | $2.30\times10^{-6}$ | [19,46] |
| Cross infection rate | $\alpha$ | Assume $1.14\times10^{-6}$ | - |
| Proportion of faeces that remains in a room | $\pi$ | 0.4 | - |
| Proportion of faeces that remains present after cleaning | $q$ | 0.1 | - |
| Time spent in unit (days) | $T_{max}$ | 108 | [26] |
| Airborne infection rate | $\omega$ | Assume $1.02\times10^{-14}$ | - |

Reproduced from Table 2 of [11].

sustained for a long period of time. Furthermore, as a number of infectious pigs enter the system, there are a large number of potential infections, which could be sufficient to sustain the infection, despite the low $R_0$ value.

This paper aims to use simulation results of interventions imposed on these models in order to propose practical interventions that could result in on farm *Salmonella* control. By investigating the mechanisms that drive *Salmonella* transmission,

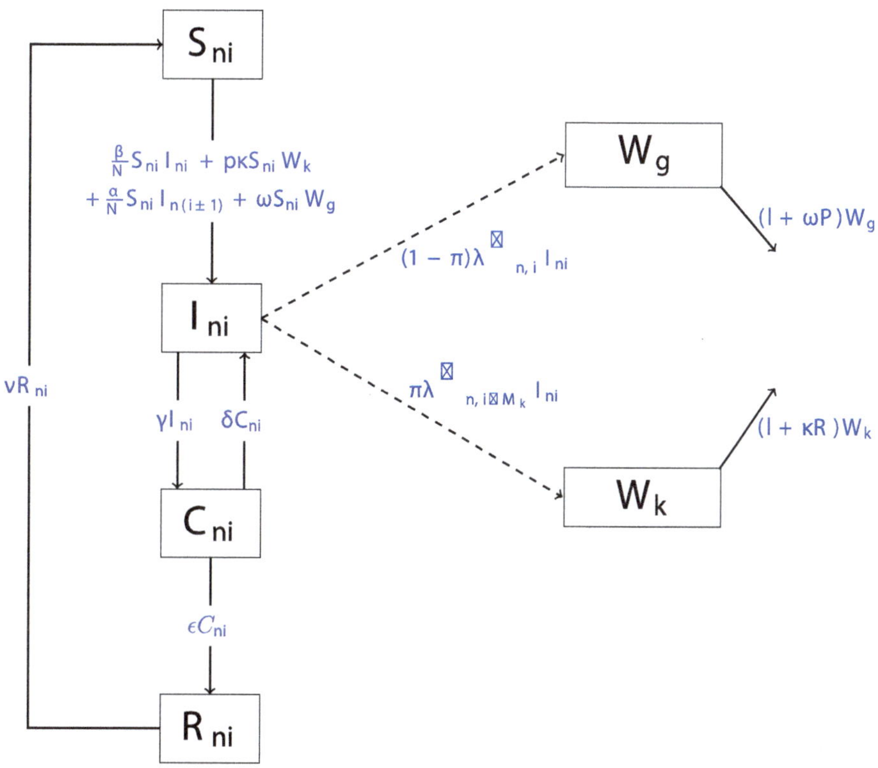


**Figure 1. Flow diagram representing transmission routes and other processes within the slatted model.** Note: $R$ denotes the number of pigs within a room, $P$ denotes the number of pigs on farm and $M_k$ denotes the set of pens within room $k$. Figure reproduced from Figure 1 of [11].

information that can inform the development of control strategies can be generated. Furthermore, potential key drivers of *Salmonella* spread are identified. Clearly, any interventions that are tested within the model must be possible to implement in practice to have any value. As such, any modification to parameters applied within the analysis are within a range that is consistent with findings within the literature. Biosecurity (measures and protocols taken to reduce the risk of disease spread on farms [13]) is becoming increasingly important as a means to reduce the risk of a disease outbreak. Biosecurity encompasses a number of aspects (people, cleaning and equipment, for example). The incorporation of cleaning and disinfection within the model of a solid floored unit enables the testing of the effect of cleaning on farm. Physiological stress (such as manure overflow, mixing/moving of pigs and bad feed [5,14]) is potentially an important factor in *Salmonella* spread as it can cause infected animals (carriers) to resume shedding the bacteria and therefore requires further investigation. The probability of becoming infected after exposure relates to transmission via the faecal-oral route, which is thought to be one of the key routes of transmission and is therefore an important parameter to analyse. Model simulations showed the amount of bacteria shed by an infectious pig was potentially a key driver of *Salmonella* transmission [11]. Further analysis of this result highlights possible interventions that could have an effect on *Salmonella* prevalence.

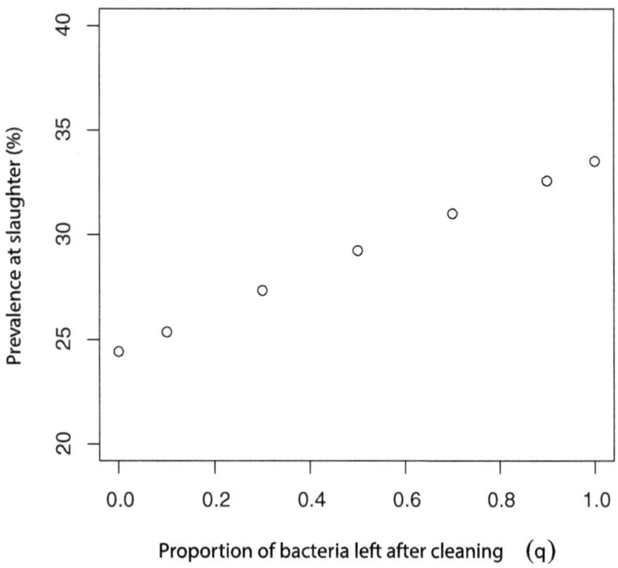


**Figure 2. Average slaughter prevalence with varying levels of efficiency of cleaning and disinfection.** Results are based on model predictions from the solid unit model. Base parameter value of $q = 0.1$.

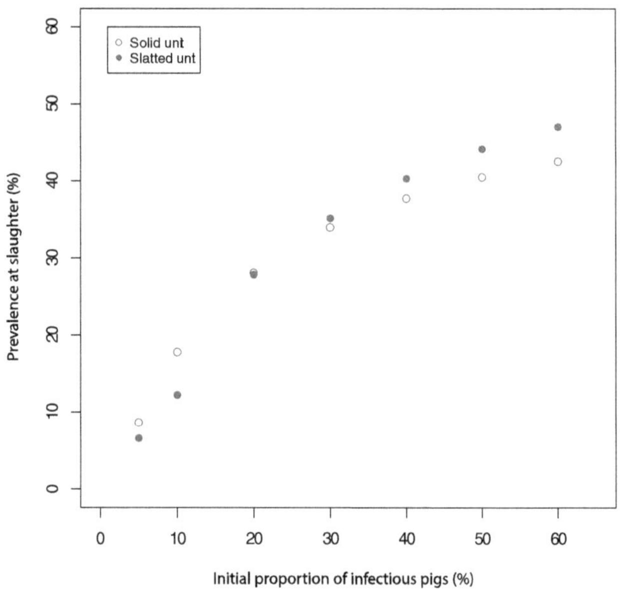


**Figure 3. Average slaughter prevalence with varying levels of infectious animals entering the unit.** Results are based on model predictions from both the slatted and solid unit models.

## Methods

We have previously developed models describing *Salmonella* transmission around two types of British pig grower-finisher farm [11]. The most typical farm structures used within the UK are 'slatted units' and 'solid units,' the difference being the type of flooring, whereby slatted flooring results in the majority of faeces falling through the slats. Each model enables the testing of a number of possible on farm interventions. Within the slatted model, we were able to compute the basic reproduction number $R_0$, which was calculated by analysing the next generation matrix, as described by [15,16]. The elements of this matrix consist of the expected number of secondary infections due to a single primary infection in a fully susceptible population, calculated class by class [17]. The value of $R_0$ is then given by the dominant eigenvalue of the next generation matrix [15]. Detailed calculations of $R_0$ can be found in Appendix 2 of [11].

Within both models, there are a total of 40 pens, each with 25 pigs per pen. Within the slatted unit, the unit is assumed to be divided into 4 rooms with 10 pens in each room (5 pens on either side of a corridor). Within the solid unit however, one large room is used, reflecting realistic practice. Although pigs can shed the bacteria within their faeces in varying amounts [14,18,19], pigs are often subclinical carriers of *Salmonella*, which consequently results in the presence of a carrier state [20,21]. This state can be defined as a pig that is infected with the bacteria in some form, within the tonsils or ileocolic lymph node for example [20], but not shedding bacteria in its faeces. As such, pigs are classed as susceptible (S), infectious (I) (and therefore excreting the bacteria), carrying (C) and recovered (R) within the models [11]. Although both infectious and carrying animals are deemed to be infected (and thus contribute to final prevalence), it is assumed that only infectious animals are capable of passing on infection.

Due to the different time scales within the model (i.e. between infections and bacterial evolution), the dynamics of the bacteria are treated deterministically in simulation. Thus, a continuous-time, semi-stochastic model is used, which differs from other studies which adopt a discrete time modelling approach [22–25]. A flow diagram representing the various types of transmission routes is shown in Figure 1, for a slatted unit. For the solid unit, only one bacterial environment is required; hence the corresponding flow diagram is not shown. The solid unit model also incorporates weekly cleaning and disinfection. Base parameter values are given in Table 1. The models incorporate direct transmission, which can occur within the same pen and between neighbouring pens via contact between susceptible and infectious pigs. Indirect transmission is also incorporated, via the faecal-oral and airborne routes. Infectious animals shed bacteria into their environment. Within the slatted unit, a proportion of faeces is assumed to fall through the slats and so a limited number of bacteria remain available for consumption; within the solid unit, any bacteria shed remain available. It was assumed that the farms operate on an all-in-all-out basis; that is, pigs enter and leave the unit as a group. Furthermore, the farm is assumed to be an exclusive grower-finisher farm and consequently weaners are assumed to be sourced from elsewhere. As such, weaners (at approximately 35 kg) enter the unit (at time 0) and are grown through to finishing in the same building. The model is run for 108 days which is the average time pigs spend within this stage of production [26]. It was assumed that the farm is emptied in its entirety, as all animals are of the same age group. Consequently, the slaughter age prevalence consists of all animals that are classed as either infectious (I) or carriers (C) at the time of slaughter. Although animal prevalence varies greatly, on average in the UK approximately 17% of weaners entering a unit are infected [27]. As such, within the model, a random 15% of pigs entering the unit are assumed to be infectious, and a further 5% considered carriers. The models were validated using data from the Zoonoses National Control Programme (ZNCP) farm visits and results from a British abattoir study. Both of which found between ≈20% to 30% of pigs to be *Salmonella* positive at slaughter, but was on average 23.4% [4]. Unfortunately data were not available for the type of unit pigs came from. As such, this prevalence must be used for both models described.

The model was run for 5,000 simulations in order to present an average end distribution of prevalence. All simulations were run in MATLAB® 7.10 running under Microsoft® Windows® on a desktop personal computer.

## Results

### The effect of cleaning and disinfection

With the incorporation of cleaning and disinfection within the solid-floored unit, the effect of this aspect of biosecurity can be analysed. Cleaning and disinfection is an important aspect of on-farm management practice, especially with regard to biosecurity. Furthermore, it is a highly intensive process and requires a large amount of time to be done efficiently. Within the model, the efficiency of cleaning could be analysed by analysis of the parameter $q$; the proportion of faeces that remained present after cleaning. A farm that cleaned to a good standard could remove 90% of the bacteria from the environment for example, compared to a farm with poor cleaning that only removed 10% of the bacteria.

The model showed that fully effective cleaning alone was not enough to eradicate *Salmonella* once infection was established. However the prevalence (proportion of animals in either infectious or carrying state) was lower if cleaning took place on farm (Figure 2), which implied that cleaning and disinfection was still a worthwhile task. This concurs with a previous study [28] which found that cleaning and disinfection reduces environmental bacteria but fails to eradicate *Salmonella* on farm. Furthermore, an on-farm study of UK pig farms found improved cleaning and disinfection on farm translated into a reduction in prevalence of approximately 10% [29]. The model predicted a reduction in prevalence of approximately 8%, which seemed a relatively good estimate for this effect.

### Proportion of infectious animals entering the unit

*Salmonella* on farm clearly has to be initiated somehow. Within the models presented here, the prevalence at slaughter was dependent on the initial number of infected animals, both infectious and carriers. With the standard number of infected pigs entering the unit (including both infectious (and therefore shedding) and carrier animals, thought to be ≈20%), an average prevalence at slaughter of approximately 24.6% and 25.4% within the slatted and solid units, respectively, was found. Within both slaughter age prevalences, the majority of animals were classed as carriers (≈15% with the standard number of infected animals), which can be assumed to be the case for all scenarios unless otherwise stated. With varying levels of infectious pigs entering the unit, it was shown that prevalence just prior to slaughter increased until approximately 60% of pigs entering the unit were infectious (Figure 3), after which any increase in the initial proportion of

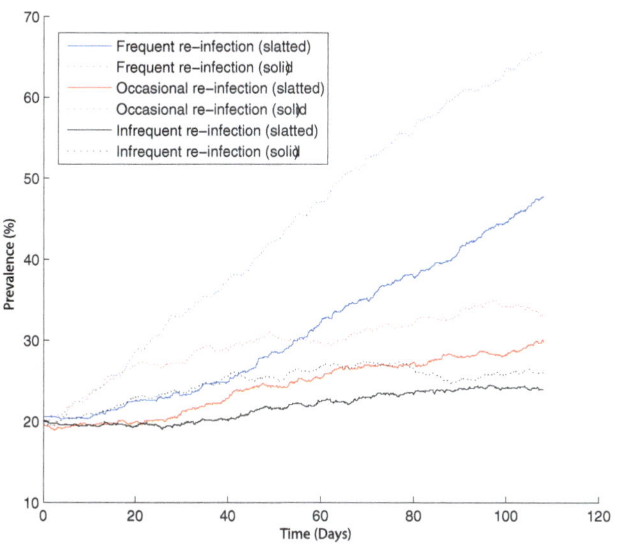


**Figure 4. Model predictions showing the effect of the rate of re-infection ($\delta$) on *Salmonella* prevalence.** One typical simulation for each rate and model. Note: $\delta = \frac{1}{14}$ for frequent re-infection, $\delta = \frac{1}{61}$ for occasional re-infection and $\delta = \frac{1}{108}$ for infrequent re-infection. Base model: infrequent re-infection.

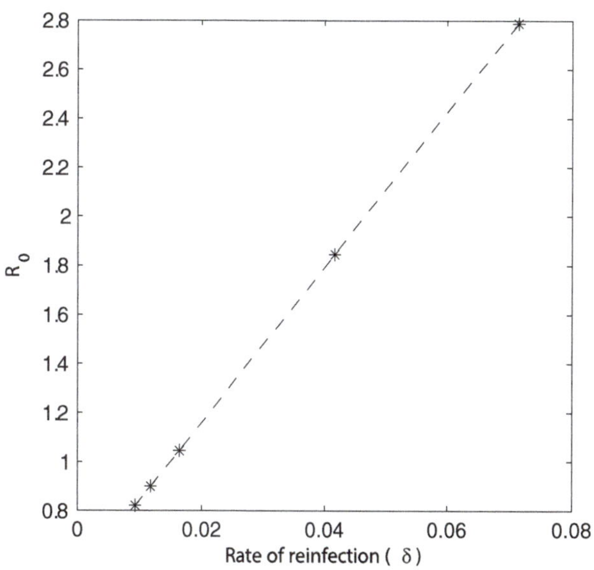


**Figure 5. The effect of the rate of re-infection ($\delta$) on $R_0$ within the slatted unit.** Base parameter value of $\delta = 0.01$.

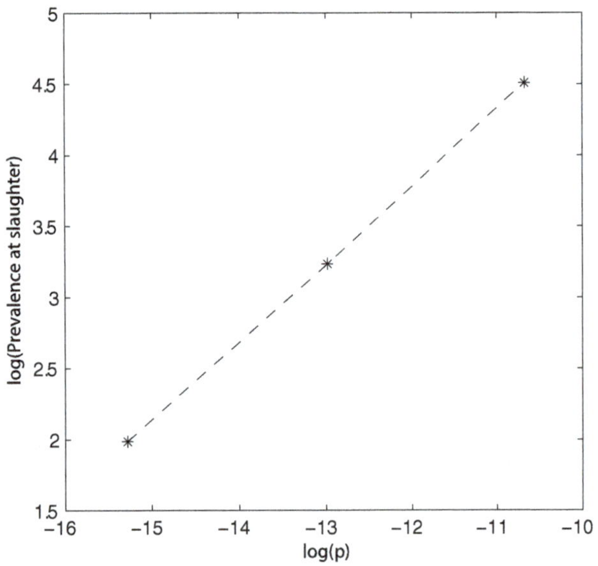


**Figure 6. The effect of the probability of infection ($p$) on slaughter prevalence within the slatted unit.** Natural logs were used, with base parameter $\log(p) = -12.98$. Although these values represent the slatted model, the trend and values are similar to the solid models.

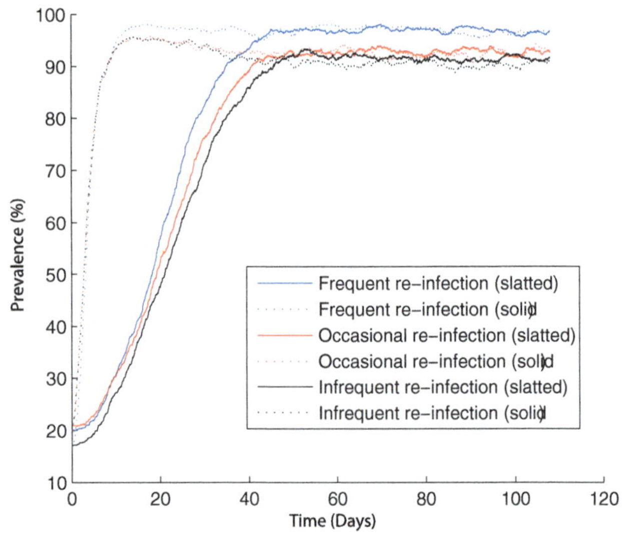


**Figure 8. The effect of the rate of re-infection ($\delta$) on *Salmonella* prevalence with high shedding.** Results are based on model predictions from both the slatted and solid unit models. One typical simulation for each rate and model. Note: $\delta = \frac{1}{14}$ for frequent re-infection, $\delta = \frac{1}{61}$ for occasional re-infection and $\delta = \frac{1}{108}$ for infrequent re-infection. Base model: infrequent re-infection.

infectious pigs entering the unit had little effect on prevalence at slaughter and was therefore not included within the Figure. This was true for both the solid and slatted floored models. To analyse how infection is able to spread when animals enter the unit, it is assumed that they are infectious as opposed to carrying, and thus the initial proportion of carriers is 0.

It was also seen that, if low levels of infectious animals enter the unit, *Salmonella* either becomes eradicated (when levels are extremely low; found via simulations with $<1\%$ of initially infectious animals) or maintained at low levels. It is important to note that this assumed that no other form of infection exists. As

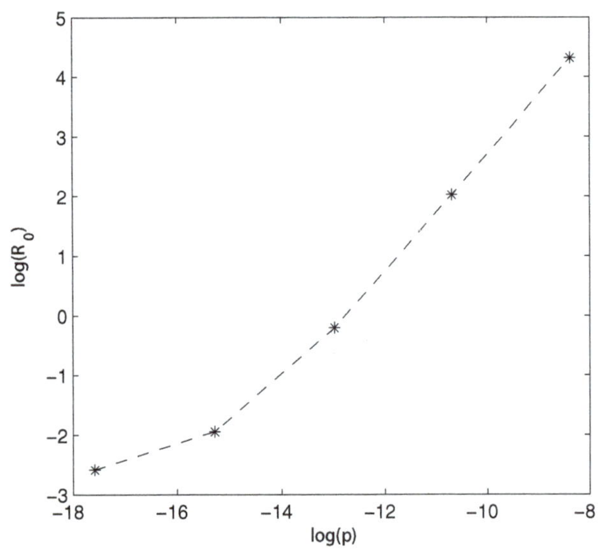


**Figure 7. The effect of the probability of infection ($p$) on $R_0$ within the slatted unit.** Natural logs were used, with base parameter $\log(p) = -12.98$.

such, this result should be taken on the side of caution, as a low number of infectious animals entering the unit may not be sufficient to ensure a low *Salmonella* prevalence, as external factors may cause additional pigs to become infected.

The prevalence at slaughter was found to be higher within the solid unit until 30% of pigs entering the unit were infectious (Figure 3). For an initial proportion of infectious pigs greater than 30%, prevalence at slaughter is higher in the slatted unit. Within the solid unit, it is possible that cleaning and disinfection on farm is insufficient to eliminate infection from the environment altogether, but can be effective at reducing levels of environmental bacteria when the environment becomes more contaminated. Within both models, at each initial prevalence level, the number of animals carrying the bacteria remained higher than those classed as infectious.

## The effect of physiological stress

It is well known that any increased stress on pigs that are infected but not shedding (carrier pigs) could cause the animals to resume shedding the bacteria in their faeces [30,31]. Clearly the majority of stress would be imposed during transport to the abattoir and during lairage, but the general movement of pigs on farm and manure overflow could also have an effect [5,14]. Within both models, it has been assumed that minimal stress is imposed on the animals on farm (corresponding to infrequent re-infection in Figure 4); increased stress that causes the animals to become re-infectious was shown to influence *Salmonella* prevalence (Figure 4), for each unit structure.

It was shown that the average on-farm prevalence increased as the rate of re-infection became greater (Figure 4). Prevalence at slaughter is higher in the solid unit than the slatted unit whatever the rate of reinfection. However, while this effect is minimal in the base model (infrequent re-infection), the difference become much more marked as the rate of re-infection increases. It was interesting to note that within the solid unit, frequent re-infection consistently

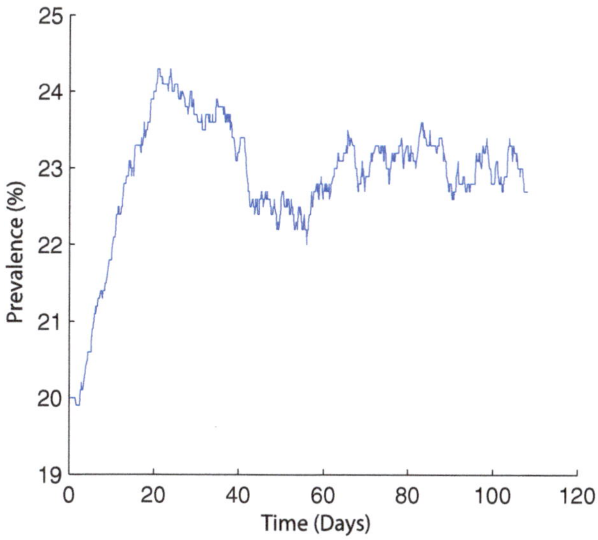


**Figure 9. One typical simulation showing the result of containing infectious pigs shedding high numbers of bacteria.** Animals are assumed to be shedding the bacteria at a rate of $\lambda = 2.25 \times 10^5$ and are contained within 1 room of a slatted unit.

appeared to only take effect after approximately 20 days, before which occasional and frequent re-infection showed similar behaviour. By differentiating the *Salmonella* status of the animals with frequent re-infection, it was shown that the numbers of both infectious and carrying animals continually increase. Thus frequent re-infection results in an increase in the number of animals capable of passing on infection, which consequently results in the increase in *Salmonella* prevalence.

It is possible that frequent re-infection caused the model to cross a threshold which resulted in the continued increase of prevalence. With the presence of occasional re-infection (after approximately 60 days), the value of the basic reproduction number $R_0$, calculated for the slatted model, became greater than 1 ($\approx 1.05$) and frequent re-infection caused $R_0$ to increase to $\approx 2.79$ (recall base value $R_0 = 0.8204$, Figure 5); this corresponds with the large increase in final prevalence (Figure 4). As such, it is important that stress or any other cause for continual re-infection is minimised in order to keep prevalence as low as possible.

## The effect of the probability of infection

One possible mode of action for *Salmonella* intervention is to reduce the probability of becoming infected after *Salmonella* exposure, which is something that could possibly be achieved via vaccination. With the base probability of infection ($p = 2.3 \times 10^{-6}$), a prevalence at slaughter of 24.6% ($\approx 14.4\%$ carriers) and 25.4% ($\approx 15.4\%$ carriers) for the slatted and solid unit respectively was found. A 10 times reduction in this probability resulted in a reduction in prevalence of a similar magnitude within both models (to $\approx 7.5\%$). Conversely, a 10 times increase in probability resulted in a prevalence of approximately 91% ($\approx 54\%$ carriers, Figure 6). Again this was consistent between both models. As such, it appeared as though, with these levels of shedding, the probability of infection had much the same impact on *Salmonella* prevalence at slaughter regardless of the structure of the unit itself. These results also highlighted a threshold that existed within the model, whereby prevalence is very sensitive to the probability of infection for $p$ between $10^{-6}$ and $10^{-5}$, but insensitive outside this range.

The probability of infection after *Salmonella* exposure had a major affect on *Salmonella* prevalence; the corresponding effect on the basic reproduction number, $R_0$, was calculated for the slatted unit. When the probability of infection was decreased 10 times, the corresponding $R_0$ value was 0.1432, compared to the base result where $R_0 = 0.8204$. The large increase in prevalence when the probability of infection was 10 times higher was reflected in the $R_0$ value, increasing to 7.59. This $R_0$ value was greater than 1 and thus a sustained infection would be expected. This increase in $R_0$ as the value of $p$ increases is shown in Figure 7.

## The effect of the amount of bacteria shed

Previous analyses of the model highlighted the importance of the shedding rate ($\lambda$) in *Salmonella* spread [11], which concurs with the findings of Lurette et al. [24]. Within both the slatted and solid unit, a ten times higher shedding rate resulted in a slaughter age prevalence of 91.2% and 90.85% (both with $\approx 54\%$ of animals classed as carriers) respectively. A potential significant difference between unit structures was highlighted from this analysis, whereby infection within the solid unit was able to spread at a much faster rate.

The finding that a number of pigs shedding high numbers of *Salmonella* in their faeces could have such a drastic effect on prevalence is important. As such, a key issue is to analyse interventions that could have an effect on disease spread even with this high rate of shedding.

**Effective interventions when shedding is high.** From simulation of the solid unit model, it was shown in [11] that with high shedding, weekly cleaning was no longer as effective in *Salmonella* control. In fact, the difference in the average *Salmonella* prevalence between farms with a high level of cleaning and those with a low level of cleaning was less than 1%. This potentially indicates that infection can become established quickly and consequently weekly cleaning of the farm is rendered inadequate. It is therefore quite possible that in order for *Salmonella* control when shedding is high, cleaning must be conducted more often in order to minimise the amount of bacteria that pigs are exposed to. This however would require further investigation.

Previously, an increase in the rate of re-infection was shown to have quite a large effect (Figure 4). However, when shedding was higher, the rate of re-infection had a much smaller effect on *Salmonella* prevalence at slaughter (Figure 8). Both models appear to have a very similar slaughter age prevalence, however the dynamics are quite different; the spread of infection was more gradual within the slatted unit. With the presence of higher shedding, the value of $R_0$ (for the slatted unit) for infrequent re-infection $\left(\delta = \frac{1}{108}\right)$ was found to be $\approx 7.60$ and became progressively higher as the rate of re-infection increased. Although $R_0$ increases, it appears that the value for infrequent re-infection was sufficiently large for any increase to have a minimal affect on prevalence, which was highlighted by simulation (Figure 8).

A point of interest was the extremely small difference in prevalence with these varying rates of re-infection; a difference of approximately 6%. This is clearly a considerable contrast with 'normal' shedding levels, where frequent re-infection resulted in a prevalence of approximately 40% higher than infrequent re-infection. When shedding was high, the majority of the population was already infected in some form, shedding or carrying, and frequently changing between states did not have a large effect. However, with 'normal' shedding levels, frequent transition between carrying and shedding (and therefore infectious) states can have a drastic affect, by increasing the number of animals that are capable of passing on the infection.

Within both models, the probability of infection was found to be an important parameter when shedding was at 'normal' levels. With the higher rate of shedding, a 10 times reduction in the probability of infection saw prevalence within both models fall to approximately 25% ($\approx$15% carriers), a reduction in prevalence of approximately two thirds. It could be concluded that the probability of becoming infected after *Salmonella* exposure is an important factor in disease spread regardless of the shedding rate, and a potential key driver of *Salmonella* transmission.

When animals shedding bacteria at a high rate (potentially "super-shedders") were randomly spread throughout the unit, prevalence at slaughter was approximately 90%. However, if all infectious pigs, shedding at a high rate, were contained within 1 room of the building (whereby infection is able to spread between pens within a room, but is less likely to spread between rooms, Figure 9) then in general, this could be enough to halt transmission as infection was unable to spread throughout the whole unit. Containing all infectious animals to 1 room limited the number of animals that were exposed to the bacteria and consequently limited *Salmonella* transmission. Although farmers cannot easily identify individual pigs that become infected, this finding could still be exploited by attempting to keep pigs in groups of pens with solid divisions, thus ensuring every effort is made to prevent contact between pens and essentially create different epidemiological groups. Clearly it is not easy to identify all infected animals due to many animals being asymptomatic. However, if the assumption that carrying animals are incapable of passing on the infection is true, then the focus should be on containing infectious pigs (i.e. those animals that are shedding the bacteria). As long as stress on farm is minimised (i.e. carrying animals rarely become re-infectious) then the presence of animals carrying the bacteria elsewhere in the unit could have a minimal effect.

## Discussion

We have analysed semi-stochastic transmission models describing the dynamics of *Salmonella* within 2 types of grower-finisher herds, as detailed in [11]. Each model enabled the assessment of different aspects of *Salmonella* dynamics. The slatted floor model allowed the calculation and analysis of the basic reproduction number $R_0$, which has not previously been analysed in this context. It was shown that $R_0$ was insufficient in determining whether the disease was eliminated quickly or able to persist. When $R_0 < 1$, the disease does not necessarily disappear (over the relevant time period). Consequently, $R_0$ calculations may not be the most effective way of examining interventions applied within this system due to the complex dynamics.

Findings from the solid unit model highlighted some key issues with cleaning and disinfection on-farm. The model found cleaning to have minimal effect on *Salmonella* prevalence when shedding was high, which is likely due to the rate at which infection was able to spread. In an attempt to counteract this high uptake of infection with the use of cleaning, it is possible that more frequent cleaning could minimise *Salmonella* spread. Although this has not been implemented here, the fact that cleaning does have some effect on reducing the prevalence was thought to confirm this supposition. However, further analysis would be needed in order to determine whether this would be feasible, beneficial and economically viable.

Separating animals that shed bacteria at a high rate was shown to have an effect on *Salmonella* prevalence, as the infection was unable to spread throughout the whole unit. Although this is a relatively simple intervention, difficulties arise in its implementation as infectious animals are hard to identify. Any advances in the ability to identify infected animals could result in an increased potential to apply this intervention effectively.

The addition of prebiotics to drinking water has been shown to be associated with a reduction in *S.* Typhimurium shedding [32]. Probiotics on the other hand have been shown to have little effect on shedding, but do show signs of reducing the presence of the bacteria internally (in the mesenteric lymph nodes for example), which implies that probiotics and prebiotics could alter the gut microflora composition to the benefit of the animal [32]. Acidification of feed has been shown to inhibit *Salmonella* growth, which results in a reduction in infection levels and consequently the amount of bacteria shed [33]. The type of food used could also have some impact on the dynamics, for example wet feed has been associated with a reduction in shedding [33]. Clearly there are a number of possible interventions that could be implemented with regard to feed, although a large factor for decision making is cost. In changing the whole system to use wet feed, it is quite possible that a large scale renovation of the unit would need to occur.

A factor that had a major impact on prevalence was the probability of infection after *Salmonella* exposure. A decrease in the probability of infection was found to be extremely influential in *Salmonella* control even when shedding was at high levels. In order for these simulated interventions to be of use, there needs to be a practical way in which such interventions can be implemented. Although it has been shown that adding antibiotics to feed can reduce the amount of *Salmonella* shed by an infected animal [34], with the presence of resistant *Salmonella* strains, the addition of the antibiotic to the resistant *Salmonella* strain can increase the quantity, duration and prevalence of faecal shedding [35]. Consequently, the use of antibiotics is somewhat controversial due to the potential increased risk of generating antimicrobial resistance, and therefore unlikely to be implemented in order to control *Salmonella*.

It would be interesting to see if vaccination would have an effect on two aspects by decreasing the amount of bacteria shed when an animal becomes infected and/or reducing the susceptibility of the animals. Although vaccinations could be useful in helping to prevent clinical salmonellosis in pigs, the capacity for a vaccination to make a contribution to reduce shedding in pigs remains unproven. Various vaccines have been developed (for example Salmoporc (IDT BIOLOGIKA) licensed live vaccine) but are not widely used on farm. It is possible that vaccination is scarcely used due to the potential for the vaccine to interfere with current control programs relying on serology [36]. Vaccination against viral infections is expected to limit the chance of bacterial infections [37] and should aim to prevent colonisation of the host and minimise the shedding of the pathogen [38]. A number of studies have been conducted that show vaccination is associated with a reduction in isolation of *Salmonella* in slaughter weight pigs with a reduction of clinical symptoms and colonization of the animal [36,39,40]. Vaccination may have a role where *Salmonella* prevalence is high, whereby piglets from vaccinated sows with high antibodies should have high maternal antibody levels. Furthermore, vaccination of sows would spread the cost of vaccination over all piglets. However, this would provide no protection during the later stages, once maternal antibodies had waned.

Physiological stress is a known factor in reactivating *Salmonella* shedding [30,31], however it is possible that carriers of the bacteria can resume shedding intermittently without any physiological stress. Within the base model, it has been assumed that animals resume shedding infrequently, which is thought to be a fair assumption, assuming best management practices and high animal welfare standards. However, changes in the rate of re-infection were shown to have a large impact on the slaughter age prevalence. Consequently, the impact of carrier pigs in the *Salmonella* transmission process may be high. As such, a future direction of this research could be to analyse the effect of carriers on the dynamics. For example, testing different proportions of infectious and carrier animals entering the unit could change the dynamics.

Both models (slatted and solid units) exhibited similar behaviour with regard to *Salmonella* prevalence at slaughter age with a number of scenarios, such as changes in the shedding rate and the probability of infection. There were nevertheless implications with regard to the application of an intervention. With the accelerated uptake of infection within the solid unit, the time at which an intervention should be applied in order to be as effective as possible may need to be during the initial uptake of infection. However, this would require further investigation.

Some issues may arise however with implementing certain interventions. For example, biosecurity practices and standards vary across the country (and elsewhere), thus biosecurity interventions will be associated with biosecurity policies within each location. Furthermore, it is possible that regional or national level actions could impact the infection of post-weaning pigs. Although the interventions described in this paper focus on pre-harvest measures to reduce *Salmonella* burden, there are other procedures that could be utilised. For example, there may be added value in developing a surveillance programme that could identify positive breeder units from negative units. This in turn could be an effective way of sourcing piglets from negative units and, in the long run, be more cost effective.

In conclusion, our study found the probability of infection after *Salmonella* exposure and the number of bacteria shed by an infectious animal to be key drivers of *Salmonella* transmission. These results should help inform the future direction of research regarding *Salmonella* transmission in pigs. Further research is required in order to identify possible measures that could be quickly and efficiently implemented to control these factors. The development of effective vaccinations or improved biosecurity measures (such as improved management practices and efficient control of sick animals) for example, may have the desired effect. Minimising physiological stress can also result in a reduction in on farm *Salmonella* prevalence. An intervention focusing on this aspect, by effective management and biosecurity practices for example, could be more realistic and easier to implement in the short term, and still provide worthwhile results.

## Acknowledgments

The authors would like to thank the two anonymous reviewers and Iain McKendrick for their comments, which have improved the presentation of this paper.

## Author Contributions

Conceived and designed the experiments: DC HC DA RC. Performed the experiments: ADCB DC. Analyzed the data: ADCB DC. Contributed reagents/materials/analysis tools: DA. Wrote the paper: ADCB DC RC.

## References

1. Popoff MY, Bockemühl J, Gheesling LL (2004) Supplement 2002 (no. 46) to the Kauffmann–White scheme. Research in Microbiology 155: 568–570.
2. Davies R, Dalziel R, Gibbens J, Wilesmith J, Ryan J, et al. (2004) National survey for *Salmonella* in pigs, cattle and sheep at slaughter in Great Britain (1999–2000). Journal of Applied Microbiology 96: 750–760.
3. VLA (2007). *Salmonella* in livestock production in GB: 2007 report.
4. DEFRA (2006) Zoonoses: Abattoir survey of foodborne pathogens in cattle, sheep & pigs in great britain 2003. Available http://www.defra.gov.uk/animalh/diseases/zoonoses/abattoir-survey.htm, Part 3 presentations. Accessed 19 June 2009.
5. Lo Fo Wong DMA, Hald T, van der Wolf PJ, Swanenburg M (2002) Epidemiology and control measures for *Salmonella* in pigs and pork. Livestock Production Science 76: 215–222.
6. Swanenburg M, Urlings HAP, Snijders JMA, Keuzenkamp DA, van Knapen F (2001) *Salmonella* in slaughter pigs: prevalence, serotypes and critical control points during slaughter in two slaughterhouses. International Journal of Food Microbiology 70: 243–254.
7. Bolton DJ, Pearce R, Sheridan JJ, McDowell DA, Blair IS (2003) Decontamination of pork carcasses during scalding and the prevention of *Salmonella* cross-contamination. Journal of Applied Microbiology 94: 1036–1042.
8. Pires SM, de Knegt L, Hald T (2011) Scientific/Technical Report submitted to EFSA Estimation of the relative contribution of different food and animal sources to human *Salmonella* infections in the European Union.
9. DEFRA (2011) Zoonoses Report: UK 2009. Available: https://www.gov.uk/government/uploads/system/uploads/attachment_data/file/69313/pb13571-zoonoses2009-110125.pdf. Accessed 6 June 2013.
10. Anonymous (2012) Annual Report on Zoonoses in Denmark 2011, National Food Institute, Technical University of Denmark.
11. Berriman ADC, Clancy D, Clough HE, Christley RM (2013) Semi-stochastic models for *Salmonella* infection within finishing pig units in the UK. In press.
12. Hethcote H (2000) The mathematics of infectious diseases. SIAM review 42: 599–653.
13. DEFRA (2008) Biosecurity guidance to prevent the spread of animal diseases. Available: http://archive.defra.gov.uk/foodfarm/farmanimal/diseases/documents/biosecurity_guidance.pdf. Accessed 6 June 2013.
14. Kranker S, Alban L, Boes J, Dahl J (2003) Longitudinal study of *Salmonella enterica* serotype Typhimurium infection in three Danish farrow-to-finish swine herds. Journal of Clinical Microbiology 41: 2282.
15. Diekmann O, Heesterbeek J (2000) Mathematical epidemiology of infectious diseases: Model building, analysis and interpretation. New York: Wiley. 320 p.
16. Heffernan J, Smith R, Wahl L (2005) Perspectives on the basic reproductive ratio. Journal of the Royal Society Interface 2: 281–293.
17. Roberts M, Heesterbeek J (2003) Mathematical models in epidemiology. Encyclopedia of Life Support Systems (EOLSS).
18. Gray JT, Fedorka-Cray PJ, Stabel TJ, Kramer TT (1996) Natural transmission of *Salmonella* choleraesuis in swine. Applied & Environmental Microbiology 62: 141.
19. Osterberg J, Wallgren P (2008) Effects of a challenge dose of *Salmonella* Typhimurium or *Salmonella* Yoruba on the patterns of excretion and antibody responses of pigs. Veterinary Record 162: 580.
20. Gray JT, Fedorka-Cray PJ, Stabel TJ, Ackermann MR (1995) Influence of inoculation route on the carrier state of *Salmonella* choleraesuis in swine. Veterinary Microbiology 47: 43–59.
21. Morgan IR, Krautil FL, Craven JA (1987) Effect of time in lairage on caecal and carcass *Salmonella* contamination of slaughter pigs. Epidemiology & Infection: 323–330.
22. Hill AA, Snary EL, Arnold ME, Alban L, Cook AJC (2007) Dynamics of *Salmonella* transmission on a British pig grower-finisher farm: a stochastic model. Epidemiology & Infection 136: 320–333.
23. Ivanek R, Snary EL, Cook AJC, Groehn YT (2004) A mathematical model for the transmission of *Salmonella* Typhimurium within a grower-finisher pig herd in Great Britain. Journal of Food Protection 67: 2403–2409.
24. Lurette A, Belloc C, Touzeau S, Hoch T, Ezanno P, et al. (2008) Modelling *Salmonella* spread within a farrow-to-finish pig herd. Veterinary Research 39: 49.
25. Soumpasis I, Butler F (2009) Development and application of a stochastic epidemic model for the transmission of *Salmonella* Typhimurium at the farm level of the pork production chain. Risk Analysis 29: 1521–1533.
26. British Pig Executive (2007) Pig Yearbook. Warwickshire, UK: MLC Economics.
27. AHVLA (2011) Interim report: Epidemiological studies of Multiresistant *Salmonella* Typhimurium in pigs.
28. Erdman MM, Harris IT, Torremorell M, Wilt VM, Hank Harris DL (2005) Occurrence of *Salmonella* serotype Typhimurium DT104 on a commercial swine farm before, during, and after depopulation and repopulation. Journal of the American Veterinary Medical Association 227: 460–466.
29. AHVLA Unpublished report.
30. Verbrugghe E, Boyen F, Van Parys A, Van Deun K, Croubels S, et al. (2011) Stress induced *Salmonella* Typhimurium recrudescence in pigs coincides with cortisol induced increased intracellular proliferation in macrophages. Veterinary Research 42.
31. Williams LP Jr, Newell KW (1970) *Salmonella* excretion in joy-riding pigs. American Journal of Public Health 60: 926–929.
32. Letellier A, Messier S, Lessard L, Quessy S (2000) Assessment of various treatments to reduce carriage of *Salmonella* in swine. The Canadian Journal of Veterinary Research 64: 27–31.
33. Blanchard P, Kjeldsen K (2003) Time to hit *Salmonella* in pigs. The Pig Journal 52: 182–194.

34. Gutzmann F, Layton H, Simkins K, Jarolmen H (1976) Influence of antibiotic-supplemented feed on occurrence and persistence of *Salmonella* typhimurium in experimentally infected swine. American Journal of Veterinary Research 37: 649.
35. Williams RD, Rollins LD, Pocurull DW, Selwyn M, Mercer HD (1978) Effect of feeding Chlortetracycline on the reservoir of *Salmonella* Typhimurium in experimentally infected swine. Antimicrobial Agents & Chemotherapy 14: 710–719.
36. Selke M, Meens J, Springer S, Frank R, Gerlach GF (2007) Immunization of pigs to prevent disease in Humans: Construction and protective efficacy of a *Salmonella enterica* serovar Typhimurium live negative-marker vaccine. Infection & Immunity 75: 2476–2483.
37. Potter A, Gerdts V, van Drunen Littel-van den Hurk S (2008) Veterinary vaccines: alternatives to antibiotics? Animal Health Research Reviews 9: 187–199.
38. Rostagno MH (2011) Vaccination to reduce *Salmonella* prevalence in pigs. Veterinary Record 169: 551–552.
39. Denagamage TN, O'Connor AM, Sargeant JM, Raji'c A, McKean JD (2007) Efficacy of vaccination to reduce *Salmonella* prevalence in live and slaughtered swine: A systematic review of literature from 1979 to 2007. Foodborne Pathogens & Disease 4: 539–549.
40. Schwarz P, Kich JD, Kolb J, Cardoso M (2011) Use of an avirulent live *Salmonella* Choleraesuis vaccine to reduce the prevalence of *Salmonella* carrier pigs at slaughter. Veterinary Record 169.
41. MLC (2006) British Pig Project: A housing blueprint for the British pig industry. Warwickshire, UK: MLC.
42. Jensen A, Baggesen D (2005) *Salmonella* infection risk associated with outdoor organic pig production. p. 8793.
43. Leek A, Callan J, Henry R, O'Doherty J (2005) The Application of Low Crude Protein Wheat-Soyabean Diets to Growing and Finishing Pigs: 2. The Effects on Nutrient Digestibility, Nitrogen Excretion, Faecal Volatile Fatty Acid Concentration and Ammonia Emission from Boars. Irish Journal of Agricultural & Food Research 44: 247–260.
44. EFSA (2010) Quantitative Microbiological Risk Assessment on *Salmonella* in Slaughter and Breeder pigs: Final Report.
45. Gray J, Fedorka-Cray P (2001) Survival and infectivity of *Salmonella* choleraesuis in swine feces. Journal of Food Protection 64: 945–949.
46. Gray J, Stabel T, Fedorka-Cray P (1996) Effect of dose on the immune response and persistence of *Salmonella* choleraesuis infection in swine. American journal of veterinary research 57: 313.

# Employing a Monte Carlo Algorithm in Newton-Type Methods for Restricted Maximum Likelihood Estimation of Genetic Parameters

**Kaarina Matilainen[1]*, Esa A. Mäntysaari[1], Martin H. Lidauer[1], Ismo Strandén[1], Robin Thompson[2]**

1 Biotechnology and Food Research, MTT Agrifood Research Finland, Jokioinen, Finland, 2 Biomathematics and Bioinformatics Department, Rothamsted Research, Harpenden, United Kingdom

## Abstract

Estimation of variance components by Monte Carlo (MC) expectation maximization (EM) restricted maximum likelihood (REML) is computationally efficient for large data sets and complex linear mixed effects models. However, efficiency may be lost due to the need for a large number of iterations of the EM algorithm. To decrease the computing time we explored the use of faster converging Newton-type algorithms within MC REML implementations. The implemented algorithms were: MC Newton-Raphson (NR), where the information matrix was generated via sampling; MC average information(AI), where the information was computed as an average of observed and expected information; and MC Broyden's method, where the zero of the gradient was searched using a quasi-Newton-type algorithm. Performance of these algorithms was evaluated using simulated data. The final estimates were in good agreement with corresponding analytical ones. MC NR REML and MC AI REML enhanced convergence compared to MC EM REML and gave standard errors for the estimates as a by-product. MC NR REML required a larger number of MC samples, while each MC AI REML iteration demanded extra solving of mixed model equations by the number of parameters to be estimated. MC Broyden's method required the largest number of MC samples with our small data and did not give standard errors for the parameters directly. We studied the performance of three different convergence criteria for the MC AI REML algorithm. Our results indicate the importance of defining a suitable convergence criterion and critical value in order to obtain an efficient Newton-type method utilizing a MC algorithm. Overall, use of a MC algorithm with Newton-type methods proved feasible and the results encourage testing of these methods with different kinds of large-scale problem settings.

**Editor:** Rongling Wu, Pennsylvania State University, United States of America

**Funding:** The study was funded by Rothamsted Research (URL: www.rothamsted.ac.uk) and MTT Agrifood Research Finland (URL: www.mtt.fi). The funders had no role in study design, data collection and analysis, decision to publish, or preparation of the manuscript.

**Competing Interests:** The authors have declared that no competing interests exist.

* E-mail: kaarina.matilainen@mtt.fi

## Introduction

Estimation of variance components (VC) by restricted maximum likelihood (REML) [1] via a Monte Carlo (MC) expectation maximization (EM) algorithm has proven a computationally attractive choice for large data sets and complex linear mixed effects models [2–4]. In such cases, it is often impossible to calculate the exact inverse of the coefficient matrix using direct methods, but it can be estimated by MC sampling methods instead. Although the idea of MC EM REML is simple, its convergence is slow, like typical for the EM algorithm. There are different ways to enhance the convergence. One possibility is to use observed information obtained by Louis' method [5], which also gives standard errors for the estimates. The MC technique can be adapted to Louis's method as well [6]. Other possibilities include Aitken's acceleration and quasi-Newton EM acceleration, as used in [7] and discussed, e.g., in [8]. However, both Louis' method and the acceleration methods require complicated calculations which may be difficult with the large-scale problems often occurring in animal breeding evaluations.

Newton-type methods are based on second derivatives and reach fast convergence in the neighbourhood of the maximum. Second derivatives with respect to all the parameters yield the information matrix, which can be used to calculate standard errors for the parameters. The Newton-Raphson (NR) method is based on the observed information matrix while Fisher's scoring uses the expected information matrix. Other Newton-type methods include average information (AI) REML, which utilizes the average of the observed and expected information matrices [9]. This is currently the most common VC estimation method used in animal breeding. Quasi-Newton methods [10], which rely on approximation of second derivatives based on the direction of the most recent step, have also been suggested and used, e.g [11]. These methods usually result in faster convergence compared to linear methods but slower convergence compared to Newton-type methods because the information matrix is replaced by an approximation.

MC techniques are useful for analyses involving complex likelihoods. Thus, the MC method has been used in the NR algorithm for generalized linear mixed effects models, e.g., by Kuk and Cheng [12], and for incomplete data, e.g., by Gauderman and Navidi [13]. More complicated models related to these examples require simulations from the conditional distribution of the missing data given the observed data with methods like Gibbs sampling.

However, the problem in animal breeding is not necessarily the complexity of the model, but rather the need to analyze large-scale data sets to obtain sufficiently accurate genetic parameter estimates. In such cases, the simple sampling method presented in García-Corts et al. [14] has shown to be practical for VC estimation in linear mixed effects models by MC EM REML [4]. Its use is also possible in Newton-type methods.

The aim of this study is to compare MC algorithms in different Newton-type methods for VC estimation of linear mixed effects models. We first introduce the AI REML and Broyden's method with MC. These methods are then compared with a sampling-based NR method (MC NR REML), where a simple approximation of second derivatives is possible from independent and identically distributed samples. Finally, we evaluate the performance of these Newton-type methods using the MC algorithm in an analysis of simulated example data.

## Materials and Methods

### Model

Consider a bivariate linear mixed effects model

$$\mathbf{y}=\mathbf{Xb}+\mathbf{Zu}+\mathbf{e}, \tag{1}$$

where $\mathbf{y}$ is a vector of observations, $\mathbf{b}$ is a vector of fixed effects, $\mathbf{u}$ is a vector of random effects, $\mathbf{e}$ is a vector of random error terms or residuals, and $\mathbf{X}$ and $\mathbf{Z}$ are design matrices for fixed and random effects, respectively. Assume that $\mathbf{u}\sim\mathcal{N}(0,\mathbf{G})$ has a covariance structure $\mathbf{G}=\mathbf{A}\otimes\mathbf{G}_0$, where $\mathbf{A}$ is a numerator relationship matrix and $\mathbf{G}_0$ is a $2\times2$ covariance matrix. Similarly, $\mathbf{e}\sim\mathcal{N}(0,\mathbf{R})$, where $\mathbf{R}=\mathbf{I}\otimes\mathbf{R}_0$ and $\mathbf{R}_0$ is a $2\times2$ covariance matrix. Thus, $\mathbf{y}\sim\mathcal{N}(\mathbf{Xb},\mathbf{V})$, where $\mathbf{V}=\mathbf{ZGZ}'+\mathbf{R}$. We assume that either both or no traits are observed.

### Methods

Let the parameter vector of covariances be $\theta=[\theta_G' \quad \theta_R']'$, where $\theta_G=vech(\mathbf{G}_0)$, $\theta_R=vech(\mathbf{R}_0)$ and *vech* is an operator changing unique elements of the matrix argument into vector form. In our case, $\theta$ is a $6\times1$ vector which contains three unique elements from both the random effect and residual covariance matrices. Newton-type methods rely on first and second derivatives of the REML likelihood function $L(\theta)$ with respect to $\theta$. For example, the NR algorithm uses the observed information matrix

$$\mathbf{H}(\theta)=\left[-\frac{\partial^2\log L(\theta)}{\partial\theta_i\partial\theta_j}\right]_{i,j=1,\ldots,6} \tag{2}$$

and the gradient vector

$$\mathbf{J}(\theta)=\left[\frac{\partial\log L(\theta)}{\partial\theta_i}\right]_{i=1,\ldots,6} \tag{3}$$

in calculating new estimates of parameters $\hat{\mathbf{q}}$ at iteration round $k$:

$$\hat{\theta}^{(k)}=\hat{\theta}^{(k-1)}-\mathbf{H}(\theta)^{-1}\mathbf{J}(\theta),$$

where information matrix $\mathbf{H}(\theta)$ and gradient vector $\mathbf{J}(\theta)$ are computed at current VC estimate $\hat{\theta}^{(k-1)}$.

First derivatives of the REML log-likelihood $\log L(\theta)$ with respect to elements in $\mathbf{G}_0$ or $\mathbf{R}_0$ can be considered simultaneously [13]. Thus, $\partial\log L(\theta)/\partial\mathbf{G}_0$ can be written as a $2\times2$ matrix which has diagonal elements $\partial\log L(\theta)/\partial\mathbf{G}_{0_{1,1}}$ and $\partial\log L(\theta)/\partial\mathbf{G}_{0_{2,2}}$ and off-diagonal elements $\frac{1}{2}\partial\log L(\theta)/\partial\mathbf{G}_{0_{1,2}}$ and $\frac{1}{2}\partial\log L(\theta)/\partial\mathbf{G}_{0_{2,1}}$:

$$\frac{\partial\log L(\theta)}{\partial\mathbf{G}_0}=-\frac{1}{2}\{q\mathbf{G}_0^{-1}-\mathbf{G}_0^{-1}(\mathbf{S}_G+\mathbf{D}_G)\mathbf{G}_0^{-1}\},$$

where $q$ is the number of levels in random effect $\mathbf{u}$, and $\mathbf{S}_G$ and $\mathbf{D}_G$ are $2\times2$ matrices with elements $\mathbf{S}_{G_{i,j}}=\mathrm{tr}(\mathbf{A}^{-1}\mathbf{C}^{\mathbf{u}_i\mathbf{u}_j})$ and $\mathbf{D}_{G_{i,j}}=\mathbf{u}_i'\mathbf{A}^{-1}\mathbf{u}_j$, respectively. Here $\mathbf{u}_i$ is a subvector of $\mathbf{u}$ corresponding to the $i^{th}$ trait in the model, and $\mathbf{C}^{\mathbf{u}_i\mathbf{u}_j}$ is the part of the inverse of the coefficient matrix of the mixed model equations (MME) corresponding to $\mathbf{u}_i$ and $\mathbf{u}_j$, $i,j=1,2$. Similarly,

$$\frac{\partial\log L(\theta)}{\partial\mathbf{R}_0}=-\frac{1}{2}\{n\mathbf{R}_0^{-1}-\mathbf{R}_0^{-1}(\mathbf{S}_R+\mathbf{D}_R)\mathbf{R}_0^{-1}\},$$

where $n$ is the number of observations, and $\mathbf{S}_R$ and $\mathbf{D}_R$ are $2\times2$ matrices with elements $\mathbf{S}_{R_{i,j}}=\mathrm{tr}(\mathbf{W}_i\mathbf{C}^{ij}\mathbf{W}_j')$ and $\mathbf{D}_{R_{i,j}}=\mathbf{e}_i'\mathbf{e}_j$. Now $\mathbf{W}_i$ is a submatrix of $\mathbf{W}=[\mathbf{X} \quad \mathbf{Z}]$ and $\mathbf{e}_i$ is a subvector of $\mathbf{e}$ corresponding to the $i^{th}$ trait, and $\mathbf{C}^{ij}$ is the part of the inverse of the coefficient matrix of MME corresponding to traits $i$ and $j$.

Matrices $\mathbf{S}_R$ and $\mathbf{S}_G$ are difficult to compute for large data sets and complex models because they require elements of $\mathbf{C}$, the inverse of the coefficient matrix of MME. These matrices can be approximated by simulating $s$ MC samples of data, i.e., $\tilde{\mathbf{y}}^h=\mathbf{Z}\tilde{\mathbf{u}}^h+\tilde{\mathbf{e}}^h$, $h=1,\ldots,s$ [14] where $\tilde{\mathbf{y}}^h$ is a vector of MC simulated observations at MC sample $h$, and $\tilde{\mathbf{u}}^h$ and $\tilde{\mathbf{e}}^h$ are simulated from their assumed normal density models using current values of the variance parameters. When the full model (1) is fitted, i.e., when MME are solved using the simulated data to obtain estimates $\hat{\tilde{\mathbf{u}}}$, element (i,j) in $\mathbf{S}_G$ can be approximated by method 1 or 2 in García-Corts et al. [16]:

$$\mathbf{S}_{G_{i,j}}^{GC1}=q\mathbf{G}_{0_{i,j}}-\frac{1}{s}\sum_{h=1}^{s}(\hat{\tilde{\mathbf{u}}}^h)_i'\mathbf{A}^{-1}\hat{\tilde{\mathbf{u}}}_j^h \tag{4}$$

or

$$\mathbf{S}_{G_{i,j}}^{GC2}=\frac{1}{s}\sum_{h=1}^{s}(\tilde{\mathbf{u}}^h-\hat{\tilde{\mathbf{u}}}^h)_i'\mathbf{A}^{-1}(\tilde{\mathbf{u}}^h-\hat{\tilde{\mathbf{u}}}^h)_j, \tag{5}$$

respectively. These formulas are also convenient for multivariate models, as shown in Matilainen et al. [4]. An increase in MC sample size $s$ will give more accurate estimates of the $\mathbf{C}$ and $\mathbf{C}^{\mathbf{uu}}$ matrices and, subsequently, more accurate estimates of the gradient.

Elements in the observed information matrix (2) require more complex calculations than those needed to calculate the gradient vector (3). Approximations are used here to avoid calculations of exact second derivatives. In the following section we present three methods applying the MC sampling scheme. The first method, which is named MC NR REML, is based on calculation of the observed information matrix by sampling. The second method, named MC AI REML, uses MC sampling in the AI REML algorithm. Finally, the third MC sampling method, named MC BM REML, is based on Broyden's method.

MC NR REML. By definition, the expected information matrix at convergence is

$$\mathrm{E}(\mathbf{H}(\theta))=\mathrm{E}(\mathbf{J}(\theta)\mathbf{J}(\theta)')=\mathrm{Var}(\mathbf{J}(\theta)).$$

Use of the MC algorithm with independent and identically distributed samples enables approximation of the information matrix by the variances of the gradients over the samples within each NR REML round. Note, however, that (4) needs to be used to compute the sampling variance of the gradients, because (5) only gives the variances of prediction error variances. Now, the information matrix is approximated by

$$\mathbf{H}(\theta) \approx \mathrm{Cov}\left(\left[\mathbf{J}^1(\theta) \quad \cdots \quad \mathbf{J}^s(\theta)\right]\right),$$

where

$$\mathbf{J}^h(\theta) = \frac{1}{2}\begin{bmatrix} \left(\mathbf{G}_0^{-1}\left[\hat{\mathbf{u}}_1^h \quad \hat{\mathbf{u}}_2^h\right]\mathbf{A}^{-1}\left[\hat{\mathbf{u}}_1^h \quad \hat{\mathbf{u}}_2^h\right]\mathbf{G}_0^{-1}\right)_{1,1} \\ 2\left(\mathbf{G}_0^{-1}\left[\hat{\mathbf{u}}_1^h \quad \hat{\mathbf{u}}_2^h\right]\mathbf{A}^{-1}\left[\hat{\mathbf{u}}_1^h \quad \hat{\mathbf{u}}_2^h\right]\mathbf{G}_0^{-1}\right)_{1,2} \\ \left(\mathbf{G}_0^{-1}\left[\hat{\mathbf{u}}_1^h \quad \hat{\mathbf{u}}_2^h\right]\mathbf{A}^{-1}\left[\hat{\mathbf{u}}_1^h \quad \hat{\mathbf{u}}_2^h\right]\mathbf{G}_0^{-1}\right)_{2,2} \\ \left(\mathbf{R}_0^{-1}\left[\hat{\mathbf{e}}_1^h \quad \hat{\mathbf{e}}_2^h\right]\left[\hat{\mathbf{e}}_1^h \quad \hat{\mathbf{e}}_2^h\right]\mathbf{R}_0^{-1}\right)_{1,1} \\ 2\left(\mathbf{R}_0^{-1}\left[\hat{\mathbf{e}}_1^h \quad \hat{\mathbf{e}}_2^h\right]\left[\hat{\mathbf{e}}_1^h \quad \hat{\mathbf{e}}_2^h\right]\mathbf{R}_0^{-1}\right)_{1,2} \\ \left(\mathbf{R}_0^{-1}\left[\hat{\mathbf{e}}_1^h \quad \hat{\mathbf{e}}_2^h\right]\left[\hat{\mathbf{e}}_1^h \quad \hat{\mathbf{e}}_2^h\right]\mathbf{R}_0^{-1}\right)_{2,2} \end{bmatrix}$$

is a gradient vector calculated based on sample $h$, $h = 1, \ldots, s$. For a $s \times 6$ matrix $\mathbf{J}$, $\mathrm{Cov}(\mathbf{J})$ returns a $6 \times 6$ matrix, where the diagonal has variance within each column in $\mathbf{J}$, and the off-diagonals contain the covariances between each two-column combinations in $\mathbf{J}$.

MC AI REML. Johnson and Thompson [17] and Gilmour et al. [9] presented AI REML noting that computation of the average of the observed and expected information matrices is easier than of either of the components:

$$\mathbf{H}(\theta) \approx \frac{1}{2}\mathbf{y}'\mathbf{P}\frac{\partial \mathbf{V}}{\partial \theta_i}\mathbf{P}\frac{\partial \mathbf{V}}{\partial \theta_j}\mathbf{P}\mathbf{y},$$

where

$$\begin{aligned} \mathrm{P} &= \mathrm{V}^{-1} - \mathrm{V}^{-1}\mathrm{X}(\mathrm{X}'\mathrm{V}^{-1}\mathrm{X})^{-1}\mathrm{X}'\mathrm{V}^{-1} \\ &= \mathrm{R}^{-1} - \mathrm{R}^{-1}\mathrm{WCW}'\mathrm{R}^{-1}. \end{aligned}$$

Define $\mathbf{F} = [\mathbf{f}_1 \quad \ldots \quad \mathbf{f}_6]$, where $\mathbf{f}_{i=1,\ldots,6}$ is

$$\begin{aligned} \mathrm{f}_i &= \frac{\partial \mathrm{V}}{\partial \theta_i}\mathrm{Py} \\ &= \mathrm{Z}\frac{\partial \mathrm{G}}{\partial \theta_i}\mathrm{G}^{-1}\hat{\mathrm{u}} + \frac{\partial \mathrm{R}}{\partial \theta_i}\mathrm{R}^{-1}\hat{\mathrm{e}} \\ &= \mathrm{Z}\left(\frac{\partial \mathrm{G}_0}{\partial \theta_i} \otimes \mathrm{A}\right)\mathrm{G}^{-1}\hat{\mathrm{u}} + \left(\frac{\partial \mathrm{R}_0}{\partial \theta_i} \otimes \mathrm{I}\right)\mathrm{R}^{-1}\hat{\mathrm{e}} \end{aligned}$$

Then

$$\begin{aligned} \mathbf{y}'\mathbf{P}\frac{\partial \mathbf{V}}{\partial \theta_i}\mathbf{P}\frac{\partial \mathbf{V}}{\partial \theta_j}\mathbf{P}\mathbf{y} &= \mathbf{F}'\mathbf{P}\mathbf{F} \\ &= \mathbf{F}'\mathbf{R}^{-1}\mathbf{F} - (\mathbf{C}\mathbf{W}'\mathbf{R}^{-1}\mathbf{F})'\mathbf{W}'\mathbf{R}^{-1}\mathbf{F} \\ &= \mathbf{F}'\mathbf{R}^{-1}\mathbf{F} - \mathbf{T}'\mathbf{W}'\mathbf{R}^{-1}\mathbf{F}, \end{aligned}$$

where

$$\mathbf{T} = \mathbf{C}\mathbf{W}'\mathbf{R}^{-1}\mathbf{F}.$$

Hence, in MC AI REML, MC sampling is needed only for estimation of first derivatives in $\mathbf{F}$, while the average information can be calculated based on current VC estimates. However, the algorithm requires additional computations to form $\mathbf{F}$, which necessitates solving $\mathbf{T}$ from the MME with data replaced by $\mathbf{f}_i, i = 1, \ldots, 6$. Thus, MME needs to be solved for each VC parameter.

MC BM REML. Broyden's method is a quasi-Newton method for numerical solution of non-linear equations [18]. It is a generalization of the secant method to multiple dimensions. Broyden's method updates the inverse of the information matrix (instead of the information matrix itself) within each round:

$$\mathbf{H}(\theta)^{-1} \approx \mathbf{B}^{(k)} = \mathbf{B}^{(k-1)} + \frac{(\Delta\theta - \mathbf{B}^{(k-1)}\Delta\mathbf{J})\Delta\theta'\mathbf{B}^{(k-1)}}{\Delta\theta'\mathbf{B}^{(k-1)}\Delta\mathbf{J}},$$

where

$$\Delta\theta = \theta^{(k-1)} - \theta^{(k-2)}$$

$$\Delta\mathbf{J} = \mathbf{J}(\theta^{(k-1)}) - \mathbf{J}(\theta^{(k-2)}).$$

Instead of using true gradients $\mathbf{J}(\theta)$ for both the update of the inverse information matrix and the update of new estimates, we used the round-to-round changes in the EM estimates [19]:

$$\Delta\mathbf{J} \approx \left(\theta^{(k-1)} - \theta_{EM}^{(k-1)}\right) - \left(\theta^{(k-2)} - \theta_{EM}^{(k-2)}\right),$$

where $\theta_{EM}^{(k-1)}$ is a vector of EM REML solutions computed from the estimates from round $(k-1)$. Apart from scaling, they are relative to the original gradients. In the beginning, $\mathbf{B}^{(0)}$ is identity matrix $\mathbf{I}$. The first update of the inverse of the information matrix is made at round $k = 2$ based on estimates from the first round $k = 1$ and initial values at $k = 0$.

This method leads to superlinear convergence although sequence $\{\mathbf{B}^{(k)}\}$ does not converge to the observed information matrix at the maximum [20]. Like with MC AI REML, MC sampling is not used to estimate the information matrix directly. Instead, the sampling variation comes in through the gradients which are used to update the inverse of the information matrix within each round.

## Analysis of test data

For this study we simulated a small data set which mimics a typical set-up in dairy cattle breeding. The two study traits resembled 305-day milk and fat yield records in 20 herds. The base generation comprised 146 unrelated sires, each of which had 1 to 10 daughters with unknown and unrelated dams. Each daughter had one observation of the two traits, and the data contained 569 observations for both traits. The pedigree comprised a total of 715 animals. Fixed herd effects and random genetic animal effects were included in the study model. Table 1 shows genetic and residual VC in $\mathbf{G}_0$ and $\mathbf{R}_0$ used to simulate 305-day milk and fat records. The simulation of observation was based on the assumed linear mixed effects model and VCs.

All the algorithms used for analyzing the data were implemented in R software [21]. First we applied analytical EM REML, NR

REML, AI REML and BM REML. For these analytical analyses we used convergence criteria based on relative squared changes in consecutive estimates, with $10^{-10}$ as the critical value. Each MC REML algorithm was tested with 20, 100 and 1000 MC samples per REML iteration round. The MC EM REML algorithm with 20 MC samples per REML round was used as a reference [4]. Estimates from round 2 of MC EM REML analysis were set as initial values for the Newton-type analyses (Table 1). For cases where Newton-type algorithms yielded estimates outside the parameter space, crash recovery was implemented by weighting the Newton-type and EM REML estimates with a weighting factor sequentially from 0.1 to 1.0 by 0.1 until the estimated VC matrices were positive-definite [15].

Convergence of an MC algorithm is difficult to identify, and so we examined the convergence performance of MC REML algorithms by continuing an additional 10 REML rounds more than required by corresponding analytical analyses. The obtained mean and relative standard deviations of the parameter estimates over the additional REML rounds are shown in Table 2. Three convergence criteria presented in the literature were then calculated for the MC AI REML algorithm. The first is a commonly used criterion, presented by Booth and Hobert [22], which is based on a change in consecutive parameter estimates relative to their standard errors. A value of 0.005 can be used as the critical value. The second criterion, by Kuk and Cheng [12], relies on the gradient vector and its variance-covariance matrix. Their stopping criterion is 90-percent quantile of a chi-square distribution with the number of parameters as degrees of freedom. This criterion attempts to stop the iteration as soon as possible. Finally, from MC AI REML round 5 onwards, convergence was also checked by a method similar to the one in Matilainen et al. [4], where the approach was to predict the parameter estimates of the next round using linear regression on previous iteration rounds. Here we took the same approach but applied the prediction method to the gradients instead of the estimates. Analyses were continued until the critical value of $10^{-10}$ as a norm for predicted round-to-round change in the gradient was reached.

## Results

Analytical EM REML converged in 401 rounds, analytical NR REML and AI REML in 5 rounds, and analytical BM REML in 11 rounds. Estimates by analytical algorithms differed by less than 3% across algorithms, as seen in Table 1. The mean and relative standard deviation for the MC REML estimates obtained from the additional 10 REML rounds after reaching the convergence point determined by corresponding analytical algorithms are given in Table 2. Due to convergence problems in MC BM REML, only results with 1000 MC samples per REML rounds are reported here(Table 2). Almost all VC estimates were in good agreement with the analytical estimates, their means deviating less than 2.5% from the analytical ones. The exceptions were estimates by MC NR REML with 20 samples and those for genetic effect by MC AI REML with 20 samples. The variability of the estimates can be seen in the relative standard deviations over the last 10 REML rounds. Round-to-round variation in the MC EM REML estimates after assumed convergence was only 0.5% for genetic VC, while MC NR REML and MC AI REML with 100 samples per REML round would still have relative standard errors of 5%–8% and 4%–5%, respectively, in the corresponding estimates.

MC REML round-to-round convergence in the genetic covariance component $\mathbf{G}_{1,2}$ is illustrated in Fig. 1. The straight lines in the figures represent the estimated genetic covariance (solid line) and estimated standard error (dashed lines) by analytical AI REML. Fig. 2 describes the relative absolute difference between estimates obtained by MC AI REML with different numbers of MC samples and the true estimate by analytical AI REML.

The standard error of the genetic covariance estimates were 7996 and 8274 by the analytical NR REML and AI REML algorithms, respectively. Standard errors were not calculated by analytical EM REML and BM REML. When calculated by MC NR REML, standard errors for the genetic covariances at REML round 10 was 11360, 8056 and 8495 with 20, 100 and 1000 MC samples per round, respectively. Corresponding, standard errors by MC AI REML at REML round 10 were 8857, 8185 and 8294 with 20, 100 and 1000 MC samples per REML round. However, it should be noted that these actual numbers of standard errors may vary from round to round due to sampling.

Of the three different convergence criteria studied for the MC AI REML algorithm, the convergence criterion presented by Booth and Hobert [22] gave average values of 0.35, 0.15 and 0.05 with 20, 100 and 1000 MC samples per MC AI REML round, respectively. This indicates the need for a huge increase in MC sample size before the critical value of 0.005 proposed by Booth and Hobert [22] can be reached. Kuk and Cheng [12], in turn, suggest stopping the iteration at MC AI REML rounds 2, 1 and 1 with 20, 100 and 1000 MC samples per round, respectively. Their criterion implies relatively small gradients after 1 or 2 steps which is probably due to large standard errors of the estimates. According to the convergence criterion in Matilainen et al. [4] using a critical value of $10^{-10}$, iteration would stop at MC AI REML rounds 101, 70 and 44 with 20, 100 and 1000 MC samples per round, respectively. Because this criterion may be too strict for practical purposes in MC REML analyses, we also checked stopping at points when the criterion gave values less than $10^{-8}$. This would mean that analyses would be stopped at MC AI REML rounds 28, 27 and 10 with 20, 100 and 1000 MC samples, respectively.

## Discussion

Whereas the MC NR REML method is easy to implement, it may require a large number of MC samples to accurately approximate the variances of first derivatives over samples. MC AI REML, in contrast, works better even with small MC sample sizes, because the AI matrix has no extra sampling noise as it depends only on variance parameters estimated in the previous round. MC AI REML rounds are computationally more demanding than MC NR REML, however, because the MME system needs to be solved

**Table 1.** Variance components used for the simulation, initial values used for the analyses and estimates by analytical EM REML.

| | $\mathbf{G}_{0_{1,1}}$ | $\mathbf{G}_{0_{1,2}}$ | $\mathbf{G}_{0_{2,2}}$ | $\mathbf{R}_{0_{1,1}}$ | $\mathbf{R}_{0_{1,2}}$ | $\mathbf{R}_{0_{2,2}}$ |
|---|---|---|---|---|---|---|
| Simulation value | 500.0 | 14.00 | 0.800 | 750.0 | 29.00 | 1.400 |
| Initial value | 350.3 | 12.18 | 0.599 | 615.8 | 21.34 | 1.061 |
| EM REML | 511.9 | 18.11 | 0.747 | 842.6 | 29.10 | 1.590 |
| NR REML | 512.1 | 18.20 | 0.730 | 842.3 | 29.02 | 1.607 |
| AI REML | 512.1 | 18.20 | 0.730 | 842.3 | 29.02 | 1.607 |
| BM REML | 512.6 | 18.08 | 0.751 | 841.9 | 29.13 | 1.586 |

The model includes three unique genetic ($\mathbf{G}_0$) and three unique residual ($\mathbf{R}_0$) (co)variance components. All values are presented in thousands.

at each MC AI REML round as many times as there are VC parameters to be estimated.

MC BM REML is computationally the least expensive of the considered methods when the number of REML rounds and the number of MC samples are kept the same. To circumvent evaluation of the information matrix, BM REML corrects the approximation of the inverse of information matrix from round to round based on the gradients. While the analytical BM REML worked reasonably well, the small data set in our study required a large MC sample size for the method to work, which indicates its sensitivity to changes in gradients from round to round. Furthermore, MC BM REML is efficient even with a fairly poor approximation to the information matrix, but extra computations are needed for standard errors after convergence has been reached.

The performance of MC NR and MC AI REML was quite similar to analytical NR and AI REML. The only clear difference was that, with small MC sample sizes, estimates by MC NR REML varied more than those by MC AI REML. With 20 MC samples, the relative standard deviations from the last 10 REML rounds by both methods were unacceptably high, although MC AI REML was better. With 100 MC samples per REML round, the standard deviations were acceptable, and estimates by MC NR REML showed approximately as much variation as the estimates of MC AI REML. Thus, the information matrix appears to be quite accurately estimated in this case. With 1000 MC samples, variation in MC NR REML and MC AI REML estimates was almost equal. Genetic covariance estimates by both methods differed on average from the true value by 5% and 2% with 100 and 1000 MC samples, respectively. Interestingly, such variation diminished when MC BM REML was applied (Fig. 1c). Why this did not happen with MC NR or MC AI REML analysis may be because the diagonals in the approximation of the inverse of the information matrix were close to unity throughout the analysis, leading to more like MC EM REML parameter estimates.

For analytical REML analysis, Newton-type REML algorithms provide much faster convergence than EM REML, leading to shorter overall solving times with small data sets. The use of MC in the algorithms speeds up convergence of Newton-type methods, but sampling variation in the estimates increases compared to MC EM REML analysis. This is due to multiplication of the gradients by the inverse of the information matrix, as seen in the increase of MC noise. If each round of iteration in Newton-type methods requires many more samples than MC EM REML, overall solving time will reduced only in case the Newton method can enhance the convergence dramatically. The solving times were not recorded in this study because they would only apply to the model and implementation used. With respect to the total number of times to solve MME along the analysis, results showed that MC EM REML with 20 MC samples and 401 EM REML rounds corresponded to MC NR REML with 100 MC samples and 80 NR REML rounds or MC AI REML with 100 MC samples and 75 AI REML rounds. Thus, the number of times required to solve MME within a REML round is $s$+1 for MC NR REML but $s+1+N_p$ for MC AI REML, where $s$ is the number of MC samples and $N_p$ is the number of VC parameters. Hence, in our example with six parameters, MC NR and MC AI REML clearly outperformed MC EM REML, especially if we consider that the analytically implemented NR REML and AI REML needed 5 REML rounds to reach convergence but EM REML needed 401 rounds.

Obtaining a fast algorithm for REML estimation requires development of a practical convergence criterion for Newton-type methods. Although convergence is the same regardless of MC sample size, MC variation affects the values of the convergence criteria. Further study is therefore needed to define a suitable critical value for genetic evaluations. Identification of a feasible convergence criterion also requires deciding which values to use as the final solutions: the average of estimates over several REML rounds or simply the estimates at the last REML round.

The performance of MC-based algorithms is the better the larger the data to be analyzed. With a large data set, the averages of the gradients for MC AI REML are more accurate also with a smaller MC sample size, which leads to more accurate moves in the EM steps of MC BM REML. Most probably the amount of MC samples needed for sufficiently accurate gradient variances in MC NR REML will also decrease somewhat. As models grow larger and more complex, the efficiency of different methods becomes more difficult to predict. Further experience is especially needed on the behaviour of MC BM REML in VC estimation of complex models. A shortfall with respect to MC AI REML is that the number of times needed to solve MME increases along with increase in the number of estimated VCs. This fact does not change even with a large data set, and so MC NR and MC EM REML may become more efficient than MC AI REML. For instance, in [4], MC EM REML was used to estimate 96 VCs in a model describing daily milk yields of dairy cows. Estimation by MC EM REML with 5 MC samples per REML round required 565 rounds. The same analysis by MC AI REML with 20 MC samples per REML round should converge in less than 25 rounds

**Table 2.** Means (relative standard deviation) of estimates over the last 10 rounds by MC REML.

| Method | $\mathbf{G}_{0_{1,1}}$ | $\mathbf{G}_{0_{1,2}}$ | $\mathbf{G}_{0_{2,2}}$ | $\mathbf{R}_{0_{1,1}}$ | $\mathbf{R}_{0_{1,2}}$ | $\mathbf{R}_{0_{2,2}}$ |
|---|---|---|---|---|---|---|
| EM 20 | 519.3 (0.5%) | 18.30 (0.5%) | 0.752 (0.4%) | 843.3 (1.1%) | 28.98 (1.0%) | 1.578 (1.0%) |
| NR 20 | 446.8 (60.8%) | 15.54 (71.8%) | 0.653 (67.5%) | 877.3 (32.4%) | 30.70 (35.4%) | 1.655 (25.3%) |
| NR 100 | 509.8 (5.4%) | 17.91 (6.4%) | 0.712 (7.7%) | 842.3 (2.6%) | 29.20 (3.4%) | 1.620 (3.3%) |
| NR 1000 | 510.9 (1.6%) | 18.18 (2.1%) | 0.730 (2.5%) | 843.3 (0.8%) | 29.04 (1.0%) | 1.607 (0.9%) |
| AI 20 | 495.5 (7.2%) | 17.44 (8.1%) | 0.689 (8.4%) | 855.3 (3.4%) | 29.57 (4.5%) | 1.632 (4.0%) |
| AI 100 | 513.4 (4.2%) | 18.20 (4.7%) | 0.729 (5.2%) | 839.9 (2.6%) | 28.93 (2.8%) | 1.602 (2.4%) |
| AI 1000 | 513.8 (1.6%) | 18.28 (1.9%) | 0.734 (1.9%) | 840.3 (0.9%) | 28.92 (1.1%) | 1.603 (0.8%) |
| BM 1000 | 502.1 (3.2%) | 17.73 (3.5%) | 0.758 (1.1%) | 852.7 (1.9%) | 29.48 (1.9%) | 1.581 (0.5%) |

The model includes three unique genetic ($\mathbf{G}_0$) and three unique residual ($\mathbf{R}_0$) (co)variance components. Values were calculated over REML rounds 402 to 411 for MC EM REML, 6 to 15 for MC NR and MC AI REML, and 12 to 21 for MC BM REML with 20, 100 or 1000 MC samples. Mean values are presented in thousands.

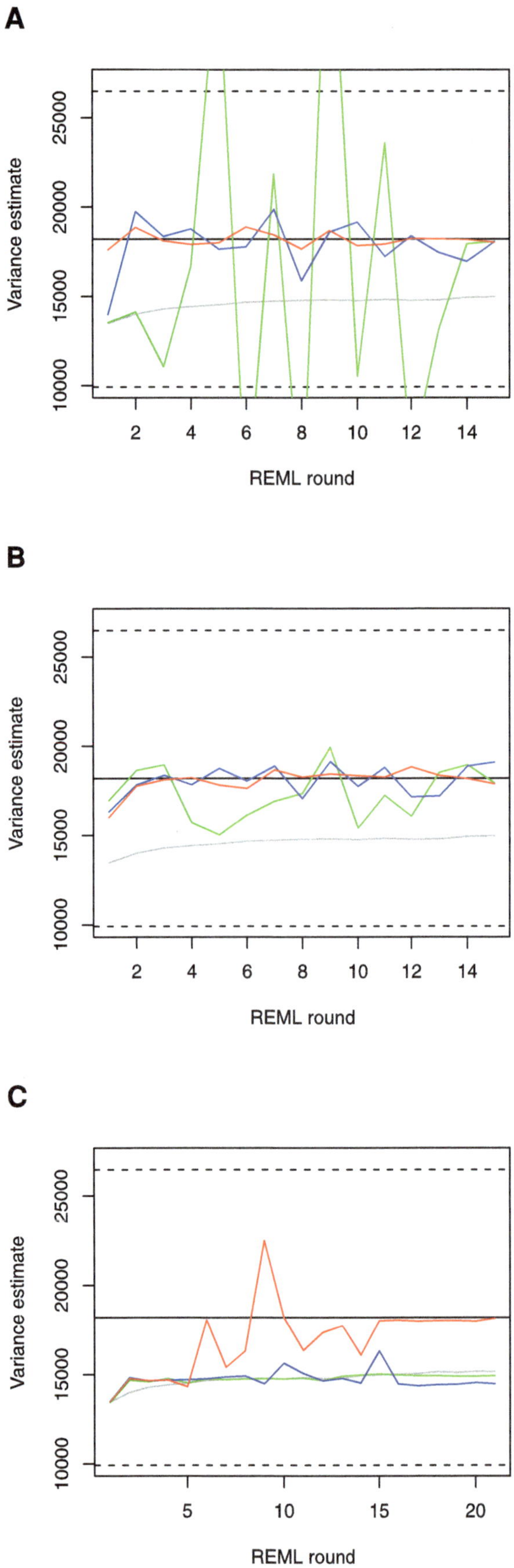


**Figure 1. Estimates of the genetic covariance component by Newton-type methods.** Analyses by MC NR REML (Figure A), MC AI REML (Figure B) and MC BM REML (Figure C) with 20, 100 and 1000 MC samples (green, blue and red line, respectively). MC EM REML with 20 MC samples is plotted as a reference (grey line). The straight lines in the figures are the estimated genetic covariance (solid line) and plus/minus one standard deviation (dashed lines) based on standard errors by analytical AI REML.

to be computationally superior over MC EM REML, given that the MME solving time is the same for both algorithms.

The estimates of the analyses presented here were weighted by corresponding EM REML estimates whenever they fell outside the parameter space. Yet, this does not guarantee convergence to the true solutions, especially with respect to Broyden's method. To avoid divergence, Broyden [18] suggested choosing a scalar multiplier, i.e., a step length that decreases the change in some gradient norm and ensures the ascent of likelihood at each step. Convergence is also guaranteed by the Wolfe conditions [10], which ensure that steps make a sufficient ascent. However, if the search direction in BM REML approximates the Newton direction well enough, the unit step length will satisfy the Wolfe conditions, as the iterates converge to the solution [10]. Based on our study, this may mean that the required MC sample size may become enormous. One way to increase the robustness of VC estimation algorithms is reparametrization of the VC matrices by Cholesky decomposition [11]. The performance of this option is worth considering in future studies.

## Conclusions

Our results show that the use of MC algorithms in different Newton-type methods for VC estimation is feasible, although there was variation in efficiency between the implementations. An efficient MC method can achieve fast convergence and short computing times for VC estimation in complex linear mixed effects models when sampling techniques are used. However,

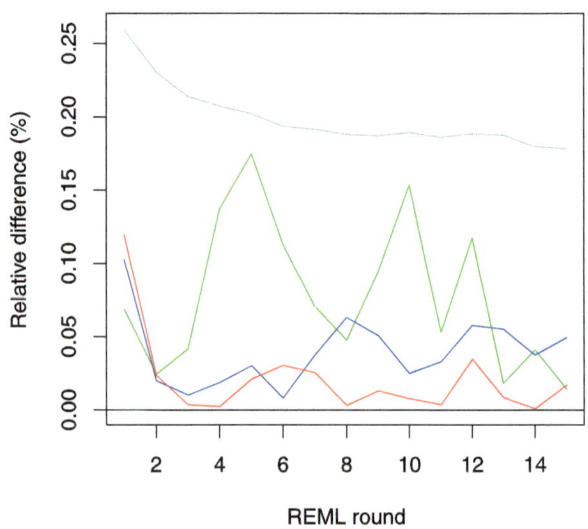


**Figure 2. Relative difference between MC AI REML estimates and the true estimate obtained by analytical AI REML.** The relative difference (%) is plotted for MC AI REML estimates with 20, 100 and 1000 MC samples (green, blue and red line, respectively) along the iteration. MC EM REML with 20 MC samples is plotted as a reference (grey line).

analysis of our small simulated data implies that the number of MC samples needed for accurate estimation is dependent on the used method. This work encourages testing the performance of the presented methods in solving large-scale problems.

## Author Contributions

Conceived and designed the experiments: KM EAM MHL IS RT. Analyzed the data: KM. Wrote the paper: KM EAM MHL IS RT.

## References

1. Patterson HD, Thompson R (1971) Recovery of inter-block information when block sizes are unequal. Biometrika 58: 545–554.
2. García-Cortés LA, Sorensen D (2001) Alternative implementations of Monte Carlo EM algorithms for likelihood inferences. Genet Sel Evol 33: 443–452.
3. Harville DA (2004) Making REML computationally feasible for large data sets: Use of Gibbs sampler. J Stat Comput Simul 74: 135–153.
4. Matilainen K, Mäntysaari EA, Lidauer MH, Strandén I, Thompson R (2012) Employing a Monte Carlo algorithm in expectation maximization restricted maximum likelihood estimation of the linear mixed effects model. J Anim Breed Genet 129: 457–468.
5. Louis TA (1982) Finding the observed information matrix when using the EM algorithm. J R Stat Soc Series B Stat Methodol 44: 226–233.
6. Wei G, Tanner M (1990) A Monte Carlo implementation of the EM algorithm and the poor man's data augmentation algorithms. J Am Stat Assoc 85: 699–704.
7. Mäntysaari E, van Vleck LD (1989) Restricted maximum likelihood estimates of variance components from multitrait sire models with large number of fixed effects. Journal of Animal Breeding and Genetics 106: 409–422.
8. Jamshidian M, Jennrich RI (1997) Acceleration of the EM algorithm by using quasi-Newton methods. J R Stat Soc Series B Stat Methodol 59: 569–587.
9. Gilmour AR, Thompson R, Cullis BR (1995) Average information REML: An efficient algorithm for variance parameter estimation in linear mixed models. Biometrics 51: 1440–1450.
10. Nocedal J, Wright SJ (2006) Numerical Optimization. New York: Springer, 2nd edition. 664 p.
11. Groeneveld E (1994) A reparametrization to improve numerical optimization in multivariate REML (co)variance component estimation. Genet Sel Evol 26: 537–545.
12. Kuk AYC, Cheng YW (1997) The Monte Carlo Newton-Raphson algorithm. J Stat Comput Simul 59: 233–250.
13. Gauderman WJ, Navidi W (2001) A Monte Carlo Newton-Raphson procedure for maximizing complex likelihoods on pedigree data. Comput Stat Data Anal 35: 395–415.
14. García-Cortés LA, Moreno C, Varona L, Altarriba J (1992) Variance component estimation by resampling. J Anim Breed Genet 109: 358–363.
15. Jensen J, Mäntysaari EA, Madsen P, Thompson R (1997) Residual maximum likelihood estimation of (co)variance components in multivariate mixed linear models using average information. J Indian Soc Agric Stat 49: 215–236.
16. García-Cortés LA, Moreno C, Varona L, Altarriba J (1995) Estimation of prediction-error variances by resampling. J Anim Breed Genet 112: 176–182.
17. Johnson DL, Thompson R (1995) Restricted maximum likelihood estimation of variance components for univariate animal models using sparse matrix techniques and average information. J Dairy Sci 78: 449–456.
18. Broyden CG (1965) A class of methods for solving nonlinear simultaneous equations. Math Comp 19: 577–593.
19. Jamshidian M, Jennrich RI (1993) Conjugate gradient acceleration of the EM algorithm. J Am Stat Assoc 88: 221–228.
20. Dennis JE, Moré JJ (1974) A characterization of superlinear convergence and its application to quasi-Newton methods. Math Comp 28: 549–560.
21. R Core Team (2012) R: A Language and Environment for Statistical Computing. R Foundation for Statistical Computing, Vienna, Austria. URL http://www.R-project.org/. ISBN 3-900051-07-0.
22. Booth JG, Hobert JP (1999) Maximizing generalized linear mixed model likelihoods with an automated Monte Carlo EM algorithm. J R Stat Soc Series B Stat Methodol 61: 265–285.

# 10

# Community-Based Management: Under What Conditions Do Sámi Pastoralists Manage Pastures Sustainably?

**Vera H. Hausner[1]*, Per Fauchald[2], Johnny-Leo Jernsletten[3]**

**1** Department of Arctic and Marine Biology, University of Tromsø, Tromsø, Norway, **2** Department of Arctic Ecology, Norwegian Institute for Nature Research, FRAM Centre, Tromsø, Norway, **3** Centre for Sámi Studies, Teorifagbygget, Tromsø, Norway

## Abstract

Community-based management (CBM) has been implemented in socio-ecological systems (SES) worldwide. CBM has also been the prevailing policy in Sámi pastoral SES in Norway, but the outcomes tend to vary extensively among resource groups ("siidas"). We asked why do some siidas self-organize to manage common pool resources sustainably and others do not? To answer this question we used a mixed methods approach. First, in the statistical analyses we analyzed the relationship between sustainability indicators and structural variables. We found that small winter pastures that are shared by few siidas were managed more sustainably than larger pastures. Seasonal siida stability, i.e., a low turnover of pastoralists working together throughout the year, and equality among herders, also contributed to more sustainable outcomes. Second, interviews were conducted in the five largest pastures to explain the relationships between the structural variables and sustainability. The pastoralists expressed a high level of agreement with respect to sustainable policies, but reported a low level of trust and cooperation among the siidas. The pastoralists requested siida tenures or clear rules and sanctioning mechanisms by an impartial authority rather than flexible organization or more autonomy for the siidas. The lack of nestedness in self-organization for managing pastures on larger scales, combined with the past economic policies, could explain why CBM is less sustainable on the largest winter pastures. We conclude that the scale mis-match between self-organization and the formal governance is a key condition for sustainability.

**Editor:** Kamal Bawa, University of Massachusetts, United States of America

**Funding:** The "Ecosystem Finnmark" and the "TUNDRA" project, both financed by the Norwegian Research Council, financed this study. The funders had no role in study design, data collection and analysis, decision to publish, or preparation of the manuscript.

**Competing Interests:** The authors have declared that no competing interests exist.

* E-mail: vera.hausner@uit.no

## Introduction

Community-based management (CBM) has been promoted as a sustainable alternative for governing common pool resources, such as pastures, forests, water and fisheries [1–3]. The endorsement of CBM is based on extensive empirical research showing that local users have successfully devised their own systems of rules and sanctions for the sustainable harvesting of resources [3,4]. Over the past two decades, policy reforms related to CBM have also been advocated by international development agencies, the European Community, and the Convention on Biological Diversity. Many of these reforms bear references to the subsidiary principle, which argues that local users have a better understanding of the socio-ecological systems (SES) they manage and greater incentives to manage their own resources sustainably; management authority, therefore, should be transferred to the lowest appropriate level [5,6]. As the experience with the devolution reforms and CBM in terms of delivering sustainable outcomes has been equivocal, the focus has shifted towards analyzing conditions likely to influence the SES [1,6,7]. Ostrom suggested to analyze the factors affecting sustainable SES before crafting sustainable solutions [6]. For CBM, this means analyzing causes that affect the local users' ability to work towards sustainable outcomes collectively.

The conventional approach has been to consider the size of the resource system, the group size and the socioeconomic heterogeneity as the key conditions for the sustainable management of SES [1,8]. In particular, small and homogenous groups of resource users have been expected to act collectively toward common goals, but the empirical evidence supporting this claim is inconclusive [8,9]. To better understand how such variables affect the likelihood of cooperation, including their interaction with other structural factors, there is a need to link these variables to explanatory mechanisms [10]. The literature related to the concept of social capital suggests that, in small communities, dense social networks and frequent interactions, should build mutual trust among resource users over time [11]. Repeated interactions build reputations for being trustworthy, and higher levels of trust make it more likely that individuals build reciprocal relationships and collaborate repeatedly. Other studies claim that trust is insignificant as a cause of cooperation, and that leadership or institutional mechanisms that require, forbid or permit and sanction specific actions result in cooperation despite low levels of trust [12,13]. Such institutions could also be informal and based on social norms, where a bad reputation and the fear of losing social bonds may encourage individuals to cooperate [10,14].

To self-organize for sustainable resource management, resource users must have a minimum degree of autonomy and secure access to the resource [4,8]. However, autonomy and the community's ability to make and enforce its own rules could be undermined by cross-scale linkages beyond the control of the local users [1,7,14].

CBM strategies that were once effective can be challenged by outside forces, resulting in overharvesting and the breakdown of cooperation in local communities [2,3,14]. Furthermore, as the spatial scale of the resource or group size increases, there arises a need for a stronger endogenous organization to enable cooperation [4]. Whereas larger cross-scale cooperation can emerge endogenously among users through their self-organization into multiple nested levels [4], there is often a need for higher levels of governance for the formal recognition of community rules, sanctioning, or conflict resolution among the users [1,5,7].

In this study, we focus on the question posed by Elinor Ostrom: "Why do some resource users self-organize to manage common pool resources sustainably and others do not?" [3,8]. This question has mainly been addressed by small-N studies or game theoretical approaches, which do not permit the testing of the relative importance of different factors in explaining sustainable outcomes. Poteete et al., therefore, call for more engagement in broadly comparative and large-N research to evaluate the generality of different hypotheses [8]. Here, we adopt a mixed methods design that combines statistical modeling and semi-structured interviews to link the relative importance of structural variables with explanatory mechanisms [8,15,16]. Such interdisciplinary approaches to sustainability have also been argued as essential to overcome some of the challenges in the Sámi pastoralist SES [17,18].

We build on a rather unique dataset which allows for comparative quantitative research on the management of reindeer pastures by Sámi pastoralists in Finnmark, Norway. The pastoralists self-organize in groups (siidas) in which the reindeer are herded together [19–21]. Usually the siida is based on kinship and is both a social and working association. The siida institution varies in form and flexibility: some pastoralists herd together all year, whereas other siidas group and regroup during the course of the year depending on seasonal pastures. The sustainability of the Sámi pastoralist SES depends on the management of both summer and winter pastures (Text S1, Figure 1). On the summer pastures one siida usually has the exclusive rights to well-defined pastures, and therefore the pastoralists usually let the reindeer roam freely to exploit the heterogeneous resources. The siida organization on winter pastures is more variable. Specifically, five large winter pastures are managed by many siidas, whereas five other pastures are managed by only a few siidas. During winter, access to grazing resources varies according to snow condition, but in order to avoid inter-mixing with neighboring herds, continuous herding with snowmobiles is often necessary in the areas with many siidas. Sustainable outcomes in such SES are likely to depend on the match or mismatch between the scale of self-organization and the scale of formal governance. It has been argued, on the other hand, that the sharing of large pastures by several groups, i.e., with fluid and overlapping boundaries between herders, adds a level of flexibility to respond to environmental variability and, therefore, is more fit for such SES [22–25]. Flexible herding and livestock mobility have also been emphasized as important for coping with climate variability in the Arctic pastoralist SES [17,26–28]. The outcomes of such flexible organizations depend on the degree of trust and the pastoralists' ability to negotiate and cooperate on a larger scale.

To explore how the organization of herders might affect sustainability, we first examined the relationships between indicators of sustainable outcomes and structural variables including: i) the area of the pastures, ii) the inequality of herders with respect to herd size, and iii) seasonal siida stability, i.e., the turnover of pastoralists working together throughout a year. Secondly, the explanatory mechanisms behind these relationships were explored by field studies on the largest winter pastures. The semi-structured interviews focused on the pastoralists' understanding of the resource situation, trust and institutional arrangements. On the basis of our results, we discuss how the CBM strategies in Norway (see Text S1) that emphasize i) the devolution of rulemaking authority to the lowest level of organization, ii) well-defined boundaries and iii) the creation of clear rules and sanction mechanisms fit the Sámi pastoralist SES.

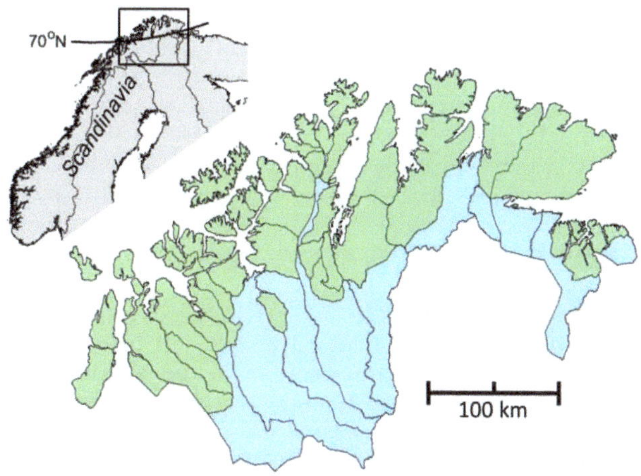


**Figure 1. Study area.** Green areas indicate the delineation of summer pasture areas and blue indicate winter pasture areas in 2003/04. Most of the summer pastures (39 of the 44) are managed by one siida. In winter many of the summer siidas are split into smaller winter siidas that share winter pastures.

## Materials and Methods

### Quantitative Analysis of Structural Variables

Data on individual herding units were mostly derived from the ecological statistics compiled annually by the Norwegian reindeer husbandry administration [29]. We analyzed data from the West and East Finnmark reindeer herding areas (Figure 1) for a ten-year period (1998 to 2007). Because we use the spatial variation to analyze the SES, we removed the noise associated with temporal changes by averaging the data for each herding unit over the ten year period (for analyses of time series; see [30]). We removed the smallest units that kept less than 20 reindeer. The final dataset comprised 292 herding units.

In Finnmark, as much as 39 of the 44 summer pastures are each managed by one siida. The remaining five summer pastures are each shared by two or three siidas. We used the delineation of siida pastures to indicate summer pasture size. In contrast, the five largest winter pastures encompass 11–21 siidas, whereas the five smallest winter pastures are managed by one to six siidas each (Figure 1). We related indicators of sustainable outcomes to structural variables in mixed linear regression models. We used three different indicators of sustainability as response variables: i) proportion of calf slaughter ii) calf body mass, and iii) livelihood income. The proportion of calves slaughtered reflects sustainable herding practices [30–32]. Slaughtering calves, instead of waiting until the animals are one year of age or older, reduces the herd size during the winter and therefore the grazing pressure and the losses due to starvation on the winter pastures. Calf body mass is the carcass mass measured at the slaughterhouses. In our study area there is a negative impact of high animal density on the quality of the summer pastures [33] and, as a consequence, a strong negative relationship between calf body mass and reindeer density [30,31]. High calf body mass reflects high level of growth during the summer, animals in good body condition and reduced vulnerability to harsh winter conditions [30,31,34,35]. The livelihood economy indicator is a measure of the economic benefits derived from the pastures, which is the sum of income from the sale of slaughtered animals and governmental disbursements through slaughter subsidies (provided by the Reindeer Husbandry Administration) and compensation for losses to predators (provided by the Directorate for Nature Management).

We used the pasture area, inequality among herders and seasonal siida stability as structural predictor variables in the analyses. We expected the pasture size to be an important determinant of sustainability. However, due to difference in the match between the pasture size and the siida organization, we expected the effect of pasture size to differ between summer and winter pastures. Inequality has been hypothesized to influence incentives for collaboration and, therefore, sustainable outcomes [8]. Previous studies of Sámi pastoral communities have documented inequalities associated with the herd size, both in terms of the access to resources and power [19,21,36]. Based on the number of reindeer possessed by each herding unit, we used the Gini coefficient as a measure of equality [37]. The Gini coefficient ranges from 0, indicating perfect equality, to 1, indicating maximal inequality, and we expected low Gini coefficients to be associated with more sustainable outcomes. The seasonal stability index reflected the turnover of pastoralists working together throughout the year. For each herding unit, the index was calculated by dividing the number of herding units cooperating year-round by the total number of herding units cooperating on either the summer or winter pastures. This index range from 0, indicating that no units herd together year-round, to 1, indicating that all pastoralists herd together year-round (i.e. stable siida partnership, Text S1). Because individual herd size is an important determinant of individual success in Sámi pastoral communities [38], we also included the individual herd size as a covariate in the analyses.

The data were analyzed by linear mixed-effect models (*lme*) using the *nlme* library in *R* [39,40]. For each of the three response variables we constructed an initial model with a given set of structural predictor variables (fixed factors). The response variables were measured on the herding unit level, giving 292 numbers of observations in each model. The initial models included six fixed factors: herd size, siida stability, winter Gini, summer Gini, winter size and summer size. Herd size and siida stability were measured on the herding unit level. The Gini coefficients and pasture sizes were measured on the winter pasture level (winter Gini and winter size) and on the summer pasture level (summer Gini and summer size). The 44 summer pastures were nested within the 10 winter pastures, i.e.; herders from different summer pastures shared the same winter pasture, while no herders from different winter pastures shared the same summer pasture. Summer pasture nested within winter pasture was accordingly included as a random factor in the models.

The degrees of freedom of the fixed factors was determined by the level at which the factor was measured. Therefore, due to different degrees of freedom, the relative importance of the fixed factors cannot be assessed by their associated *t* and *P*-values. To compare the relative importance of the different coefficients, we standardized each variable to a mean equal to zero and a standard deviation equal to one. The effect of a coefficient $b_1$ from a predictor X1 with standard deviation $s_1$, on a response variable Y with standard deviation $s_Y$, should be interpreted as follow: An increase of $s_1$ in X1 results, on average, in an increase in $b_1 * s_Y$ in Y.

We used the Akaike's Information Criterion (AIC) to investigate how model simplification affected the relationships revealed by the initial models [41]. From a pool of candidate models containing all the possible combinations of the six fixed factors, we selected the model with the lowest AIC. The random structure was held constant for all candidate models.

## Interview Inquiry of Explanatory Mechanisms

The quantitative analyses showed that unsustainable outcomes were more common in the five largest winter pastures indicating that collective action problems might be more pronounced in these areas. Accordingly, we selected the five largest winter pasture areas for the interview inquiry (for design and methods see http://www.ecologyandsociety.org/vol16/iss3/art4/appendix3.html, Accessed 2012 Nov 12, [30]). We asked: Q1. Is there agreement among pastoralists about sustainability policies? Q2. Is there trust among pastoralists? Q3. What is the role of institutional arrangements? In most cases, we coded the interviews on three ordinal levels. "Not Available" (NA) was used in all cases where the variables could not be coded due to unclear answers or no response. The number of respondents (N), therefore, varied for each variable.

**Q1. Is there agreement among pastoralists about sustainability policies?.** One hypothesis for the lack of community solutions is the occurrence of disagreements about the resource situation among the pastoralists. Reduction of the number of reindeer and pastoralists has been the major policy goal in the last 20 years [30,42]. Agreement on the overall goal that reindeer numbers should be adjusted to pasture capacity to increase the condition of the animals, therefore, was a major question (Q1.1). The policies also aim to change herding practices to increase the production per hectare, particularly by reducing the herd sizes (Q.1.2) and stimulating calf slaughter (Q1.3) [30,42]. Lastly, meat production for market and the maximization of economic income has not traditionally been the aim of Sámi pastoralists, and a lack of adaptation could, therefore, be due to their inclination for non-market values (Q.1.4) [19,21,36]. We coded policy agreement on three levels: disagree, partly agree and agree.

**Q2. Is there trust among pastoralists?.** Trusting other pastoralists to reciprocate is assumed to be essential for cooperation [8], particularly in those systems in which the flexible use of pastures demands continuous negotiations between the pastoralists [22–24]. We coded trust on three levels for both the summer and winter pastures (Q2.1 and 2.2). Trust was coded as "no" for pastoralists keeping their herds away from neighboring pastoralists or claims of negative reciprocity, such as suspicious losses of reindeer, changes of earmarks or keeping reindeer too long on others pastures. Trust was coded as weak when the pastoralists responded that they communicate but are unsure about the others' intentions or disagree about their decisions. On the summer pastures, we additionally coded trust as weak for claims of other pastoralists not contributing sufficiently to labor or financial costs. Full trust of other pastoralists was coded as strong in situations where the pastoralists reported good communication and positive reciprocity, such as informing other pastoralists of lost reindeer and returning the reindeer or letting them remain on their pastures for a period if needed.

**Q3. What is the role of institutional arrangements?.** We hypothesized that outcomes on the five largest pastures could be explained by a mismatch between the scale of self-organization (the siidas) and the scale of formal governance (i.e. the need for well-defined boundaries) or, a lack of sufficient autonomy for the pastoralists to make their own decisions. Voluntary programs for establishing customary tenures have been initiated by the administration, but formalizing customary rights through independent courts have also been discussed (see Text S1, [28,43]). We therefore asked about the need for customary tenures on the winter pastures and whether such a division should be based on negotiations or formalized by a court (Q3.B). The need for clearly defined rights on the winter pastures depends on the ability of the users themselves to negotiate and resolve conflicts related to pasture use. Local arenas that allow for rapid conflict resolution may increase the levels of trust and the degree of cooperation among the users [4]. We therefore asked about the existing conflict resolution mechanisms (No, Negotiation, or Formalized) (Q3.A). The high degree of participation in rulemaking has been shown to increase the likelihood of sustainable management [44]. However, to solve the problems of collective action among users, the higher levels of governance must themselves be trustworthy [11]. We coded the pastoralists' trust to the multilevel co-management system as no, weak or strong levels of trust (Q3.C). Lastly, we asked if more autonomy is needed at the level of the summer siidas (Q3.D).

### Ethics Statement

Our standards for ethics were approved by the Norwegian Social Science Data Services and included a letter of information about the study before the start of the interview to secure voluntary participation and a perusal of the final transcript of the interview, if requested. We informed about the survey, the purpose and how data would be stored and used. Due to abuse of written statements in the past, Sámi elders are reluctant to provide signatures of any kind. Thus we obtained verbal consent from the participants. This procedure was approved by the Norwegian Social Science Data Services, which check the ethics and is the only institution that issue official licenses to all social science studies in Norway.

## Results

### Statistical Analyses

The size of the winter pasture had, relative to the other structural variables, a strong negative effect on all the sustainable outcome variables (Table 1; Figure 2A, 2B, 2C). The effect was however, only marginally significant for the proportion of calf slaughter ($P = 0.048$) and not significant for calf body mass ($P = 0.066$). However, model simplification based on AIC retained the effect of winter pasture size in all the analyses. The effects were significant ($P<0.05$) in all the selected models, suggesting that winter pasture size was an important factor explaining sustainable outcomes. In contrast, the effects of summer pasture size were relatively weak and variable, and this factor was consistently removed by the model selection procedure (Table 1). There was a significant positive relationship between the seasonal siida stability and the proportion of calf slaughter (Table 1; Figure 2D). This result shows that herders, on average, slaughtered a higher proportion of calves when they were organized into the same siida throughout the year. With respect to the calf body mass, the Gini coefficient on summer pastures had a negative effect (Table 1; Figure 2E), indicating that equality in herd size among herders on the summer pastures was associated with higher slaughter weights of calves. Finally, we found a positive relationship between herd size and income and between herd size and the proportion of calf slaughter (Table 1).

### Interview Inquiry

The interviews suggested that the variables used as outcome variables in the statistical analyses were in agreement with the pastoralists' own views on sustainability (Q1, Figure 3). There was strong support for an adjustment of reindeer numbers to the pasture capacity in which higher weights are preferable to keeping large herds. However, most of the pastoralists indicated the need for others to reduce their herds; i.e. either small owners who have alternative income or large owners who keep very large herds. The majority preferred to slaughter the surplus calves before moving on to the winter pastures to reduce the risk of losses and to improve production outcomes. The focus on production was also strongly supported by the pastoralists, who reported meat production as their primary goal. Most of the respondents argued that the costs of reindeer pastoralism are high, thus to cover the expenses for snowmobiles and housing, they needed to think in terms of production and income. This does not mean, however, that non-market factors were unimportant. The pastoralists claimed that the prevailing cause for keeping large herds was to gain informal influence (71%) and to access the best winter pastures (66%).

There was a clear difference in the levels of trust for the summer and winter pastures (Q2, Figure 4). Most summer pastures were managed by one siida group in which a high degree of trust was based on strong family ties and a long history of collaboration. A lack of trust was more evident on the winter pastures, as 52% of the respondents are suspicious to their neighbors, and only 19% have a strong degree of trust for their neighboring herdsmen. According to the older pastoralists, there have always been flexible boundaries on the winter pastures, potentially leading to tension between neighbors, but they formerly had a core tenure associated with their winter siida groups that was generally respected by the other pastoralists. The flexible boundaries were necessary for the adaptation to shifting pasture conditions, such as icing or deep snow, that limit the access to lichen. Many of those who experienced conflicts explained that the 1978 Reindeer Husbandry Act named the winter pastures as "common pastures" which, according to the respondents, had not been the case previously. The system of customary tenures was consequently not respected by some of the pastoralists who expanded their herds into these areas (see also Text S1). Keeping large herds bestows influence and access to the best winter pastures, providing incentives for some pastoralists to increase their herds.

The lack of conflict resolution mechanisms for the winter pastures was striking (Q3.A, Figure 5). The small herders explained that the only options are to move away from the large herds to avoid losing too many reindeer. In conflict areas, the pastoralists spend most of their time herding reindeer on snowmobiles to avoid the mixing of herds. Traditionally, conflicts have been solved through negotiation, and some respondents referred to informal leaders, "siida isit," who used to confer with neighbors on the winter pastures. There is a special police unit, the reindeer police, who mitigate and resolve conflicts among the pastoralists related to such activities as gathering, separation, marking and slaughtering. However, the pastoralists considered it inappropriate to report anyone to the police, and some pastoralists explained that such a report could bring retribution and "losses" of reindeer. The respondents also argued that neither the police nor the administration provide any possibility of solving major conflicts regarding the pastures because the grazing rights are not formalized.

Most of the pastoralists preferred to divide the winter pastures into customary tenures, but they disagreed on the use of voluntary agreements versus formalization through the courts. There were

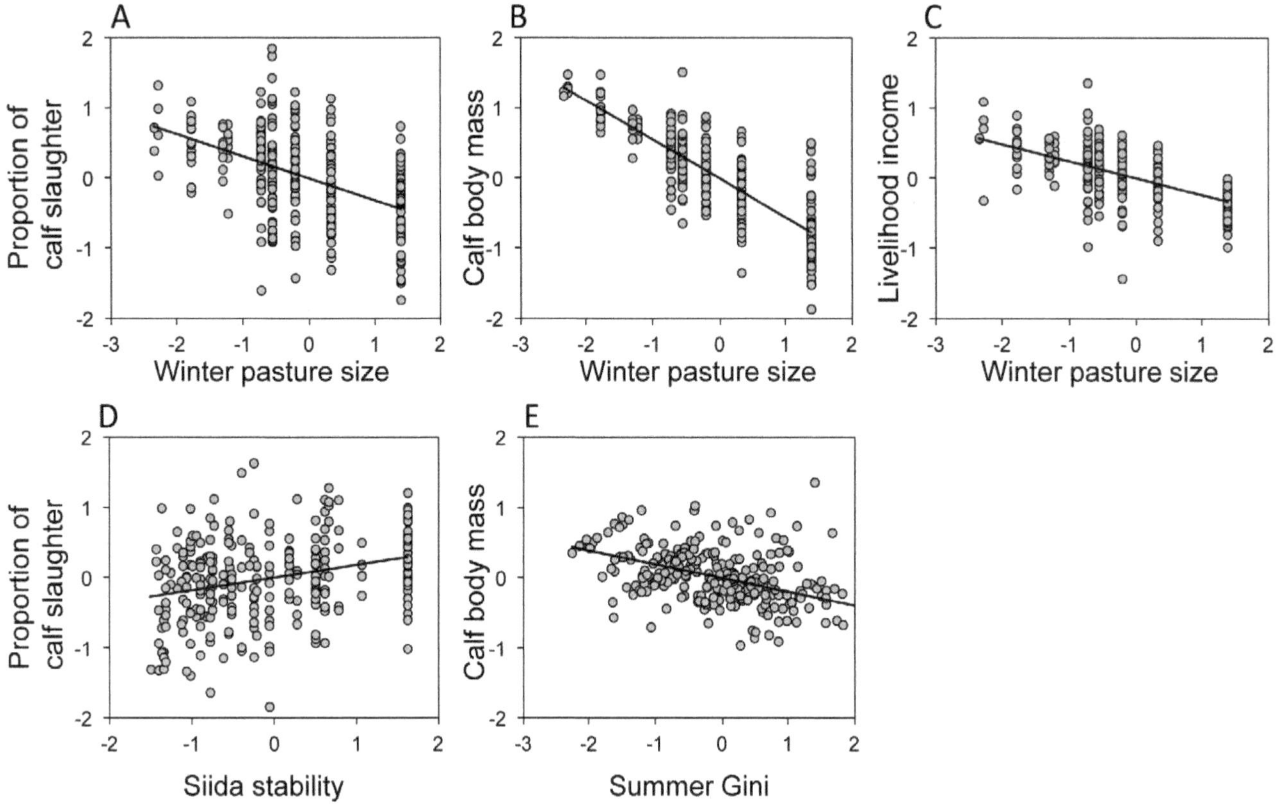


**Figure 2. Relationships between structural variables and sustainability indicators selected by AIC.** Partial residual plots between (A) winter pasture area and proportion of calf slaughter, (B) winter pasture area and slaughter weights of calves, (C) winter pasture area and livelihood income, (D) seasonal siida stability and proportion of calf slaughter, and (E) inequality in herd size on the summer pasture (Gini coefficient) and slaughter weight of calves. Each point is the average value of a herding unit from 1998 to 2007. The partial residuals were computed from the initial mixed effect model, relating the sustainability indicators to six structural predictor variables (see Table 1). To allow comparisons of effect sizes, all variables were standardized to mean equal to zero and standard deviation equal to one.

pastoralists who claimed that newly rich pastoralists do not respect the customary tenures, so these respondents only options was to use the court to claim pasture rights. Others argued that the voluntary division of pastures does not work due to overlapping use by different siida groups. Only 12% did not want to divide the winter pastures into smaller tenures related to siida groups. A few warned that such divisions would ruin the traditional culture, solidarity and flexible use of the pastures and move reindeer pastoralism toward ranching.

As much as 88% perceived the governance system as a necessary third party for formalized conflict resolution, monitoring and enforcement. The lack of impartial and fair treatment due to kinship or friendship among the co-management boards was the main source of distrust. Recent decisions associated with total allowable reindeer numbers, weighing programs, and the processes behind the division of winter pastures had also caused discontent with the administration. Most pastoralists argued for more impartiality, either by using co-management boards from neighboring regions or by transferring authority to the governor.

The pastoralists did not support increased autonomy for the summer siidas. The primary arguments against devolution were the need for regulating the competition between siidas on the winter pasture and the protection against encroachment. Those who partly agreed argued that self-determination could work under the condition that appropriate rules or customary tenures were established on the winter pastures. For some, the transfer of rulemaking authority to the summer siidas was undesirable, as the elites have too much power in the local boards, whereas others maintained that the reindeer are privately owned and they cannot tell their relatives to slaughter their herds. Many pastoralists perceived the 2007 Act that devolved the rulemaking authority to the siidas as a repudiation of liability, as the governing bodies caused the situation. The generous allocation of herding licenses and subsidies allowed too many pastoralists to accumulate herds on shared pastures during the 1980s. Combined with the definition of winter pastures as commons in the 1978 Act, which some pastoralists interpreted as open access, new herds and expansionism have resulted in crowding on the winter pastures.

## Discussion

Our results suggest that the match between the scale of self-organization and the scale of formal governance is a key condition for sustainable outcomes in a SES. The CBM was designed to delegate the rulemaking authority to the siidas, but the challenge remains in how the users self-organize at the higher nested levels, i.e. winter pastures. In our case, the collective action problems emerged on the large winter pastures shared by several competing siidas. Our findings were based on the combination of quantitative analyses of data from official databases and interview inquiries, linking structural variables with explanations of sustainable outcomes [3,10,15]. In the quantitative analyses, we found winter pasture size to be the key variable influencing sustainable outcomes (Table 1, Figure 2). The selection of the five largest

**Table 1.** Mixed-effect models of sustainability outcomes.

| | | | | | |
|---|---|---|---|---|---|
| **A) Proportion of calf slaughter** | | | | | |
| **Fixed effects** | **Estimate** | **S.E.** | **DF** | **t-value** | **P** |
| Herd size | 0.134 | 0.042 | 246 | 3.22 | 0.002 |
| Stability | 0.181 | 0.063 | 246 | 2.86 | 0.005 |
| Winter size | −0.314 | 0.131 | 7 | −2.39 | 0.048 |
| −/Summer size | 0.233 | 0.157 | 32 | 1.49 | 0.147 |
| −/Winter Gini | −0.220 | 0.108 | 7 | −2.04 | 0.081 |
| −/Summer Gini | −0.032 | 0.091 | 32 | −0.35 | 0.728 |
| **Random effects** | | | | **StDev** | **nObs** |
| Among winter pastures | | | | 0.10 | 10 |
| Among summer pastures (within winter pastures) | | | | 0.60 | 44 |
| Residuals | | | | 0.59 | 292 |
| **B) Calf body mass** | | | | | |
| **Fixed effects** | **Estimate** | **S.E.** | **DF** | **t-value** | **P** |
| −/Herd size | −0.078 | 0.030 | 246 | −2.55 | 0.011 |
| −/Stability | −0.106 | 0.049 | 246 | −2.18 | 0.030 |
| Winter size | −0.558 | 0.256 | 7 | −2.17 | 0.066 |
| −/Summer size | −0.117 | 0.224 | 32 | −0.52 | 0.605 |
| −/Winter Gini | 0.158 | 0.200 | 7 | 0.79 | 0.456 |
| Summer Gini | −0.370 | 0.094 | 32 | −3.93 | 0.000 |
| **Random effects** | | | | **StDev** | **nObs** |
| Among winter pastures | | | | 0.07 | 10 |
| Among summer pastures (within winter pastures) | | | | 0.61 | 44 |
| Residuals | | | | 0.58 | 292 |
| **C) Livelihood income** | | | | | |
| **Fixed effects** | **Estimate** | **S.E.** | **DF** | **t-value** | **P** |
| Herd size | 0.811 | 0.023 | 246 | 35.60 | 0.000 |
| −/Stability | 0.064 | 0.035 | 246 | 1.80 | 0.073 |
| Winter size | −0.242 | 0.071 | 7 | −3.40 | 0.012 |
| −/Summer size | 0.071 | 0.089 | 32 | 0.80 | 0.432 |
| −/Winter Gini | −0.075 | 0.060 | 7 | −1.25 | 0.252 |
| −/Summer Gini | 0.026 | 0.052 | 32 | 0.49 | 0.626 |
| **Random effects** | | | | **StDev** | **nObs** |
| Among winter pastures | | | | 0.00 | 10 |
| Among summer pastures (within winter pastures) | | | | 0.35 | 44 |
| Residuals | | | | 0.32 | 292 |

(A) proportion of calf slaughter (B) slaughter weights of calves, and (C) livelihood income. The response variables were modeled with respect to (fixed factors): herd size, seasonal siida stability, pasture size (summer and winter size) and the inequality in herd size (summer and winter Gini). Summer pasture nested within winter pasture were modeled as random factors. To allow comparisons of effect sizes, all variables were standardized to mean equal to zero and standard deviation equal to one prior to the analyses.

−/Fixed effects removed by model simplification based on AIC.

winter pastures for the interview inquiries allowed us to analyze the explanatory mechanisms behind this result. The interviewed pastoralists agreed on the sustainability policies, but had a low degree of trust to other herders on the winter pastures and no conflict resolution mechanisms for pasture use among the siida groups. As a result, the Sámi pastoralists did not call for more autonomy at the lowest scale but instead requested customary tenures or an impartial authority to devise clear rules and sanctioning mechanisms. The multilevel co-management system has not filled this institutional gap [19,42,45,46] and lacks sufficient trust to act as a neutral, objective and impartial authority.

The effect of pasture size on sustainability differed between seasonal pastures. Large winter pastures had a negative effect on

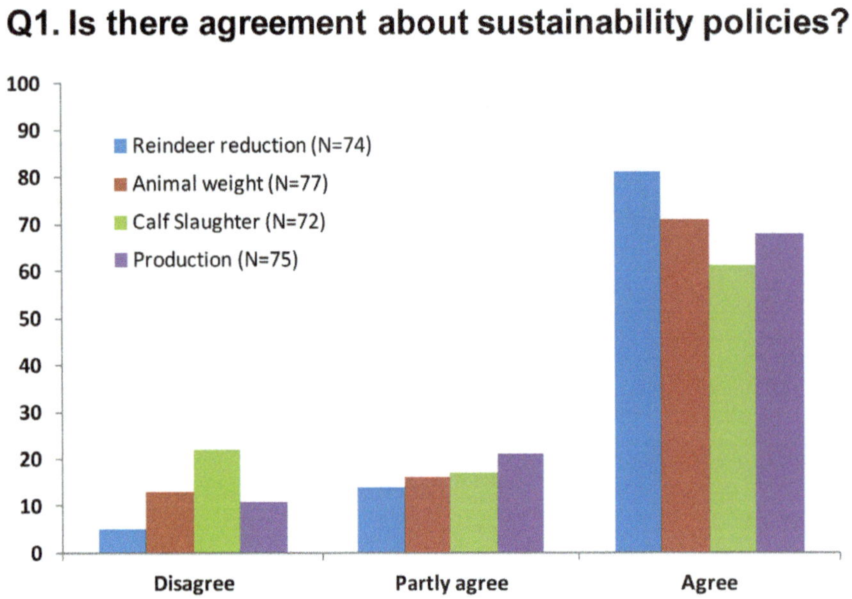


**Figure 3. Agreement about sustainability policies.** Percentages of the pastoralists who disagreed, partly agreed and agreed with sustainability policies.

the herders' livelihood income, the proportion of calf slaughter and the slaughter weights of calves. No such relationships were evident on the summer pastures. Most summer pastures have well-defined boundaries, with usually one siida group having the exclusive rights to the pasture. Accordingly, the size of the summer pasture only had small effects on the sustainability indicators. These differences between seasonal pastures were also evident in the interview inquiry (Figure 3). On the five largest winter pastures, the low level of trust also corresponds with the negative reciprocity among herders as well as the higher costs of monitoring and herding reindeers on snowmobiles. Open conflicts are usually avoided as the outcome could be retribution, and for small herders, the best strategy is usually to move away from large herds to avoid losses of reindeer. The seasonal difference in the effect of pasture size, combined with the seasonal difference in trust (and cooperation) among herders, do support the view that a mis-match between the scale of self-organization and formal governance on the winter pasture is among the major reasons behind unsustainable outcomes in the SES. Unknown confounders, such as systematic differences in the biophysical conditions or herding practices, could have biased the quantitative analyses, and in particular the estimated effect of winter pasture size. However, the field study suggests that the effects of such biases were small. The reindeer herders do not ascribe to alternative herding practices or sustainability goals (Figure 3). They report low degree of trust and the weak institutional arrangements on the largest winter pastures as their major challenges (Figures 4 and 5).

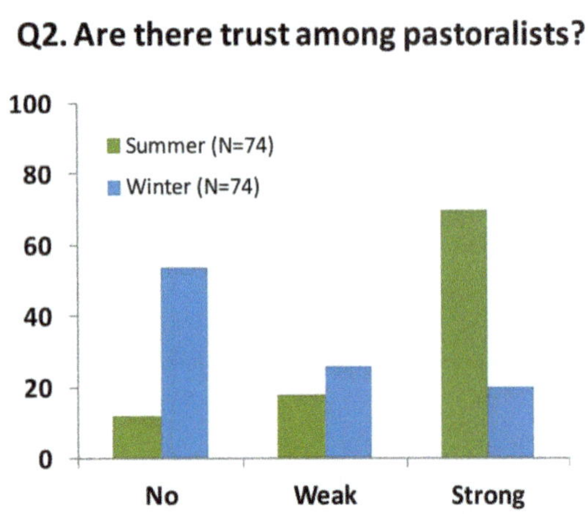


**Figure 4. Trust among pastoralists.** Percentages of the pastoralists who expressed no, weak and strong degrees of trust for other pastoralists sharing winter and summer pastures.

Our assertion that collective action problems are crucial for understanding this pastoral SES, is also corroborated by the fact that a high stability in the social organization of herders and a high degree of equality among herders, in addition to small winter pasture size, had a positive effect on some of the sustainability indicators. Specifically, the quantitative analyses showed that pastoralists belonging to the same siida throughout the year slaughtered a higher proportion of calves compared to herders that changed the siida group seasonally. Calves are generally slaughtered before the herds move into the winter pastures, and calf slaughter will, as explained by the herders in the interviews, spare the winter pastures and reduce the risk of losses of weak animals under unfavorable winter conditions (see also [31,32,47]). On the other hand, calf slaughter will reduce the herd size and thereby the herders' influence and ability to acquire pasture during winter. The indicator was therefore expected to be particularly sensitive to the degree of cooperation among herders. The equality with respect to herd size was important for the slaughter weight of calves, which indicate how well reindeer density is regulated by the siidas. Finally, the effect of winter pasture size was also evident for the economic benefits derived from the pastures. Controlled for herd size, small winter pastures was the only factor that contributed positively to income. The lower income on the largest winter pastures could be explained by strategies associated with non-market values [36,38,48], but in our interviews the pastoralists perceive production as a primary goal, which is necessary to cover the high expenses of reindeer herding [17,19,34].

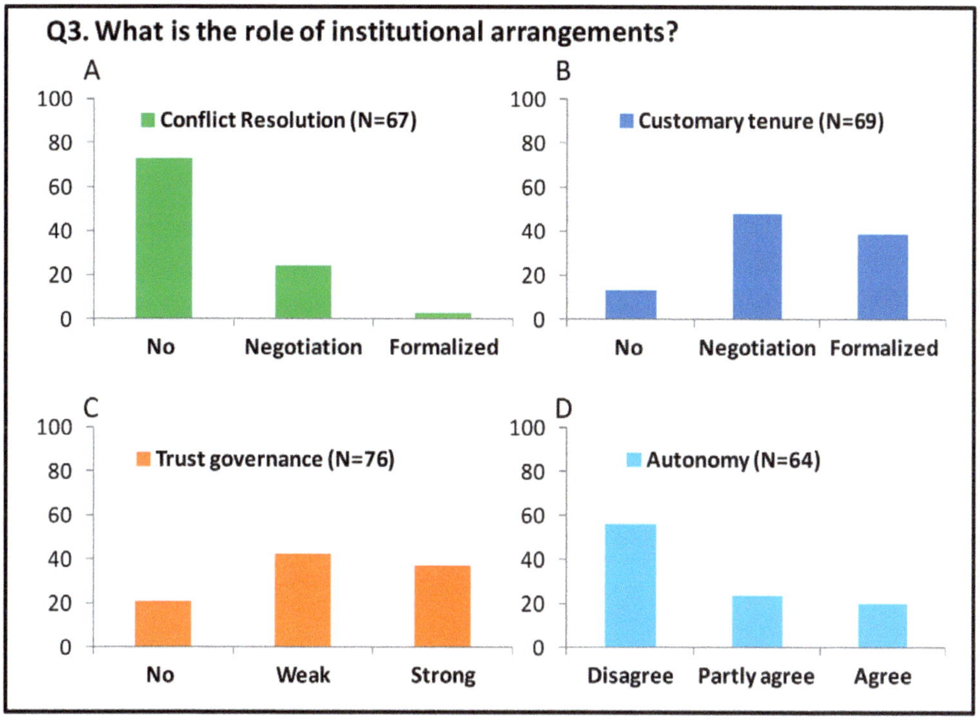


**Figure 5. Institutional arrangements.** The pastoralists' perceptions of institutional arrangements. A) Percentages of the pastoralists who reported no, negotiated and formalized conflict resolution mechanisms. B) Percentages of the pastoralists who responded negatively to the idea of dividing the pastures into customary tenures or responded positively to the idea of establishing customary tenures, either by voluntary agreements or court formalization. C) Percentages of the pastoralists who expressed no, weak and strong degrees of trust in the current governance regime. D) Percentages of the pastoralists who disagreed, partly agreed and agreed that more autonomy is needed to achieve sustainability.

As CBM is often based on the subsidiary principle, the nestedness of social organization is easy to ignore when policies are put into practice [2,3,5]. The mismatch of CBM with an appropriate scale has also been suggested as one of the reasons behind the persistent decline of common pastures in the drylands of Africa [23]. A multilevel co-management system regulates macro-scale mobility in Sámi pastoral ecosystems but has failed to regulate micro-scale use on the higher nested levels. The prescription has been more devolution to users, or "enforced self-organization," as local boards now have full responsibility for crafting rules for the sustainable use of all of the seasonal pastures. As our results show, this reform toward self-determination is likely to work better when the winter pasture size is smaller. Large pastures require an impartial third party to regulate the competition between the siidas, as was also confirmed by another survey showing that only 1/5 of the pastoralists think they will be able to manage these pastures themselves, i.e., without external assistance [43]. The importance of this mismatch between governance and the siida organization is corroborated by the comparison with other regions in Norway, where pastoralists in general manage their smaller pastures more sustainably [19,31,46]. Similar collective action problems have also been observed in northernmost Finland, where herding cooperatives self-organize to manage pastures without impartial authorities to deal with conflicts and power inequalities [49].

Critics of CBM have also argued that the strong emphasis on formalized rules, rights and sanctioning mechanisms does not fit the flexible social organization needed in mobile pastoral systems [23]. High climatic variability and an unpredictable supply of resources demand flexible resource-sharing arrangements, which cannot be as easily delineated as secure tenure rights [17–19]. Although they acknowledge the need for flexibility under adverse weather conditions, most Sámi pastoralists prefer delineation of borders on the basis of customary rights to tenures [46]. There are, however, reciprocal norms for sharing pastures, particularly in the rim zones of the siida tenures. These norms typically depend on prior agreement [46], but the timing and duration of stay are also factors, along with the degree of trust between neighbors. The call for customary rights in the five largest winter pastures could be explained by the low levels of trusts. Trust is a key condition for negotiations and flexible use on winter pastures, which may have eroded as number of siidas and herds have increased.

CBM focus on the capabilities of pastoralist groups to establish their own rules, monitoring and sanctioning mechanisms for managing common pastures [4,23]. The fit of CBM to Sámi pastoralist ecosystems depends on the scale of resource sharing, as well as the setting in which rulemaking authority has been transferred. Historically, the "big-push" economic policies rapidly modernized the Sámi pastoral ecosystems from a system of low living conditions and the use of simple technology to an SES of higher welfare, marked integration and the use of snowmobiles and ATVs (see Text S1, [19,30]). The increase in number of pastoralists and herd expansion has subsequently resulted in pressures on the customary tenure systems, which have limited the adaptability of the SES [19,20,30,46]. The challenges have been met by CBM policies that devolve rulemaking authority to the lowest level of organization. We found CBM to have a better fit to such a SES when pastoralists cooperate on smaller scales and when there are higher stability in siida partnership. The lack of trust and self-organization among siidas on larger scales often need higher level of governance for devising clear rules, conflict resolution and sanctioning mechanisms [1,5,7]. The co-management board did not fill this institutional gap which may explain the pastoralists request for well-defined boundaries, and clearer rules for pasture use devised by an impartial authority.

## Supporting Information

**Text S1 The text provides a more detailed description of community-based management of Sámi pastoral ecosystems in Norway.**


## Acknowledgments

We especially thank the Sàmi pastoralists that participated in the interviews.


## Author Contributions

Analyzed the data: VH PF JLJ. Wrote the paper: VH PF.

## References


1. Agrawal A (2001) Common property institutions and sustainable governance of resources. World Dev 29: 1649–1672.
2. Berkes F (2007) Community-based conservation in a globalized world. PNAS 104: 15188–15193.
3. Ostrom E (2007) A diagnostic approach for going beyond panaceas. PNAS 104: 15181–15187.
4. Ostrom E (1990) Governing the commons: The evolution of institutions for collective action. Cambridge: Cambridge University Press.
5. Brondizio ES, Ostrom E, Young OR (2009) Connectivity and the Governance of Multilevel Social-Ecological Systems: The Role of Social Capital. Annu Rev Environ Resour 34: 253–278.
6. Berkes F (2010) Devolution of environment and resources governance: trends and future. Environ Conserv 37: 489–500.
7. Young OR (2002) The institutional dimensions of environmental change: fit, interplay, and scale. Cambridge, Mass.: MIT Press.
8. Poteete AR, Janssen M, Ostrom E (2010) Working together: collective action, the commons, and multiple methods in practice. Princeton, N.J.: Princeton University Press.
9. Agrawal A, Gibson CC (1999) Enchantment and disenchantment: the role of community in natural resource conservation. World Dev 27: 629–649.
10. Ostrom E (2010) Analyzing collective action. Agric Econ 41: 155–166.
11. Putnam RD (1993) Making Democracy Work: Civic traditions of modern Italy. Princeton: Princeton University Press.
12. Raymond L (2006) Cooperation without trust: Overcoming collective action barriers to endangered species protection. Pol Stud J 34: 37–57.
13. Cook KS, Hardin R, Levi M (2005) Cooperation without trust? New York: Russell Sage Foundation.
14. Cudney-Bueno R, Basurto X (2009) Lack of Cross-Scale Linkages Reduces Robustness of Community-Based Fisheries Management. PLoS ONE 4(7): e6253. doi:10.1371/journal.pone.0006253.
15. Lieberman ES (2005) Nested analysis as a mixed-method strategy for comparative research. Am Polit Sci Rev 99: 435–452.
16. Agrawal A, Chhatre A (2011) Strengthening causal inference through qualitative analysis of regression residuals: explaining forest governance in the Indian Himalaya. Environ Plan A 43: 328–346.
17. Forbes B, Bölter M, Müller-Wille L, Hukkinen J, Müller F, et al (2006) Reindeer management in northernmost Europe : linking practical and scientific knowledge in social-ecological systems. Berlin: Springer.
18. Pape R, Löffler J (2012) Climate change, land use conflicts, predation and ecological degradation as challenges for reindeer husbandry in Northern Europe: What do we really know after half a century of research? AMBIO 41: 421–434. http://dx.doi.org/10.1007/s13280-012-0257-6. Accessed 2012 Nov 12.
19. Riseth JA, Vatn A (2009) Modernization and Pasture Degradation: A Comparative Study of Two Sami Reindeer Pasture Regions in Norway. Land Econ 85: 87–106.
20. Kalstad JKH (1999) Reindriftspolitikk og samisk kultur- en uløselig konflikt? Et studie av reindriftstilpasninger of moderne reindriftspolitikk. Kautokeino: Nordic Sámi Institute.
21. Paine R (1994) Herds of the Tundra: a portrait of Saami reindeer pastoralism. London: Smithsonian Institution Press.
22. Beyene F (2010) Interclan Cooperation in a Risky Pastoral Ecology: Some lessons from Eastern Ethiopia. Hum Ecol 38: 555–565.
23. Turner MD (2011) The new pastoral development paradigm: Engaging the realities of property institutions and livestock mobility in dryland Africa. Soc Nat Resour 24: 469–484.
24. Niamir-Fuller M (1998) The resilience of pastoral herding in Sahelian Africa. In: Berkes F, Folke C, editors. Linking social and ecological systems: management practices and social mechanisms for building resilience. Cambridge: Cambridge University Press. 250–284.
25. Robinson LW, Berkes F (2010) Applying Resilience Thinking to Questions of Policy for Pastoralist Systems: Lessons from the Gabra of Northern Kenya. Hum Ecol 38: 335–350.
26. Forbes BC, Stammler F, Kumpula T, Meschtyb N, Pajunen A, et al. (2009) High resilience in the Yamal-Nenets social-ecological system, West Siberian Arctic, Russia. PNAS 106: 22041–22048.
27. Brännlund I, Axelsson P (2011) Reindeer management during the colonization of Sami lands: A long-term perspective of vulnerability and adaptation strategies. Glob Env Change 21: 1095–1105.
28. Sara MN (2011) Land usage and siida autonomy. Arctic Review of Law and Politics 2011/2.
29. Anonymous (2010) Ressursregnskap for reindriftsnæringen 08/09 (Ecological statistics of reindeer husbandry). Alta: Reindeer Husbandry Administration. http://www.reindrift.no/asset/1627/1/1627_1.pdf. Accessed 2012 Nov 12.
30. Hausner V, Fauchald P, Tveraa T, Pedersen E, Jernsletten J-L, et al. (2011) The ghost of development past: the impact of economic security policies on Saami pastoral ecosystems. E & S 16: 4. http://dx.doi.org/10.5751/ES-04193-160304. Accessed 2012 Nov 12.
31. Tveraa T, Fauchald P, Yoccoz NG, Ims RA, Aanes R, et al. (2007) What regulate and limit reindeer populations in Norway? Oikos 116: 706–715.
32. Kojola I, Helle T (1993) Calf harvest and reproductive rate of reindeer in Finland. J Wildl Man 57: 451–453.
33. Bråthen KA, Ims RA, Yoccoz NG, Fauchald P, Tveraa T, et al. (2007) Induced shift in ecosystem productivity? Extensive scale effects of abundant large herbivores. Ecosystems 10: 773–789.
34. Helle TP, Jaakkola LM (2008) Transitions in herd management of semi-domesticated reindeer in northern Finland. Ann Zool Fenn 45: 81–101.
35. Olofsson A, Danell Ö, Åhman B, Forslund P (2011) Carcass records of autumn-slauthered reindeer as indicator of long-term changes in animal condition. Rangifer 31: 7–20.
36. Johannesen AB, Skonhoft A (2011) Livestock as Insurance and Social Status: Evidence from Reindeer Herding in Norway. Env Res Econ 48: 679–694.
37. Gini C (1921) Measurement of Inequality and Incomes. The Economic Journal 31: 124–126.
38. Næss MW, Bårdsen BJ (2010) Environmental Stochasticity and Long-Term Livestock Viability-Herd-Accumulation as a Risk Reducing Strategy. Hum Ecol 38: 3–17.
39. Pinheiro J, Bates D, DebRoy S, Sarkar D and the R Development Core Team (2011) nlme: Linear and Nonlinear Mixed Effects Models. R package version 3.1–100.
40. The R Development Core Team (2011) R: A language and environment for statistical computing. R Foundation for Statistical Computing,Vienna, Austria.
41. Burnham KP, Anderson DR (2002) Model selection and multimodel inference: a practical information-theoretic approach. New York: Springer Inc.
42. Ulvevadet B, Hausner V (2010) Incentives and regulations to reconcile conservation and development: Thirty years governance of Sami pastoral ecosystems in Finnmark, Norway. J of Env Man 92: 2794–2802.
43. Ravna Ø, Olli JJ (2011) Sedvanerettslige oppfatninger om arealbruk blant reindriftsutøvere: Herunder oppfatninger om reintallsfastsetting og om behov for en rettsinstans som er egnet for å ordne beitebruk i reindriftsområder. Diedut. Kautokeino: Sami University College.
44. Persha L, Agrawal A, Chhatre A (2011) Social and Ecological Synergy: Local Rulemaking, Forest Livelihoods, and Biodiversity Conservation. Science 331: 1606–1608.
45. Ulvevadet B (2008) Management of reindeer husbandry in Norway - power-sharing and participation. Rangifer 28: 53–78.
46. Ulvevadet B (2011) Problems and challenges for user participation: The system of user participation in reindeer husbandry in Norway. Rangifer 31: 161–182.
47. Tveraa T, Fauchald P, Henaug C, Yoccoz NG (2003) An examination of a compensatory relationship between food limitation and predation in semi-domestic reindeer. Oecologia 137: 370–376.
48. Riseth JA (2006) Sami reindeer herd managers: why do they stay in a low-profit business? Brit Food J 108: 541–559.
49. Laakso AM (2008) The shadow field of reindeer management: A case study from Finland. Acta Borealia 25: 138–159.

# 11

# Mapping Large-Area Landscape Suitability for Honey Bees to Assess the Influence of Land-Use Change on Sustainability of National Pollination Services

**Alisa L. Gallant[1]*, Ned H. Euliss Jr.[2], Zac Browning[3]**

**1** United States Geological Survey, Earth Resources Observation and Science Center, Sioux Falls, South Dakota, United States of America, **2** United States Geological Survey, Northern Prairie Wildlife Research Center, Jamestown, North Dakota, United States of America, **3** Browning's Honey Company, Inc., Jamestown, North Dakota, United States of America

## Abstract

Pollination is a critical ecosystem service affected by various drivers of land-use change, such as policies and programs aimed at land resources, market values for crop commodities, local land-management decisions, and shifts in climate. The United States is the world's most active market for pollination services by honey bees, and the Northern Great Plains provide the majority of bee colonies used to meet the Nation's annual pollination needs. Legislation requiring increased production of biofuel crops, increasing commodity prices for crops of little nutritional value for bees in the Northern Great Plains, and reductions in government programs aimed at promoting land conservation are converging to alter the regional landscape in ways that challenge beekeepers to provide adequate numbers of hives for national pollination services. We developed a spatially explicit model that identifies sites with the potential to support large apiaries based on local-scale land-cover requirements for honey bees. We produced maps of potential apiary locations for North Dakota, a leading producer of honey, based on land-cover maps representing (1) an annual time series compiled from existing operational products and (2) a realistic scenario of land change. We found that existing land-cover products lack sufficient local accuracy to monitor actual changes in landscape suitability for honey bees, but our model proved informative for evaluating effects on suitability under scenarios of land change. The scenario we implemented was aligned with current drivers of land-use change in the Northern Great Plains and highlighted the importance of conservation lands in landscapes intensively and extensively managed for crops.

**Editor:** Dennis vanEngelsdorp, University of Maryland, United States of America

**Funding:** The United States Geological Survey; the United States Department of Agriculture, Farm Service Agency; and the United States Department of Agriculture, Natural Resources Conservation Service, Conservation Effects Assessment Project—Wetlands jointly supported this project. Other than policy review by the United States Geological Survey and a general editorial review the authors solicited from a colleague at the United States Department of Agriculture's Farm Service Agency, the funders had no role in study design, data collection and analysis, decision to publish, or preparation of the manuscript.

**Competing Interests:** Zac Browning is a co-owner and operator of Browning's Honey Company, Inc. There are no patents, products in development or marketed products to declare.

* E-mail: gallant@usgs.gov

## Introduction

Maintaining human prosperity and quality-of-life standards requires landscapes that sustainably deliver ecosystem services valued by society. Decisions that improve the sustainability of landscapes call for information on the simultaneous responses of multiple ecosystem services to changes in land use and climate as a more comprehensive approach to safeguard against unintended, and potentially negative, outcomes of land management or policy actions [1]. Changes in land use and climate clearly alter how ecosystems function within any given landscape [2], affecting all ecosystem services, from supporting to cultural, provisioning, and regulating services [3]. Socioeconomic factors cause dynamic, often unforeseen, fluctuations and shifts in land use, so adaptive modeling systems that are quantitatively linked to monitoring activities provide both the feedback on landscape responses needed to inform decision makers and the basis for continued model improvement [2].

Socioeconomic factors are exerting considerable influence on decisions affecting land use in the Northern Great Plains of North America, a region of major agricultural production [4] and the most important area for production of continental waterfowl populations [5], [6]. The area's significance for conservation in the United States has influenced various federal, state, and nongovernmental entities to protect lands in the region from agricultural development. The U.S. Department of Agriculture administers various programs that target delivery of ecosystem services, through traditional agricultural crop programs to voluntary conservation programs. The Conservation Reserve Program (CRP) [7] is the largest of the latter and has received considerable recognition from an array of conservation interests [8], [9], [10]. The program enables landowners to enroll or re-enroll eligible farmland for 10- or 15-year contracts during specified sign-up periods or on a continuous basis for priority conservation practices, such as to support pollinators and wetlands.

Euro-American settlement of the Northern Great Plains brought conversion of vast tracts of grassland to cropland, and while grasslands continue to be converted to cropland, periods in recent decades have seen some reversal of that land-change dynamic [11], [12], [13], [14], [15]. Satellite imagery revealed a notable return of cropland to grassland in years following inception of the CRP [11], as landowners realized increased land values resulting from enrollment in the program [16]. Land conversion since the mid-2000s has favored cropland over grassland, likely influenced by the convergence of technological advances in agriculture, a push for renewable energy from biofuels, and rising commodity prices [17], [18], as well as crop insurance and disaster relief programs [12].

As incentives have risen for landowners to maximize crop acreages, Congress has reduced total allowable acreage enrollments for the CRP. The highest cap allocated for the program was 39.2 M acres (15.9 M ha; note, we report here the measurement units used in the legislation and federal reports) during 2002–2007 [19]. From 2008–2013 the cap was lowered about 20% to 32 M acres (12.9 M ha) [20]. Actual acres enrolled in CRP nationally have dropped more than 27% since the peak of the program in 2007. Reductions in enrollment in the Northern Great Plains exceeded national rates. North Dakota's peak enrollment was 338.9 K acres (137.1 K ha) in 2007, dropping 53% by 2013. The latest Farm Bill [21] specifies further, annual decrements of total allowable acres, from 27.5 M acres (11.1 M ha) in 2014 to 24 M acres (9.7 M ha) by 2017.

Concurrent interests in achieving national energy independence [22], new technology (e.g., genetically modified crops) [23], and strong economic incentives for production of agricultural crops in recent years have resulted in many private landowners converting expiring CRP acreage back into agricultural production, especially for biofuel and feedstock crops. The impact of a land-use shift of this magnitude on the ecosystem services delivered from the Northern Great Plains has been the subject of recent investigations on carbon sequestration [8], [9], wildlife habitat provisioning [10], [24], sediment reduction [9], and others. Several studies underway in the Northern Great Plains are evaluating the importance of the region for meeting the needs of national pollination services, an industry now valued at $29B (USD) annually, with about $19B attributable to honey bees [25]. No quantitative evaluations have been conducted, however, to assess how future land-use shifts may affect national pollination services.

There have been increasing concerns about declines in pollinators over past decades [26], [27], but it was the major loss of honey bees to Colony Collapse Disorder in 2007 that brought pollination services into national focus and caught the attention of the U.S. Congress and the public [28], [29]. Approximately 70% of the world's flowering plants rely on pollinators [30]. In the United States, the agricultural industry depends heavily on the pollination services of honey bees, a commercially managed insect [31], and the Nation is the world's most active market for their pollination services [32]. Numbers of managed honey bee colonies in the United States have declined by 60% since the 1940s [29], [33], while the proportion of crops in the agricultural industry that rely on their pollination services has increased [33]. In recent years rates of overwintering losses in colonies have been around 30% in the United States [34], much higher than the 18% loss rate beekeepers anticipate on average [35], and there is a concomitant perception of waning bee health [36]. Annual reports dating back to the mid-1980s also show a downward trend in national honey production [37], [38], [39], [40], [41], [42], [43], [44], [45].

A principal contributor to declines in honey bees (as well as native bees) is the loss of floral sources of pollen and nectar in the landscape [30], [46], [47]. Worldwide, conversion of land cover has been extensive, typically to support agriculture [48]. These shifts, along with trends in using synthetic fertilizers rather than rotations of nitrogen-fixing cover types (e.g., clover, alfalfa) that are beneficial to bees, challenge bees to find sufficient floral sources [30]. Understanding how land-use change affects honey bees and other pollinators is one element needed to inform decisions affecting development of sustainable management paradigms for delivering ecosystem services to society.

The Northern Great Plains carries great importance for U.S. honey production and pollination services. Commercial beekeepers from states as far away as coastal areas transport honey bees to the prairies each summer to produce honey, a crop valued at more than $256.5M nationally in 2011, with about 50% of the contribution coming from the Northern Great Plains [49]. The region typically produces most of the Nation's honey [37], [38], [39], [40], [41], [42], [43], [44], [45], but the importance of the area extends far beyond the honey crop. In 2011, over one million colonies of bees were maintained in the region during the summer, representing nearly 40% of the total number of colonies in the United States. Essentially, all of these bees are transported throughout the Nation at other times of the year to pollinate a large variety of crops, often making multiple stops in different geographic locations. Although there have been no quantitative assessments, honey bees that spend the summer in the Northern Great Plains are thought to provide 60–85% of the pollination needs for the entire Nation (Dr. Jeff Pettis, U.S. Department of Agriculture-Agricultural Research Service, personal communication). The summer locations beekeepers use for honey production in the Northern Great Plains have tremendous influence on the nutritional status and health of honey bees because foods, especially pollen, collected and stored in hives during the summer are the primary sources for nutrition during winter and other times when natural foods are unavailable. Land-cover types with abundant supplies of polyfloral and nutritious pollen are known to positively influence honey bee health through improved immune system function [50].

Although CRP lands in the Northern Great Plains have abundant polyfloral sources of pollen for pollinators, there is little information available to quantify the magnitude of that ecosystem service across large geographic extents or its contribution to national pollination services. Government policies and programs that target land resources exert influence on land-management decisions at the local level. Decisions are made parcel by parcel, but tend to accumulate to regional patterns of change [51] and can have substantial influence on ecosystem services over broad scales. Agencies that manage land resources or administer policies and programs focused on land resources seek consistent ways to evaluate outcomes of land management for ecosystem services. Ideally, these evaluations also could be conducted in a proactive context to inform decisions about policies, programs, and management actions. Within the context of developing sustainable management paradigms for landscapes, ecosystem services that are provided in one geographic location but are dependent upon conditions from other locations require special attention because they are subject to different interests and support groups [52]. Provisioning national pollination services for agricultural crops is a good example because most beekeepers are migratory and transport honey bees from pastoral locations used for honey production to distant locations for pollinating a variety of agricultural crops [53].

We describe here a prototype model we developed to quantify the influence of land use on honey bees in the Northern Great Plains to inform decisions affecting conservation programs that target delivery of national pollination services. Our objectives were to (1) develop an approach to map regional-scale habitat suitability for honey bees based on their nutritional needs from the local landscape, (2) evaluate the efficacy of available land-cover data for monitoring effects of interannual land-cover change on landscape suitability for honey bees, and (3) demonstrate how our approach can be used to assess outcomes for honey bees from drivers of land-cover and land-use change. These objectives are the foundation for a module on honey bees in a larger, integrated assessment tool developed to evaluate effects of policies, programs, climate, and other drivers of land change simultaneously across multiple ecosystem services [54].

## Materials and Methods

We prototyped our approach for North Dakota, the Northern Great Plains state with the highest production of honey among all 50 states in most years [37], [38], [39], [40], [41], [42], [43], [44], [45]. We identified local landscape criteria for honey bees and designed a spatially explicit model to locate places in the North Dakota landscape that could meet these land-cover requirements. We used available land-cover data to map interannual changes in landscape suitability, then applied our model to these maps as well as to maps we developed to fit a scenario based on recent drivers of land-use change.

### Landscape criteria for honey bees

We specified landscape criteria based on industry standards professional beekeepers use to place hives in the Northern Great Plains. These criteria consider the potential of sites to produce honey crops and to provide the nutrition needed to ensure the health of honey bees. We developed landscape criteria to support apiaries of 100 hives (Table 1, "original criteria"). This is approximately twice the number of hives a commercial beekeeper would maintain in an average apiary setting for this part of the country; however, competition for locations to place hives in the summer is high, and it is common to find multiple apiaries of 40–48 or more hives placed within overlapping forage ranges by multiple beekeepers. In many cases the cumulative stocking rate for a local landscape may exceed 100 hives.

### Modeling approach

Apiaries in the Northern Great Plains typically are maintained in grassland settings within proximity of roads so hives can be transported. Therefore, we first identified grassland settings sufficient in size to place commercial numbers of hives. We merged information from two sources to develop a map of grassland cover. We used proprietary data on lands with graminoid/herbaceous land treatments enrolled in the CRP [7]. These data were delineated by the Farm Service Agency with orthoimagery of high spatial resolution (typically one to several meters), and boundaries were very accurate for our type of application. For non-CRP lands we relied on grassland cover mapped from satellite data by the North Dakota Gap Analysis Program (NDGAP) [55]. The NDGAP data had lower spatial resolution (30 m) and only a fair level of mapping accuracy for grasslands (somewhat better than 50%), but provided wall-to-wall data on land cover with sub-classes that allowed us to exclude saline and sand prairie grasslands, which we deemed unlikely to provide good floral conditions for honey bees. We merged NDGAP and CRP grasslands into a single map, then fragmented the resulting grassland areas by intersecting them with the road network for North Dakota [56] to measure accessibility for transporting hives.

North Dakota requires beekeepers to register locations of their apiaries by the "quarter section" designation from the United States public land survey system [57]. We used the areal extent of a quarter section, approximately 65 ha (160 ac in the U.S. land system), as a minimum size threshold for eliminating grassland polygons that were less likely to support 100 bee colonies. We also eliminated grassland polygons that were too far from road access via information from a national gridded layer representing Euclidean distance to the nearest road [58], [59]. The road network across North Dakota largely is laid out as a grid of north-south and east-west routes, with road spacings of a mile (1620 m; although, many farms also have field access roads every half-mile or quarter-mile). We rationalized that grassland areas more than a mile from the nearest public road would be infeasible for a beekeeper to place hives; thus, we removed grassland areas having centroids (approximate centers) farther than 1620 m (1 mi) from a road. Only 0.6% of the area of the state was farther than a mile from the nearest public road, and most of this area was associated with badlands, river breaks, and large expanses of water (e.g., reservoirs). In total this criterion removed of 0.2% of the grassland centroids from further consideration as potential apiary sites. Remaining grassland polygons were at least 65 ha in size and had edges within reasonable proximity to road access. We considered the locations of centroids of these remaining polygons as potential apiary sites, though in reality an apiary could be sited anywhere within a grass polygon.

We needed to quantify acreages of desired land-cover types (per Table 1, "original criteria") within forage distance from potential apiary sites. Forage distance for honey bees varies based on floral availability, landscape structure, and ambient conditions, such as from wind (e.g., [60], [61], [62]), but beekeepers typically look for sites providing good cover types within a distance of about 3.2 km from the hive [63], defining a forage area of 3255 ha around an apiary. We developed a separate map layer for each land-cover class of interest, determining that we needed to use a common spatial resolution of 10 m across all layers to retain certain important features (see section on "Compiling annual maps of land cover").

Conceptually, we wanted to move a circular forage window across North Dakota to summarize land-cover classes at each potential apiary location. This is a challenge for geospatial analytical software at the 10-m spatial resolution we used because each land-cover layer is well over 47,000 rows by 77,100 columns, and a circular forage window would encompass about 325,500 cells and require geometric calculations to determine which cells fell within the forage radius around an apiary centroid. We overcame these challenges with a modified moving-window approach (Figure 1). First, we applied a 10×10 rectangular moving window over each 10-m binary map to calculate the proportion per hectare of the respective land-cover type, then saved the results at a 100-m (one ha) cell size, yielding output layers having 0.01 times the number of cells of the original, binary layers. For example, if the original maps showed that a 10-cell×10-cell area contained 30 cells of grassland, 35 cells of non-suitable cover (such as corn and roads), 20 cells of alfalfa, and 15 cells of sunflowers, the resulting ha-sized cell would carry the information of 30% grassland, 20% alfalfa, 15% sunflowers, and 35% other (although the information for each cover type would be stored as a separate map layer, as shown in Figure 1). Second, we applied a 57×57 rectangular moving window (dimensions needed to survey acreage comparable to the forage area around an apiary) across each 100-m raster layer to calculate the amount of each cover type within forage distance from the center of the window, advancing the window one cell (100 m) at a time. We avoided the compute-intensive, circular moving window with a square window of the closest possible areal match to the size of the desired forage

**Table 1.** Landscape criteria used to identify potential locations for apiaries that would ensure sources of pollen and nectar throughout the growing season in the Northern Great Plains.

| **Typical timing of flowering for major beneficial cover types in the Northern Great Plains** |
|---|
| Deciduous trees/shrubs[a]: Second half of May |
| Canola: First 3 weeks in June |
| CRP[b] grassland and comparable mixed grass/forb: 3rd week of June through end of growing season (sometime in September) |
| Oilseed sunflowers: 3rd week of July through end of August |
| Second-crop[c] alfalfa: Mid-July through end of growing season (sometime in September) |
| **Original criteria:** |
| ≥65 ha (160 ac[d]) deciduous trees/shrubs |
| ≥130 ha (320 ac) CRP[b] grassland or comparable mixed grass/forb cover |
| ≥65 ha (160 ac) alfalfa |
| ≥65 ha (160 ac) oilseed sunflowers |
| ≥65 ha (160 ac) canola |
| presence of surface water[e] |
| **Relaxed criteria:** |
| ≥130 ha (320 ac) CRP[b] grassland or comparable mixed grass/forb cover |
| ≥130 ha (320 ac) any combination of alfalfa, oilseed sunflowers, and canola |

[a]Occurrence of deciduous trees and shrubs in the Northern Great Plains generally is limited to windbreaks, fence lines, riparian corridors, and landscaping around houses.
[b]Land enrolled in the Conservation Reserve Program. Of particular interest for honey bees are grasslands rich in leguminous species because they offer long-flowering land cover with good sources of pollen and nectar.
[c]The first cutting of alfalfa for hay typically is done prior to blooming, but flowers are present following re-growth between subsequent cuttings.
[d]Acres (ac) are the unit of measure for the property system in the United States and are provided here because they relate to multiples of standard crop field sizes.
[e]Bees collect water for evaporative cooling in the hive, so our criteria recognize this landscape requirement in addition to sources of pollen and nectar.

area. Given the imprecision of land-cover data and forage behavior by bees (influenced by wind, air temperature, and floral density, in addition to land cover type), having a round, rather than square, window was not worth the extra processing burden for a statewide application. Third, we summarized acreages across land-cover types within forage distances around all 100-m cells that had been identified as potential apiary locations to determine which sites met the areal thresholds for required land-cover types (per Table 1).

## Compiling annual maps of land cover

We compiled annual land-cover maps from multiple sources, as no single source provided all information needed to assess landscape suitability for bees. We encountered expected differences in data vintages, resolutions, and origins (see Table S1 for information sources), as well as product quality. We used the Cropland Data Layer produced annually by the National Agricultural Statistics Service (NASS) [64] for information on alfalfa (*Medicago sativa*), canola (a group of cultivars developed from *Brassica napus* and *B. campestris*), and sunflowers (*Helianthus annuus*), the three crop types needed for our analysis. All three types have been distinguished annually in map products in North Dakota since 2002, prior to which alfalfa was aggregated with grassland hay crops not nutritious for bees. We used information on woody cover types from the same two sources we used for grassland cover, the CRP and NDGAP products. The dataset for CRP-enrolled lands contained information on the years for which the conservation treatments were under contract, which enabled us to update our land-cover maps annually for both woody and grassland parcels. The NDGAP data we used for the remainder of the landscape was a single, static map derived from a collection of Landsat imagery spanning 1992–1998. We amended the NDGAP information with the relatively high-resolution wetland cover delineated by the National Wetlands Inventory (NWI) [65], as bees need sources of water and wetlands typically are poorly represented in land-cover products generated from the mid-resolution satellite data used by NDGAP [66]. The NWI map was derived from high resolution orthoimagery, but similar to the NDGAP product, represented a static depiction of wetlands derived from a collection of different orthophoto dates mosaicked across North Dakota.

The annual maps we assembled spanned 2002, the earliest year for which we could distinguish all three crop types of interest, through 2006, the latest year for which we had annual data on CRP-enrolled lands (although, we later received updated data on CRP enrollments that enabled the example application we describe in the next section). We filled the remainder of the landscape each year with data from the static NDGAP and NWI product sources. We processed all geospatial data in an equal-area map projection (following parameters used by the U.S. Geological Survey for the conterminous United States, see [67]) to maintain appropriate areal properties. All land-cover information was converted to raster layers of 10-m cell size, the spatial resolution needed to retain features from the finest-grained sources, the CRP and NWI data. We developed binary maps ("1" for presence and "0" for absence) for each land-cover type of interest, including alfalfa, canola, oilseed sunflowers, grasslands (CRP grasslands and NDGAP [non-saline and non-sandy] grasslands), wetlands, and woodlands (CRP and NDGAP woodlands).

## Assessing outcomes from a scenario of land change

We used a scenario that captured contemporary drivers of change in the Northern Great Plains to evaluate if our model could reflect influences from programs and policies on landscape

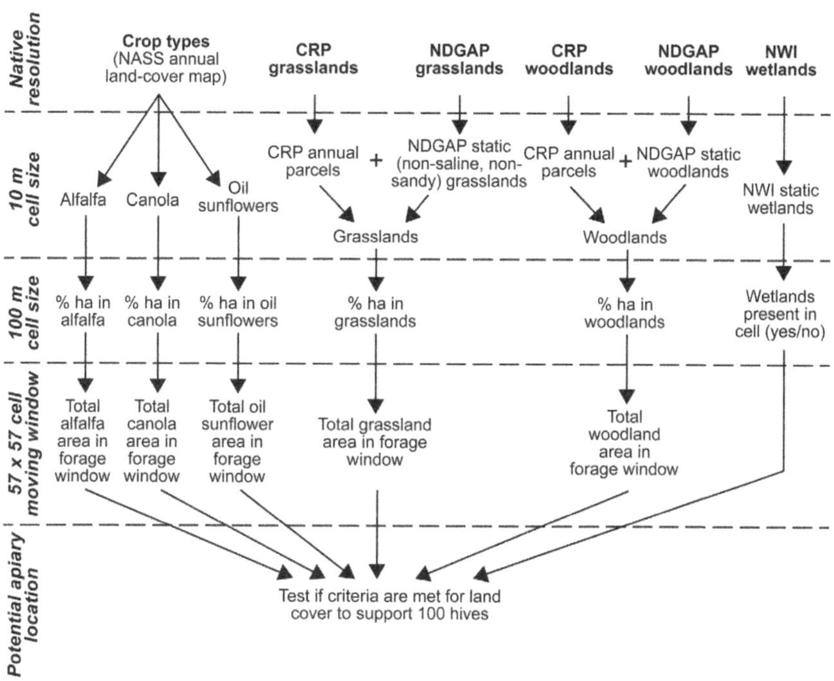


**Figure 1. Geospatial processing schematic for compiling annual land-cover maps.** We extracted land-cover components needed to meet criteria shown in Table 1 and saved them as separate gridded layers at 10-m cell resolution to retain needed spatial detail. We then summarized the proportion of area per hectare for each land-cover type (except wetlands, for which we recorded only presence/absence) and saved results as separate gridded layers at 100-m cell resolution to reduce the dimensions of the data layers. We used a moving filter of 57×57 cells to tally land-cover elements over an area comparable with the forage zone for honey bees, and queried the tallies for grassland locations we previously had identified as potential sites for placing apiaries. Figure abbreviations: NASS – National Agricultural Statistics Service, CRP – Conservation Reserve Program, NDGAP – North Dakota GAP Analysis Program, NWI – National Wetlands Inventory.

suitability for honey bees. Owners of agricultural lands in North Dakota have been avid participants in the CRP. Nationwide, approximately 28 million acres (11.3 M ha) under CRP contract were set to expire between 2007 and 2010 [68]. During that same time frame, production pressures for corn-based ethanol and soybean biodiesel, the two major crops used to produce biofuels in the United States [69], [70], were on the rise [71], [72] from requirements for substantial increases in national use of renewable fuels [22]. In addition, commodity prices for corn and soybeans have been trending upwards since the mid-2000s [73]. We therefore compiled land-cover maps for 2002 and 2010 to capture conditions before and after the convergence of these pivotal influences of land-use change.

Initial results from the analysis of landscape suitability (discussed in the "Results" section) motivated us to modify some of the source information for this scenario application. We used CRP land-cover information, as before, but changed the second source of data on grasslands and woody cover from NDGAP to the 2001 National Land Cover Database (NLCD) [74]. The NLCD offered notably better mapping accuracy [75] than the NDGAP product [55] (Table S2), though did not provide information distinguishing sandy and saline grassland subtypes. For cropland information we extracted pixels from NASS maps for 2002 and 2010 across all crop types to obtain general extent of cropland for each of the years, and used specific information for canola and sunflowers to distinguish those two cover types. The accuracy for alfalfa on the NASS cropland maps was not very high; total acreages represented for alfalfa departed considerably from statistics reported from ground-based surveys conducted by NASS. Mapped alfalfa was 16x the acreage reported from ground-based surveys for 2002 and 5x the acreage reported for 2010 [76]. In addition, the distribution of alfalfa pixels tended to have a "shotgun" appearance, with many isolated pixels scattered across the landscape, rather than occurring in contiguous patches as would be expected for agricultural fields.

We developed our own maps of potential alfalfa distribution with remotely sensed data from the Landsat Enhanced Thematic Mapper sensor obtained through the Web Enabled Landsat Data (WELD) online distribution site [77], [78]. WELD data are terrain corrected and spectrally calibrated to top-of-atmosphere reflectance. We used the data layer for the Normalized Difference Vegetation Index, a data transformation that highlights photosynthetic activity in plants [79]. We targeted the WELD composite for October because it is a month in which alfalfa remains green in the Northern Great Plains after other cover types (with the exception of some grasses) have senesced for the season. For 2010 we used the WELD composite for November to fill any data gaps in the October composite. The year 2002 was exceptionally cloudy for North Dakota overpasses, so we substituted data from 2003, with the rationale that alfalfa is managed on a four- to eight-year rotation cycle in the Northern Great Plains [80], therefore minimizing potential differences in the distribution of alfalfa between 2002 and 2003. We filled data gaps in the WELD October composite for 2003 with information from the WELD Autumn composite for that year, as the November composite was contaminated with widespread clouds and snow. To generate distributions of alfalfa we simply selected thresholds of green biomass response that resulted in relatively comparable total acreages reported from the 2002 and 2010 NASS ground-based surveys. We had no means to determine the accuracy of our results, but were satisfied we had usable examples for the scenario application, given total acreage tallies and the fact our maps had

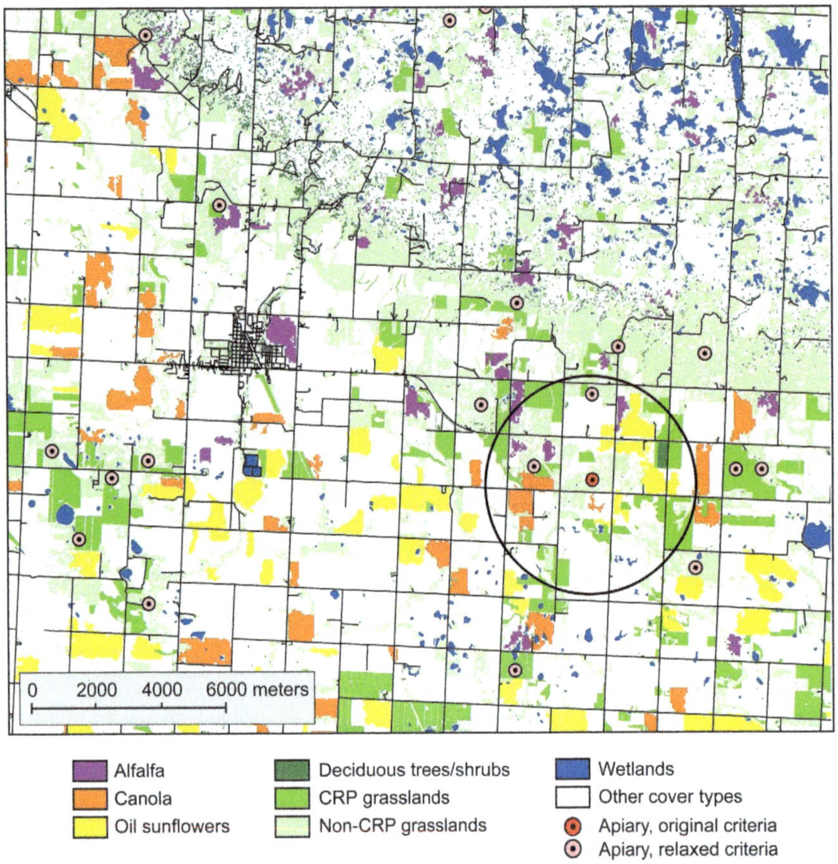


**Figure 2. Predicted apiary locations resulting from original versus relaxed criteria for 2006.** The original criteria resulted in only one apiary site identified (circle shows forage zone for this apiary), but relaxed criteria resulted in many more potential sites. Beekeepers often install apiaries in close proximity to one another in good landscapes such that forage zones are overlapping, as would be the case in this figure.

much more contiguous patches of alfalfa than did the NASS cropland maps. Still, our alfalfa maps also captured grassy areas that remained green late in the year, such as along drainage courses, in other depressional areas, and in urban settings. We make no claim as to how well our land-cover compilations represented actual North Dakota land-cover patterns for 2002 and 2010 (other than the CRP data would have been fairly accurate), but our maps did provide total acreages of key land-cover types in amounts approximating those reported by NASS.

## Results

### Output from the modeling approach

A relatively sparse number of locations met our initial land-cover criteria. We found the requirements likely were challenged by limitations in the source information we used for the annual land-cover maps. Less than half the woodlands in the landscape actually had been mapped correctly as woodlands in the NDGAP product [55], which made it hard to meet the acreage criterion for early season flowering shrubs and trees. Meeting certain crop requirements also was difficult. For example, in 2006, only 20% of the alfalfa was mapped accurately in the cropland data layer [81], meaning that 80% of the alfalfa in the actual landscape was not classified as alfalfa in the crop map. We relaxed the land-cover criteria to overcome these limitations in the source data (see Table 1, "Relaxed criteria"), then re-ran the landscape model. The model was able to find many additional locations that could meet the relaxed criteria (Figure 2), though these locations might not support apiaries as large as the 100 colonies our original criteria were defined to support.

North Dakota requires that all apiaries be registered with the state by quarter-section legal units, so we obtained (from the North Dakota Department of Agriculture, Beekeepers Association) and mapped annual registered locations of apiaries for visual comparison with potential apiary locations produced by our model (Figure 3). We did not perform a quantitative comparison between modeled and registered sites because registered apiary sites can contain any number of beehives (based on the discretion of the beekeeper), and our model was parameterized to identify landscapes that could support large apiaries. Also, once a beekeeper registers an apiary location, they tend to re-register that same location over time, regardless if they stock it with bees, as most beekeepers do not own the property on which their hives are located and are motivated to maintain ongoing agreements with landowners. To validate this assumption we constructed a time series of registered apiary locations for North Dakota from 1981 (90 registered sites) to 2010 (10,054 registered sites). We found that the total number of sites increased every year (e.g., Figure 3B-C shows registered sites in 2002 and 2010, respectively), and rarely did any site drop from the registry. We selected 2002 to compare the distribution of locations identified with our model and registered locations for apiaries (Figure 3) because that year had the highest rates of mapping accuracy in the cropland maps for the three crop types of interest. An informal visual comparison

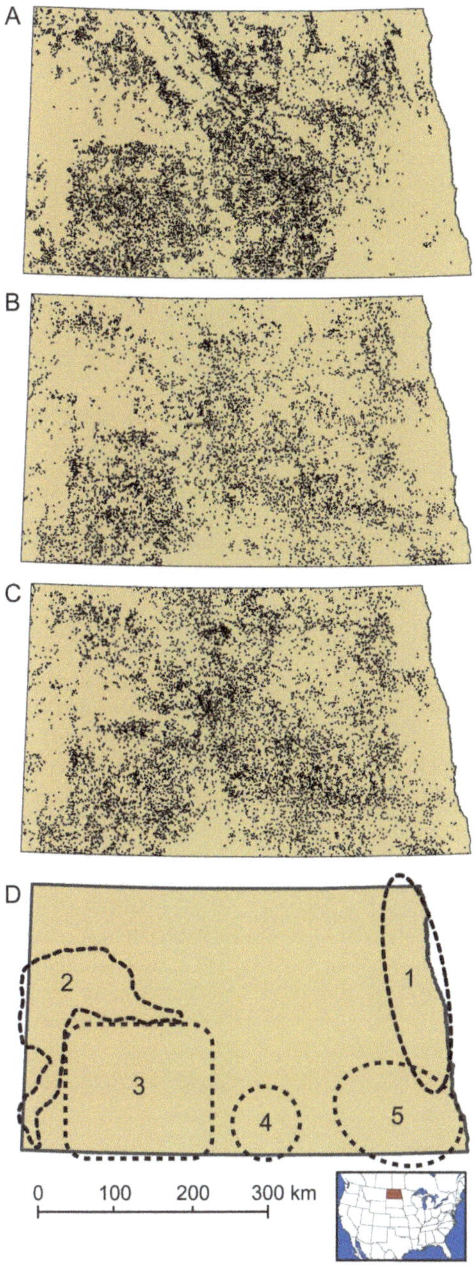


**Figure 3. Predicted versus registered apiary locations (note: 1 dot = 1 apiary location).** The cropland data we used had the highest mapping accuracies for the three crops of interest in 2002 (metadata on accuracy for all years is available through [82]). The distribution of sites we predicted to support large apiaries in 2002 (A) shares general similarities with apiaries registered for 2002 (B). Distribution of registered sites has grown more dense over time, as shown by a map of sites registered for 2010 (C). Hand-delineated areas of particular agreement (1, 2, and 3) and disagreement (4 and 5) of point distribution patterns between predicted and registered apiaries in 2002 (D), as described in the Discussion.

revealed general similarities in statewide patterns of predicted locations and locations registered by beekeepers, but a notable difference is our model did not identify many locations for large apiaries in the southeastern part of North Dakota, though many apiary sites have been registered there (Figure 3D, area 5).

## Annual maps of land cover

Our annual series predicting potential apiary locations revealed a legacy of patterns from the mosaic of Landsat imagery used by NASS to develop the crop-type maps, particularly for years 2002–2004 (Figure 4). Despite such issues associated with satellite image boundaries, many suitable locations were identified throughout the central swath of the state in all years. In contrast, we found few locations in the eastern, southeastern, and central-west parts of the state for any year (Figure 3D, areas 1, 2, and 5). Interestingly, our model consistently predicted a dense distribution of potential locations in south-central North Dakota that is not echoed by the record of registered apiaries (Figure 3D, area 4).

## Assessing outcomes from a scenario of land change

An important motivation to evaluate effects of land change on honey bees is to provide information on how resource manage-

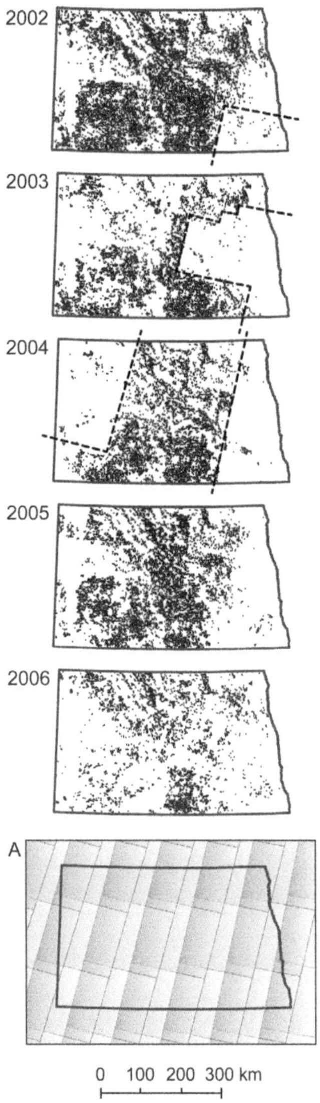


**Figure 4. Effects of mosaicking satellite images for mapping crop types on subsequent application of resulting crop maps (note: 1 dot = 1 apiary location).** Annual predicted apiary sites for 2002–2006 and Landsat scene boundaries (map A) from (side-lapping) north-south orbital paths. The influence of satellite scene boundaries on the crop-type maps derived from the set of Landsat overpasses needed to provide statewide coverage of cloud-free data are most apparent in our apiary predictions for 2002–2004 (note dashed lines). The various dates of imagery pieced together to provide statewide coverage of data often represent differences in phenological timing or other environmental conditions, making it difficult for image analysts or algorithms to detect comparable cover types across scenes [66]. This example highlights limitations of current large-area land-cover products for applications such as ours that rely on local-scale landscape components.

ments programs and policies can influence pollination services. We highlighted predicted locations that depended on CRP-enrolled lands to meet criteria for siting apiaries (Figure 5). Landowners throughout North Dakota participate in the CRP, but the benefits for honey bees may best be realized in the eastern/southeastern part of the state, where our model found few locations outside of CRP-enrolled lands to meet the criteria for siting large apiaries.

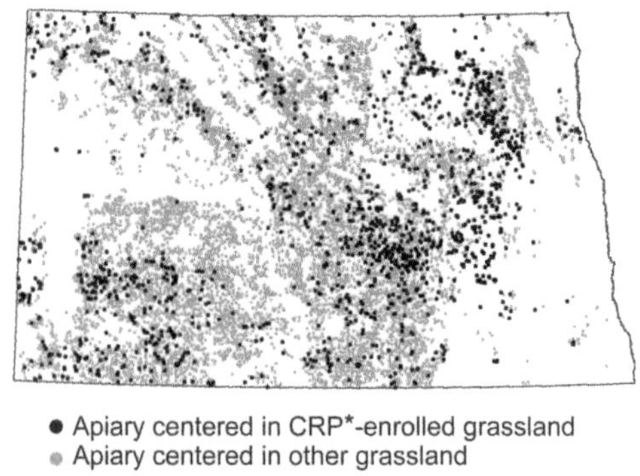


**Figure 5. Influence of the Conservation Reserve Program (CRP).** In this map we distinguish potential apiary sites that met criteria only because CRP-enrolled grasslands provided a suitable place for locating hives. This example shows how our model could be used to distinguish contributions to apiary-friendly landscapes from different sources of land management.

Our results for the scenario application representing pre- and post-incentive years to expand production of biofuel crops revealed a reduction in landscape suitability for large apiaries between 2002 and 2010 (Figure 6). The model identified 10% fewer locations for large apiaries in 2010 than in 2002 across all potential grassland centroids, but the decrease in centroids located in CRP parcels was 60%. The number of CRP-enrolled polygons available as potential apiary centroids in North Dakota declined by about 15% between 2002 and 2010, but geographically this decrease represented 24% less surface area enrolled in the program in North Dakota. Coupled with changes in the surrounding land cover, much of the CRP land remaining in 2010 was unable to meet the local landscape criteria for large apiaries. This was most evident in eastern North Dakota, but substantial reduction in CRP-related sites was evident throughout most of the state.

## Discussion

Our approach to assess landscape suitability for honey bees provides the foundation for a module in a larger, integrated framework to assess multiple ecosystem services simultaneously to inform decisions on land management and policy, as well as evaluate outcomes from government conservation investments [54]. Competing drivers of change in the agricultural industry can have substantial influence on the Nation's ability to sustain the annual demand for pollination services by honey bees. For example, conservation programs supported by the Food, Conservation, and Energy Act of 2008 [20], which generally amended the conservation programs from earlier farm legislation (e.g., the CRP was in the Food Security Act of 1985 [83], have encouraged increased acreage of floral sources beneficial for bees. Conversely, the Energy Independence Security Act of 2007 [22] laid requirements for considerable increases in the volume of biofuels produced over the next decade, which currently encourages expansion of acreage in corn and soybeans, crops that provide little nutritional value for bees in the Northern Great Plains. The model we developed enables assessing outcomes from such competing influences on landscape suitability for honey bees.

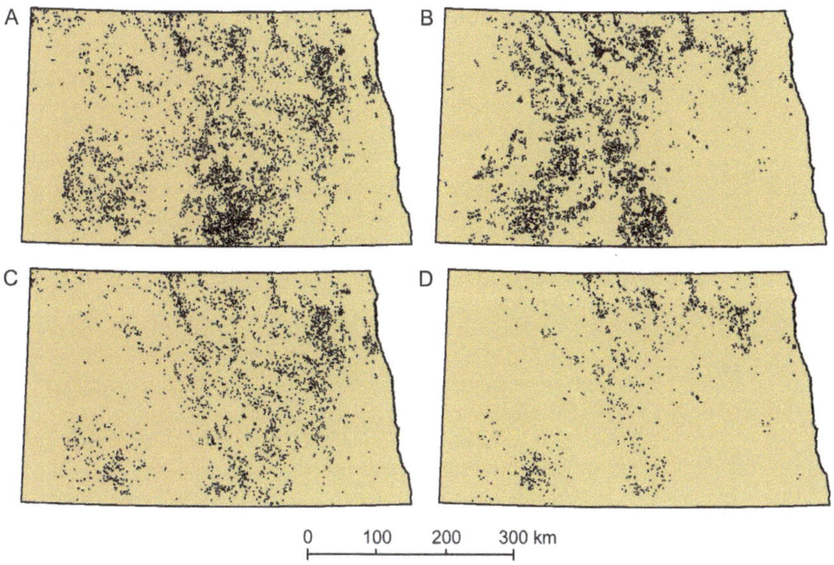


**Figure 6. Results from a scenario of land-use change (note: 1 dot = 1 apiary location).** A scenario influenced by high commodity prices and incentives to expand production of biofuel crops resulted in a loss of local landscapes suitable for apiaries. The distribution of all grassland centroids meeting criteria to support large apiaries in 2002 (A), prior to land-change incentives, is more extensive than in 2010 (B), subsequent to incentives. The same is true for the distribution of CRP centroids that met the apiary criteria in 2002 (C) versus 2010 (D).

## Output from the modeling approach

Informal visual comparison of the distribution of locations predicted by our model for 2002 with registered locations of apiaries for that year exhibited some general similarities (Figure 3). Both maps depict ample landscape opportunities for apiaries throughout the central swath of the state. Both maps agree the arid, erosional landscapes of the badlands and river breaks in western North Dakota and the crop types of the wall-to-wall cropland in the Lake Agassiz Plain in eastern North Dakota hold lower value for honey bees (Figure 3D, areas 1 and 2, respectively), and the maps show a strong pattern of agreement for the distribution of good apiary locations in the Missouri Plateau (Figure 3D, area 3). The maps differ most in the southeastern part of the state (Figure 3D, area 5), where our model found few locations for supporting large apiaries, yet many apiary sites are registered. It is likely these apiaries support far fewer colonies than in more floristically diverse landscapes elsewhere in the state, though this has not always been the case. According to beekeepers who have operated in southeastern North Dakota for many decades, it was rare to see corn or soybean crops in this area a decade ago, but now there is extensive cropland planted in corn, soybeans, and wheat, which are not very nutritious for bees (although, soybean varieties grown elsewhere in the United States can be beneficial for honey bees at times). In previous decades, southeastern North Dakota had many more acres of CRP, alfalfa, sunflowers, and canola (Doug Ruby, Ruby's Apiaries, personal communication). Data from the Farm Service Agency show that land enrollment in the Conservation Reserve Program for southeastern counties of North Dakota totaled 103,856 ha in 1990, rose to 163,472 ha by 2000, then dropped to 135,553 ha by 2010, and continues to drop (120,874 ha by 2013) [84]. Maps produced by the U.S. Department of Agriculture for the agricultural census series show the distribution of alfalfa has been decreasing in southeastern North Dakota since at least 1978 (though we note there is no indication canola or oilseed sunflowers were grown in much quantity in this part of the state during the past few decades [85], [86], [87], [88], [89]. Migratory beekeepers have little flexibility for shifting their operations to new regions of North Dakota to accommodate changes in land cover and instead adjust the sizes and numbers of colonies per apiary to compensate for diminished acres of forage. Given the current cropland types grown wall-to-wall in southeastern North Dakota, the benefits of land enrolled in conservation programs there can be especially important for supporting honey bees, an outcome highlighted by our model results (Figure 6). Beekeepers in that area have experienced sharp declines in honey yields as a result of these changing cropping trends, and have been required either to reduce per-site hive allotments or to abandon sites (Doug Ruby, Ruby's Apiaries, personal communication). In contrast, our model found more locations for supporting apiaries than were reflected by the distribution of registered sites in the southern part of the Prairie Coteau Slope region (see Figure 3D, area 4). The distribution of registered apiaries is not very dense in this area, but our model found plenty of locations suitable for apiaries across all years (Figure 4) because the area had many CRP-enrolled parcels, abundant acreage in oilseed sunflowers, and moderate acreages in alfalfa and mixed grass prairie.

Results from our analyses helped us identify several ways to refine our model. First, the approach we developed to identify potential apiary locations was successful in implementing local landscape requirements across a statewide extent, but the all-or-nothing recipe for 100 hives is limiting. A shift from a model based completely on expert opinion to one developed statistically from data on number of hives and honey production relative to landscape setting would enable predicting suitability across a range of apiary sizes, as well as contribute information towards understanding how land-cover changes in the Northern Great Plains could affect the supply of honey bee colonies for national pollination needs. Second, we also should incorporate seasonal influences from weather into our model because of effects on plant flowering and nectar flow. Even if land-cover types were held constant across years, different weather conditions would provide substantially more or substantially less pollen and nectar because of variations in moisture and growing degree days. Third, we used

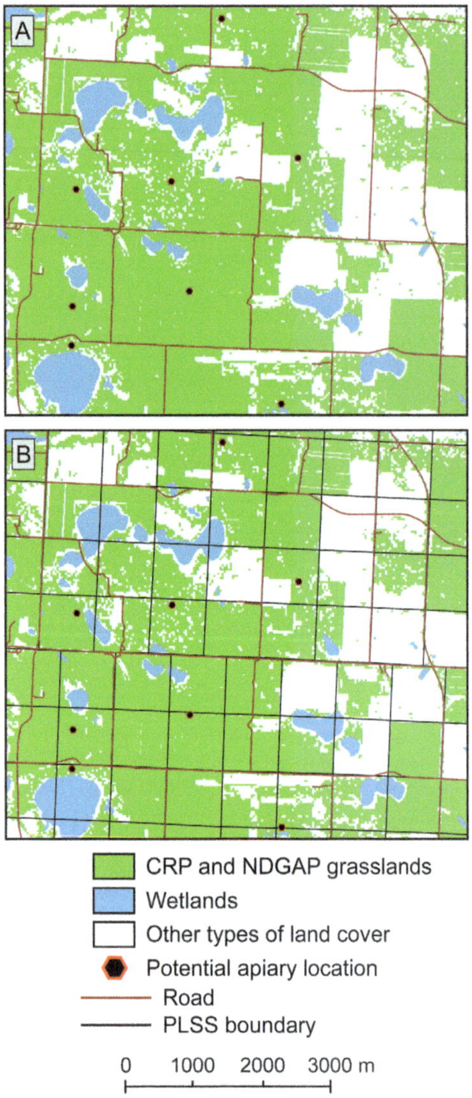


**Figure 7. Approach we used versus future algorithm modification for identifying potential grassland sites for placing apiaries.** We identified centroids (grassland locations for potential apiary sites) based on criteria for minimum extent of grassland acreage and proximity to roads (A). Density of potential site locations would increase if, instead of using grassland centroids, we applied the spatial framework of the U.S. Public Land Survey System used for registering apiary locations (B). We could maintain our same criteria for grassland extent and road access, but could have multiple apiaries within a large grassland patch, a situation that often occurs in the actual landscape. Figure abbreviations: CRP – Conservation Reserve Program, NDGAP – North Dakota GAP Analysis Program, PLSS – Public Land Survey System.

centroid locations within grassland patches that met criteria for areal extent and proximity to roads to identify potential apiary sites. For future model refinement we could retain the areal and proximity requirements for grasslands, but instead use centroids from the legal mapping units (quarter sections) that beekeepers register. This would densify the number of point locations around which we could analyze land-cover components (Figure 7) and would better mimic the concentration of overlapping forage areas we have observed *in situ* with commercial apiaries.

## Using available land-cover data to monitor annual changes in landscape suitability for honey bees

The data we assembled to compile a land cover time series for North Dakota were a mix of annual and static information, and the vintages of the static sources were not always well matched to the annual sources. The Northern Great Plains have had low rates of annual land-cover conversion (typically a fraction of a percent of the total area per year) [11] during the years of our analysis and in preceding decades, providing some reassurance our assembled landscapes were relatively representative of this rather stable region of the United States. The principal land-change dynamic in the Northern Great Plains during this period was the influence of the CRP [11], for which we did have temporally dynamic data, as well as annual changes in crop types, for which we had the annual NASS data. One implicit assumption in our approach, however, was that grasslands were suitably rich in season-long floral sources. We know of no data on the floral components of land cover (except for isolated, local cases) and currently have no means to refine this assumption.

For initial trials of our model we used the NDGAP land-cover product because it offered an array of grassland and woodland subclasses that allowed us to discard those we expected to be poor for bees. The downside to using this product was relatively low rates of mapping accuracy for grassland and woodland classes. For that reason we used non-cropland land-cover information from the NLCD for our subsequent example application because NLCD had a better rate of accuracy for mapping woodland and grassland cover (Table S2) [75], [90], [91]. In addition, the NLCD provided land-cover products for 1992, 2001, and 2006 [74], which we could more easily crosswalk (than the single NDGAP product) to the vintages of our other land-cover information, and the NLCD continues to provide new national maps of land cover.

Until recently, annual mapping of land cover for the United States has been untenable at mid-resolution spatial scales. Improvements in technology and free access to Landsat data [92] have advanced this capability, as evidenced by national-scale annual products from NASS [64]. The methods employed to classify land-cover types across extensive geographic areas, however, produce maps better suited for applications that summarize outcomes at regional scales, rather than applications like ours that rely on local-scale accuracy. The results we generated based on annual land-cover maps compiled from NASS, NLCD, and NDGAP data inherited patterns of inaccuracies from those maps (e.g., Figure 4) that underscore limitations in the current state of operational development of large-area land-cover products. These maps fall short of the accuracy and precision needed for local-scale applications such as ours, an observation that has been noted elsewhere (e.g., [93]). Our algorithm relies on tallying land-cover components within a 3.2-km radius around each potential apiary site throughout the state. Local errors in individual pixels therefore can have strong influence on our results. We conclude that current operationally produced land-cover maps do not provide sufficient local accuracy for monitoring landscape suitability for honey bees. This finding underscores the need for those conducting similar habitat assessments to exercise caution in reporting shifts in landscape suitability and to evaluate if they signal mapping inaccuracies, rather than true landscape change.

There are legitimate reasons for the errors in operational-scale land-cover products. Spatiotemporal ambiguities in the spectral characteristics of land cover make it difficult to optimize a classification system for all cover types at once for all places in the landscape [66], [94], as demonstrated by image pattern effects on our annual results (Figure 4). For this reason, NASS focuses on

**Table 2.** Comparison of acreage planted in 2002 and 2010 in North Dakota for corn and soybeans, the major crop types used to produce biofuels in the United States, versus other cover types that provide important sources of pollen and nectar for bees.

| Commodity | Hectares planted in 2002 (reported acres[c]) | Hectares planted in 2010 (reported acres[c]) | Change in hectares (change in acres[c]) |
|---|---|---|---|
| Alfalfa[a] | 44,515 (110,000) | 32,375 (80,000) | −12,140 (−30,000) |
| Canola[a] | 526,095 (1,300,000) | 518,000 (1,280,000) | −8095 (−20,000) |
| Sunflowers[a] | 554,420 (1,370,000) | 358,150 (885,000) | −196,270 (−485,000) |
| CRP-enrolled lands[b] | 1,346,345 (3,326,883) | 1,100,510 (2,719,413) | −245,835 (−607,470) |
| Corn[a] | 497,765 (1,230,000) | 829,610 (2,050,000) | 331,845 (820,000) |
| Soybeans[a] | 1,080,515 (2,670,000) | 1,659,220 (4,100,000) | 578,705 (1,430,000) |

[a]Data are from probability-based surveys of farmers conducted by the U.S. Department of Agriculture [76].
[b]Data are total acres enrolled, provided online by the U.S. Department of Agriculture [97].
[c]Acres are the unit of measure for the property system in the United States and are reported in the literature we accessed for U.S. agricultural statistics.

optimizing their annual land-cover algorithms for a handful of important crops, though their maps do distinguish numerous other crop types. NASS assesses mapping accuracy for all crop types, but not for other classes of land cover. Hence, although the availability of annually updated, wall-to-wall maps of land cover is attractive for a growing population of users, the NASS maps are best employed for applications centered on the land-cover classes NASS maps with high accuracy. The metadata for the NASS cropland data layers we used (via [95]) showed the accuracy for mapping alfalfa in 2002, for example, ranged from 19% to 95% for different parts of North Dakota, and for 2004, ranged from 40% to 97%. The accuracy for mapping canola across North Dakota ranged from 88% to 97% in 2002 and from 86% to 99% in 2004. Likewise, the accuracy for mapping sunflowers varied from 80% to 99% in 2002 and 63% to 99% in 2004. The best rates of mapping accuracy at the statewide level across all years were achieved in 2002 (77% accuracy for alfalfa, 93% for canola, and 90% for sunflowers). However, even in this year when 77% percent of the land area planted in alfalfa was correctly mapped as alfalfa, the commission rate for that cover type was 55%, meaning that over half the time a pixel was called "alfalfa" it was not really alfalfa on the ground. This type of error likely contributed to the great density of locations we identified as suitable for large apiaries in 2002 compared with other years (Figure 4). We found far fewer locations that met our landscape criteria in 2006, which likely was more an artifact of a change in mapping methods used by NASS than a change in crop distribution. The NASS product for 2006 is much cleaner in appearance than in previous years, but does not provide improved mapping accuracy (alfalfa in 2006 had 80% omission error and 39% commission error; see [82]). One way NASS might improve mapping accuracy for alfalfa is to extend the range of seasonal dates of Landsat imagery they use. Alfalfa, along with some grasses, in the Northern Great Plains is one of the earliest cover types to become green in the spring, and remains green into the fall after other cover types have senesced.

### Using the model to assess outcomes for honey bees from a scenario of land-cover change

Although we found that the land-cover products currently available lack sufficient local accuracy to enable monitoring of actual landscape suitability for honey bees, our model does achieve the goal to provide a way to evaluate how different policies, programs, and market incentives could influence land cover and alter landscape suitability for honey bee colonies. Land-cover maps can be developed to meet various scenario criteria (e.g., [96]), and our model then could be applied to assess how distribution patterns of apiary locations would be affected.

The scenario we employed showed marked differences in landscape suitability prior and subsequent to agricultural and energy drivers currently influencing change in the Northern Great Plains (Figure 6). The landscapes we assembled met reported abundances of crop types important for bees in amounts comparable to those reported by NASS from ground-based sources. Ground-based statistics for 2002 and 2010 indicated decreases in acreages for alfalfa, canola, sunflowers, and CRP-enrolled land, and corresponding increases in acreages of corn and soybeans, the two major crop types associated with biofuels (Table 2) [98]. Further corroboration for shifts in land cover that affected landscape suitability is offered by the U.S. Department of Agriculture's findings that native grasslands were converted to croplands in the Northern Great Plains at a rate of about 1% (311,610 ha) per year from 1997–2007 [12], and that acreage in principal crops for North Dakota increased 117,765 ha between 2002 and 2010 [71], [99]. Also, as of 2010, North Dakota was among the top ten U.S. states for production of ethanol and biodiesel, and the amount of corn used for production of biofuels has increased sharply since 2002 [100].

## Conclusions (and Future Directions)

We developed a modified moving-window approach to tally land-cover components within local-scale proximity to potential apiaries for statewide or larger regional assessments of landscape suitability for honey bees. Our work responds to a need to estimate effects from various scenarios of land change that can result from major drivers such as market values, shifts in climate, or how national programs and policies for land resources are implemented. Because the approach is sensitive to local-scale information, results based on existing land cover time series generated from remotely sensed data may reveal more about limitations in land-cover products than about actual land-cover change. Our model therefore currently is more valuable for assessing outcomes from different scenarios of land-use change than for monitoring actual changes in landscape suitability. New approaches for monitoring land-cover change with remotely sensed data are emerging (e.g., [101], [102]) that eventually will enable our model to be used as a monitoring tool in addition to a scenario-assessment tool.

Our model was parameterized based on expert opinion for a set of land-cover criteria sufficient in floral sources to support apiaries of approximately 100 hives. These criteria highlighted the best landscapes for bees in North Dakota, but is not very representative of the range of numbers of hives commercial beekeepers typically place in apiaries. Having demonstrated a method that successfully implements local-scale criteria across a large geographic extent, our future efforts will focus on developing a more sophisticated model to provide predictions covering a more realistic spectrum of apiary sizes and environmental conditions. This would be a move towards estimating the number of colonies that could be contributed from the Northern Great Plains to meet national pollination demands under different land-change scenarios.

Our motivation to develop an approach to assess landscape suitability for honey bees is to support a larger, integrated effort to evaluate multiple ecosystem services in concert. This capability is needed to provide information for land resource agencies and policy makers on potential outcomes for ecosystem services from drivers of change. Our initial efforts already have highlighted the importance of conservation lands for honey bees in landscapes extensively cropped in cover types having little pollen or nectar value for bees. Incorporating our model into the larger, integrated framework of ecosystem service models [54] will provide a more balanced assessment to support decisions related to policy, programs, and other drivers of land-use and land-cover change.

## Supporting Information

**Table S1 Information on source data.**

**Table S2 Accuracy of grassland and woody land cover in maps from the National Land Cover Database and the North Dakota GAP Analysis Program.**

## Acknowledgments

We benefitted from reviews by S. Hyberg, J. Walker, D. vanEngelsdorp (PLOS ONE academic editor), and two anonymous reviewers on earlier versions of this paper.

## Author Contributions

Conceived and designed the experiments: NHE ALG ZB. Performed the experiments: ALG. Analyzed the data: ALG. Contributed reagents/ materials/analysis tools: ALG NHE. Wrote the paper: ALG NHE.

## References


1. Euliss NH Jr, Smith LM, Wilcox DA, Browne BA (2008) Linking ecosystem processes with wetland management goals: Charting a course for a sustainable future. Wetlands 28: 553–562.
2. Euliss NH Jr, Smith LM, Liu S, Feng M, Mushet DM, et al. (2010) The need for simultaneous evaluation of ecosystem services and land use change. Environ Sci Technol 44: 7761–7763.
3. Millennium Ecosystem Assessment (2005) Ecosystems and human well-being: Synthesis. Washington DC: Island Press. 137 p.
4. National Agricultural Statistics Service (2009) 2007 Census of agriculture – Summary and state data. U.S. Department of Agriculture, Volume 1 Geographic Area Series, Part 51, AC-07-A-51. 739 p.
5. Batt BDJ, Anderson MG, Anderson CD, Caswell FD (1989) The use of prairie potholes by North American ducks. In: van der Valk AG, editor. Northern prairie wetlands. Ames: Iowa State University Press. pp. 204–227.
6. Smith AG, Stoudt JH, Gallop JB (1964) Prairie potholes and marshes. In: Lindiska JP, editor. Waterfowl tomorrow. Washington DC: U.S. Fish and Wildlife Service. pp 39–50.
7. Farm Service Agency (2008) Conservation Reserve Program (CRP) benefits: water quality, soil productivity and wildlife estimates. U.S. Department of Agriculture Fact Sheet. Available: http://www.fsa.usda.gov/Internet/FSA_File/factsheet_crp_bennies.pdf. Accessed 2012 July 20.
8. Euliss NH Jr, Gleason RA, Olness A, McDougal RL, Murkin HR, et al. (2006) North American prairie wetlands are important nonforested land-based carbon storage sites. Sci Total Environ 36: 179–188.
9. Gleason RA, Laubhan MK, Euliss NH Jr (2008) Ecosystem services derived from wetland conservation practices in the United States Prairie Pothole Region with an emphasis on the USDA Conservation Reserve and Wetland Reserve Programs. U.S. Geological Survey Professional Paper 1745. 58 p.
10. Haufler JB (2005) Fish and wildlife benefits of Farm Bill conservation programs: 2000–2005 update. Wildl Soc Tech Rev 05–02.
11. Auch RF, Sayler KL, Napton DE, Taylor JL, Brooks MS (2011) Ecoregional differences in late-20$^{th}$-century land-use and land-cover change in the U.S. Northern Great Plains. Great Plains Res 21: 231–243.
12. Claassen R, Carriazo F, Cooper JC, Hellerstein D, Ueda K (2011) Grassland to cropland conversion in the Northern Plains – the role of crop insurance, commodity, and disaster programs. U.S. Department of Agriculture, Economic Research Report EER-120. 77 p.
13. Gutmann MP, Parton WJ, Cunfer G, Burke IC (2005) Population and environment in the U.S. Great Plains. In: Entwisle B, Stern PC, editors. Population, land use, and environment: research directions: National Academies Press. pp. 84–105.
14. Hartman MD, Merchant ER, Parton WJ, Gutmann MP, Lutz SM, et al. (2011) Impact of historical land-use changes on greenhouse gas exchange in the U.S. Great Plains, 1883–2003. Ecol. Appl. 21: 1105–1119.
15. Sylvester KM, Brown DG, Deane GD, Kornak RN (2013) Land transitions in the American plains: multilevel modeling of drivers of grassland conversion (1956–2006). Agric Ecosyst Environ 168: 7–15.
16. Wu JJ, Lin H (2010) The effect of the Conservation Reserve Program on land values. Land Econ 86: 1–21.
17. Hellerstein D, Malcolm S (2011) The influence of rising commodity prices on the Conservation Reserve Program. U.S. Department of Agriculture, Economic Research Service ERR-110.
18. Stubbs M (2007) Land conversion in the Northern Plains. Washington, D.C.: Congressional Research Service Report for Congress RL33950. 13 p.
19. Farm Security Act of 2002 (U.S. Public Law 107-171). Available: http://www.gpo.gov/fdsys/pkg/PLAW-107publ171/html/PLAW-107publ171.htm. Accessed 2014 March 19.
20. Food, Conservation, and Energy Act of 2008 (U.S. Public Law 110-246). Available: http://www.gpo.gov/fdsys/pkg/PLAW-110publ246/html/PLAW-110publ246.htm. Accessed 2014 March 19.
21. Agricultural Act of 2014 (U.S. Public Law 113-79). Available: http://www.gpo.gov/fdsys/pkg/BILLS-113hr2642enr/pdf/BILLS-113hr2642enr.pdf. Accessed 2014 March 19.
22. Energy Independence Security Act of 2007. Available: http://www.gpo.gov/fdsys/pkg/PLAW-110publ140/html/PLAW-110publ140.htm. Accessed 2014 March 19.
23. Fernandez-Cornejo J (2012) Biotechnology and seed use for major U.S. crops. In: Osteen C, Gottlieb J, Vasavada U, editors. Agricultural resources and environmental indicators, 2012 Edition. U.S. Department of Agriculture, Economic Research Service, Economic Information Bulletin 98. pp. 18–20.
24. Balas CJ, Euliss NH Jr, Mushet DM (2012) Influence of conservation programs on amphibians in the prairie pothole region. Wetlands 32: 333–345.
25. Calderone NW (2012) Insect pollinated crops, insect pollinators and US agriculture: Trend analysis of aggregate data for the period 1992–2009. PLoS One 7(5): e37235. doi:10.1371/journal.pone.0037235.
26. National Research Council (2007) Status of pollinators in North America. Washington DC: The National Academies Press. 307 p.
27. Johnson R (2007) Recent honey bee colony declines. Congressional Research Service Report for Congress, RL33938.
28. Skokstad E (2007) The case of the missing hives. Science 316: 970–972.
29. Pettis JS, Delaplane KS (2010) Coordinated responses to honey bee decline in the USA. Apidologie DOI: 10.1051/apido/2010013.
30. Spivak M, Mader E, Vaughan M, Euliss NH Jr (2011). The plight of the bees. Environ Sci Technol 46: 34–38.
31. Morse RA, Calderone NW (2000) The value of honey bees as pollinators of U.S. crops in 2000. Bee Cult 128: 1–15.
32. Rucker RR, Thurman WN, Burgett M (2012) Honey bee pollination markets and the internalization of reciprocal benefits. Am J Agric Econ 94: 956–977.
33. Aizen MA, Harder LD (2009) The global stock of domesticated honey bees is growing slower than agricultural demand for pollination. Curr Biol 19: 1–4.
34. vanEngelsdorp D, Caron D, Hayes J, Underwood R, Henson M, et al. (2012) A national survey of managed honey bee 2010–2011 winter colony losses in the USA: results from the Bee Informed Partnership. J Apic Res 51: 115–124.
35. vanEngelsdorp D, Jayes J Jr, Underwood RM, Pettis JS (2010) A survey of honey bee colony losses in the United States, Fall 2008 to spring 2009. J Apic Res 49: 7–14.
36. Johnson %, Corn ML (2013) Bee health: background and issues for Congress. Congressional Research Service Report, R43191.
37. U.S. Department of Agriculture, National Agricultural Statistics Service. Honey: final estimates for 1986–1992. Statistical Bulletin Number 912. Available: http://usda.mannlib.cornell.edu/MannUsda/viewDocumentInfo.do?documentID = 1520. Accessed 2014 March 17.
38. U.S. Department of Agriculture, National Agricultural Statistics Service. Honey: final estimates 1993–1997. Statistical Bulletin Number 956. Available:

http://usda.mannlib.cornell.edu/MannUsda/viewDocumentInfo.do?documentID=1520. Accessed 2014 March 17.
39. U.S. Department of Agriculture, National Agricultural Statistics Service. Honey: final estimates 1998–2002. Statistical Bulletin Number 992. Available: http://usda.mannlib.cornell.edu/MannUsda/viewDocumentInfo.do?documentID=1520. Accessed 2014 March 17.
40. U.S. Department of Agriculture, National Agricultural Statistics Service. Honey: final estimates 2003–2007. Statistical Bulletin Number 1025. Available: http://usda.mannlib.cornell.edu/MannUsda/viewDocumentInfo.do?documentID=1520. Accessed 2014 March 17.
41. U.S. Department of Agriculture, National Agricultural Statistics Service. Honey report 2009. Released 27 February 2009. Available: http://usda.mannlib.cornell.edu/MannUsda/viewDocumentInfo.do?documentID=1191. Accessed 2014 March 17.
42. U.S. Department of Agriculture, National Agricultural Statistics Service. Honey report 2010. Released 26 February 2010. Available: http://usda.mannlib.cornell.edu/MannUsda/viewDocumentInfo.do?documentID=1191. Accessed 2014 March 17.
43. U.S. Department of Agriculture, National Agricultural Statistics Service. Honey report 2011. Released 25 February 2011. Available: http://usda.mannlib.cornell.edu/MannUsda/viewDocumentInfo.do?documentID=1191. Accessed 2014 March 17.
44. U.S. Department of Agriculture, National Agricultural Statistics Service. Honey report 2012. Released 30 March 2012. Available: http://usda.mannlib.cornell.edu/MannUsda/viewDocumentInfo.do?documentID=1191. Accessed 2014 March 17.
45. U.S. Department of Agriculture, National Agricultural Statistics Service. Honey report 2013. Released 18 March 2013. Available: http://usda.mannlib.cornell.edu/MannUsda/viewDocumentInfo.do?documentID=1191. Accessed 2014 March 17.
46. Naug D (2009) Nutritional stress due to habitat loss may explain recent honeybee colony collapses. Biol Conserv 142: 2369–2372.
47. Di Pasquale G, Salignon M, Le Conte Y, Belzunces LP, Decourtye A, et al. (2013) Influence of pollen nutrition on honey bee health: do pollen quality and diversity matter? PLoS One 88(8): e72016. DOI: 10.1371/journal.pone.0072016.
48. Gallant AL, Klaver RW, Casper GS, Lannoo MJ (2007) Global rates of habitat loss and implications for amphibian conservation. Copeia 2007: 967–979.
49. National Agricultural Statistics Service (2012) Honey Report ISSN:1949–1492.
50. Alaux A, Ducloz F, Crauser D, Le Conte Y (2010) Diet effects on honeybee immunocompetence. Biol Lett doi:10.1098/rsbl.2009.0986.
51. Auch RF, Drummond MA, Sayler KL, Gallant AL, Acevedo W (2012) An approach to assess land-cover trends in the conterminous United States (1973–2000). In: Giri C, editor. Remote sensing of land use and land cover. Boca Raton: CRC Press. pp. 351–368.
52. Tuvendal M, Elmqvist T (2011) Ecosystem services linking social and ecological systems: River brownification and response of downstream stakeholders. Ecol Soc 16(4): 21. Available: http://dx.doi.org/10.5751/ES-04456-160421. Accessed 2013 November 2.
53. U.S. Department of Agriculture (1980) Beekeeping in the United States. Washington DC: Agriculture Handbook 335. 193 p.
54. Euliss NH Jr, Smith LM, Liu S, Duffy WG, Faulkner SP, et al. (2011) Integrating estimates of ecosystem services from conservation programs and practices into models for decision makers. Ecol Appl 21: S128–S134.
55. Strong LL, Sklebar TH, Kermes KE (2005) A GAP analysis of North Dakota, final report. Jamestown: U.S. Geological Survey. 118 p.
56. Broome FR, Godwin LS (2003) Partnering for the people: improving the U.S. Census Bureau's MAR/TIGER database. Photogramm Eng Remote Sens 69: 1119–1123.
57. Stewart LO (1935) Public land surveys. Reprinted 1976. Minneapolis: The Meyers Printing Co. 202 p.
58. Watts RD (2005) Distance to nearest road in the conterminous United States. U.S. Geological Survey: Fact Sheet 2005–3011.
59. Watts RD, Compton RW, McCammon JH, Rich CL, Wright SM, et al. (2007) Roadless space of the conterminous United States. Science 316: 736–738.
60. Beekman M, Ratnieks FLW (2000) Long-range foraging by the honey-bee, *Apis mellifera* L. Funct Ecol 14: 490–496.
61. Hagler JR, Mueller S, Teuber LR, Machtley SA, Van Deynze A (2011) Foraging range of honey bees, *Apis mellifera*, in alfalfa seed production fields. J Insect Sci 11: 1–12.
62. Steffan-Dewenter I, Kuhn A (2003) Honeybee foraging in differentially structured landscapes. Proc R Soc Lond B Biol Sci 270: 569–575
63. Dadant MG (1946) The apiary. In: Grout RA, editor. The hive and the honeybee: a new book on beekeeping to succeed the book Langstroth on the hive and the honeybee. Hamilton: Dadant & Sons. pp. 290–299.
64. Boryan C, Yang Z, Mueller R, Craig M (2011). Monitoring US agriculture: the US Department of Agriculture, National Agricultural Statistics Service, Cropland Data Layer Program. Geocarto Int 2011: 1–18.
65. Wilen BO, Bates MK (1995) The US Fish and Wildlife Service's National Wetlands Inventory project. Vegetatio 118: 153–169.
66. Gallant AL (2009) What you should know about land-cover data. J Wildl Manage 73: 796–805.
67. Multi-Resolution Land Characteristics Consortium (MRLC): National Land Cover Database, Frequently asked questions. Available http://www.mrlc.gov/faq_dau.php. Accessed 2012 July 20.
68. Cowan T (2010) Conservation Reserve Program: status and current issues. Washington DC: Congressional Research Service Report for Congress 7-5700, RS21613. 11 p.
69. U.S. Government Accountability Office (2009) Biofuels – potential effects and challenges of required increases in production and use. Washington DC: GAO report GAO-09-446. 175 p.
70. Manuel J (2007) Battle of the biofuels. Environ Health Perspect 115(2): A92–A95.
71. Stubbs M (2009) Agricultural conservation issues in the 111th Congress. Washington DC: Congressional Research Service Report for Congress 7-5700, R40692. 25 p.
72. Thompson W, Meyer S, Green T (2010) The U.S. biodiesel use mandate and biodiesel feedstock markets. Biomass Bioenergy 34: 883–889.
73. National Agricultural Statistics Service (2011) Agricultural Prices. Washington DC: Report ISSN:1937–4216.
74. Homer C, Fry J (2012) The National Land Cover Database. U.S. Geological Survey Fact Sheet 2012–3020.
75. Wickham JD, Stehman SV, Fry JA, Smith JH, Homer CG (2010) Thematic accuracy of the NLCD 2001 land cover for the conterminous United States. Remote Sens Environ 114: 1286–1296.
76. National Agricultural Statistics Service: Quick Stats. Available: http://quickstats.nass.usda.gov. Accessed 2013 January 30.
77. U.S. Geological Survey: WELD: Web-Enabled Landsat Data. Available: http://weld.cr.usgs.gov. Accessed 2013 January 30.
78. Roy DP, Ju J, Kline K, Scaramuzza PL, Kovalsky V, et al. (2010) Web-enabled Landsat data (WELD): Landsat ETM+ composited mosaics of the conterminous United States. Remote Sens Environ 114: 35–49.
79. Pettorelli N, Vik JO, Mysterud A, Gaillard J-M, Tucker CJ, et al. (2005) Using the satellite-derived NDVI to assess ecological responses to environmental change. Trends Ecol Evol 20: 503–510.
80. Undersander D, Becker R, Cosgrove D, Cullan E, Doll J, et al. (2004) Alfalfa management guide. North Central Region Extension Publication NCR547. 51 p.
81. U.S. Department of Agriculture, National Agricultural Statistics Service: Cropland Data Layer metadata. Available: http://www.nass.usda.gov/research/Cropland/metadata/meta.htm. Accessed 2012 August 20.
82. U.S. Department of Agriculture, National Agricultural Statistics Service: CropScape - Cropland Data Layer. Available: http://nassgeodata.gmu.edu/CropScape. Accessed 2012 August 17.
83. Food Security Act of 1985 (U.S. Public Law 99-198). Available: http://history.nih.gov/research/downloads/PL99-198.pdf. Accessed 2014 March 19.
84. CRP enrollment and rental payments by county, 1986-2013. U.S. Department of Agriculture, Farm Service Agency. Available: http://www.fsa.usda.gov/FSA/webapp?area=home&subject=copr&topic=crp-st. Accessed 2014 March 11.
85. U.S. Department of Commerce, Bureau of the Census (1985) 1982 census of agriculture - Volume 2 Subject Series. U.S. Department of Commerce Graphic Summary 189 p.
86. U.S. Department of Commerce, Bureau of the Census (1990) 1987 census of agriculture - Volume 2 Subject Series, Part 1, Agricultural Atlas of the United States. 223 p.
87. U.S. Department of Agriculture, National Agricultural Statistics Service. 1992 Census of Agriculture, Agricultural Atlas of the United States. Available: http://www.agcensus.usda.gov/Publications/1992/Agricultural_Atlas/Maps/index.asp. Accessed 2014 March 19.
88. U.S. Department of Agriculture, National Agricultural Statistics Service (2005) 2002 Agricultural atlas of the United States - Volume 2 Subject Series, Part 1.
89. U.S. Department of Agriculture, Census of Agriculture. 2007 Census Ag Atlas Maps. Available: http://www.agcensus.usda.gov/Publications/2007/Online_Highlights/Ag_Atlas_Maps/Crops_and_Plants. Accessed 2014 March 19.
90. Stehman SV, Wickham JD, Smith JH, Yang L (2003) Thematic accuracy of the 1992 National Land-Cover Data for the eastern United States: statistical methodology and regional results. Remote Sens Environ 86: 500–516.
91. Wickham JD, Stehman SV, Smith JH, Yang L (2004) Thematic accuracy of the 1992 National Land-Cover Data for the western United States. Remote Sens Environ 91: 452–468.
92. U.S. Geological Survey (2008) Imagery for everyone. U.S. Geological Survey Technical Announcement, 21 April 2008. Available: http://landsat.usgs.gov/documents/USGS_Landsat_Imagery_Release.pdf. Accessed 2012 July 20.
93. Thogmartin WE, Gallant AL, Knutson MG, Fox TJ, Suarez MJ (2004) Commentary: a cautionary tale regarding use of the National Land Cover Dataset 1992. Wildl Soc Bull 32: 970–978.
94. Sohl TL, Gallant AL, Loveland TR (2004) The characteristics and interpretability of land surface change and implications for project design. Photogramm Eng Remote Sens 70: 439–448.
95. U.S. Department of Agriculture, National Agricultural Statistics Service (2009) 2006 North Dakota Cropland Data Layer metadata. Available: http://www.nass.usda.gov/research/Cropland/metadata/metadata_nd06.htm#7. Accessed 2012 August 20.
96. Sohl TL, Sleeter BM, Sayler KL, Bouchard MA, Reker RR, et al. (2012) Spatially explicit land-use and land-cover scenarios for the Great Plains of the United States. Agric Ecosyst Environ 153: 1–15.

97. U.S. Department of Agriculture, Farm Service Agency. Conservation programs. Available: http://www.fsa.usda.gov/FSA/webapp?area = home&subject = copr& topic = crp-st. Accessed 2013 January 30.
98. Schnepf R (2010) Agriculture-based biofuels: overview and emerging issues. Congressional Research Service Report 7-5700, R41282. 33 p.
99. U.S. Department of Agriculture, National Agricultural Statistics Service (2002) Acreage. Report ISSN1949-1522. 41 p.
100. U.S. Department of Agriculture, National Agricultural Statistics Service (2012) Acreage. Report ISSN 1949-1522. 40 p.
101. Zhu Z, Woodcock CE, Olofsson P (2012) Continuous monitoring of forest disturbance using all available Landsat imagery. Remote Sens Environ 122: 75–91.
102. Zhu Z, Woodcock CE (2014) Continuous change detection and classification of land cover using all available Landsat data. Remote Sens Environ 144: 152–171.

# 12

# Impact of Stakeholders Influence, Geographic Level and Risk Perception on Strategic Decisions in Simulated Foot and Mouth Disease Epizootics in France

**Maud Marsot[1], Séverine Rautureau[2], Barbara Dufour[3], Benoit Durand[1]***

**1** Université Paris-Est, French Agency for Food, Environmental and Occupational Health and Safety (Anses), Laboratoire de Santé Animale, Epidemiology Unit, Maisons-Alfort, France, **2** Animal Health Unit, French General Directorate for Food (DGAL), Ministry of Agriculture, Food, Fisheries and Rural Affairs, Paris, France, **3** EPIMAI, Alfort National Veterinary School (ENVA), USC Anses, Maisons-Alfort, France

## Abstract

Comparison of control strategies against animal infectious diseases allows determining optimal strategies according to their epidemiological and/or economic impacts. However, in real life, the choice of a control strategy does not always obey a pure economic or epidemiological rationality. The objective of this study was to analyze the choice of a foot and mouth disease (FMD) control strategy as a decision-making process in which the decision-maker is influenced by several stakeholders (government, agro-food industries, public opinion). For each of these, an indicator of epizootic impact was quantified to compare seven control strategies. We then determined how, in France, the optimal control strategy varied according to the relative weights of stakeholders and to the perception of risk by the decision-maker (risk-neutral/risk-averse). When the scope of decision was national, whatever their perception of risk and the stakeholders' weights, decision-makers chose a strategy based on vaccination. This consensus concealed marked differences between regions, which were connected with the regional breeding characteristics. Vaccination-based strategies were predominant in regions with dense cattle and swine populations, and in regions with a dense population of small ruminants, combined with a medium density of cattle and swine. These differences between regions suggested that control strategies could be usefully adapted to local breeding conditions. We then analyzed the feasibility of adaptive decision-making processes depending on the date and place where the epizootic starts, or on the evolution of the epizootic over time. The initial conditions always explained at least half of the variance of impacts, the remaining variance being attributed to the variability of epizootics evolution. However, the first weeks of this evolution explained a large part of the impacts variability. Although the predictive value of the initial conditions for determining the optimal strategy was weak, adaptive strategies changing dynamically according to the evolution of the epizootic appeared feasible.

**Editor:** Glenn Francis Browning, The University of Melbourne, United States of America

**Funding:** The authors have no support or funding to report.

**Competing Interests:** The authors have declared that no competing interests exist.

* E-mail: benoit.durand@anses.fr

## Introduction

Foot and mouth disease (FMD), often considered a prototype of epizootic disease, is an infectious disease affecting cloven-hoofed animals, in particular cattle, sheep, pigs and goats. FMD represents a major concern for developed countries, because of its impact on trade and the economic losses that may result of it. Following the successes obtained for Rinderpest, the Food and Agriculture Organization of the United Nations (FAO) and the World Organisation for Animal Health (OIE) recently launched a global strategy to eradicate FMD worldwide. In many countries of the European Union (EU), annual vaccination of cattle against FMD was performed from the 1960's. This measure, associated with the stamping-out of infected farms and the restriction of animal movements in infected areas, allowed the successful control of all endemic foci at the end of the 1980's. After a risk-benefit analysis, this success led the EU to prohibit all vaccination after 1991. Emergency vaccination remained however possible for epizootic control, but this tool was not used in recent epizootics: member states preferred using preemptive cull of dangerous contacts herds (except for the Dutch specific context, in 2001). However, the development of new vaccines and serological DIVA tests (that can differentiate vaccinal and natural immunity) could allow extending the use of vaccine [1–3]. Besides basic mandatory measures, fixed by EU regulations (stamping-out and disinfection in infected farms, restrictions of animal movements…), the panel of available measures for the design of FMD control strategies is relatively rich and may combine or not, according to the susceptible species, preemptive cull, suppressive vaccination (vaccinated animals are culled at the end of the epizootic like in Dutch in 2001) and protective vaccination (vaccinated animals are not culled but are subjected to movement restrictions).

Simulation models have been used to analyze the influence of the environment (e.g. farms density and aggregation) and farming practices (e.g. those that induce movements of animals between farms) on the spread of an FMD epizootic, and the efficacy of control measures. For example, Green *et al.* [4] analyzed the respective impact of animal movements and of local spread on the size and geographic extent of an epizootic; and Orton *et al.* [5]

investigated the efficacy of movement control measures. Tildesley *et al.* [6] compared emergency ring vaccination strategies in various epidemiological contexts and under logistical constraints. A similar study was devoted to the comparison of preemptive culling strategies [7]. Comparison of control strategies against simulated FMD epizootics allows determining optimal strategies according to their epidemiological and/or economic impacts [8,9]. Studied impacts may be first epidemiological: in the study of Martinez *et al.* [10], for example, depopulation and vaccination of farms in contact with outbreaks decreased the number of infected herds, compared to the stamping-out of outbreaks alone. This result was confirmed by Tildesley *et al.* [11], who also showed that preemptive slaughter alone was sometimes unable to control an epizootic. Some studies consider at the same time the economic and epidemiologic consequences FMD epizootics: Schoenbaum *et al.* [12], for example, showed that the early ring vaccination resulted in lower government costs and outbreak duration in fast-spreading scenarios. However, studied impacts are also often only economic [13]: for Garner *et al.* [14], the stamping-out strategy was, on average, the less costly, while Tomassen *et al.* [15] showed that the implementation of alternative strategies (ring-vaccination, ring-culling) could reduce economic impacts. Studied economic impacts may be limited to direct costs (e.g. public costs) or also analyze the impact of an epizootic (and of control measures) on the whole economy (indirect costs, [16,17]). Indeed, the consequences of FMD epizootics are not limited to the agricultural sector or to agro-food industries; other economic sectors are affected such as tourism [18]. Part of the economic impacts result from the reactions of consumer (or, more generally, of public opinion) against the control measures [19]. More generally, FMD epizootics may have a social impact due to the reactions of public opinion against the control measures, as reported by media. This may lead, for example, to the mobilization against the massive slaughter of animals induced by preemptive slaughter strategies. Even if the problem of the social acceptability of massive slaughter is obvious, the social impact of this type of strategy is not considered in literature, except in some studies on the 2001 epizootic that underlined the importance of public opinion [20,21].

Decision-making for the control of epizootic diseases does not always obey a pure scientific rationality, based on economic or epidemiological criteria. These decisions remain basically political: in France, in 2007, the government chose to focus the vaccination efforts against bluetongue in the areas that had been the most severely affected the preceding year. This decision was clearly not optimal on an epidemiological or economic point of view; however, it allowed satisfying the breeders who had suffered most the preceding year. Rationalizing the choice of a control strategy against FMD epizootics should take into account several criteria in the decision-making process: the direct economic impact (e.g. public costs), the indirect economic impact (e.g. the export losses for agro-food industries, the shortfall for the tourism sector), and social impacts that may induce a distrust of public control of animal diseases.

The objective of this study was to analyze the choice of a FMD control strategy as a decision-making process in which the decision-maker is influenced by several lobbies, whose interests may be divergent. These lobbies take an active part in the decision-making process and are considered stakeholders in this process. The choice made by the decision-maker depends on the relative weights of these stakeholders (or, alternatively, on the relative influence of the corresponding lobbying groups). Three stakeholders were considered: the government (that represents public finances), the public opinion, and the agro-food industries (that represents the economic sectors directly or indirectly affected by FMD epizootics). We determined how, in France, the optimal control strategy would vary according to the relative weights of stakeholders and to the perception of risk by the decision-maker (either risk-neutral or risk-averse). Two geographic levels were compared for the decision-making process: a single decision taken at the national level, or independent decisions taken in each region. We finally analyzed the feasibility of adaptive decision-making processes that could allow adapting the choice of the control strategy depending on the date and place where the epizootic starts, or depending on the evolution of the epizootic over time.

## Materials and Methods

### Disease transmission, detection and control models

We used the simulation model of FMD epizootics developed by Rautureau *et al.* [22]. This tool allows simulating realistic FMD epizootic dynamics in France, based upon actual data on French holdings (N = 390,565 including farms, markets, dealers and slaughterhouses) and on animal movements of cattle [23] and swine [24]. The basic unit was the holding, consisting of at most three mono-specific animal batches: cattle, small ruminants and swine. Five individual exclusive health states were considered: exposed and susceptible (denoted S), infected but not infectious (denoted L), subclinically infectious (denoted I), clinically infectious (denoted J) and immune (denoted R). Inside a holding, we assumed that animals of a given species batch were exposed to three forces of infection: the within-batch force of infection, the between-batch force of infection and the environmental force of infection. The latter two forces were based upon the within-batch force of infection, decreased according to species-specific biosecurity parameters, which represented the levels of separation between species in usual farm organization and management. Three modes of transmission between holdings were considered: live animal movements (inducing potentially nationwide epizootics), direct transmission by contacts between herds on pastures; and indirect transmission by contaminated vehicles, materials or fomites. Animal movements were simulated according to records stored in a national database. The health state of animals moved from one holding to another was assumed to be random. The direct between-herd force of infection was modelled as a fraction of the within-herd force of infection (decreased by a biosecurity parameter), whereas the indirect between-herd force of infection was modelled as a fraction of the within-herd environmental force of infection (also decreased by a biosecurity parameter). The disease detection model was based upon an explicit representation of the awareness of farmers and veterinarians (i.e. their ability to recognize and report FMD cases) that evolved dynamically when these actors observe diseased animals [22]. Two categories of actors were assumed to be involved in disease surveillance: the farmer who manages the herds within a holding, and the veterinarian responsible for several holdings in a given area. In the model, two events contributed to a disease report: (i) the farmer observes clinical signs and decides whether it is necessary to call the veterinarian, (ii) the veterinarian recognizes the disease and reports a suspected FMD case to the authorities. The probabilities of these events occurring were modeled according to an actor-specific awareness level. Actor awareness was represented by a state variable that increased daily, according to the observations of the actor during the day. Therefore, the time to detection of the first case was not a parameter of the model but an output variable, the value of which resulted from the dynamics of the disease-reporting process. Parameterization of the disease transmission and detection models varied according to the species of the affected animals: infected cattle and swine are known to be highly contagious and to exhibit clear clinical signs, whereas infected small ruminants are known to be less contagious, clinical signs being also lighter. Parameter values were fixed based on literature values or on expert opinions (transmission and detection parameters were fixed using the Delphi method) [22].

The disease control model was based on the stamping-out (SO) strategy: slaughter and disinfection of reported outbreaks, epidemiological inquiry (including contact tracing), and movement restriction in contact herds and in a 10 km radius area around reported outbreaks. Six supplementary control strategies were defined (Table 1) that combined preemptive slaughter and emergency vaccination. Preemptive slaughter corresponded to the slaughter of all herds in a radius of 1 km around reported outbreaks, and of herds identified "at risk" by epidemiological enquiries conducted in outbreaks. Vaccination was either suppressive (the vaccinated animals were culled at the end of the epizootic and their carcasses were destroyed) or protective (the vaccinated animals were not culled at the end of the epizootic but were subjected to movement restrictions). In three of these control strategies, measures were species-specific and varied according to the economic value of the animals (high for cattle and breeding pigs *vs* low for small ruminants and non-breeding pigs, Table 1). To ensure the completion of simulations, it was assumed that, whatever the control strategy, an emergency protective vaccination campaign (PV control strategy in Table 1) was undertaken if the number of reported outbreaks exceeded 500. Such epizootics are termed below "uncontrolled" epizootics, as opposed to "controlled" epizootics, for which no systematic vaccination was necessary. The post-epizootic serological surveys (necessary to FMD-free status recovery) were simulated according to EU regulations (Council directive 2003/85/EC). Human and material resources dedicated to control measures were considered limited, with a maximum of 3 farm depopulations (reported outbreaks having priority over preemptive culls), 5 disinfections and 5 epidemiological inquiries (in affected farms) per day and department; 1000 vaccinated animals per day and veterinarian; and 8000 serological analyses per day. If several departments were affected, their resources were pooled. Resources dedicated to each control task (depopulation, disinfection, epidemiological inquiries, vaccination) were considered independent (because of the different qualifications required for each task), and were thus not prioritized.

## Quantification of epizootic impact for stakeholders

The indicator of epizootic impact for government was the total public expense, based on the following unitary costs [22]:

- depopulation and disinfection operations, compensations paid to farmers: 2,500€ per animal for cattle, 900€ for breeding pigs, 500€ for non-breeding pigs and for small ruminants;
- vaccine and vaccination operations: 5.8€ per vaccinated animal;
- laboratory analyses: 15.4€ per test.

The indicator chosen to quantify the impact of an epizootic on public opinion was the total number of slaughtered herds (either in reported outbreaks, or in herds submitted to preemptive slaughter). Finally, the impact of an epizootic on agro-food industries was quantified by the total export losses caused by bans. Assuming that the OIE zoning principle would be applied using the NUTS (Nomenclature of Territorial Units for Statistics) level 2 (i.e. the 21 French administrative "regions"), and based on OIE and EU regulations, the duration and extent of export bans were fixed as follows:

- exports to EU countries: ban on the exports of meat and dairy products originating from each of the affected regions, ending one week after the completion of the post-epizootic serological surveys,
- exports to non-EU countries: ban on the exports of meat and dairy products originating from each of the affected regions, ending 3 months after the last depopulation if no emergency vaccination is performed, and 6 months after the last depopulation or vaccination otherwise.

For each region, daily export losses that would be caused by an export ban were estimated using the 2010 annual statistics. Meat and dairy products export statistics were obtained from the national institute of statistics and economic studies (INSEE) for 2010 and for each French region, according to the destination of exports: EU *vs* non-EU countries. For a given epizootic, the total export losses were the product of the total daily exports from the affected regions to EU and non-EU countries, by the corresponding ban durations.

**Table 1.** Simulated FMD control strategies.

| Code | Name | Preemptive slaughter | | Emergency vaccination | | |
|---|---|---|---|---|---|---|
| | | Species | Radius | Species | Radius | Cull |
| SO | Stamping-out | - | - | - | - | - |
| PV | Protective vaccination | - | - | All | 10 km | No |
| PS | Preemptive slaughter | All | 1 km | - | - | - |
| SV | Suppressive vaccination | - | - | All | 1 km | Yes |
| SPV | Selective protective vaccination | - | - | Cattle, breeding pigs | 10 km | No |
| SPS | Selective preemptive slaughter | Small ruminants, non-breeding pigs | 10 km | - | - | - |
| SPSV | Selective preemptive slaughter and protective vaccination | Small ruminants, non-breeding pigs | 1 km | Cattle, breeding pigs | 10 km | No |

Note: Whatever the strategy, basic control measures are applied: slaughter and disinfection of reported outbreaks, contact tracing, movements restriction in contact herds and in a 10 km radius area around reported outbreaks.

## Decision making

For each stakeholder, the choice of a control strategy was modelled according to the perception of risk by the decision-maker. Risk-neutral decision-makers were defined as decision-makers who base their decision on the average of the predicted impact: they would thus choose the control strategy associated with the lowest mean impact. Conversely, risk-averse decision makers rather want to avoid severe epizootic impacts, even if the mean impact is not the lowest. They were defined as decision-makers who base their decision on the variance of the predicted impact: they would thus choose the control strategy associated with the lowest standard deviation of the epizootic impact.

Epizootic impacts (public costs, slaughtered herds and export losses) were first separately computed for each of the three stakeholders (respectively: government, public opinion and agro-food industries) and the optimal control strategy was determined for risk-neutral and risk-averse decision-makers. However, in a real situation, the decision would probably not favour a single stakeholder, but rather result from a compromise between them. Such a collective decision was modelled as a multi-criteria decision, using a simple additive method. Each stakeholder-specific impact was first recalibrated as follows: $X_{i,s} = (x_{i,s} - x_{SO,s}) / x_{SO,s}$, where $X_{i,s}$ is the recalibrated impact of the $i$th simulated epizootic for the stakeholder $s$; $x_{i,s}$ is the simulated impact of the epizootic $i$ for the stakeholder $s$; and $x_{SO,s}$ is the mean impact of a simulated epizootic for stakeholder $s$, when the basic and mandatory stamping-out (SO) strategy is applied. The recalibrated impact of the epizootic $i$ ($X_{i,s}$) thus represented the deviation from the mean impact when the stamping-out strategy was applied, the unit being also the average impact when the SO strategy was used.

The "collective impact" $Y_i$ of a the $i$th simulated epizootic was then defined as a linear function of the three recalibrated impacts: $Y_i = w_g\ X_{i,g} + w_p\ X_{i,p} + w_a\ X_{i,a}$, where the subscripts $g$, $p$ and $a$ correspond to the government, the public opinion and the agro-food industry, respectively; and $w_g$, $w_p$ and $w_a$ are the weights of the three stakeholders in the collective decision; or, alternatively, those of the corresponding lobbies in the decision-making process. Finally, for a given triple ($w_g$, $w_p$, $w_a$), the optimal strategy was the one that minimized the mean (for a risk-neutral decision-maker) or the variance (for a risk-averse decision-maker) of the collective impact.

## Optimal decision-making for fixed control strategies

The breeding structures are known to show strong regional variations in France (Fig. 1), and large variations of epizootic impacts were expected according to the region in which the virus was introduced. Simulations were thus stratified by region. Fifty herds were randomly selected among those housing at least one susceptible animal (cattle, small ruminant or pig) in each of the 21 French regions, as well as a random calendar date for virus introduction (a single animal being initially infected, the species of which was randomly chosen). In the following, we define a scenario as a unique situation corresponding to a herd in which the virus is initially introduced, a date for this introduction, and a control strategy. A total number of 7,350 scenarios were thus defined (50 herds and dates of virus introduction, 21 regions, and 7 control strategies). In this part of the study, the aim was to quantify the average impact of FMD epizootics (according to the control strategy and to the region where the virus is initially introduced) rather than the variability of the impacts for a given scenario (as defined above). Therefore, a single simulation was run for each scenario (*i.e.* 50 per region). In these simulations, once started, FMD epizootics could either remain confined to the region where the virus had been introduced, or spread to the neighbouring areas, or further afield. For a given simulation and a given impact, impact value was the total of the impact in the region where the virus had been introduced, plus the impacts in the other affected regions. For each region and each control strategy, we computed the average impact of simulated epizootics starting in that region (and possibly affecting other regions). In order to compute the country-level mean impact of an FMD epizootic, an assumption had to be made about the risk of virus introduction in each region. We assumed that region-specific risks of introduction were proportional to the number of herds in each region. These numbers were thus used as weights to calculate the average impact of an FMD epizootic in France, as well as the corresponding variance. Optimal strategies were identified for each stakeholder and for the two types of risk perception by decision-makers. Collective impacts were computed for varying combinations of stakeholder weights: the weight of the government ($w_g$) was fixed to a constant value of 1 whereas the weights of agro-food industry ($w_a$) and of public opinion ($w_p$) independently varied between $10^{-3}$ and $10^3$. The corresponding optimal strategies were separately determined for risk-neutral and for risk-averse decision-makers, and the variations of the optimal strategy according to stakeholder weights were plotted.

## Feasibility of decision-making for adaptive control strategies

Two types of adaptations were considered: the choice of a fixed strategy according to the place where the epizootic starts (the "geographic" adaptation), and the dynamical adaptation of the strategy according to how the epizootic evolves (the "time" adaptation). In both cases, we decomposed the variability of impacts into an explained part (by geography and/or by time), and an unexplained part that results from the intrinsic variability of epizootics. The feasibility of adaptive control strategies was considered proportional to the explained part of the impact variability. This part of the study was restricted to the three regions where livestock density was the highest (regions D, F and T, Fig. 1).

To study the feasibility of a geographic adaptation of the control strategy, 350 scenarios were defined: 50 herds were randomly chosen in regions D, F ant T, as well as random dates for virus introduction, and the 7 control strategies were associated to each of these 50 herds. Because the aim of this part of the study was to analyze the variability of epizootic impacts, 30 simulations were run for each scenario (changing the seed of the random number generator). A total of 10,500 simulations were thus executed and the impacts were computed for each of the three stakeholders. The variability of the log-transformed impacts was analyzed separately for each control strategy, using linear mixed models in which the scenario (coded as a single categorical variable) was treated as a random effect, and without fixed effects. The part of variance explained by the scenario was computed, the intrinsic variability of epizootics was the residual variance.

To study the feasibility of a time adaptation of the control strategy, 70 scenarios were defined, that combined the 7 control strategies and 10 herds. These were randomly chosen in regions D, F and T, and a randomly chosen virus introduction date was associated with each of them. A total of $5^5$ simulations were run for each scenario as follows. Starting from the date of virus introduction, 5 simulations were first run (changing the seed of the random number generator) and stopped when the first infected premise was detected, at day D0. Each of these 5 situations at D0 was successively used as the starting situation for 5 new simulations, which were started at D0 and stopped at D7. A total of 25 distinct situations were thus obtained at D7. The same operation was recursively performed at D14, D28 and D56.

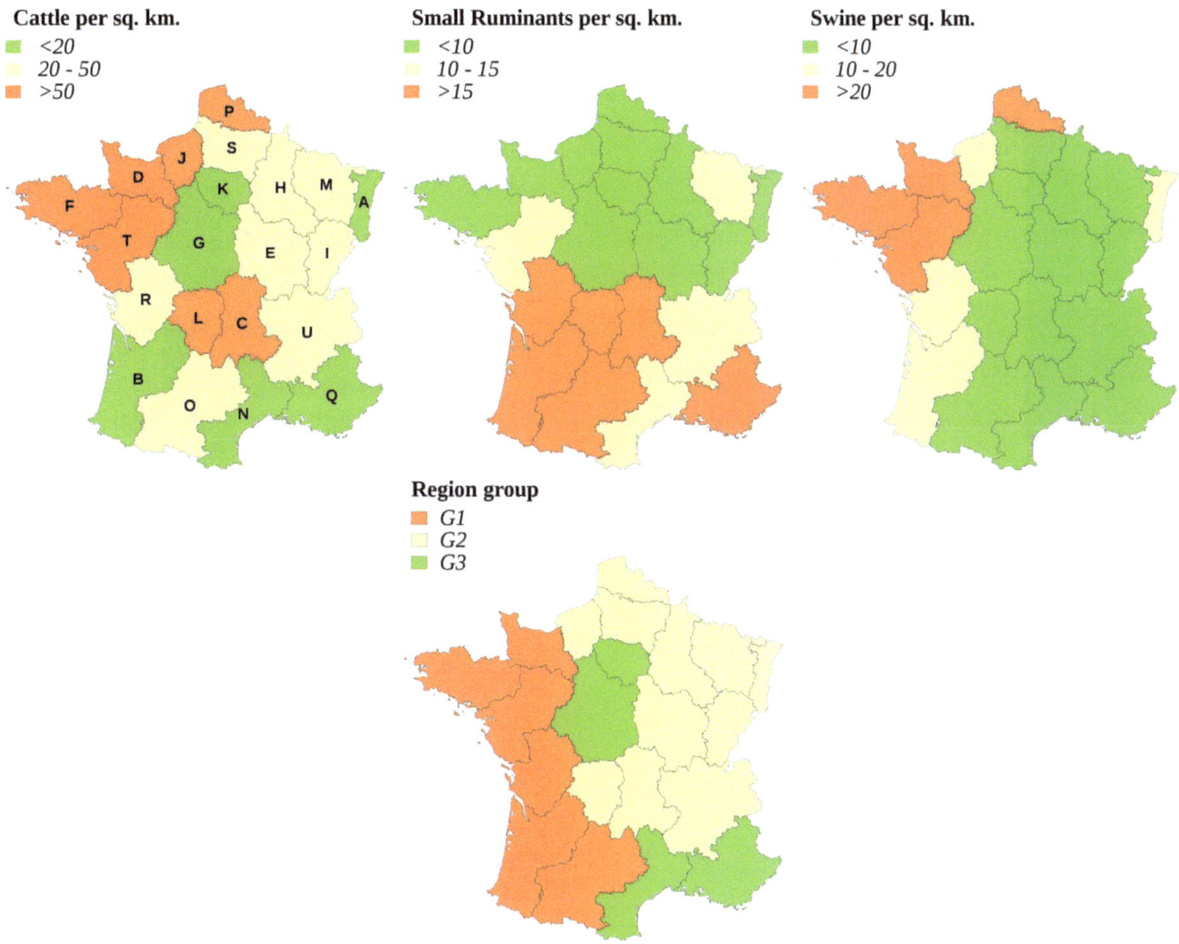


**Figure 1. Typology of French regions according to the animal densities of cattle, swine and small ruminants.** Group G1: high cattle and pig density (upper class), high density of small ruminants (upper class) associated with a low or medium cattle and swine density. Group G2: high or medium density of cattle and/or swine associated with a similar or lower density of small ruminants (same class or lower class). Group G3: low densities of cattle and swine. Capital letters on the first map are used to identify the French regions: A: Alsace, B: Aquitaine, C: Auvergne, D: Basse-Normandie, E: Bourgogne, F: Bretagne, G: Centre, H: Champagne-Ardenne, I: Franche-Comté, J: Haute-Normandie, K: Ile-de-France, L: Limousin, M: Lorraine, N: Languedoc-Roussillon, O: Midi-Pyrénées, P: Nord-Pas-de-Calais, Q: Provence-Alpes-Côte d'Azur, R: Poitou-Charentes, S: Picardie, T: Pays-de-Loire, U: Rhône-Alpes.

Globally, a total number of 218,750 (70 scenarios $\times 5^5$ simulations) were thus executed and the impacts were computed for each of the three stakeholders. The log-transformed impacts were separately analyzed for each control strategy, using linear mixed models. The scenario (place and date where the epizootic starts) was treated as a random effect. The D0–D7, D7–D14, D14–D28, and D28–D56 evolutions were treated as recursively nested random effects, the D0–D7 evolution being also nested in the initial conditions. There was no fixed effect. The part of variance explained by the D0–D7, D7–D14, D14–D28 and D28–D56 evolutions were computed; the intrinsic variability of epizootics was the residual variance.

## Results

### General results

The vast majority of simulated epizootics could be controlled without the help of an emergency protective vaccination campaign. At the national level, the proportion of controlled epizootics was high whatever the control strategy (Fig. 2): 96% for the basic stamping-out strategy and higher for the strategies with vaccination, reaching 99% for the protective vaccination (PV) strategy. The worst strategy was the selective preemptive slaughter (SPS) strategy: an emergency protective vaccination campaign was necessary to epizootic control in 15% of simulations. It was also the strategy for which the average number of infected premises was the highest, whereas this average number was the lowest for the strategies with vaccination. The national-level average duration of epizootics (number of days between the first outbreak report and the end of the post-epizootic surveys) and their average geographic extent (represented by the number of herds in the restricted areas) were similar for most of the strategies, and a little higher for the SPS strategy (Fig. 2.). The average number of slaughtered herds was high for the SPS and for the suppressive vaccination (SV) strategies, but for very different reasons: because of a greater mean number of infected premises for the SPS strategy, and because of the slaughter of vaccinated animals for the SV strategy. The corresponding compensations, paid to farmers, represented the main public costs, which were thus on average higher for the SPS and SV strategies (Fig. 2). Average total export losses were higher for the SPS strategy, and lower for the strategies with vaccination, despite the longer ban durations when protective vaccination is used. This was due to a shorter average duration of epizootics when vaccination was used, associated to a lower geographic extent of epizootics (the number of regions submitted to export bans being then lower). Furthermore, the long duration of bans mainly affected exports to non-EU countries, which represented only 22% of exports for meat products, and 23% for dairy products. Most of the export losses were thus due to EU bans (Fig. 2).

## Decisions of national-scope

When the decision was taken at the national-level, whatever their perception of risk, decision-makers wishing to minimize the public costs or the number of slaughtered herds chose the protective vaccination strategy (Table 2). Decision-makers wishing to minimize export losses also chose a strategy based upon vaccination: the suppressive vaccination strategy for risk-neutral decision-makers and the selective protective vaccination strategy for risk-averse decision-makers (Table 2). The corresponding export losses were, on average, €33 million higher for a risk-averse than for a risk-neutral decision-maker. However, in 21% of epizootics (on average, at the national level), the export losses were higher for the risk-neutral than for the risk-averse decision-maker.

After a collective decision-making process, the chosen strategy remained the protective vaccination strategy if the weight of the government was the strongest, whatever the risk perception of the decision-maker (Fig. 3, lower-left quadrant in both diagrams). It was also the case if the weight of public opinion was the strongest (Fig. 3, upper-left quadrant in both diagrams). Conversely, if the weight of the agro-food industries in the decision process was the strongest (Fig. 3, lower-right quadrant of both diagrams), the optimal strategy became the selective protective vaccination for risk-averse decision-makers, as well as for risk-neutral decision-makers when the relative weight of agro-food industries was moderate. When the relative weight of agro-food industries was high and the decision-maker was risk-neutral, the optimal collective decision became the suppressive vaccination strategy. Finally, when the weight of the government was weaker than that of public opinion and of agro-food industries (Fig. 3, upper right quadrant), the chosen strategy was the protective vaccination

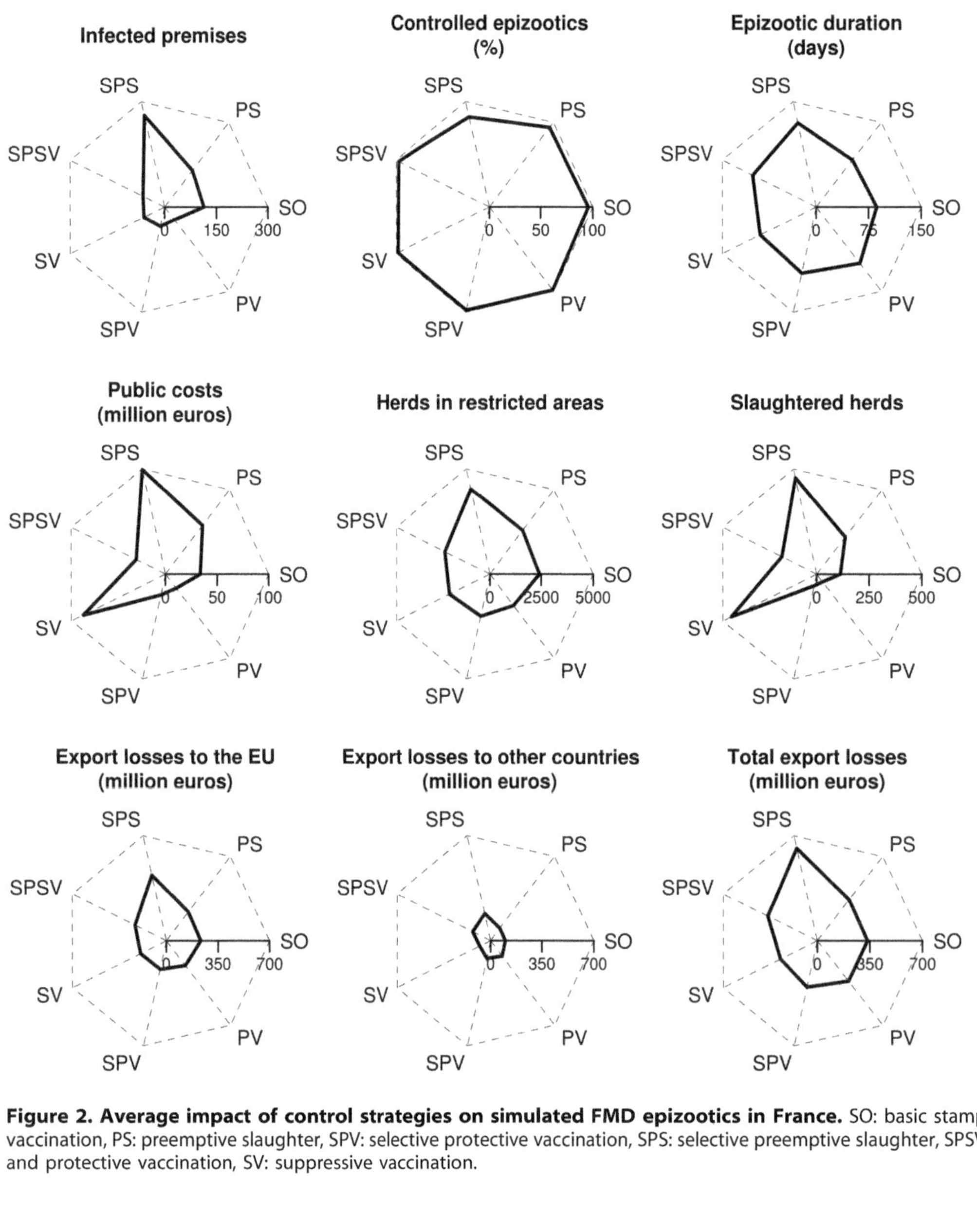


**Figure 2. Average impact of control strategies on simulated FMD epizootics in France.** SO: basic stamping-out strategy, PV: protective vaccination, PS: preemptive slaughter, SPV: selective protective vaccination, SPS: selective preemptive slaughter, SPSV: selective preemptive slaughter and protective vaccination, SV: suppressive vaccination.

**Table 2.** Average impact of simulated FMD epizootics according to the risk perception of decision makers and to the geographic level at which the control strategy is chosen and applied.

| Impact | Decision-makers risk perception | Geographic decision level | | National/Regional |
|---|---|---|---|---|
| | | National mean (sd[1]) | Regional mean (sd) | Ratio for mean (sd) |
| Public costs[5] | Risk neutral | 16.7 (33.7)[2] | 16.0 (33.5) | 0.95 (1.00) |
| | Risk averse | 16.7 (33.7)[2] | 16.3 (33.4) | 0.97 (0.99) |
| Export losses[5] | Risk neutral | 275.7 (502.9)[3] | 250.0 (457.2) | 0.91 (0.91) |
| | Risk averse | 308.3 (456.2)[4] | 277.4 (430.8) | 0.90 (0.94) |
| Slaughtered herds | Risk neutral | 45.2 (107.0)[2] | 45.0 (106.8) | 1.00 (1.00) |
| | Risk averse | 45.2 (107.0)[2] | 45.3 (106.6) | 1.00 (1.00) |

[1]Standard deviation.
[2]Optimal strategy: protective vaccination.
[3]Optimal strategy: suppressive vaccination.
[4]Optimal strategy: selective protective vaccination.
[5]Million €.
Note: National level: a single decision is taken and applied to the whole country. Regional level: specific decisions are taken and applied in each region.

when the public opinion had a stronger weight than agro-food industries, and selective protective vaccination or suppressive vaccination otherwise.

## Decisions of regional-scope: decisions taken by each stakeholder

The independent choice of optimal strategies in each region induced a reduction of average national-level public costs of approximately 5%, and a very low reduction of the standard deviation of public costs (Table 2). No significant reduction of the mean and standard deviation was observed for the average number of slaughtered herds. Conversely, a 10% reduction was observed for the export losses, both for the mean value and for the standard deviation.

For risk-neutral decision-makers (Fig. 4), the strategy allowing minimizing public costs was always based on protective vaccination in western regions where herd density is the highest (e.g. regions D, F and T). It was the basic stamping-out strategy in the other regions (except region G where the selective preemptive slaughter was optimal). To minimize the number of slaughtered herds, the optimal strategy was always based on protective vaccination (PV or SPV), except in three regions for which the stamping-out strategy was preferred. To minimize export losses, the optimal strategy was based on preemptive slaughter (PS or SPS) in 16 of 21 regions, selective protective vaccination (SPV) being optimal in a single region and suppressive vaccination (SV) in two regions. To minimize public costs and slaughtered herds, the optimal strategy was different for risk-averse and risk-neutral decision makers in six regions (indicated by stars in Fig. 5), where the protective vaccination was preferred (PV or SPV). Finally, to

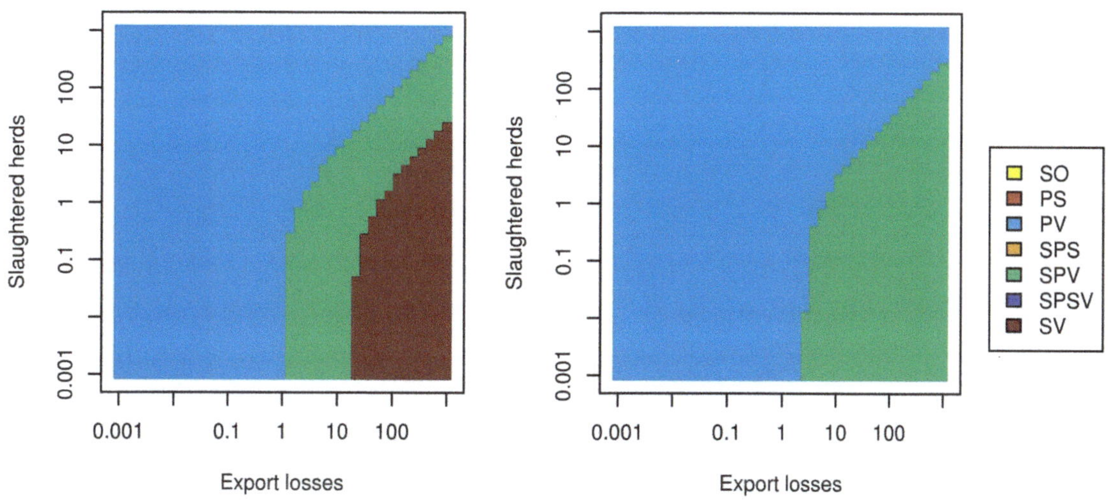


**Figure 3. Optimal national control strategies according to the risk-perception and to the respective weights of stakeholders.** Optimal control strategy decided at the national level by a risk-neutral decision-maker (left) or by a risk-averse decision-maker (right), according to the respective weights of stakeholders in decision-making: government (indicator: public costs, constant weight of 1), agro-food industry (indicator: export losses, x-axis), public opinion (indicator: number of slaughtered herds, y-axis). SO: basic stamping-out strategy, PV: protective vaccination, PS: preemptive slaughter, SPV: selective protective vaccination, SPS: selective preemptive slaughter, SPSV: selective preemptive slaughter and protective vaccination, SV: suppressive vaccination.

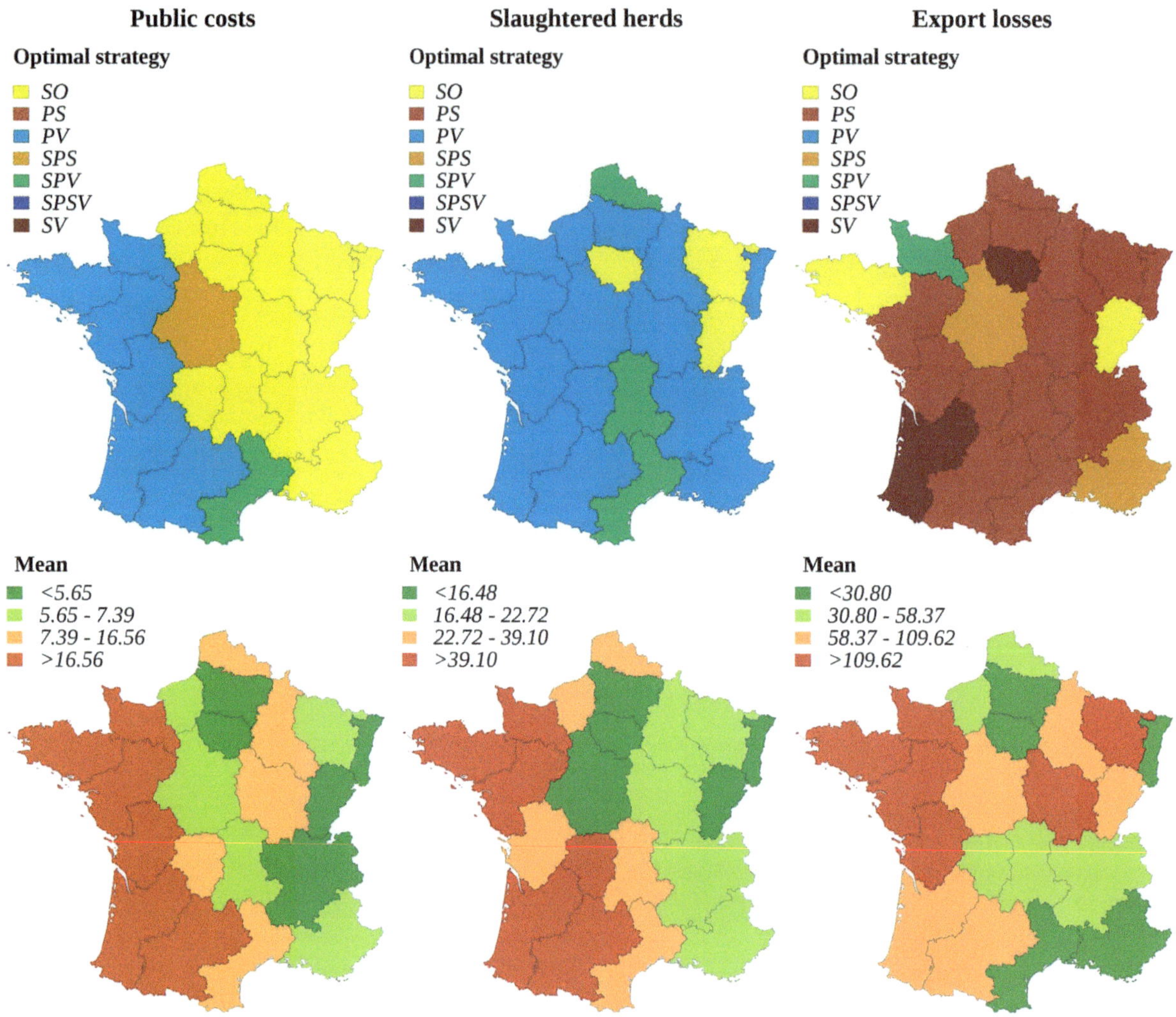


**Figure 4. Optimal regional control strategies decided by risk-neutral decision-makers, and corresponding mean impact for each stakeholder.** Optimal control strategies decided at the regional level by risk-neutral decision-makers, and corresponding mean impact of FMD epizootics for each stakeholder: government (indicator: public costs), agro-food industry (indicator: export losses), and public opinion (indicator: number of slaughtered herds). SO: basic stamping-out strategy, PV: protective vaccination, PS: preemptive slaughter, SPV: selective protective vaccination, SPS: selective preemptive slaughter, SPSV: selective preemptive slaughter and protective vaccination, SV: suppressive vaccination. Public costs and export losses are given in million €. Class bounds were fixed using the quartiles of the distributions.

minimize export losses, strategies based on protective vaccination (PV or SPV) were more often optimal (5 regions) for risk-averse decision makers, and strategies based on preemptive slaughter (PS or SPS) were less often optimal (12 of 21 regions) (Fig. 5).

## Decisions of regional-scope: collective decision-making

The result of a collective decision-making process showed various and sometimes complex patterns according to the region and to the relative weights of the three stakeholders (Fig. 6 and 7). In region I (see Fig. 1 for the location of regions), for a risk-neutral decision-maker, the optimal strategy was always the stamping-out strategy, whatever the weights of stakeholders (Fig. 6). In region D, the optimal strategy was always the protective vaccination strategy (whatever the relative weights if stakeholders) for a risk-averse decision-maker (Fig. 7). More complex patterns were obtained in other regions such as in region H: for a risk neutral decision-maker, five of the seven studied strategies could become optimal according to the relative weights of stakeholders (Fig. 6). Figures 8 and 9 map the optimal strategy for a risk-neutral and a risk-averse decision-maker respectively and for specific combinations of stakeholder weights (indicated by crosses in figures 6 and 7). The region-specific decision patterns of risk-neutral decisions-makers shown in Fig. 6 could be classified in 3 groups:

- Group G1 (regions B, D, F, O, R and T): the decision pattern was divided in two areas. The protective vaccination (PV) strategy was optimal when the weight of the government or of the public opinion exceeded that of the agro-food industries. It was replaced by another strategy when the agro-food industries dominated: selective protective vaccination (region D), stamping-out (region F), suppressive vaccination (region B) and/or preemptive slaughter (regions O, R, T). The regions that compose this group were characterized by either a high density of cattle and swine (regions D, F and T), or a high density of small ruminants associated with a low or medium density of cattle (regions B, O, R, Fig. 1).
- Group G2 (regions A, C, E, H, J, L, P, S and U): the decision pattern was divided in three areas. The stamping-out strategy

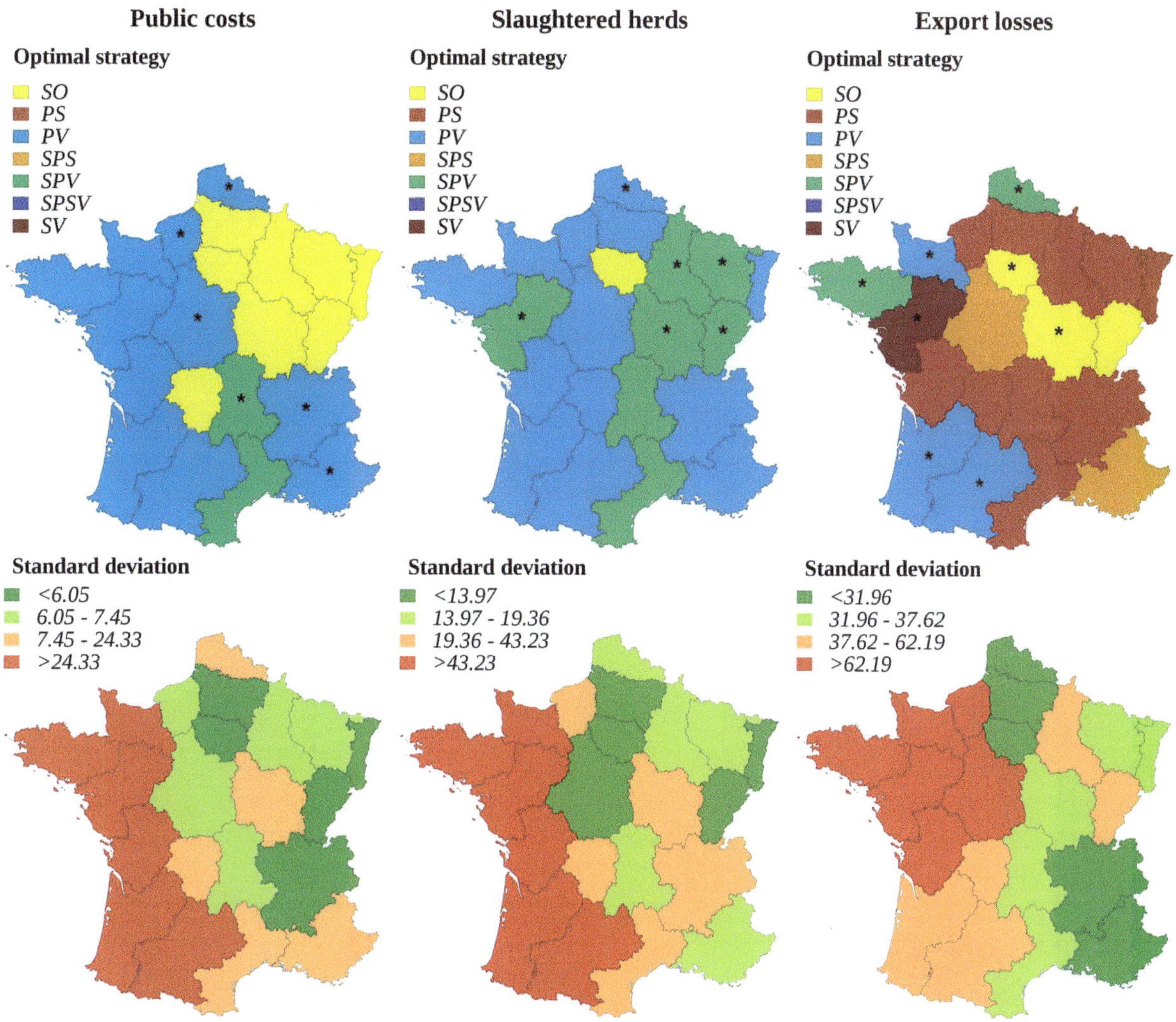


**Figure 5. Optimal regional control strategies decided by risk-averse decision-makers, and corresponding standard deviation of the impact.** Optimal control strategies decided at the regional level by risk-averse decision-makers, and corresponding standard deviation of the impact of FMD epizootics for each stakeholder: government (indicator: public costs), agro-food industry (indicator: export losses), and public opinion (indicator: number of slaughtered herds). SO: basic stamping-out strategy, PV: protective vaccination, PS: preemptive slaughter, SPV: selective protective vaccination, SPS: selective preemptive slaughter, SPSV: selective preemptive slaughter and protective vaccination, SV: suppressive vaccination. Public costs and export losses are given in million €. Class bounds were fixed using the quartiles of the distributions. Regions where risk-neutral and risk-averse decision-makers disagree are indicated by stars.

was optimal when the government's weight was the strongest, or when the agro-food industries and the public opinion had similar weights. When the public opinion dominated the decision, a vaccination-based strategy was optimal, whereas if the weight of agro-food industries was the strongest, the preemptive slaughter strategy was preferred. The regions that compose this group were characterized by high or medium cattle and/or swine densities, the density of small ruminants being always lower than the density of cattle (Fig. 1). Regions M and I also shared these characteristics.

- Group G3: regions G, N and Q showed various decision patterns according to stakeholders' weights. They were characterized by low cattle and swine densities, possibly associated with high (region Q) or medium (region N) densities of small ruminants.

This classification of decision patterns remained globally valid for a risk-averse decision-maker (Figs. 7 and 9), with a wider use of vaccination-based strategies, and a less frequent use of preemptive slaughter. In the G1 group, the protective vaccination strategy was always optimal for risk-averse decision-makers in regions B, D and O, whereas the decision pattern remained similar to that of risk-neutral decision-makers in the other regions (F, R and T). The decision pattern was also conserved for the G2 group in regions A, H, L, M and S. When agro-food industries dominated the decision, preemptive slaughter was sometimes replaced by the stamping-out strategy or by selective protective vaccination (regions E and P). Similarly, when the government's weight was the strongest, the stamping-out strategy was sometimes replaced by a strategy based on protective vaccination (PV or SPV in regions C, J, P and U). For the G3 group, the decision patterns obtained for risk-averse decision-makers were similar to these of risk-neutral decision-makers.

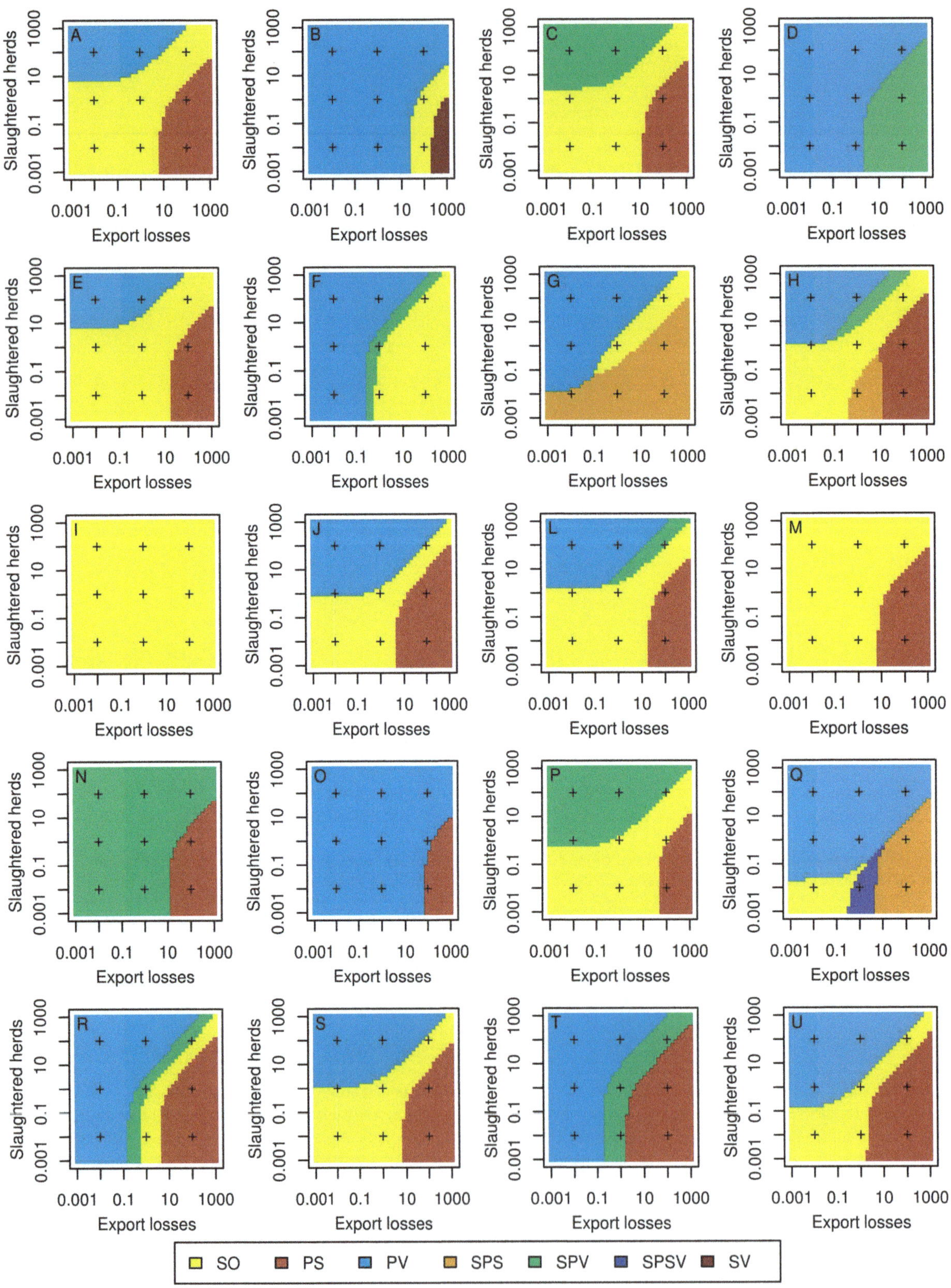


**Figure 6. Optimal regional control strategies decided by risk-neutral decision-makers, according to the respective weights of stakeholders.** Optimal control strategies decided at the regional level by risk-neutral decision-makers, according to the respective weights of stakeholders in decision-making: government (indicator: public costs, constant weight of 1), agro-food industry (indicator: export losses, x-axis), and public opinion (indicator: number of slaughtered herds, y-axis). SO: basic stamping-out strategy, PV: protective vaccination, PS: preemptive slaughter, SPV: selective protective vaccination, SPS: selective preemptive slaughter, SPSV: selective preemptive slaughter and protective vaccination, SV: suppressive vaccination. Plus signs indicate weights combinations for which maps are drawn in figs. 5a and 5b. Capital letters indicate the region as in fig. 3a.

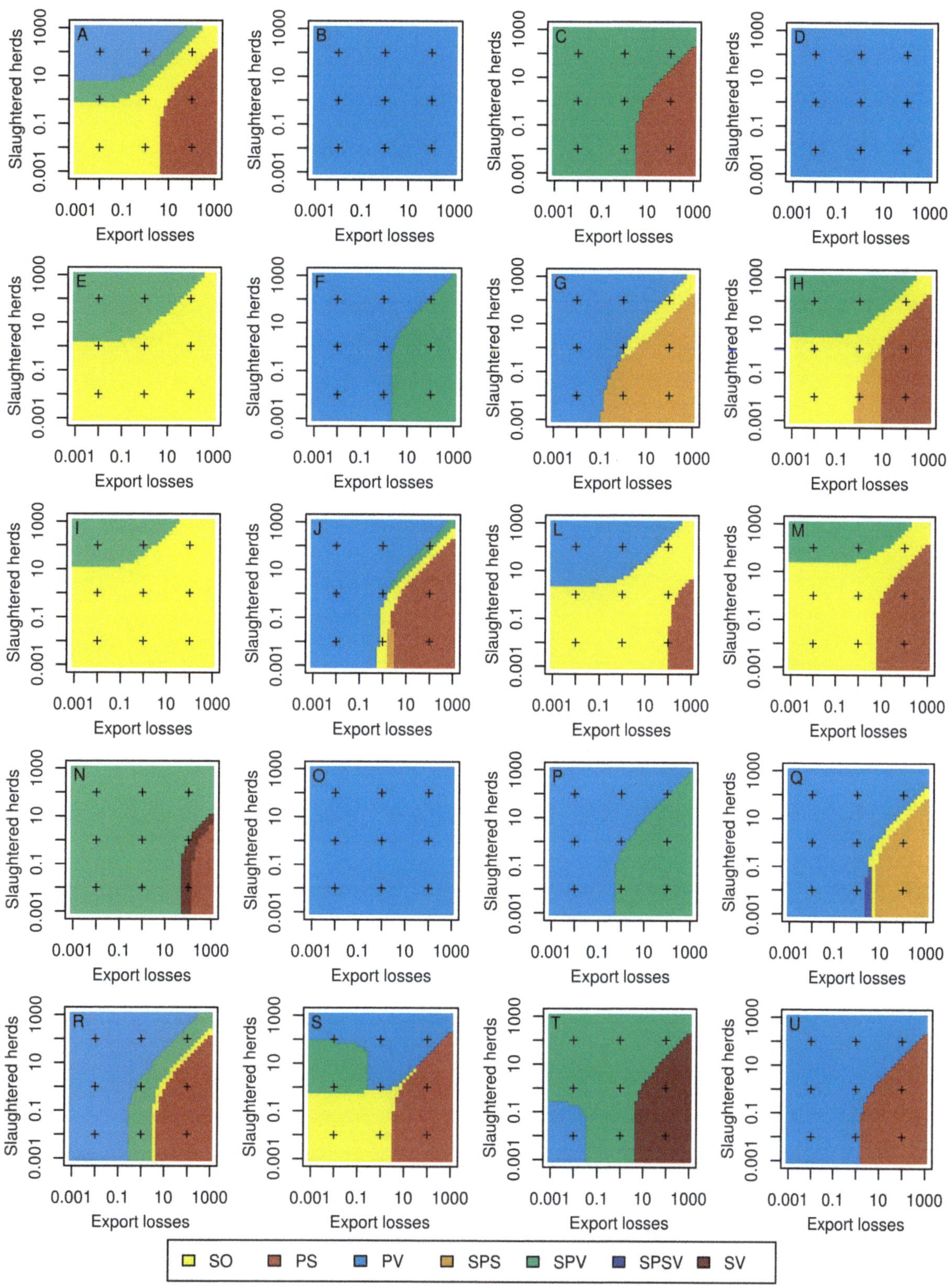


**Figure 7. Optimal regional control strategies decided by risk-averse decision-makers, according to the respective weights of stakeholders.** Optimal control strategies decided at the regional level by a risk-averse decision-makers, according to the respective weights of stakeholders in decision-making: government (indicator: public costs, constant weight of 1), agro-food industry (indicator: export losses, x-axis), and public opinion (indicator: number of slaughtered herds, y-axis). SO: basic stamping-out strategy, PV: protective vaccination, PS: preemptive slaughter, SPV: selective protective vaccination, SPS: selective preemptive slaughter, SPSV: selective preemptive slaughter and protective vaccination, SV: suppressive vaccination. Plus signs indicate weights combinations for which maps are drawn in figs. 5a and 5b. Capital letters indicate the region as in fig. 3a.

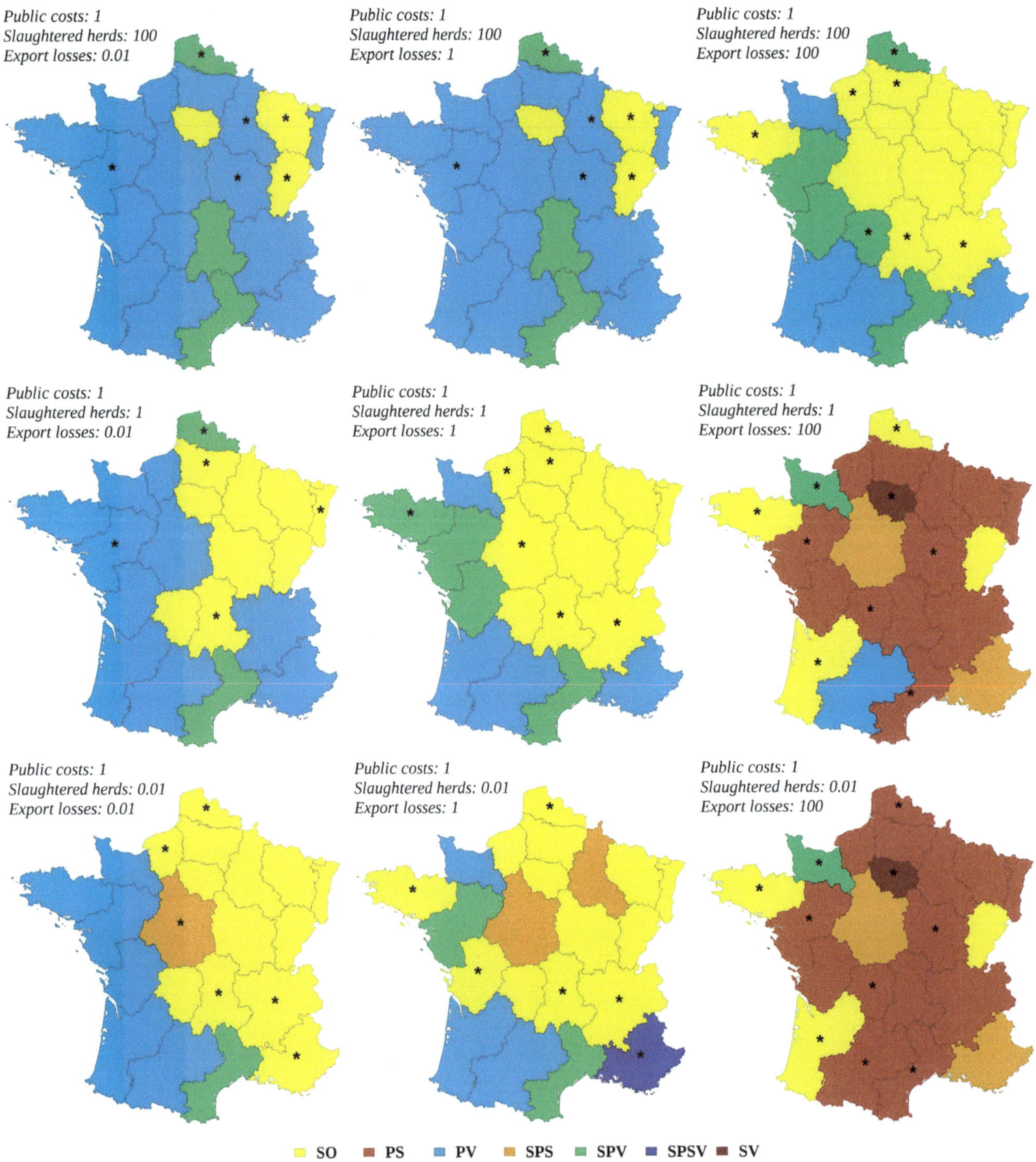


**Figure 8. Optimal regional control strategies decided by risk-neutral decision-makers, for specific combinations of stakeholders weights.** Optimal control strategies decided at the regional level by risk-neutral decision-makers, for specific combinations of stakeholders weights in decision-making: government (indicator: public costs), agro-food industry (indicator: export losses), and public opinion (indicator: number of slaughtered herds). SO: basic stamping-out strategy, PV: protective vaccination, PS: preemptive slaughter, SPV: selective protective vaccination, SPS: selective preemptive slaughter, SPSV: selective preemptive slaughter and protective vaccination, SV: suppressive vaccination. Regions where risk-neutral and risk-averse decision-makers disagree are indicated by stars.

## Feasibility of decision-making for adaptive control strategies

The initial conditions (time and place where the simulated epizootic begins) always explained at least half of the variance of impacts, whatever the stakeholder and the control strategy. The intrinsic variance of epizootics (the residual variance) remained however significant, the lowest value (*i.e.* the situation with the best predictability) being 25% of the total variance for export losses

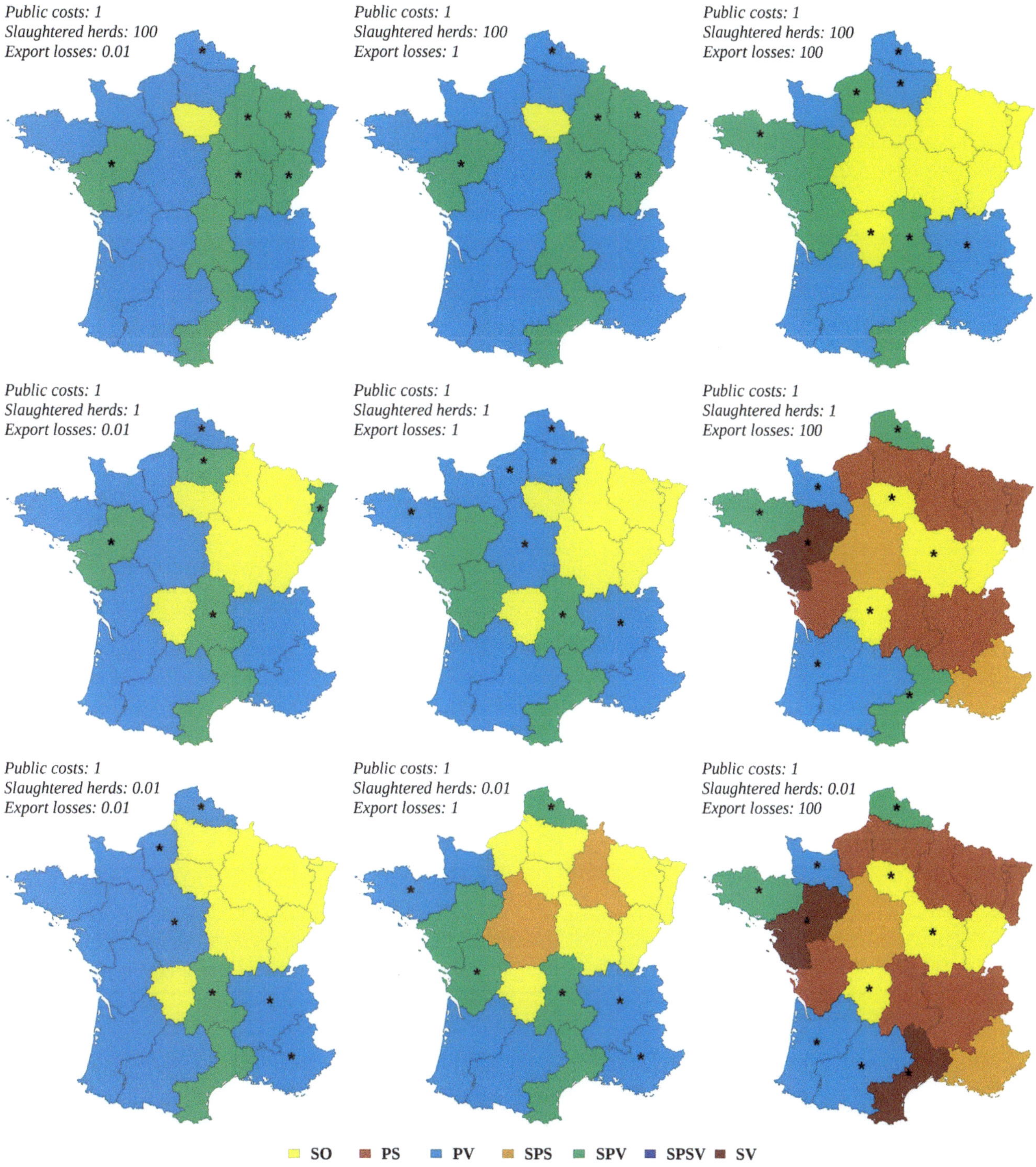


**Figure 9. Optimal regional control strategies decided by risk-averse decision-makers, for specific combinations of stakeholders weights.** Optimal control strategies decided at the regional level by risk-averse decision-makers, for specific combinations of stakeholders weights in decision-making: government (indicator: public costs), agro-food industry (indicator: export losses), and public opinion (indicator: number of slaughtered herds). SO: basic stamping-out strategy, PV: protective vaccination, PS: preemptive slaughter, SPV: selective protective vaccination, SPS: selective preemptive slaughter, SPSV: selective preemptive slaughter and protective vaccination, SV: suppressive vaccination. Regions where risk-neutral and risk-averse decision-makers disagree are indicated by stars.

with the protective vaccination strategy (Table 3). Besides the effect of the initial conditions, when the evolution of the epizootic was taken into account, the proportion of residual variance was the lowest (the predictability being thus better) for the protective vaccination strategy and for the selective protective vaccination strategy (Table 4). After removal of the variance attributable to the initial conditions and with the PV strategy, the proportion of residual variance was 2% for the public costs and for the number of slaughtered herds. It was 9% for the export losses, the proportion of residual variance being globally higher for this type of impact. The highest proportion of residual variance was observed for the selective preemptive slaughter and for the stamping-out strategies: >10% for the public costs and for the number of slaughtered herds, and >30% for the export losses (Table 4). For the public costs and for the number of slaughtered herds, the proportion of variance attributable to the evolution of the epizootics during successive periods was often higher for the first two periods: D0–D7 and D7–D14. The opposite situation was observed for the export losses, the last two periods having a stronger impact on the overall variance: D14–D28 and D28–D56 (Table 4).

## Discussion

The choice of a FMD control strategy was analyzed as a collective decision-making process in which three types of stakeholders are involved: the government, the agro-food industries and the public opinion. Seven simulated control strategies, the most realistic to implement, were compared and two geographic levels were considered for the decision scope: national-scope decisions made once for the whole country, and regional-scope decisions independently made in each region. Two types of decision-makers were distinguished according to their perception of risk: risk-neutral and risk-averse decision-makers (risk-seeker decision makers being considered unrealistic for the control of an epizootic disease). The main factor influencing the optimality of a strategy was the type of stakeholder, with opposite effects according to the decision scope: whereas national-scope optimal strategies were based on vaccination whatever the stakeholder, this consensus concealed a strong heterogeneity when the decision scope was regional. Besides the stakeholder and the geographic level, the risk perception by the decision-maker only induced minor variations on the optimal decision. Vaccine was more often used by risk-averse decision-makers and preemptive slaughter less often. However the type of optimal strategy (vaccination or preemptive slaughter) was often the same for risk-neutral and risk-averse decision-makers.

In this study, we used a realistic model of FMD spread, which was forced by actual data about herd populations and animal movements. Differences between species for clinical signs and infectiousness were taken into account. As in other studies [9,19,25], the model accounted for the heterogeneity of the within-herd virus diffusion according to the species housed, and to the efficacy of the biosecurity barriers between species inside a farm [26–28]. The disease detection process was explicitly represented in the model. Therefore, the time to first detection was not a parameter of the model, but a stochastic output. No correlation was observed between time to detection and epizootic duration or size (number of reported outbreaks). Despite the costs for government were quantified relatively precisely, a simplified cost model was used. For example, losses borne by farmers (e.g. induced by movement restrictions) were not taken into account. A more comprehensive cost model may be found in Backer *et al.* [29]. Herd culls performed for welfare reasons were not included in the model. This could have induced an under-estimation of the social impact of the control strategies. Indeed, the indicator we chose to quantify the social impact was simple. This impact may also be influenced, for example, by the total duration of the epizootic and the extent of the area submitted to restrictions of animal and human movement. It may also be linked to the impact on agro-food industries, which could induce a slowdown in economic activity. The indicator we chose for the impact on agro-food industries (export losses) was also rough: the production losses, and the shortfall for other economic sectors were not considered, for example, as well as the lower market acceptance of products of vaccinated animals [29]. However, indicators were either considered separately (for each stakeholder) or recalibrated before to be considered jointly (in collective decisions). Therefore, assuming that in each region, the chosen indicators are proportional to the real impacts, the preceding limits of these indicators should not bias significantly the geographic variations of optimal strategies against FMD epizootics.

When the decision scope was national, vaccination-based strategies (both preventive and suppressive) showed the best epidemiological effectiveness as they allowed lowering the number of infected premises. Oppositely, this number was higher for strategies based on preemptive slaughter, the ability of which to control epizootics was weaker. This result may be explained by the fact that, when preemptive slaughter was applied, some infected

**Table 3.** Feasibility of fixed control strategies adapted to the time and place where a simulated FMD epizootic begins.

| | Control strategy | | | | | | |
|---|---|---|---|---|---|---|---|
| | **SO** | **PS** | **PV** | **SPS** | **SPV** | **SPSV** | **SV** |
| Direct costs | | | | | | | |
| Initial conditions | 1.4 (52%) | 1.7 (59%) | 0.8 (55%) | 3.3 (58%) | 0.8 (53%) | 1.3 (58%) | 1.2 (53%) |
| Residual variance | 1.3 (42%) | 1.2 (41%) | 0.6 (45%) | 2.4 (42%) | 0.7 (47%) | 0.9 (42%) | 1.1 (47%) |
| Export losses | | | | | | | |
| Initial conditions | 0.4 (60%) | 0.6 (64%) | 0.3 (75%) | 0.8 (60%) | 0.3 (68%) | 0.3 (68%) | 0.3 (58%) |
| Residual variance | 0.3 (40%) | 0.3 (36%) | 0.1 (25%) | 0.5 (40%) | 0.1 (32%) | 0.2 (32%) | 0.2 (42%) |
| Slaughtered herds | | | | | | | |
| Initial conditions | 1.1 (42%) | 1.3 (57%) | 0.6 (51%) | 1.9 (55%) | 0.6 (50%) | 0.9 (55%) | 1.0 (52%) |
| Residual variance | 1.0 (48%) | 1.0 (43%) | 0.6 (49%) | 1.6 (45%) | 0.6 (50%) | 0.8 (45%) | 0.9 (48%) |

Note: The table gives, for a given impact and for a specific control strategy, the parts of variance explained by the initial conditions (place and date where the epizootic begins) and by the intrinsic variability of epizootics (residual variance). Impacts correspond to three stakeholders: government (indicator: public costs), agro-food industry (indicator: export losses), and public opinion (indicator: number of slaughtered herds).SO: basic stamping-out strategy, PV: protective vaccination, PS: preemptive slaughter, SPV: selective protective vaccination, SPS: selective preemptive slaughter, SPSV: selective preemptive slaughter and preventive vaccination, SV: suppressive vaccination. Impact indicators have been log-transformed.

**Table 4.** Feasibility of adaptive control strategies dynamically updated according to the evolution of simulated FMD epizootics.

| | Control strategy | | | | | | |
|---|---|---|---|---|---|---|---|
| | **SO** | **PS** | **PV** | **SPS** | **SPV** | **SPSV** | **SV** |
| Direct costs | | | | | | | |
| D0–D7 period | 0.134 (28%) | 0.180 (34%) | 0.067 (30%) | 0.329 (26%) | 0.114 (39%) | 0.153 (43%) | 0.114 (29%) |
| D7–D14 period | 0.134 (28%) | 0.120 (23%) | 0.069 (31%) | 0.166 (13%) | 0.076 (26%) | 0.056 (16%) | 0.093 (24%) |
| D14–D28 period | 0.094 (19%) | 0.111 (21%) | 0.059 (26%) | 0.282 (22%) | 0.063 (22%) | 0.079 (22%) | 0.090 (23%) |
| D28–D56 period | 0.074 (15%) | 0.075 (14%) | 0.024 (11%) | 0.258 (21%) | 0.030 (10%) | 0.055 (15%) | 0.068 (17%) |
| Residual variance | 0.049 (10%) | 0.040 (8%) | 0.004 (2%) | 0.219 (17%) | 0.009 (3%) | 0.016 (5%) | 0.026 (7%) |
| Export losses | | | | | | | |
| D0–D7 period | 0.015 (12%) | 0.013 (12%) | 0.004 (13%) | 0.051 (15%) | 0.006 (15%) | 0.009 (16%) | 0.009 (29%) |
| D7–D14 period | 0.015 (12%) | 0.017 (15%) | 0.005 (18%) | 0.030 (9%) | 0.006 (15%) | 0.006 (11%) | 0.011 (12%) |
| D14–D28 period | 0.021 (17%) | 0.025 (23%) | 0.009 (34%) | 0.073 (21%) | 0.012 (29%) | 0.013 (25%) | 0.026 (29%) |
| D28–D56 period | 0.032 (27%) | 0.028 (26%) | 0.07 (26%) | 0.082 (24%) | 0.012 (28%) | 0.015 (28%) | 0.026 (29%) |
| Residual variance | 0.038 (32%) | 0.025 (23%) | 0.003 (9%) | 0.108 (31%) | 0.006 (14%) | 0.010 (19%) | 0.017 (19%) |
| Slaughtered herds | | | | | | | |
| D0–D7 period | 0.114 (31%) | 0.114 (29%) | 0.057 (37%) | 0.221 (28%) | 0.079 (39%) | 0.112 (41%) | 0.081 (26%) |
| D7–D14 period | 0.081 (22%) | 0.081 (22%) | 0.036 (23%) | 0.116 (14%) | 0.047 (23%) | 0.043 (16%) | 0.074 (24%) |
| D14–D28 period | 0.072 (19%) | 0.088 (24%) | 0.039 (26%) | 0.181 (23%) | 0.046 (23%) | 0.061 (22%) | 0.077 (25%) |
| D28–D56 period | 0.060 (16%) | 0.061 (16%) | 0.018 (12%) | 0.158 (20%) | 0.023 (11%) | 0.043 (16%) | 0.057 (18%) |
| Residual variance | 0.044 (12%) | 0.033 (9%) | 0.003 (2%) | 0.122 (15%) | 0.007 (4%) | 0.014 (5%) | 0.023 (7%) |

Note: The table gives, for a given impact and for a specific control strategy, the parts of variance attributable to the epizootic evolution during specific periods, and to the intrinsic variability of epizootics (residual variance). The part of variance imputable to the initial conditions (place and date where the epizootic starts) is not considered here. Impacts correspond to three stakeholders: government (indicator: public costs), agro-food industry (indicator: export losses), and public opinion (indicator: number of slaughtered herds). SO: basic stamping-out strategy, PV: protective vaccination, PS: preemptive slaughter, SPV: selective protective vaccination, SPS: selective preemptive slaughter, SPSV: selective preemptive slaughter and protective vaccination, SV: suppressive vaccination. Impact indicators have been log-transformed.

premises were not identified: contact tracing could not be performed (as epidemiological inquiries were only performed in reported outbreaks), which indirectly favoured disease spread. In the particular case of selective preemptive slaughter, preemptive cull was targeted on animals with a low economic value, which were also the most abundant (small ruminants and non-breeding pigs): the strong decrease in the density of susceptible animals (only cattle and reproductive pigs remained) also induced a decrease of the number of diseased animals seen by farmers and veterinarians. The disease detection process was thus slowed in infected premises and the time to detection was lengthened, whereas disease spread was slowed but not stopped. These negative effects of preemptive slaughter were particularly marked in densely populated livestock areas (DPLA), where they often led to a saturation of available resources and to a loss of control over the epizootic course. The weight of DPLAs in the computation of national average impacts explained the predominance of vaccination in national-scope decisions. Because it allowed reducing the number of slaughtered herds, vaccination-based strategies were optimal for public opinion. For the same reason, vaccination allowed reducing the amount of compensations paid to farmers, and was also optimal for the government in national-scope decisions: the cost of vaccination was compensated by the decrease of the number of slaughtered herds (except for suppressive vaccination). The use of vaccine induced prolonged export bans, and was thus *a priori* not favourable to agro-food industries. However, such prolonged bans mainly affected exports to non-EU countries. As they represented <25% of total exports, vaccination-based strategies were also optimal for agro-food industries in national-scope decisions.

The average public costs and export losses were significantly reduced when the optimal decision was independently determined in each region, and the optimal strategies appeared much more variable than in national-scope decisions. Each of the 7 considered control strategies turned out to be optimal in at least one region, for a specific stakeholder (or for a specific combination of stakeholders' weights in a collective decision process). If vaccination-based strategies remained optimal in most regions for public opinion, the epizootics reduction of ban duration allowed preemptive slaughter-based strategies to be optimal in many regions for agro-food industries. Besides, from the government's point of view, vaccination and even preemptive slaughter induced significant extra-costs that were not always necessary to epizootic control: the basic stamping-out strategy was thus also optimal in many regions for the government.

The variations of stakeholders' weights in collective decisions induced various patterns, linked to the regional densities of cattle, swine and small ruminants. When they started in areas with high cattle and swine densities, epizootics were often severe and difficult to control. It was also the case when the density of small ruminants was high, whereas the density of cattle was medium or low. Because small ruminants show in general few clinical signs, infected premises were identified late, and dense populations of small ruminants allowed a prolonged disease spread, at a low level. In both types of areas (regions of group G1), vaccination-based strategies were optimal, except when the weight of agro-food industries was the strongest. In areas with a high or medium cattle density, infected premises were, on average, identified more quickly, even if the density of small ruminants was also high or medium. In these regions (regions of group G2), the stamping-out procedure became optimal when the government's weight was the strongest.

The variability of the optimal strategy between regions, and the connection between collective decision patterns and regional animal densities suggest that FMD control strategies could be usefully adapted to the local breeding conditions. To investigate the feasibility of such adaptive strategies, the proportion of impact variance attributable to the local conditions was estimated. If this proportion was high, adaptive strategies should be feasible: a carefully chosen set of variables describing the local conditions could be identified, that would allow determining the optimal strategy. In most cases, approximately 50% of variance impact was explained by the local conditions: simulated FMD epizootics thus showed a significant intrinsic variability and the predictive value of the initial conditions for determining the optimal strategy was not expected to be very high. A similar approach was used to investigate the feasibility of adaptive strategies that could be dynamically changed according to the evolution of the epizootic: for the public costs and for the number of slaughtered herds, knowing how the epizootic has evolved during the first weeks should allow predicting its total impact, and thus permit to adapt the control strategy.

In this study, it was assumed that control strategies were decided at the beginning of the epizootics and never changed afterwards. Such situations are not realistic: during real epizootics, decision-makers have to re-evaluate daily the chosen control strategy, and to decide whether this strategy has to be maintained or changed. Our results suggested that a steering tool could be designed for the control of FMD epizootics, based on the combination of a control panel and of a decision procedure based on control panel indicators.

## Acknowledgments

The authors would like to thank the following scientists and veterinarians for their excellent help and expert opinion: Pascal Hendrikx, Labib Bakkali Kassimi, Yves Leforban, François Moutou, Xavier Rosières, Bernard Toma, Gina Zanella and Stephan Zientara.

## Author Contributions

Conceived and designed the experiments: B. Durand MM SR B. Dufour. Performed the experiments: MM B. Durand. Analyzed the data: MM B. Durand. Contributed reagents/materials/analysis tools: B. Durand B. Dufour. Wrote the paper: MM B. Durand.

## References

1. Barnett P, Garland AJM, Kitching RP, Schermbrucker CG (2002) Aspects of emergency vaccination against foot-and-mouth disease. Comparative Immunology, Microbiology and Infectious Diseases 25: 345–364.
2. Leforban Y (2002) How predictable were the outbreaks of foot and mouth disease in Europe in 2001 and is vaccination the answer? Revue scientifique et technique (International Office of Epizootics) 21: 549–556, 539.
3. Toma B, Moutou F, Dufour B, Durand B (2002) Ring vaccination against foot-and-mouth disease. Comparative Immunology, Microbiology and Infectious Diseases 25: 365–372.
4. Green DM, Kiss IZ, Kao RR (2006) Modelling the initial spread of foot-and-mouth disease through animal movements. Proceedings of the Royal Society B: Biological Sciences 273: 2729–2735.
5. Orton RJ, Bessell PR, Birch CPD, O'Hare A, Kao RR (2012) Risk of foot-and-mouth disease spread due to sole occupancy authorities and linked cattle holdings. PLoS ONE 7.
6. Tildesley MJ, Savill NJ, Shaw DJ, Deardon R, Brooks SP, et al. (2006) Optimal reactive vaccination strategies for a foot-and-mouth outbreak in the UK. Nature 440: 83–86.
7. Tildesley MJ, Bessell PR, Keeling MJ, Woolhouse MEJ (2009) The role of pre-emptive culling in the control of foot-and-mouth disease. Proceedings of the Royal Society B: Biological Sciences 276: 3239–3248.
8. Ferguson NM, Donnelly CA, Anderson RM (2001) Transmission intensity and impact of control policies on the foot and mouth epidemic in Great Britain. Nature 413: 542–548.
9. Keeling MJ, Woolhouse MEJ, Shaw DJ, Matthews L, Chase-Topping M, et al. (2001) Dynamics of the 2001 UK foot and mouth epidemic: Stochastic dispersal in a heterogeneous landscape. Science 294: 813–817.
10. Martínez-López B, Perez AM, Sánchez-Vizcaíno JM (2010) A simulation model for the potential spread of foot-and-mouth disease in the Castile and Leon region of Spain. Preventive Veterinary Medicine 96: 19–29.
11. Tildesley MJ, Smith G, Keeling MJ (2012) Modeling the spread and control of foot-and-mouth disease in Pennsylvania following its discovery and options for control. Preventive Veterinary Medicine 104: 224–239.
12. Schoenbaum MA, Disney WT (2003) Modeling alternative mitigation strategies for a hypothetical outbreak of foot-and-mouth disease in the United States. Preventive Veterinary Medicine 58: 25–52.
13. Kobayashi M, Carpenter TE, Dickey BF, Howitt RE (2007) A dynamic, optimal disease control model for foot-and-mouth disease: I. Model description. Preventive Veterinary Medicine 79: 257–273.
14. Garner MG, Lack MB (1995) An evaluation of alternate control strategies for foot-and-mouth disease in Australia: A regional approach. Preventive Veterinary Medicine 23: 9–32.
15. Tomassen FHM, De Koeijer A, Mourits MCM, Dekker A, Bouma A, et al. (2002) A decision-tree to optimise control measures during the early stage of a foot-and-mouth disease epidemic. Preventive Veterinary Medicine 54: 301–324.
16. Bates TW, Carpenter TE, Thurmond MC (2003) Benefit-cost analysis of vaccination and preemptive slaughter as a means of eradicating foot-and-mouth disease. American Journal of Veterinary Research 64: 805–812.
17. Mahul O, Durand B (2000) Simulated economic consequences of foot-and-mouth disease epidemics and their public control in France. Preventive Veterinary Medicine 47: 23–38.
18. Thompson D, Muriel P, Russell D, Osborne P, Bromley A, et al. (2002) Economic costs of the foot and mouth disease outbreak in the United Kingdom in 2001. OIE Revue Scientifique et Technique 21: 675–687.
19. Backer JA, Hagenaars TJ, Nodelijk G, van Roermund HJW (2012) Vaccination against foot-and-mouth disease I: Epidemiological consequences. Preventive Veterinary Medicine 107: 27–40.
20. Haydon DT, Kao RR, Kitching RP (2004) The UK foot-and-mouth disease outbreak - The aftermath. Nature Reviews Microbiology 2: 675–681.
21. Kitching RP, Hutber AM, Thrusfield MV (2005) A review of foot-and-mouth disease with special consideration for the clinical and epidemiological factors relevant to predictive modelling of the disease. Veterinary Journal 169: 197–209.
22. Rautureau S, Dufour B, Durand B (2012) Structuring the passive surveillance network improves epizootic detection and control efficacy: A simulation study on foot-and-mouth disease in France. Transboundary and Emerging Diseases 59: 311–322.
23. Rautureau S, Dufour B, Durand B (2011) Vulnerability of animal trade networks to the spread of infectious diseases: A methodological approach applied to evaluation and emergency control strategies in cattle, France, 2005. Transboundary and Emerging Diseases 58: 110–120.
24. Rautureau S, Dufour B, Durand B (2012) Structural vulnerability of the French swine industry trade network to the spread of infectious diseases. Animal 6: 1152–1162.
25. Bates TW, Thurmond MC, Carpenter TE (2003) Description of an epidemic simulation model for use in evaluating strategies to control an outbreak of foot-and-mouth disease. American Journal of Veterinary Research 64: 195–204.
26. Bouma A, Dekker A, De Jong MCM (2004) No foot-and-mouth disease virus transmission between individually housed calves. Veterinary Microbiology 98: 29–36.
27. Eblé P, De Koeijer A, Bouma A, Stegeman A, Dekker A (2006) Quantification of within- and between-pen transmission of foot-and-mouth disease virus in pigs. Veterinary Research 37: 647–654.
28. Honhold N (2006) The impact of farm gate biosecurity on the transmission of FMD in UK in 2001. Session of the Research Group of the Standing Technical Committee of EuFMD. Paphos, Cyprus.
29. Backer J, Bergevoet R, Hagenaars T, Bondt N, Nodelijk G, et al. (2009) Vaccination against foot-and-mouth disease - differentiating strategies and their epidemiological and economic consequences Wageningen UR.

# Investigating the Role of State and Local Health Departments in Addressing Public Health Concerns Related to Industrial Food Animal Production Sites

**Jillian P. Fry[1,2]*, Linnea I. Laestadius[2,3], Clare Grechis[2,3], Keeve E. Nachman[1,2,3], Roni A. Neff[1,2,3]**

**1** Department of Environmental Health Sciences, Johns Hopkins Bloomberg School of Public Health, Baltimore, Maryland, United States of America, **2** Center for a Livable Future, Johns Hopkins Bloomberg School of Public Health, Baltimore, Maryland, United States of America, **3** Department of Health Policy & Management, Johns Hopkins Bloomberg School of Public Health, Baltimore, Maryland, United States of America

## Abstract

***Objectives:*** Evidence of community health concerns stemming from industrial food animal production (IFAP) facilities continues to accumulate. This study examined the role of local and state health departments in responding to and preventing community-driven concerns associated with IFAP.

***Methods:*** We conducted semi-structured qualitative interviews with state and county health department staff and community members in eight states with high densities or rapid growth of IFAP operations. We investigated the extent to which health concerns associated with IFAP sites are reported to health departments, the nature of health departments' responses, and barriers to involvement.

***Results:*** Health departments' roles in these matters are limited by political barriers, lack of jurisdiction, and finite resources, expertise, and staff. Community members reported difficulties in engaging health departments on these issues.

***Conclusions:*** Our investigation suggests that health departments frequently lack resources or jurisdiction to respond to health concerns related to IFAP sites, resulting in limited engagement. Since agencies with jurisdiction over IFAP frequently lack a health focus, increased health department engagement may better protect public health.

**Editor:** Joan A. Caylà, Public Health Agency of Barcelona, Spain

**Funding:** The authors have no support or funding to report.

**Competing Interests:** The authors have declared that no competing interests exist.

* E-mail: jfry@jhsph.edu

## Introduction

Industrial food animal production (IFAP), the dominant meat, dairy, and egg production method in the US, involves housing large numbers of animals in close quarters [1]. Between 1959 and 2007, mean farm size increased over 2,300 percent for hogs and over 30,000 percent for broilers (chickens sold for meat). During this time, the number of farms in the US with swine and broilers decreased by 95.9 and 98.5 percent, respectively [2], [3]. In addition to a dramatic shift to fewer, larger operations, IFAP sites are geographically concentrated, as Figure 1 illustrates for US hog production. The impacts of IFAP are much more widespread than the map indicates, however, because other types of IFAP are concentrated in different areas of the country. Producing large numbers of animals in close proximity results in the concentration of massive amounts of animal waste in small geographic areas, leading to practices such as over-application to fields as fertilizer, storing waste in sheds or large cesspits or lagoons, and transporting waste from overburdened regions [4], [5]. Regulations exist to prevent animal waste from entering surface water, but the US Environmental Protection Agency has noted that many facilities are failing to comply with regulations [6]. Also, there is much variation at the state level regarding regulatory stringency and enforcement [7], [8]. The regulatory term Concentrated Animal Feeding Operation (CAFO) is used to describe these facilities, but we use the term IFAP, which includes CAFOs, as there is concern that sites that use these methods but fail to meet the technical CAFO definition due to size and/or structure can still contribute to serious environmental public health problems.

Our study seeks to understand the role of state and local health departments (HDs) in addressing citizen health concerns related to IFAP facilities in the US. We also explore potential barriers that may limit agency activities aimed at addressing these issues.

Documented air emissions from IFAP include hydrogen sulfide [9], particulate matter [9], endotoxins [10], ammonia [11], allergens [12], and volatile organic compounds [13], [14]. Exposure to these emissions has been associated with multiple respiratory, cardiovascular, and neurological outcomes for IFAP workers [15], [16]. Community health effects largely parallel those from occupational exposure, with growing evidence suggesting that residential proximity to IFAP facilities also increases respiratory health risks [13], [17–19]. Studies link chronic exposure to animal production odors to headaches, nausea, upset stomach, mood disorders, and sleep problems [20–23]. Water is also

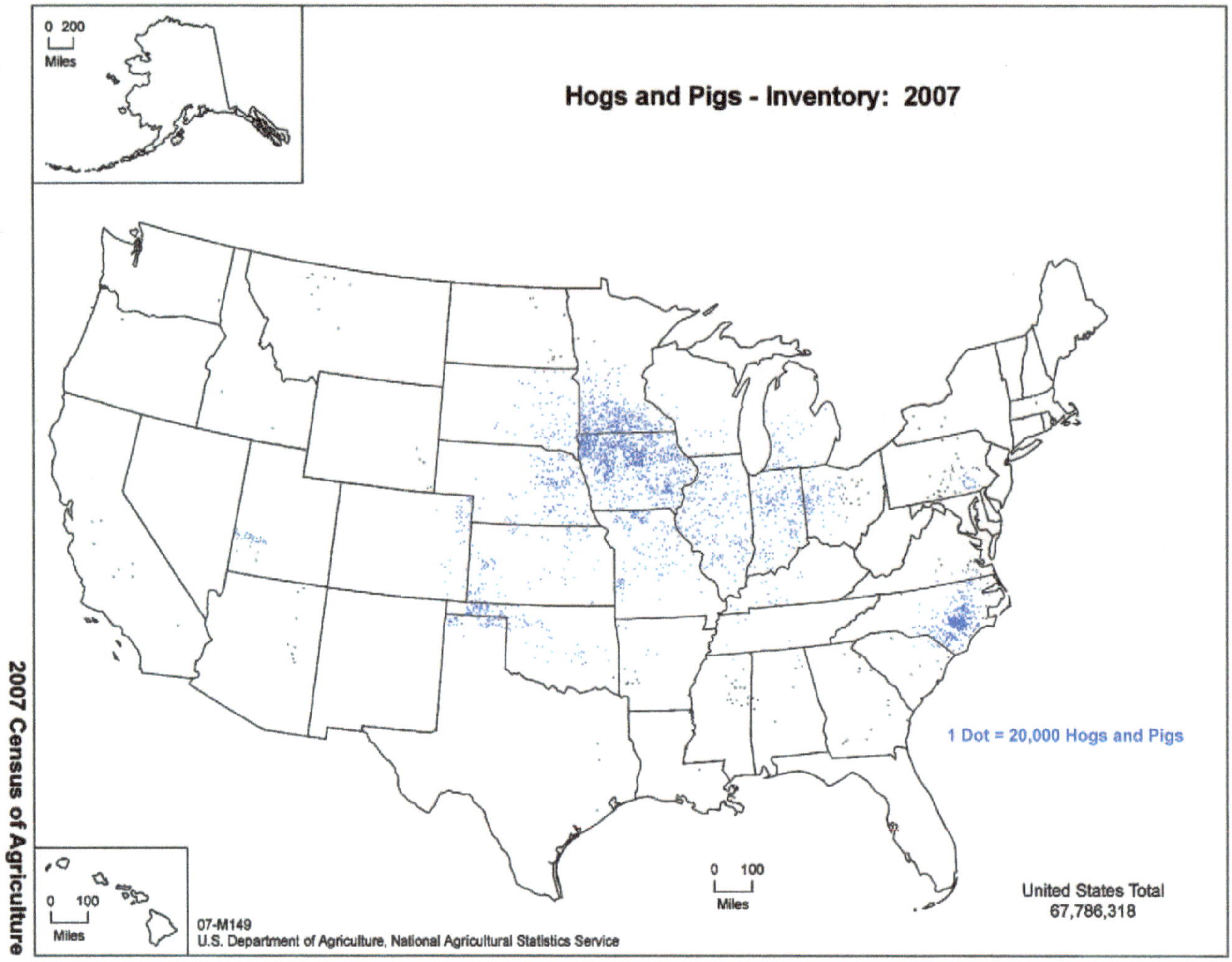


**Figure 1. Geographic concentration of hog and pig production in the US.** Figure Legend: 2007 Census of Agriculture data from the United States Department of Agriculture (USDA) shows the geographic concentration of hog and pig production in the US. Source: USDA Census of Agriculture; 2007 Census Ag Atlas Maps www.agcensus.usda.gov/Publications/2007/Online_Highlights/Ag_Atlas_Maps/.

a concern; rural residents who rely on private wells for their water supply are at particular risk as animal waste contaminants, including nitrates, pathogens, pharmaceuticals, metals, and hormones, can leach into ground water [24], [25].

Studies in two states have shown that IFAP facilities are disproportionately sited in low-income communities with high-percentage minority populations [26–28]. These populations are at increased risk of respiratory conditions such as asthma, and are already significantly less likely to have health insurance than higher-income white populations [29], [30], thus increasing both their vulnerability to the health consequences of IFAP pollution and the challenge of obtaining medical advice and treatment.

In some cases, HDs have funded or performed investigations to collect data on IFAP health effects [28], [31–34]. On the other hand, two of the authors (RN and JF) witnessed an individual at a public meeting state his belief that the health department would surely know about any health concerns posed by an IFAP facility and alert the public to the situation. This was striking because in this case the HD had little to no involvement in the decision to permit the facility in question, nor with monitoring or regulation of IFAP sites in the area. This led to the realization that research was needed to examine the extent to which human health concerns associated with IFAP are reported to HDs, and how agencies with IFAP operations in their jurisdiction respond to public concerns about health problems that may be caused by these operations.

Environmental health concerns related to air and water quality have long been key to the conception of public health in the US [35]. The 10 essential public health services, which include monitoring, diagnosing and investigating community health hazards and informing/educating community members, suggest that some aspects of IFAP would fall within public health agency mandates [36]. Environmental health responsibilities, however, are increasingly shared between Departments of Health and Departments of Environment, resulting in challenges to co-ordination and data sharing [37]. A report by the National Council of State Legislatures indicates that primary state-level regulatory authority over IFAP facilities falls largely with Departments of Environment and Natural Resources and that health departments generally have little or no role in these activities [8], which raises concerns that regulations may not be designed to protect human health.

Another critical factor to consider when examining the HD role in addressing health impacts from IFAP is that many local and state HDs in the US have suffered significant and sustained resource reductions, in part attributable to the recession and weak economic recovery in the early 2000s [38], [39]. Limited staff and resources hamper state HDs' ability to investigate non-communicable disease clusters [40]. Additionally, lower rates of illness and better health outcomes have been linked to more comprehensive HD programs and higher levels of resources, including funding and staff [41–44]. Identifying and publicizing the implications of HD resource levels and program characteristics is important to effectively addressing the limitations and gaps that may be experienced by HD programs.

## Methods

Due to the exploratory nature of the research, we determined that a qualitative approach to data collection and analysis was most appropriate. We performed semi-structured interviews with state- and county-level HD employees in eight states, and with a community member in each of those states. We used USDA Census of Agriculture hog inventory data to rank all US counties in two ways: 1) 2007 county hog inventory at operations with 1,000 or more hogs, and 2) increase in hog inventory at 1,000+ head operations between 2002 and 2007. Many hogs produced in intensive settings or a large increase in the number of hogs may lead community members or others to request that HDs become involved with IFAP. We sought to increase the chance that the relevant hog operations were near residential areas by ranking the top sixty counties from each list by population density using 2000 Census data. We selected the top fifteen counties by population density from the two lists and, due to overlap, eighteen counties in eight states remained. To ensure that no state was disproportionately represented, we limited the sample to a maximum of two counties per state. The final sample comprised fourteen counties in eight states.

To identify appropriate HD personnel to interview, we performed background research using health department websites and contact with the Association of State and Territorial Health Officials (ASTHO). We contacted HD staff who led environmental health divisions within an HD or the head of the HD if there was no division focused on environmental health. At times, HD staff we contacted referred us to others in the department because they could better answer our questions. In two instances, we found that the county HD did not address environmental health concerns. In these cases, we included the county level department that handled environmental health concerns, generally an environmental services department, instead of the HD. To avoid unnecessary complication, we include these agencies with the county HDs in the results except where they differed substantially.

We identified community members through Internet research on involved organizations in each state and by asking known contacts who work on IFAP issues for suggestions. Recognizing that speaking to only one community member per state was limiting, we nonetheless felt it important to complement HD comments with community perspectives. We sought affected or potentially affected residents who had been active in addressing IFAP issues in their state, as opposed to lawyers, environmental scientists, or others who were involved professionally but not directly affected by IFAP operations where they live. When possible, we interviewed community members who were from the included counties. As these individuals were often members of community groups working on IFAP concerns, they spoke both as individuals and as representatives of their groups. We conducted all interviews by telephone.

The questionnaire included mostly open-ended and some closed-ended questions (See Appendix S1 for survey instruments). We sent background questions to HD staff members prior to the interview so information on budgets and workforce could be compiled beforehand. We did not use the terms IFAP or CAFO in the interviews. Before beginning each interview, we read participants a confidentiality statement. The Johns Hopkins School of Public Health Institutional Review Board (IRB) determined the study was exempt from IRB oversight.

After data collection was complete, notes from six interviews were double coded by JF and LL through an inductive coding process using *HyperRESEARCH 3.0.3* (ResearchWare, Randolph, MA). We then jointly discussed codes, developed a uniform codebook, and applied it to the remaining interviews. Due to the qualitative nature of the study and small, non-representative sample, we provide limited numeric information from the data to avoid giving the impression that the findings are generalizable to a larger population [45].

## Results

We conducted telephone interviews with staff from thirteen county HDs and eight state HDs, plus eight community members between November 2010 and January 2012. One county HD contacted declined to participate.

### Health Departments

**Contacts made by community members.** HD staff reported receiving calls about IFAP facilities generally a few times a year to a few times a month, with some reporting that they have not been contacted about issues related to IFAP. We read participants a list of the most common health concerns associated with IFAP, as identified in the literature. Of these, the most common reasons for calls were odor, water quality concerns, respiratory health, general health concerns, and stress. Individual residents, rather than physicians or organized groups, usually made these calls. Almost all HD staff noted that the frequency of related calls had either remained constant or decreased over the past few years. A few county HD staff cited the passage and/or enforcement of ordinances and regulations that limited new facility construction or improved waste management practices as an explanation for the reduction in call volume. One county HD staff member noted that the volume of reports changed "when the state actually started enforcing feedlot rules." Two county-level HD staffers in the same state indicated that the number had increased in recent years, and some state HD staff members said call volume generally increased when new facilities were being proposed or built.

**Response to calls.** Overall, staff indicated that HDs do not have a prescribed response to reported concerns about IFAP facilities, and most counties and states described their response as a case-by-case process. Few HD staff reported keeping formal records of calls. In almost all instances, regardless of the type of concern, county and state HD staff reported that they refer residents with concerns about IFAP to another agency (Table 1). Most indicated that this was due to their agencies' lack of regulatory authority over animal production farms. Specific referrals generally depended on what agency had regulatory authority over IFAP in that particular state, with referrals to Departments of Environment and Natural Resources most common. Staff from a few counties and states said they might speak to the IFAP facility operator, but noted that they could only request voluntary changes. A state HD staff member noted, "the best we can do is bring it to the operator's attention and hope they

**Table 1.** Referrals reported by State and County Health Department staff in response to concerns related to IFAP.

| |
|---|
| State Departments of Agriculture |
| Soil Conservation Districts |
| Cooperative Extensions |
| State Departments of Environment/Natural Resources/Pollution Control |
| State HD referral to county HD and vice versa |

take care of it voluntarily," and a county HD staff member said, "we have no control over manure spreading or manure management, all we can do is consult with the farmer to try to work with them."

**Additional involvement.** We also queried HD staff about their involvement with IFAP issues more generally. Staff from eight counties and three states reported little to no involvement; others' activities are described below.

About half of the county and state HD staff reported that their agencies collaborated with other agencies and government entities at times on IFAP issues, for example through communicating with them or visiting problematic IFAP facilities along with other agencies if the situation warranted their attention. Most reported that they had not engaged with citizen groups active on IFAP issues, and that they were not aware of any such citizen groups in their regions.

Most county and state interviewees reported no involvement in decision-making regarding animal agriculture. A small number said they had been involved with regulatory issues such as zoning, siting requirements, and setback requirements (distance between a building and the property edge). A few state HD staff members said they serve on advisory committees or provide input to those making decisions. Only two county agency staff members mentioned involvement with permitting IFAP facilities; these counties were atypical in that their environmental health programs were located within their environmental services departments rather than HDs.

About half the states reported collecting some IFAP-relevant data in the past (e.g. well testing, counting flies, air monitoring). These instances appeared to have been rare, however, and primarily motivated by IFAP facilities with particularly poor standards rather than routine data collection. Additionally, the vast majority of county and state HD staff interviewed had not provided any educational services to residents regarding potential health effects of animal agriculture. Efforts mentioned by staff that did provide information included informally explaining IFAP health effects to community members and creating an online document reviewing research and opinions on IFAP and health.

**Barriers to action.** State and county HD staffers reported multiple barriers to addressing concerns about intensive animal agriculture. Lack of jurisdiction over IFAP was the most frequently reported. In almost all states and counties examined, odor and air quality issues were reported to be outside the jurisdiction not only of HDs, but also of other state and county agencies, including Departments of Environment and Agriculture, due to a lack of regulation. One state has an ordinance pertaining to hydrogen sulfide, and the interviewee said they are often contacted by residents in other states asking what they can do about air quality problems because those states do not have air quality regulations relevant to IFAP emissions. Some interviewees also mentioned resource and infrastructure challenges. More than half of the state HD staff mentioned these as preventing them from undertaking a formal study to assess an issue like well water quality, and one county stated that they are trying hard enough to handle their own mandated programs with limited resources. In addition, HD flexibility in allocating environmental health funds varied. About half of HD staff interviewed indicated that they had no or quite limited flexibility in allocating these funds. No HD received funds specifically designated for supporting IFAP related efforts or to train staff in this area. Levels of expertise also hindered response, with one state HD staffer noting that they simply do not have the expertise needed to have their own regulatory program on IFAP. Another state HD staff member felt the current science on IFAP and public health is insufficient and does not justify health agencies providing guidance on IFAP for other agencies. A few county HD staff members stated that they do not believe that odor is associated with health effects, and one county HD staff member mentioned that the agency with regulatory authority over IFAP will investigate spills, but "if it's just smell, they don't worry about it."

Many HD staffers at both the state and county level also described political barriers to addressing the issue, including industry political influence in the state or county, agriculture's importance for economies in rural areas, and intentional efforts by legislators and administrators to avoid enforcing IFAP regulations. A county HD staffer explained that their early efforts to pass an ordinance to regulate IFAP facilities failed in part because so many people in the area make an income from farming. Another respondent said a state law had been changed to thwart local efforts to control CAFO siting, invoking the strategy of pre-emption, and a county-level interviewee described state legislators pressuring agencies to avoid looking into the issue. Several state HD staff members indicated that their state governments feel that the financial incentive to have large farms outweighs other concerns, and one noted that they had to be extremely careful with any statements they make on the issue of IFAP and health because anything perceived to be detrimental to industry could result in a lawsuit. Another noted that his department has to "protect health while still allowing breathing space for [industry] to 'make a buck'."

**Needed resources.** We provided HD staff with a list of possible resource needs and asked them to indicate any they needed to more effectively address health concerns stemming from IFAP (Table 2). The most commonly indicated need was for educational materials that could be distributed to the public on possible health effects of IFAP facilities. A few county HD staff members indicated that they had no need for any additional resources, primarily because this was not an issue they planned to address.

## Community Members

All community members interviewed expressed concerns about IFAP-related health impacts. Community members reported a wide range of activities, both personally and with community groups, to address IFAP concerns (Table 3), including contacting local and state HDs and local boards of health. As Table 3 shows, community members and organizations have been assuming roles that would traditionally fall under HD purview, including environmental monitoring, educating and providing information to the community, seeking information from other government agencies, and partnering with researchers. Communication with other local and national level groups addressing IFAP was common.

No community members reported having had an interaction with an HD about an IFAP issue that resolved their concerns.

**Table 2.** Resources indicated by HD staff as needed to address IFAP concerns (ranked by number of HDs indicating need).

| |
|---|
| 1. Educational materials for distribution |
| 2. Increased funding for department |
| 3. Training for staff on issues relevant to animal production farms |
| 4. Updated information from researchers on health effects of concern |
| 5. Environmental quality tracking tools |
| 6. More staff |
| 7. Different political climate |
| 8. Funding specifically for animal production farm activities |
| 9. Connections to experts |

Generally they stated that HDs were not engaged with this issue. Some community respondents indicated that HD staff had attended meetings on this topic or showed an initially encouraging response; however, they said that the HDs ultimately took no action. One community member thought that the HD was placing the burden of proof on community members. Community members who said they were unsuccessful in having their concerns addressed by contacting HDs generally did not persist in contacting them, and instead moved on to other agencies and approaches. Despite this, almost all community members interviewed indicated that there should be a role for HDs in addressing this issue complementary to that of permitting agencies.

Community interviewees stated that the number of concerns reported to agencies was likely limited by agriculture's influence in small rural communities. One suggested that everyone in the county is related to someone who works in agriculture and asked, "Who is willing to rat out their cousin?" Another described social stigma associated with expressing concerns about IFAP and noted several instances of harassment against individuals who had done so.

Community members recognized many of the themes expressed by HD staff regarding barriers to HD response. They stated that perceived barriers included politics and animal agriculture's economic importance, limited agency resources, and active efforts by agricultural interests to minimize regulatory attention to IFAP facilities. Many community members perceived an inappropriate connection between government agencies, legislators, and industry. One community member noted that "they're not even working for us anymore; it's like they're working for industry." Another community member suggested that county HD staff members are privately very concerned about IFAP and health, but publicly refrain from addressing it due to "political posturing."

## Discussion

Evidence suggests that contaminated air, soil, and water near IFAP facilities can pose significant public health threats. Despite that, this analysis found that the sampled HDs received few contacts from the public on IFAP issues, and engaged little on the topic. Significant barriers to engagement were identified.

The low call volume to HDs should probably not be interpreted as a lack of public concern, but rather as a function of the unique political and social circumstances surrounding IFAP in many rural communities. Community members commented that when they call HDs and are referred to other agencies because animal agriculture is said not to be under the HDs' jurisdiction, they often stop contacting the HD about IFAP concerns. Community members also reported social pressures that appear to limit the voicing of IFAP-related concerns.

Due to regulations relevant to IFAP, jurisdiction is almost exclusively within Departments of Agriculture and Environment/Natural Resources. As human health is generally not a primary part of the mission of these agencies, if it is included at all, HD involvement even in an advisory capacity can be valuable. Unfortunately, many barriers in addition to jurisdiction are currently preventing HDs from engaging on this issue, including resource constraints, lack of expertise, and political pressure.

This is the first study to examine formally the role of HDs in responding to community concerns arising from IFAP operations. We used qualitative methods to engage with both HD staff and community members to gain context regarding their respective views on roles, responsibilities, and responsiveness. Due to the use

**Table 3.** Actions taken by community members/groups in response to IFAP concerns.

| |
|---|
| Contacting local and state agencies (departments of health, environment, natural resources, and agriculture), state legislators, and governors |
| Forming community organizations and seeking advice from other organizations that have been working on this issue |
| Holding public events and conducting public education |
| Reading scientific literature on IFAP and health |
| Organizing meetings |
| Attending permit hearings and reviewing permit applications |
| Conducting own environmental monitoring |
| Requesting samples to be taken |
| Media advocacy |
| Advocating for state legislation |
| Forming a group to hire a lawyer |
| Contacting local police |
| Collaborating with researchers |
| Collaboration with other national and regional groups |
| Serving on local boards (e.g. board of health) |

of a small sample that may not be representative of all US regions impacted by IFAP we cannot extrapolate findings to all counties and states with these types of operations. Additionally, the opinions and perceptions of the individual HD staff members interviewed likely influenced results. While there was frequently only one staff member identified as relevant at each HD, it is possible that other staffers may have provided substantively different responses. Despite these considerations, we anticipate that the results of our investigation have identified common themes in agricultural communities.

For community members in particular, the sample was limited and likely does not reflect the full range of views within the surveyed regions. Views identified here may not, for example, be representative of the views of rural community members who have not actively sought to address IFAP issues. Additionally, inclusion of others involved with these issues, such as staff at nonprofit environmental organizations instead of community members, would likely result in different responses. Despite this, we believe their inclusion adds further context that aids in understanding the perspectives of HD staff on issues related to IFAP operations.

## Conclusion

It is critical for affected communities, healthcare providers, policy makers, and public health professionals to be aware of the limited engagement on this issue by HDs. We found that HD staff reported inadequate resources and/or jurisdictional authority needed to monitor and regulate IFAP facilities; as a result, IFAP facilities that pose a threat to public health may go unnoticed and unaddressed by HDs. While giving HDs a formal role has the potential to improve public health protection, HDs could play a more significant role even with no change in jurisdictional authority if resources and the political landscape changed. HDs with IFAP operations in their county or state should be provided with resources such as training, educational materials, and increased funding. These resources could be provided through organizations working on IFAP and public health issues, state legislatures, or from the federal government as part of the national oversight of CAFO permitting programs. It should be noted, however, that even with these resources or a change in jurisdictional authority, political barriers will likely remain a significant challenge to fully addressing IFAP and public health.

There are many potential directions for future research that can build on these findings. For example, a questionnaire informed by these results could be developed that would allow researchers to study a much larger sample of HD staff and/or community members. A larger study would facilitate more robust comparisons between states and regions and also allow investigators to explore additional factors such as which characteristics increase the likelihood that an HD will take on a more proactive role or how race, ethnicity, or socio-economic status of affected communities impact these situations. Other groups, such as staff of agencies with jurisdiction over IFAP and environmental organizations, could also be included. We hope this study will serve to include IFAP in the wider debate on resource limitations faced by HDs and their role in protecting environmental public health and spur additional research on this topic.

## Supporting Information

**Appendix S1** Health Department Personnel and Community Members Questionnaires.

## Acknowledgments

We thank the health department staff members and community residents who participated in the study.

## Author Contributions

Conceived and designed the experiments: JF LL CG KN RN. Performed the experiments: JF LL CG KN RN. Analyzed the data: JF LL. Wrote the paper: JF LL KN RN.

## References

1. US Department of Agriculture (2009) The Transformation of U.S. Livestock Agriculture: Scale, Efficiency, and Risks. Economic Information Bulletin No. (EIB-43) 46 p. Available: http://www.ers.usda.gov/publications/eib-economic-information-bulletin/eib43.aspx. Accessed 2012 Aug 8.
2. US Department of Agriculture. Census of Agriculture Historical Archive. Available: http://agcensus.mannlib.cornell.edu/AgCensus/censusParts.do?year=1959. Accessed 2012 Aug 8.
3. US Department of Agriculture. Census of Agriculture Quick Stats. Available: http://quickstats.nass.usda.gov/?source_desc=CENSUS. Accessed 2012 Aug 8.
4. Graham JP, Nachman KE (2010) Managing waste from confined animal feeding operations in the United States: The need for sanitary reform. J Water Health 8: 646–670.
5. US Department of Agriculture (2000) Manure Nutrients Relative to the Capacity of Cropland and Pastureland to Assimilate Nutrients: Spatial and Temporal Trends for the United States. Available: http://www.nrcs.usda.gov/wps/portal/nrcs/detail/national/technical/nra/rca/?&cid=nrcs143_014126. Accessed 2012 Aug 8.
6. US Environmental Protection Agency (2012) FY 2012 Office of Enforcement and Compliance Assurance (OECA) Final National Program Manager (NPM) Guidance. Available: http://nepis.epa.gov/Exe/ZyPDF.cgi?Dockey=P100B2HE.pdf. Accessed 2012 Aug 8.
7. Centner TJ, Newton GL (2011) Reducing CAFO permitting requirements. J Anim Sci 89: 4364–4369. doi:10.2527/jas.2011-4256.
8. National Conference of State Legislatures (2008) Concentrated Animal Feeding Operations A Survey of State Policies. Available: http://www.ncifap.org/bin/k/w/CAFO_State_Survey_Introduction.pdf. Accessed 2012 Aug 8.
9. Schiffman SS, Bennett JL, Raymer JH (2001) Quantification of odors and odorants from swine operations in North Carolina. Agr Forest Meteorol 108: 213–240.
10. Schenker MB, Christiani D, Cormier Y, et al. (1998) American Thoracic Society: Respiratory Health Hazards in Agriculture. Am J Resp Crit Care Med 158: S1–S76.
11. Wilson SM, Serre ML (2007) Examination of atmospheric ammonia levels near hog CAFOs, homes, and schools in Eastern North Carolina. Atmos Environ 41: 4977–4987.
12. Williams DA, Breysse P, McCormack M, Diette G, McKenzie S, et al. (2011) Airborne cow allergen, ammonia and particulate matter at homes vary with distance to industrial scale dairy operations: an exposure assessment. Environ Health 10: 72. doi:10.1186/1476-069X-10-72.
13. Schinasi L, Horton RA, Guidry VT, Wing S, Marshall SW, et al. (2011) Air Pollution, Lung Function, and Physical Symptoms in Communities Near Concentrated Swine Feeding Operations. Epidemiology 22: 208–215.
14. Trabue S, Scoggin K, Li H, Burns R, Xin H, et al. (2010) Speciation of volatile organic compounds from poultry production. Atmos Environ 44: 3538–3546.
15. Mitloehner FM, Calvo MS (2008) Worker health and safety in concentrated animal feeding operations. J Agric Saf Health 14: 163–187.
16. Donham K (2010) Community and occupational health concerns in pork production: A review. J Anim Sci 88: e102-e111.
17. Thu K, Donham K, Ziegenhorn R, Reynolds S, Thorne P, et al. (1997) A control study of the physical and mental health of residents living near a large-scale swine operation. J Agric Saf Health 3: 13–26.
18. Merchant JA, Naleway AL, Svendsen ER, Kelly KM, Burmeister LF, et al. (2005) Asthma and farm exposures in a cohort of rural Iowa children. Environ Health Perspect 113: 350–356.
19. Radon K, Schulze A, Ehrenstein V, van Strien RT, Praml G, et al. (2007) Environmental exposure to confined animal feeding operations and respiratory health of neighboring residents. Epidemiology 18: 300–308.
20. Schiffman SS, Miller EA, Suggs MS, Graham BG (1995) The effect of environmental odors emanating from commercial swine operations on the mood of nearby residents. Brain Res Bull 37: 369–375.
21. Avery RC, Wing S, Marshall SW, Schiffman SS (2004) Odor from industrial hog farming operations and mucosal immune function in neighbors. Arch Environ Health 59: 101–108.
22. Nimmermark S (2004) Odour influence on well-being and health with specific focus on animal production emissions. Ann Agr Environ Med 11: 163–173.

23. Horton RA, Wing S, Marshall SW, Brownley KA (2009) Malodor as a trigger of stress and negative mood in neighbors of industrial hog operations. Am J Public Health 99: S610–615.
24. Osterberg D, Wallinga D (2004) Addressing externalities from swine production to reduce public health and environmental impacts. Am J Public Health 94: 1703–1708.
25. Burkholder J, Libra B, Weyer P, Heathcote S, Kolpin D, et al. (2007) Impacts of waste from concentrated animal feeding operations on water quality. Environ Health Perspect 115: 308–312.
26. Wing S, Cole D, Grant G (2000) Environmental injustice in North Carolina's hog industry. Environ Health Perspect 108: 225–231.
27. Wilson SM, Howell F, Wing S, Sobsey M (2002) Environmental injustice and the Mississippi hog industry. Environ Health Perspect 110: 195–201.
28. Mirabelli MC, Wing S, Marshall SW, Wilcosky TC (2006) Race, poverty, and potential exposure of middle-school students to air emissions from confined swine feeding operations. Environ Health Perspect 114: 591–596.
29. US Centers for Disease Control and Prevention. National Surveillance for Asthma - United States, 1980–2004. Available: http://minorityhealth.hhs.gov/templates/content.aspx?ID = 6170. Accessed 2012 Aug 8.
30. US Census Bureau (2010) Income, Poverty, and Health Insurance Coverage in the United States: 2009. Available: http://www.census.gov/prod/2010pubs/p60-238.pdf. Accessed 2012 Aug 8.
31. Thu K (2002) Public Health Concerns for Neighbors of Large-Scale Swine Production Operations. J Agric Saf Health 8: 175–184.
32. Missouri Department of Health and Senior Services (2003) Health Consultation Final Report on Exposure Investigation Findings: Valley View Concentrated Animal Feeding Operations. Available: http://health.mo.gov/living/environment/hazsubstancesites/pdf/HC%20CAFO%20Final.pdf. Accessed 2012 Aug 8.
33. Wing S, Wolf S (2000) Intensive livestock operations, health, and quality of life among eastern North Carolina residents. Environ Health Perspect 108: 233–238.
34. Wing S (2002) Social responsibility and research ethics in community-driven studies of industrialized hog production. Environ Health Perspect 110: 437–444.
35. Resnick BA, Zablotsky J, Burke TA (2009) Protecting and promoting the nation's health: The environmental public health workforce as a critical component. J Public Health Manag Pract 15: S40–S45.
36. American Public Health Association. 10 Essential Public Health Services. Available: http://www.apha.org/programs/standards/performancestandardsprogram/resexxentialservices.htm. Accessed 2012 Aug 8.
37. Johns Hopkins Bloomberg School of Public Health (2008) The Northeast/Mid-Atlantic Profile of Environmental Public Health Practice. Available: http://www.jhsph.edu/ecehp/Profile_Report_Part_I_final_9-16-08.pdf. Accessed 2012 Aug 8.
38. National Association of County and City Health Officials (2012) Local Health Department Job Losses and Program Cuts: Findings from the January 2012 Survey. Available: www.naccho.org/topics/infrastructure/lhdbudget/upload/Research-Brief-Final.pdf. Accessed 2012 Aug 8.
39. Association of State and Territorial Health Officials (2012) Budget Cuts Continue to Affect the Health of Americans: March 2012 Update. Available: http://www.aahd.us/wp-content/uploads/2012/04/ASTHO-Budget-Cuts-Impact-Research-0312.pdf. Accessed 2012 Aug 8.
40. Juzych NS, Resnick B, Streeter R, Herbstman J, Zablotsky J, et al. (2007) Adequacy of State Capacity to Address Noncommunicable Disease Clusters in the Era of Environmental Public Health Tracking. Am J Public Health 97: S163–S169. doi:10.2105/AJPH.2006.096453.
41. Zablotsky Kufel JS, Resnick BA, Fox MA, McGready J, Yager JP, et al. (2011) The Impact of Local Environmental Health Capacity on Foodborne Illness Morbidity in Maryland. Am J Public Health 101: 1495–1500. doi:10.2105/AJPH.2011.300137.
42. Erwin PC, Greene SB, Mays GP, Ricketts TC, Davis MV (2011) The association of changes in local health department resources with changes in state-level health outcomes. Am J Public Health 101: 609–615. doi:10.2105/AJPH.2009.177451.
43. Kanarek N, Stanley J, Bialek R (2006) Local public health agency performance and community health status. J Public Health Manag Pract 12: 522–527.
44. Hyde JK, Shortell SM (2012) The Structure and Organization of Local and State Public Health Agencies in the U.S. A Systematic Review. Am J Prev Med 42: S29–S41. doi:10.1016/j.amepre.2012.01.021.
45. Maxwell JA (2010) Using Numbers in Qualitative Research. Qual Inq 16: 475–482.

# 14

# Percutaneous Disc Decompression with Nucleoplasty–Volumetry of the Nucleus Pulposus Using Ultrahigh-Field MRI

**Richard Kasch[1]*, Birger Mensel[2], Florian Schmidt[1], Wolf Drescher[3], Ralf Pfuhl[4], Sebastian Ruetten[5], Harry R. Merk[1], Ralph Kayser[1]**

**1** Clinic and Outpatient Clinic for Orthopedics and Orthopedic Surgery, University Medicine Greifswald, Greifswald, Germany, **2** Department of Diagnostic Radiology and Neuroradiology, University Medicine Greifswald, Greifswald, Germany, **3** Department of Orthopedic and Trauma Surgery, RWTH Aachen University, Aachen, Germany, **4** Leibnitz Institute for Farm Animal Biology, Dummerstorf, Germany, **5** Department of Spine Surgery and Pain Therapy, Center for Orthopedics and Traumatology, St. Anna-Hospital Herne, Herne, Germany

## Abstract

***Purpose:*** To evaluate changes in nucleus pulposus volume as a potential parameter for the effects of disc decompression.

***Methods:*** Fifty-two discs (T8 to L1) were extracted from 26 pigs and separated into thoracic (T8 to T11) and thoracolumbar discs (T12 to L1). The discs were imaged using 7.1 Tesla ultrahigh-field magnetic resonance imaging (MRI) with acquisition of axial T2-weighted turbo spin-echo sequences for determination of baseline and postinterventional nucleus pulposus volumes. Volumes were calculated using OsiriX® (http://www.osirix-viewer.com). After randomization, one group was treated with nucleoplasty, while the placebo group was treated with an identical procedure but without coblation current. The readers analyzing the MR images were blinded to the kind of procedure performed. Baseline and postinterventional volumes were compared between the nucleoplasty and placebo group.

***Results:*** Average preinterventional nucleus volume was 0.799 (SD: 0.212) ml. Postinterventional volume reduction in the nucleoplasty group was significant at 0.052 (SD: 0.035) ml or 6.30% ($p<0.0001$) (thoracic discs) and 0.082 (SD: 0.042) ml or 7.25% ($p = 0.0078$) (thoracolumbar discs). Nucleoplasty achieved volume reductions of 0.114 (SD: 0.054) ml or 14.72% (thoracic) and 0.093 (SD: 0.081) ml or 11.61% (thoracolumbar) compared with the placebo group.

***Conclusions:*** Nucleoplasty significantly reduces thoracic and thoracolumbar nucleus pulposus volumes in porcine discs.

**Editor:** Laxmaiah Manchikanti, University of Louisville, United States of America

**Funding:** AArthroCare Deutschland, Remscheid, provided the technical equipment needed for nucleoplasty. The authors received for their project the nucleoplasty Convenience Pack (DLR SpineWand and sterile 17-gauge Crawford needle (6") with mandrin) from ArthroCare Germany. No additional external funding received for this study. The funders had no role in study design, data collection and analysis, decision to publish, or preparation of the manuscript.

**Competing Interests:** The authors have read the journal's policy and have the following conflicts: ArthroCare Deutschland, Remscheid, provided the technical equipment needed for nucleoplasty. The authors received for their project the nucleoplasty Convenience Pack (DLR SpineWand and sterile 17-gauge Crawford needle (6") with mandrin) from ArthroCare Germany. There are no further patents, products in development or marketed products to declare.

* E-mail: richard.kasch@uni-greifswald.de

## Introduction

Magnetic resonance imaging (MRI) has gained an important role as a noninvasive tool in biomedical research. No harmful side effects are known, and its excellent soft tissue contrast makes it the imaging procedure of choice for examinations of the spine [1,2]. While 1.5 and 3 Tesla MRI have become established in clinical routine [3], high-resolution 7 Tesla ultrahigh-field MRI is becoming increasingly available for answering more specialized questions [4]. Data on disc morphology and the effects of intradiscal therapy are still limited. MRI is well suited for providing such data, allowing measurement of intradiscal volume [5–6].

In the past a variety of different intradiscal procedures were used for treating symptomatic disc prolapse [7–10], many of which have now been abandoned [11]. Various studies have reported the clinical results of these treatments [8] [10] [12–14], and the effectiveness of some procedures has been demonstrated in high-quality studies [7] [10] [9] [15]. For many procedures, however, the mechanism of action remains to be demonstrated in an experimental setting.

More than 10 years ago, in July 2000, nucleoplasty was approved in the USA by the Food and Drug Administration (FDA) as a treatment for symptomatic disc prolapse [16]. Since then several valid studies have demonstrated its clinical effectiveness in treating the lumbar spine [16], and it is now considered safe and reliable [17]. Chen et al. experimentally demonstrated that nucleoplasty works by reducing pressure [18], and its histological effect on disc tissue has been characterized as well [19–20]. Experience with other intradiscal procedures suggests that the clinical effectiveness of nucleoplasty is due to its volume-reducing effects [21]. To the best of our knowledge, a study experimentally investigating volume reduction after nucleoplasty has not been published in the English literature.

This study investigated the question of whether nucleoplasty has volume-reducing effects on biomechanically different spinal segments in pigs - the thoracic and thoracolumbar spine [22–23] - and whether these effects can be demonstrated by volumetry of the nucleus pulposus using in vitro data sets acquired by 7 Tesla MRI.

## Materials and Methods

### Specimens

Our study included 53 ex vivo discs (T8- L1) from 26 freshly slaughtered "German native breed" pigs (mean age, 12.7 months; range, 4–54 months). We differentiated between thoracic (T8 to T11) and thoracolumbar junction discs (T12 to L1).

Discs were assigned randomly to either the nucleoplasty group or the placebo group, and the experiment was performed within 24 hours of slaughter. Discs showing damage (from slaughter, transport, etc.) on gross inspection or MRI were excluded from analysis (n = 1).

### Imaging

Imaging was performed on a 7.1 Tesla MR imager (ClinScan, Bruker Bioscan GmbH, Ettlingen, Germany). Discs were placed on 1-channel surface coils with the ventral side down and examined. Before treatment, a mark was placed on each disc to match preoperative and postoperative disc positions relative to the coil. The entire nucleus pulposus was imaged (Fig. 1A) using axial, gapless, T2-weighted turbo spin-echo sequences (Fig. 1B-D) (repetition time (TR), 2000 ms; echo time (TE), 42 ms; slice thickness, 0.7 mm; field of view (FOV), 45×45 mm; 512×512 pixels). The acquisition time was 7:20 min.

### Volumetry

MR images were viewed and processed using the OsiriX® software (version 3.6.1, 32-bit, http://www.osirix-viewer.com) [24]. Volumetry of the nucleus pulposus was performed semi-automatically. A radiologist manually outlined the junction of the nucleus pulposus and the anulus fibrosus, first cranially and then caudally, in the axial image stack (Fig. 1B-D). The software then generated the missing outlines between the cranial and the caudal slices, and each slice was afterwards corrected manually. OsiriX® was used to calculate the nucleus pulposus volume by multiplying all outlined areas of the slices with the slice thickness. Volumes were measured and calculated in the same way before and after nucleoplasty or placebo treatment (Fig. 1A-D).

### Nucleoplasty

We used the ArthroCare System 2000 (Arthrocare Deutschland, Remscheid, Germany) with control unit, foot switch, and Convenience Pack (DLR SpineWand and sterile 17-gauge Crawford needle (6") with mandrin). Appropriate to the manufacturer's instructions, the ex vivo coblation current, in the nucleoplasty group (Fig. 1D), was applied in 6 positions for 10 sec each to create 6 channels with an application field of 360°. In the placebo group (Fig. 1C) the identical procedure was performed but without application of current. Discs were randomized to nucleoplasty or placebo treatment. The volumes of the discs treated with and without application of coblation current were then compared to determine the effectiveness of nucleoplasty.

### Data Selection

A total of 52 discs were included. Unpaired samples – randomly selected, independent value pairs – were assigned to one of two groups: thoracic discs and thoracolumbar junction discs, which have different biomechanical characteristics. Independent value pairs were selected randomly to minimize the potential for systematic bias (global disc disease in an individual pig, for example). Thus, only one functional lumbar spinal unit (vertebra-disc-vertebra) (T8/T9 and T9/T10) was considered per spine. The two groups (therapy versus placebo group) were formed using the SAS randomization program. For the thoracic spine, 18 discs were assigned to the therapy group and 18 to the placebo group. For the thoracolumbar spine, each group was assigned 8 discs.

### Statistics

The data were assessed with the Wilcoxon test for independent (nonparametric) samples using SAS 9.1 TS (XP_PRO Windows NT Server, Cary, North Carolina, USA). Data are given as absolute values and standard deviations. Calculated differences were considered statistically significant at $p<0.05$.

## Results

There were no complications during the interventions. All porcine discs were found to be normal, showing no abnormalities or degenerative changes. In both postoperative groups we were able to track the placebo/nucleoplasty channel created in the center of the disc (Fig. 1C-D). The average nucleus volume for all 52 examined discs was 0.799 (SD: 0.212) ml.

The results for thoracic discs were as follows. The average baseline volume was 0.754 (SD: 0.203) ml in the nucleoplasty group (n = 18) and 0.786 (SD: 0.219) ml in the placebo group (n = 18), showing no statistically significant difference ($p = 0.130$) (Fig. 2). Thoracic discs treated with nucleoplasty showed a significant nucleus pulposus volume reduction (pre- versus postprocedure) of 0.052 (SD: 0.035) ml or 6.30% ($p<0.0001$). Placebo-treated thoracic discs showed a significant volume increase (due to instrument manipulation) of 0.062 (SD: 0.053) ml or 8.42% ($p = 0.0002$) (Table 1, Fig. 3).

In the thoracic group, nucleoplasty decreased nucleus volume by 0.114 (SD: 0.054) ml or 14.72% compared with the placebo group (Table 2, Fig. 4).

The average baseline volume of the larger thoracolumbar junction nuclei was 0.850 (SD: 0.200) ml for the nucleoplasty group (n = 8) and 0.881 (SD: 0.214) ml for the placebo group (n = 8), showing no significant difference ($p = 0.461$) (Fig. 2). For the thoracolumbar junction group as well, a significant post-nucleoplasty volume reduction (pre- versus postprocedure) of 0.082 (SD: 0.042) ml or 7.25% ($p = 0.0078$) was demonstrated. In the placebo group, a nonsignificant volume increase (due to instrument manipulation) of 0.011 (SD: 0.082) ml or 4.36% ($p = 0.547$) was measured (Table 1, Fig. 3).

In the thoracolumbar junction group, nucleoplasty decreased nucleus volume by 0.093 (SD: 0.081) ml or 11.61% compared with the placebo group (Table 3, Fig. 4).

## Discussion

Different studies have demonstrated the effectiveness of minimally invasive intradiscal procedures in treating symptomatic disc prolapse [7–8] [10] [12] [16,25] [26]. In order to gain long-term acceptance, however, clinical treatment methods require more than simply evidence-based data evaluation. Their effectiveness must also be measurable in experimentally verifiable

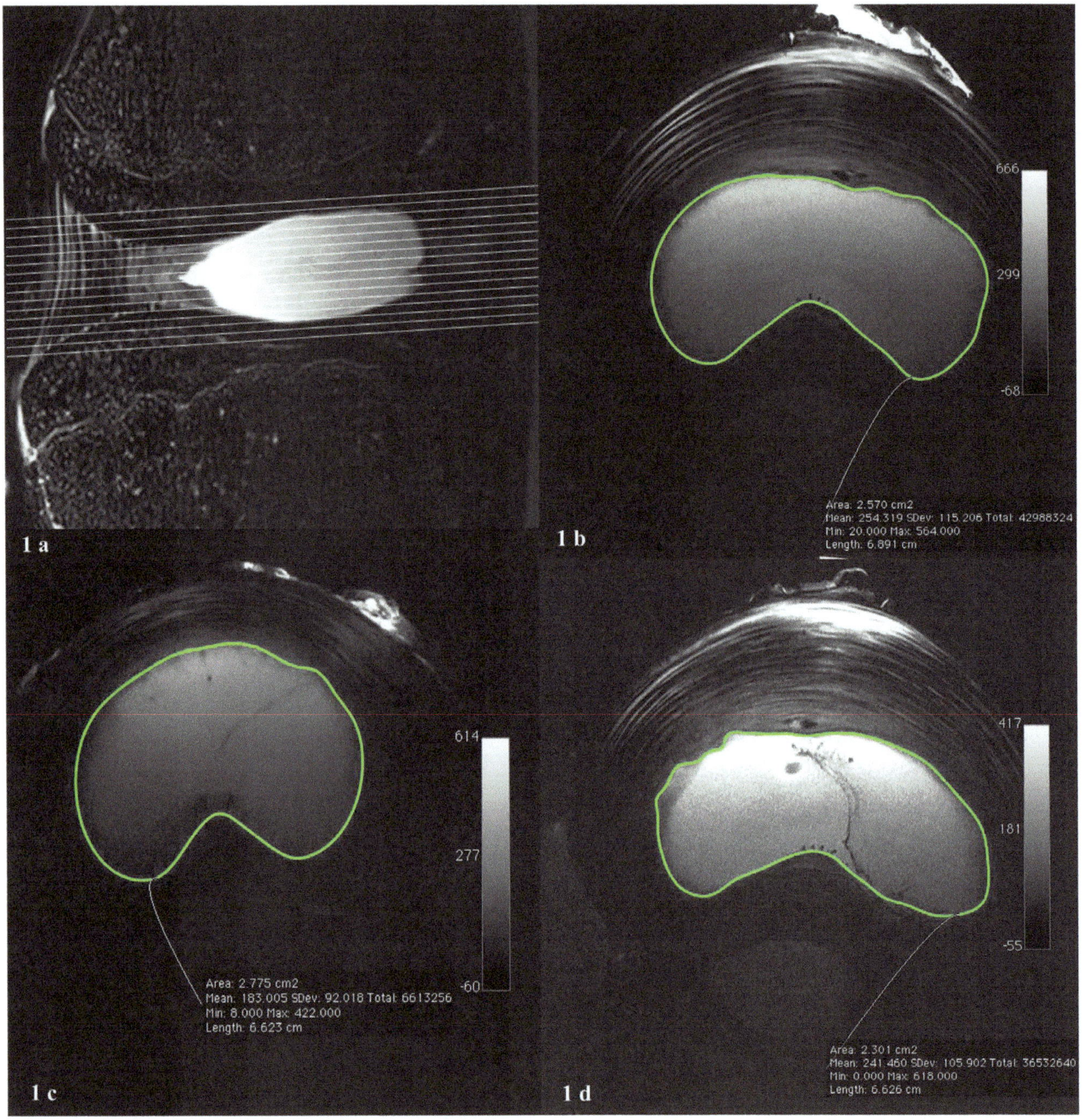


**Figure 1. T2-weighted sagittal and axial images of ex vivo discs of pigs, depicting the hyperintense nucleus pulposus and the surrounding inhomogeneous hypointense anulus fibrosus.** a: Sagittal image of lumbar motion segments showing axial stack. The axial images were acquired without gaps, parallel to the disc. b: Preoperative image. The green line outlines the nucleus pulposus area for volume calculation. c: Postinterventional image of the placebo group. Visible coblation channel at the 7–8 o'clock position on the left. d: Postinterventional image of nucleoplasty group. Visible coblation channel at the 4–5 o'clock position on the right; the channel is hypointense in the nucleus pulposus and hyperintense in the anulus fibrosus.

models [27]. For chemonucleolysis or percutaneous laser disc decompression (PLDD), with a level of evidence II-2 for short- and long-term pain relief, [8] the mechanism of action has now been clarified in an experimental setting [14] [28–31]. For nucleoplasty, however, before this study, this was not the case.

Coblation employs an electrolyte-rich medium to generate a plasma field of highly ionized particles with enough energy to break the molecular bonds in soft tissue, so that tissue is vaporized and escapes through the introducer needle [32]. It is a technology that has been successfully established in a number of therapeutic fields, not only those involving the musculoskeletal system [33–34].

The risk of indirect injury to nerve structures close to the discs has been discussed repeatedly. Temperatures of 60 to 65°C or even higher can be reached at a distance of 3–4 mm from the probe [35]. Considering that neurodegeneration begins at 45°C [36], this is not to be taken lightly. Several experiments, however,

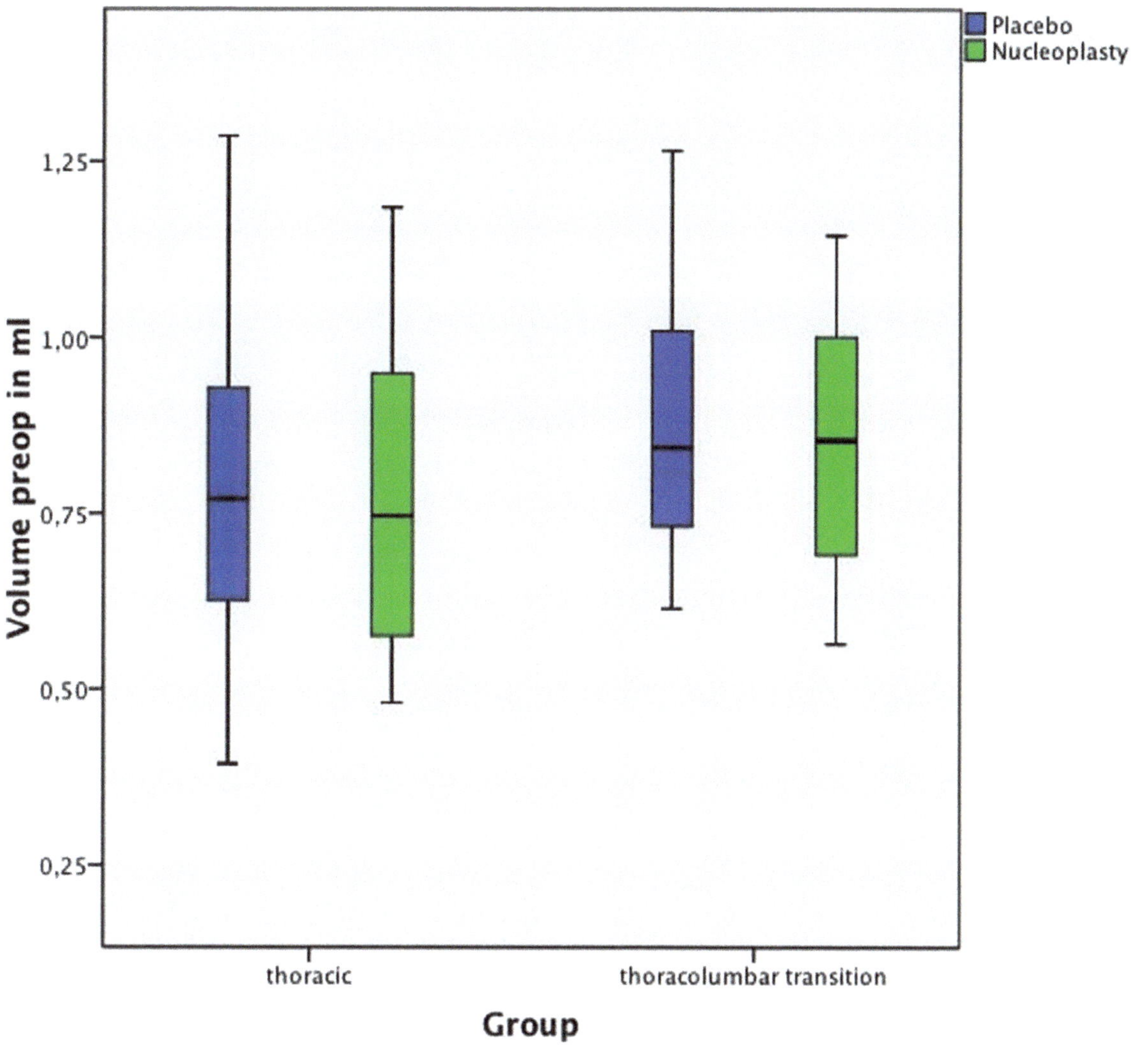


**Figure 2. Initial nucleus pulposus volumes in the independent thoracic (placebo group n = 18, nucleoplasty group n = 18) and thoracolumbar (placebo group n = 8, nucleoplasty group n = 8) groups in a box-whisker plot, presenting the 25% (lower box end), 50% (marking in the box) and 75% quartiles (lower and upper box end).** The whiskers represent the smallest and largest values in the 1.5 × interquartile range. The group differences were not significant - p = 0.130 and p = 0.461.

performed both in vivo and in vitro using discs from pigs and sheep, have demonstrated that any tissue damage produced by coblation will be confined to the plasma field surrounding the electrode [19–20] [36].

Some intradiscal procedures have proven to be more effective when they are restricted to discs that have not lost too much height or volume due to degeneration [37]. Intradiscal electrothermal therapy (IDET) offers, significant relief in one-half of chronic discogenic low back pain patients [10] and, has been demonstrated to be more effective than placebo only when height loss of the discs in the segment treated is less than 20% [38]. With height loss of 50% or more, no therapeutic effect can be demonstrated at all [14]. The efficacy of most intradiscal procedures is tied to a particular disc volume, limiting their usefulness in cases of height loss or disc degeneration. For this reason our study only included discs that were free of all signs of degenerative damage in the preinterventional MRI.

Chen et al. demonstrated experimentally that nucleoplasty involves a reduction in pressure within the disc [18]. Using three human cadavers they also showed that the intradiscal pressure reduction achieved by nucleoplasty on lower thoracic and lumbar discs was dependent on the degree of disc degeneration. Applying monopolar radiofrequency in an in vivo sheep model Podhaysky et al. also demonstrated an intradiscal pressure reduction that was still measurable 4 weeks later [37]. The demonstrated pressure effects are attributable to the reduction in nucleus pulposus volume we show in our study. To our knowledge, such quantitative volume effects have not been published before.

The mechanism of action of PLDD can be described as volume reduction – which also results in intradiscal pressure reduction [39–41]. Volumetry of organs and organ systems is widely used and well established in medical research and has found a prominent place in routine clinical examinations as well. MR data sets are used to outline the target volume manually, semiautomatically, or fully automatically. This, along with the remaining image parameters (slice thickness, gap), is used to calculate the volume. Volumetry is thus used in the diagnosis of neurodegenerative diseases, cardiac diseases, and stroke as well as in the planning and follow-up of operative and nonoperative cancer treatment [42–45].

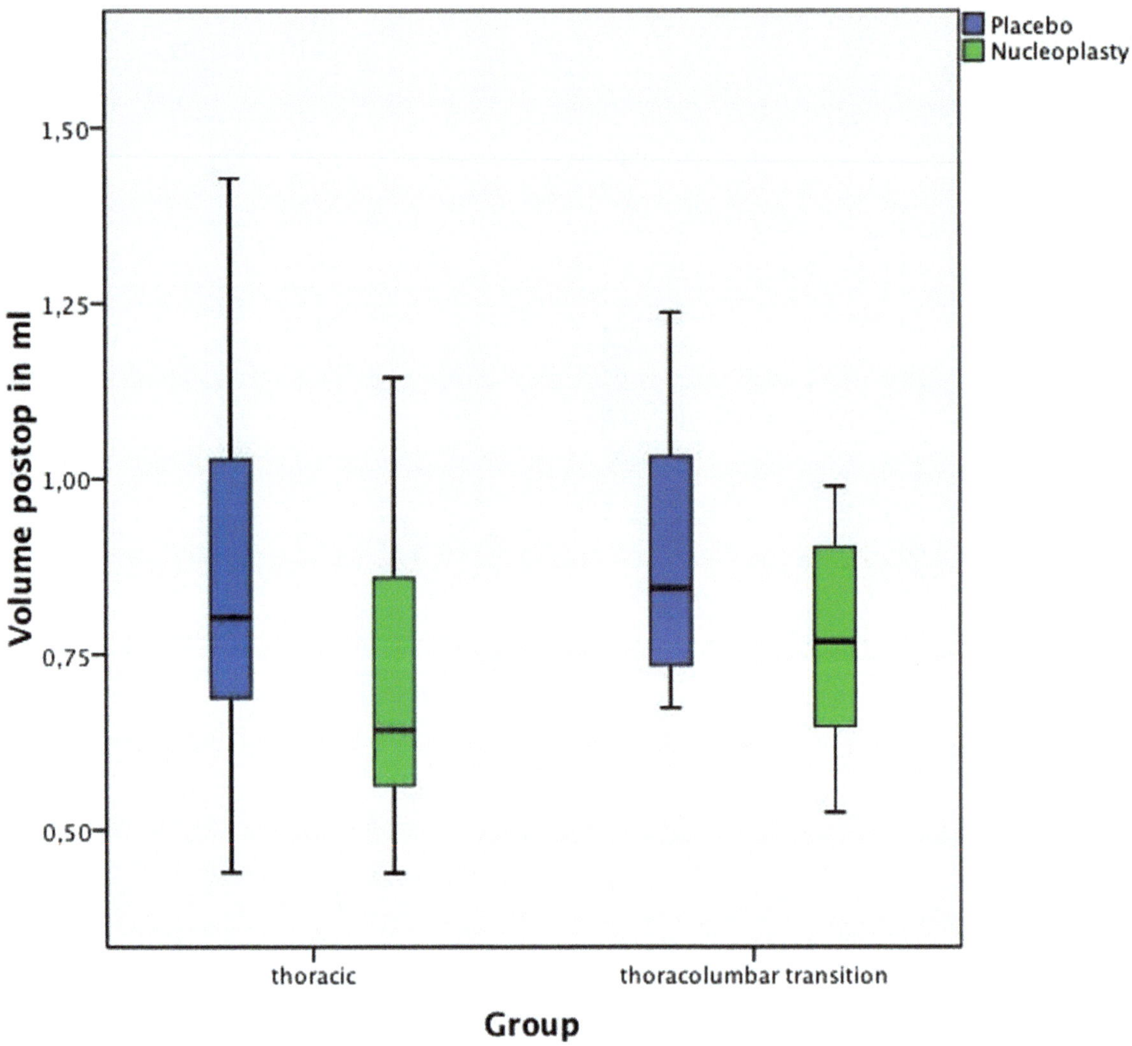


**Figure 3. Postoperative volumes in the independent placebo and nucleoplasty groups for the two spinal segments investigated.**

Tissue vaporization with reduction of the nucleus pulposus volume appears to be the mechanism of action by which nucleoplasty reduces pressure in intact discs as used in our study. Chen et al. demonstrated this volume-reducing effect qualitatively by performing postoperative histological examinations [19] without providing any observations on the quantitative extent of ablation or its effects. Case et al. also showed a correlation between pressure and volume changes in the disc [21].

Our experimental data show that, in a placebo-controlled setting, nucleoplasty can be used to reduce the initial volume of nucleus pulposus by a statistically highly significant degree. We attribute the volume increase in the placebo group to the fact that

**Table 1.** Summary of intradiscal volume changes in the thoracic (n = 36) and thoracolumbar (n = 16) spine: nucleoplasty versus placebo.

| Cohort | n | Absolute volume change | | | | Relative volume change | |
|---|---|---|---|---|---|---|---|
| Thoracic discs | | Nucleoplasty group (n = 18) | P | Placebo group (n = 18) | p | Nucleoplasty group (n = 18) | Placebo group (n = 18) |
| Total | 36 | −0.052 ml (SD: 0.035) | <0.0001 | 0.062 ml (SD: 0.053) | 0.0002 | −6.30% | 8.42% |
| Thoraco-lumbar discs | | Nucleoplasty group (n = 8) | P | Placebo group (n = 8) | p | Nucleoplasty group (n = 8) | Placebo group (n = 8) |
| Total | 16 | −0.082 ml (SD:0.042) | 0.0078 | 0.011 ml (SD:0.082) | 0,547 | −7.25% | 4.36% |

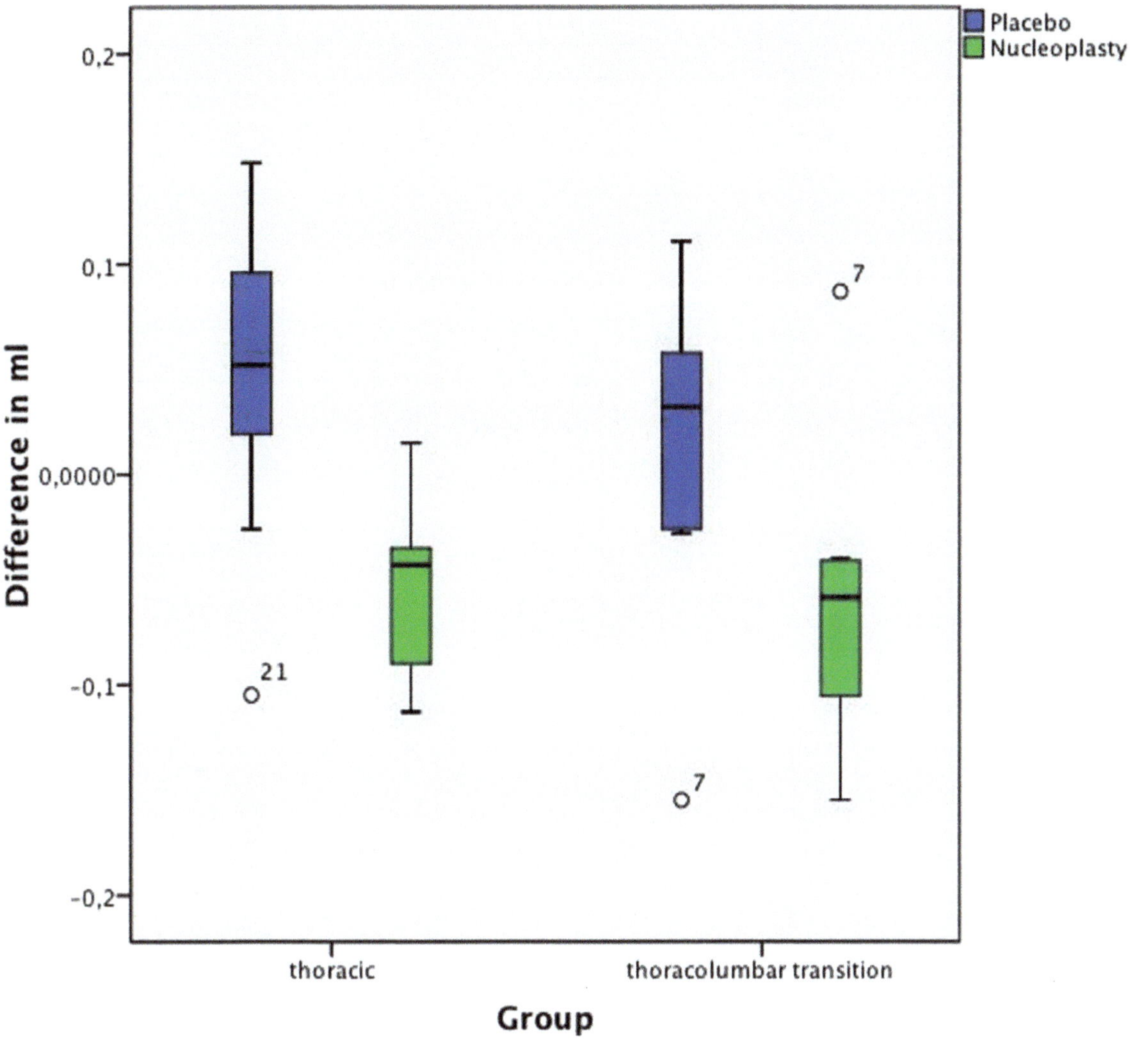


**Figure 4. Differences between initial and postoperative volumes for placebo and nucleoplasty groups.** Comparison of the values of the thoracic group and thoracolumbar group shows a volume increase in the placebo group and a volume reduction in the nucleoplasty group (○-markers denote "mild" outliers within the 1.5× and 3× interquartile ranges).

the nucleus was pushed aside and compressed by insertion of the Spine Wand. Conversely, in the nucleoplasty group, the compressed volume was reduced with the coblation current.

Our study examined regions of the porcine spine that bear the closest anatomical resemblance to the human spine [46–48]. First we quantified the volume-reducing effects of nucleoplasty on the lower thoracic spine and on the thoracolumbar junction to the upper lumbar spine, corroborating this effect by comparison with the placebo group. We demonstrated that nucleoplasty reduced initial nucleus pulposus volumes and also that these volume reductions – (T8- T11 = 14.72%) and (T12 to L1 = 11.61%) – were significant compared to the placebo procedure. Although the porcine disc model we used is very similar to the human spine, our experimental results do not allow us to draw any conclusions regarding the volume reduction in humans. This is precluded because porcine and human discs are not fully identical and because we did not measure the amount of material removed.

While the thoracic discs of the thoracic spine were smaller than the discs of the thoracolumbar junction, the volume reduction achieved by nucleoplasty was similar in the two groups. It can be assumed that decompression involves the entire anulus fibrosus and not the herniated disc portion alone [28]. Intradiscal pressure

**Table 2.** Analysis of intradiscal volume changes in the thoracic spine for independent pairs (n = 18).

| Thoracic discs | Average (mean) | Median | SD | Range | Interquartile range | p |
|---|---|---|---|---|---|---|
| Δ V absolute in ml | 0.114 | 0.093 | 0.054 | 0.207 | 0.076 | <0.0001 |
| Δ relative in % | −14.725 | −15.090 | 6.747 | 25.980 | 6.670 | <0.0001 |

**Table 3.** Analysis of intradiscal volume changes in the thoracolumbar spine for independent pairs (n = 8).

| Thoraco-lumbar discs | Average (mean) | Median | SD | Range | Interquartile range | p |
|---|---|---|---|---|---|---|
| Δ V absolute in ml | 0.093 | 0.115 | 0.081 | 0.264 | 0.073 | <0.05 |
| Δ relative in % | −11.610 | −11.880 | 8.280 | 27.240 | 8.480 | <0.05 |

decreases in proportion to volume reduction [21]. The proteins within the nucleus pulposus exert an osmotic force that produces a continual flow of water into the disc [30]. Nucleoplasty, like PLDD, denatures some of the proteins, thus reducing the reperfusion effect in a way that contributes to the long-term effectiveness of the procedure [19] [41].

## Conclusion

Our study demonstrates that nucleoplasty has a volume-reducing effect on the nucleus pulposus of the thoracic and thoracolumbar spine. While this effect was demonstrated in an experimental setting and remains to be verified in patients, our results suggest that the potential benefit likely to be achieved in the clinical setting is of interest and deserves to be pursued further.

## Acknowledgments

The authors would like to thank ArthroCare Deutschland, Reimscheid, Germany for providing the technical equipment needed for nucleoplasty and Mr. Stefan Hadlich from the Department of Diagnostic Radiology and Neuroradiology at the University of Greifswald, Germany for his technical assistance in the compilation of MRI data sets.

## Author Contributions

Conceived and designed the experiments: R. Kasch R. Kayser HM BM. Performed the experiments: R. Kasch FS BM R. Kayser RP. Analyzed the data: R. Kasch FS BM R. Kayser WD SR. Contributed reagents/materials/analysis tools: RP. Wrote the paper: R. Kasch R. Kayser FS SR WD HM.

## References

1. McGirt MJ, Eustacchio S, Varga P, Vilendecic M, Trummer M, et al. (2009) A prospective cohort study of close interval computed tomography and magnetic resonance imaging after primary lumbar discectomy: factors associated with recurrent disc herniation and disc height loss. Spine (Phila Pa 1976) 34: 2044–2051.
2. Arun R, Freeman BJ, Scammell BE, McNally DS, Cox E, et al. (2009) 2009 ISSLS Prize Winner: What influence does sustained mechanical load have on diffusion in the human intervertebral disc?: an in vivo study using serial postcontrast magnetic resonance imaging. Spine (Phila Pa 1976) 34: 2324–2337.
3. Weintraub MI, Khoury A, Cole SP (2007) Biologic effects of 3 Tesla (T) MR imaging comparing traditional 1.5 T and 0.6 T in 1023 consecutive outpatients. J Neuroimaging 17: 241–245.
4. Krug R, Stehling C, Kelley DA, Majumdar S, Link TM (2009) Imaging of the musculoskeletal system in vivo using ultra-high field magnetic resonance at 7 T. Invest Radiol 44: 613–618.
5. Pfirrmann CW, Metzdorf A, Elfering A, Hodler J, Boos N (2006) Effect of aging and degeneration on disc volume and shape: A quantitative study in asymptomatic volunteers. J Orthop Res 24: 1086–1094.
6. Violas P, Estivalezes E, Briot J, Sales de Gauzy J, Swider P (2007) Quantification of intervertebral disc volume properties below spine fusion, using magnetic resonance imaging, in adolescent idiopathic scoliosis surgery. Spine (Phila Pa 1976) 32: E405–412.
7. Kallewaard JW, Terheggen MA, Groen GJ, Sluijter ME, Derby R, et al. (2010) 15. Discogenic low back pain. Pain Pract 10: 560–579.
8. Singh V, Manchikanti L, Benyamin RM, Helm S, Hirsch JA (2009) Percutaneous lumbar laser disc decompression: a systematic review of current evidence. Pain Physician 12: 573–588.
9. Manchikanti L, Derby R, Benyamin RM, Helm S, Hirsch JA (2009) A systematic review of mechanical lumbar disc decompression with nucleoplasty. Pain Physician 12: 561–572.
10. Helm S, Hayek SM, Benyamin RM, Manchikanti L (2009) Systematic review of the effectiveness of thermal annular procedures in treating discogenic low back pain. Pain Physician 12: 207–232.
11. Nordby EJ, Wright PH, Schofield SR (1993) Safety of chemonucleolysis. Adverse effects reported in the United States, 1982–1991. Clin Orthop Relat Res: 122–134.
12. Steppan J, Meaders T, Muto M, Murphy KJ (2010) A metaanalysis of the effectiveness and safety of ozone treatments for herniated lumbar discs. J Vasc Interv Radiol 21: 534–548.
13. Choy DS (2004) Percutaneous laser disc decompression: a 17-year experience. Photomed Laser Surg 22: 407–410.
14. Freeman BJ, Fraser RD, Cain CM, Hall DJ, Chapple DC (2005) A randomized, double-blind, controlled trial: intradiscal electrothermal therapy versus placebo for the treatment of chronic discogenic low back pain. Spine (Phila Pa 1976) 30: 2369–2377; discussion 2378.
15. Levin JH (2009) Prospective, double-blind, randomized placebo-controlled trials in interventional spine: what the highest quality literature tells us. Spine J 9: 690–703.
16. Gerges FJ, Lipsitz SR, Nedeljkovic SS (2010) A systematic review on the effectiveness of the Nucleoplasty procedure for discogenic pain. Pain Physician 13: 117–132.
17. Birnbaum K (2009) Percutaneous cervical disc decompression. Surg Radiol Anat 31: 379–387.
18. Chen YC, Lee SH, Chen D (2003) Intradiscal pressure study of percutaneous disc decompression with nucleoplasty in human cadavers. Spine (Phila Pa 1976) 28: 661–665.
19. Chen YC, Lee SH, Saenz Y, Lehman NL (2003) Histologic findings of disc, end plate and neural elements after coblation of nucleus pulposus: an experimental nucleoplasty study. Spine J 3: 466–470.
20. Lee MS, Cooper G, Lutz GE, Doty SB (2003) Histologic characterization of coblation nucleoplasty performed on sheep intervertebral discs. Pain Physician 6: 439–442.
21. Case RB, Choy DS, Altman P (1995) Change of intradisc pressure versus volume change. J Clin Laser Med Surg 13: 143–147.
22. Kim YJ, Bridwell KH, Lenke LG, Rhim S, Kim YW (2007) Is the T9, T11, or L1 the more reliable proximal level after adult lumbar or lumbosacral instrumented fusion to L5 or S1? Spine (Phila Pa 1976) 32: 2653–2661.
23. Vernon-Roberts B, Fazzalari NL, Manthey BA (1997) Pathogenesis of tears of the anulus investigated by multiple-level transaxial analysis of the T12-L1 disc. Spine (Phila Pa 1976) 22: 2641–2646.
24. OsiriX® (2010) Version 3.6.1, 32-bit, Available: http://www.osirix-viewer.com.
25. Zhu H, Zhou XZ, Cheng MH, Shen YX, Dong QR (2011) The efficacy of coblation nucleoplasty for protrusion of lumbar intervertebral disc at a two-year follow-up. Int Orthop.
26. Bokov A, Skorodumov A, Isrelov A, Stupak Y, Kukarin A (2010) Differential treatment of nerve root compression pain caused by lumbar disc herniation applying nucleoplasty. Pain Physician 13: 469–480.
27. Mayer HM (2008) Reviewer's comment concerning "Percutaneous cervical nucleoplasty treatment in the cervical disc herniation" (Jian Li, et al. MS-no: ESJO-D-08–00079R2). Eur Spine J 17: 1670.
28. Dang L, Liu Z (2010) A review of current treatment for lumbar disc herniation in children and adolescents. Eur Spine J 19: 205–214.
29. Choy DS, Altman P (1995) Fall of intradiscal pressure with laser ablation. J Clin Laser Med Surg 13: 149–151.
30. Schenk B, Brouwer PA, van Buchem MA (2006) Experimental basis of percutaneous laser disc decompression (PLDD): a review of literature. Lasers Med Sci 21: 245–249.
31. Kutschera HP, Lack W, Buchelt M, Beer R (1998) Comparative study of surface displacement in discs following chemonucleolysis and lasernucleotomy. Lasers Surg Med 22: 275–280.
32. Chen SS, Humphrey JD (1998) Heat-induced changes in the mechanics of a collagenous tissue: pseudoelastic behavior at 37 degrees C. J Biomech 31: 211–216.
33. Timms MS, Temple RH (2002) Coblation tonsillectomy: a double blind randomized controlled study. J Laryngol Otol 116: 450–452.
34. Yetkinler DN, Greenleaf JE, Sherman OH (2002) Histologic analysis of radiofrequency energy chondroplasty. Clin Sports Med 21: 649–661, viii.

35. Nau WH, Diederich CJ (2004) Evaluation of temperature distributions in cadaveric lumbar spine during nucleoplasty. Phys Med Biol 49: 1583–1594.
36. Kapural L, Mekhail N, Hicks D, Kapural M, Sloan S, et al. (2008) Histological changes and temperature distribution studies of a novel bipolar radiofrequency heating system in degenerated and nondegenerated human cadaver lumbar discs. Pain Med 9: 68–75.
37. Podhajsky RJ, Belous A, Johnson K, Maul DH, Finch PM (2007) Effects of monopolar radiofrequency heating on intradiscal pressure in sheep. Spine J 7: 229–234.
38. Pauza KJ, Howell S, Dreyfuss P, Peloza JH, Dawson K, et al. (2004) A randomized, placebo-controlled trial of intradiscal electrothermal therapy for the treatment of discogenic low back pain. Spine J 4: 27–35.
39. Choy DS, Diwan S (1992) In vitro and in vivo fall of intradiscal pressure with laser disc decompression. J Clin Laser Med Surg 10: 435–437.
40. Choy DSJ, Thorton J, Moser F (1992) The relationship of lumbar disc volume to body size. J Clin Laser Med Surg 10: 433–434.
41. Choi JY, Tanenbaum BS, Milner TE, Dao XV, Nelson JS, et al. (2001) Theramal, mechanical, optical, and morphologic changes in bovine nucleus pulposus induced by Nd:YAG (lambda = 1.32 microm) laser irradiation. Lasers Surg Med 28: 248–254.
42. Baghi M, Mack MG, Hambek M, Bisdas S, Muerthel R, et al. (2007) Usefulness of MRI volumetric evaluation in patients with squamous cell cancer of the head and neck treated with neoadjuvant chemotherapy. Head Neck 29: 104–108.
43. Groschel K, Hauser TK, Luft A, Patronas N, Dichgans J, et al. (2004) Magnetic resonance imaging-based volumetry differentiates progressive supranuclear palsy from corticobasal degeneration. Neuroimage 21: 714–724.
44. Karlo C, Reiner CS, Stolzmann P, Breitenstein S, Marincek B, et al. (2010) CT- and MRI-based volumetry of resected liver specimen: comparison to intraoperative volume and weight measurements and calculation of conversion factors. Eur J Radiol 75: e107–111.
45. Lotjonen JM, Jarvinen VM, Cheong B, Wu E, Kivisto S, et al. (2008) Evaluation of cardiac biventricular segmentation from multiaxis MRI data: a multicenter study. J Magn Reson Imaging 28: 626–636.
46. Lundin O, Ekstrom L, Hellstrom M, Holm S, Sward L (1998) Injuries in the adolescent porcine spine exposed to mechanical compression. Spine (Phila Pa 1976) 23: 2574–2579.
47. Alini M, Eisenstein SM, Ito K, Little C, Kettler AA, et al. (2008) Are animal models useful for studying human disc disorders/degeneration? Eur Spine J 17: 2–19.
48. Cotterill PC, Kostuik JP, D'Angelo G, Fernie GR, Maki BE (1986) An anatomical comparison of the human and bovine thoracolumbar spine. J Orthop Res 4: 298–303.

# 15

# Efficacy of a Vaccine Formula against Tuberculosis in Cattle

**Germinal J. Canto Alarcon[1]*, Yezenia Rubio Venegas[1], Luis Bojorquez Narvaez[2], Oscar E. Pizano Martínez[3], Leticia García Casanova[3], Susana Sosa Gallegos[1], Alejandro Nava Vargas[3], Andrea M. Olvera Ramírez[1], Feliciano Milian Suazo[1]**

**1** Facultad de Ciencias Naturales, Universidad Autónoma de Querétaro, Querétaro, Qro., México, **2** Productora Nacional de Biológicos Veterinarios (PRONABIVE), Distrito Federal, México, **3** Centro Nacional de Investigación Disciplinaria en Fisiología y Mejoramiento Animal, INIFAP, Ajuchitlán, Qro., México

## Abstract

"Test-and-slaughter" has been successful in industrialized countries to control and eradicate tuberculosis from cattle; however, this strategy is too expensive for developing nations, where the prevalence is especially high. Vaccination with the Calmette-Guérin (BCG) strain has been shown to protect against the development of lesions in vaccinated animals: mouse, cattle and wildlife species. In this study, the immune response and the pathology of vaccinated (BCG-prime and BCG prime-CFP-boosted) and unvaccinated (controls) calves were evaluated under experimental settings. A $10^6$ CFU dose of the BCG strain was inoculated subcutaneously on the neck to two groups of ten animas each. Thirty days after vaccination, one of the vaccinated groups was boosted with an *M. bovis* culture filtrate protein (CFP). Three months after vaccination, the three groups of animals were challenged with $5\times10^5$ CFU via intranasal by aerosol with a field strain of *M. bovis*. The immune response was monitored throughout the study. Protection was assessed based on immune response (IFN-g release) prechallenge, presence of visible lesions in lymph nodes and lungs at slaughter, and presence of bacilli in lymph nodes and lung samples in histological analysis. Vaccinated cattle, either with the BCG alone or with BCG and boosted with CFP showed higher IFN-g response, fewer lesions, and fewer bacilli per lesion than unvaccinated controls after challenge. Animals with low levels of IFN-g postvaccine-prechallenge showed more lesions than animals with high levels. Results from this study support the argument that vaccination could be incorporated into control programs to reduce the incidence of TB in cattle in countries with high prevalence.

**Editor:** Stephen J. Turner, University of Melbourne, Australia

**Funding:** These authors have no support or funding to report.

**Competing Interests:** The authors have declared that no competing interest exist.

* E-mail: gcanto07@uaq.mx

## Introduction

Bovine tuberculosis (bTB) is an infectious disease of cattle and other animal species, including man [1]. The importance of bTB is that it represents a risk to public health, causes serious economic losses to the livestock industry worldwide due to animal disposal, carcass confiscation, premature culling, low production, poor reproductive performance, and because it represents a constraint for international trade of animals [2], [3].

Mexico, as many countries around the world, counts with a national program to control and eradicate tuberculosis from cattle. This program is based on the strategy of "test-and-slaughther" where animals reacting to the tuberculin tests are eliminated. Even though this strategy has been successful in reducing the incidence of the disease in some industrialized countries [4], in developing countries, where the prevalence of the disease is especially high, the success has been limited. In Mexico, as in most of Latin America, tuberculosis concentrates in populations of dairy cattle, where "test-and-slaughter" is unfeasible because elimination of animals represents less milk for the human population and there is a cost for the producer due to lack of compensation for disposal of animals and the cost of importing replacements from abroad. Therefore, the use of a vaccine to control and, in the long run, eradicate bTB could be an important addition to control programs already established [5], [6].

Studies evaluating the efficacy of vaccines for tuberculosis in cattle have focused on the use of the bacillus Calmette–Guerin (BCG) strain, an attenuated strain of *M. bovis* used in humans since 1921 [7]. BCG is the only licensed vaccine against human TB; therefore, it is likely that any vaccine or vaccination strategy in cattle is based on this strain [8]. As a matter of fact, numerous studies have already shown that BCG induces a significant level of protection against development of lesions in animals vaccinated and challenged with virulent strains [9], [10], [11], [12], [13]. Subcutaneous and oral administrations of BCG, the use of BCG in a lipid matrix, and different BCG strains have shown protection [14], [15]. It has been shown that: low doses of BCG [$10^3$–$10^6$ colony forming units (CFU)] induce better protection than higher doses [5], pre-exposure to environmental *Mycobacterium* can negatively affect vaccine efficacy [16], vaccination of neonatal calves induced a higher level of immunity than that observed in calves vaccinated at 5–6 months of age [11], [12], [17], and that boosting with culture filtrate proteins (CFP) of *M. bovis*, plus adjuvant induce a better Th1 response. Conclusions of these studies are that vaccine protection occur by reducing the number

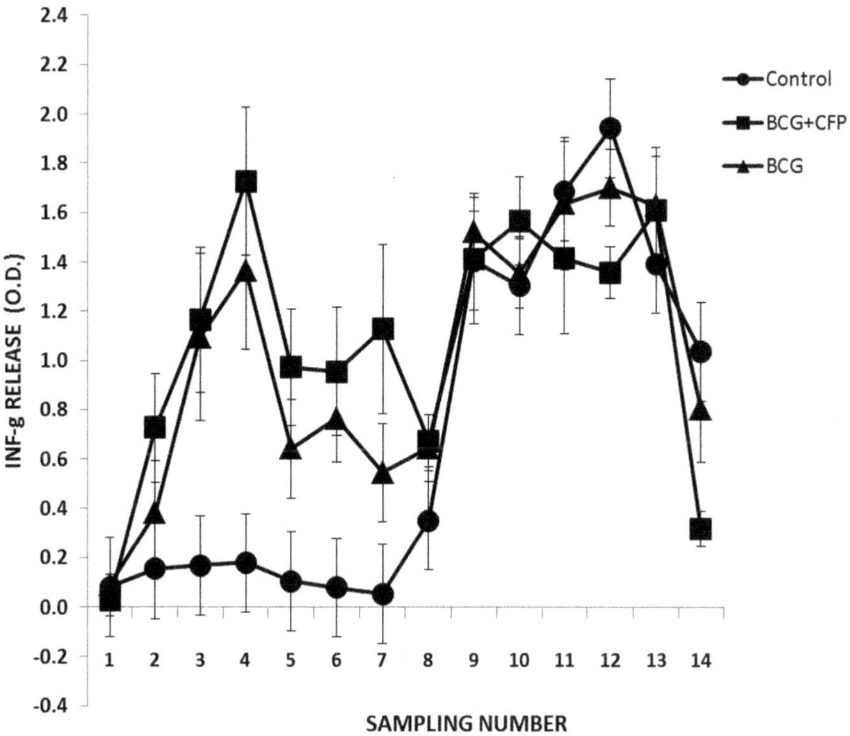


**Figure 1. IFN-g response after vaccination and challenge.** Plasma from whole-blood cultures of vaccinated (BCG and BCG+CFP) and unvaccinated (control) animals were stimulated *in vitro* with bovine PPD and avian PPD and assayed for IFN-g. Challenge was delivered in sampling number 8. Sampling was performed at days 0, 15, 40 and then about every two weeks until the end of the experiment.

of animals with lesions, the number of lesions per animal and bacteria load per lesion [11], [12].

Vaccination against tuberculosis has been used in humans for many years; however, it has never been officially used in control programs in cattle because the immune response interferes with the tuberculin test, which in most countries is the only official test to estimate prevalence. Nevertheless, on the one hand, it has been shown that only a few vaccinated animals react to the comparative cervical test, since BCG induces response to both avian and bovine-PPD [11], [12], and, on the other hand, this limitation is irrelevant in countries with no eradication programs or countries with large populations of dairy cattle where the tuberculin test is rarely used. More than that, nowadays there are at least two antigens, ESAT6 and CFP10 that can differentiate between TB infected and TB vaccinated animals [18]. In Mexico, the bTB Control and Eradication Program has been successful in beef cattle, reducing the prevalence to less than 0.5% in about 85% of the national territory; however, in dairy cattle the program has failed since only about 30% of the farms have some partial participation in the program [19]; therefore, tuberculin testing is not performed in a routine bases. Nevertheless, dairy farmers around the country are willing to participate more in the bTB National Program as far as herds are not labeled as quarantined and animals reacting to the tuberculin tests are not removed. As a matter of fact, a recent survey in dairy regions showed that more than 90% of the farmers would be willing to participate in the program if a vaccine is used. Therefore, the aim of this study was to evaluate the efficacy of a BCG vaccine formula against bovine tuberculosis under experimental settings to provide to the National Program with a new tool against this problematic disease.

## Materials and Methods

### Experimental Animals

Thirty, five to six months old, 150 kg weight calves free of TB coming from a TB-free area were included in the study. Animals were divided into three groups and allocated in isolation in high-security confinement units, fed a diet based on grain and forage with free access to water during the experimental period. All procedures received approval of the institutional committee for the protection of animals used in research (*Comité del Instituto Nacional de Investigaciones Forestales, Agrícolas y Pecuarias para el Cuidado de los Animales usados en Investigación*).

### Vaccine Strain

The *M. bovis* BCG Phipps strain was used. This strain was selected because it induced the best protection against challenge in a mouse model where 10 BCG substrains were compared [20]. Bacteria were grown in Dorset-Henley medium supplemented with 0.25% (wt/vol) glucose. Cultures were incubated at 37°C for four weeks. Bacteria were then harvested and shacked continuously at 200 rpm/2 h. Bacterial count was determined by serial dilution on Stonebrink media tubes supplemented with sodium piruvate. CFU were determined after 21 days of incubation. Vaccine was then adjusted to contain $10^6$ bacteria per ml and stored at $-70$°C until use.

### Culture Filtrate Proteins (CFP)

*M. bovis* CFP were prepared as follows. *M. bovis* AN5, the strain used to produce the PPD for skin testing cattle in México was grown as a pellicle in Dorset Henley medium for 8–10 weeks. Cultures supernatant was sterile-filtered twice through a 0.22 µm filter and tangentially filtered using a 10 kDa molecular weight

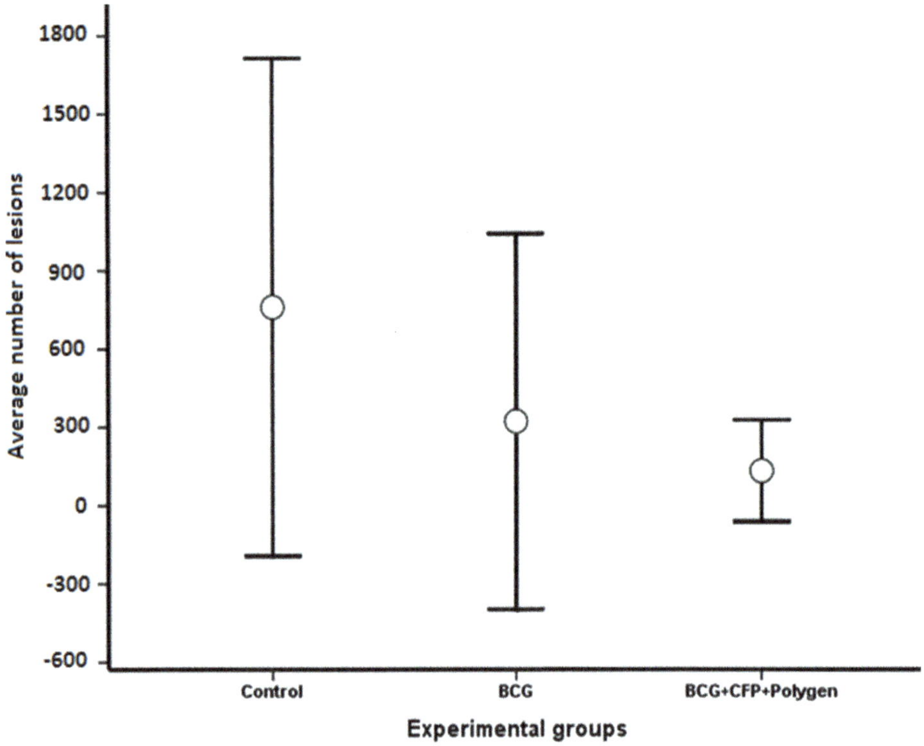


**Figure 2. Total number of lesions per group in calves vaccinated (BCG and BCG+CFP) and unvaccinated (control) against tuberculosis.** This number includes lesions from lymph nodes (mediastinal and trachebronchial) and lungs.

cut-off filter (Pelicon Membrane, Millipore, Inc., USA). The CFP was stored at 4°C until use. Protein content of the culture filtrate was determined using a Biuret Reagent, and protein adjusted to 200 µg/ml (P2B010A05, membrane type: PBGC). CFP was prepared by mixing *M. bovis* CFP with Polygen TM (MVPLaboratories, Inc.). The adjuvant was then added to a final concentration of 10% (v/v).

## Study Design and Vaccination

Animals were allocated throughout a completely randomized experimental design into three groups: the control, unvaccinated group, the BCG vaccinated and the BCG vaccinated and CFP boosted group. Vaccinated groups were inoculated subcutaneously on the neck with a dose of $10^6$ colony forming units (CFU) of the BCG strain in 2 ml diluent. The boosted group was vaccinated with the BCG strain and four weeks later with a *M. bovis* CFP, 400 µg/ml, plus polygen.

## *M. bovis* Challenge

The challenge inoculums consisted of a mid-log-phase of *M. bovis* grown in Stonebrink media supplemented with piruvate. The challenge strain was obtained from the lymph node of a dairy cow

**Table 1.** Average number of visible lesions per affected organ in vaccinated (BCG and BCG+CFP) and unvaccinated (control) calves challenged with a field strain of *M. bovis*.

| Organ | Group | Average number of lesions | Standar deviation | $IC_{95\%}$ | Min. | Max. |
|---|---|---|---|---|---|---|
| **Mediastinal lymph node** | Control | 316[a] | 432 | 7;625 | 0 | 1000 |
| | BCG | 147[a] | 406 | −144;437 | 0 | 1300 |
| | BCG+CFP | 48[a] | 76 | −6;102 | 0 | 250 |
| **Tracheobronchial lymph node** | Control | 351[a] | 473 | 12;689 | 0 | 1000 |
| | BCG | 114[a] | 218 | −41;270 | 0 | 550 |
| | BCG+CFP | 20[b] | 31 | −2;42 | 0 | 100 |
| **Lungs** | Control | 91[a] | 130 | −1;184 | 2 | 437 |
| | BCG | 64[a] | 155 | −50;172 | 0 | 500 |
| | BCG+CFP | 61[a] | 154 | −46;174 | 0 | 500 |

**Table 2.** Average number of *M. bovis* bacilli per 100 fields, 40× lens in vaccinated (BCG and BCG+CFP) and unvaccinated (control) calves challenged with a field strain of *M. bovis*.

| Group | Average number of bacilli per 100 fields, 40× lens | Estándar deviation | $IC_{95\%}$ | Min. | Max. |
|---|---|---|---|---|---|
| Control | 15a | 30.04 | −6.5;36.5 | 0 | 75 |
| BCG | 6a | 12.85 | −3.2;15.2 | 0 | 6 |
| BCG+CFP | 2.8a | 3.2 | 0.52;5.1 | 0 | 20 |

in 2010. To harvest the bacilli, bacteria were pelleted by centrifugation at 750 *g* and washed twice with phosphate-buffered saline solution (PBS, 0.01 M, pH 7.2). Bacilli were then perfectly homogenized in PBS and shacked by glass beads continuously at 200 rpm for 1 h. The homogenate was sterile-filtered twice throughout a 40 Watman filter and diluted to the established doses ($5\times10^5$ CFU) for use as inoculum in 2 ml of PBS. Bacilli counting were performed by serial dilution in Stonebrink media plates. Calves were challenged by aerosol as described by Palmer *et al.*, [21]. Briefly, previous to challenge, animals were inoculated 0.25 mg/kg of Xylazine hydrochloride 10% IM, and then restrained in a head chute. The challenge inoculum was delivered by nebulization into a mask covering the nostrils and mouth. The nebulization apparatus consisted of a compressed air tank and a commercially available aerosol delivery system (Equine Aeromask, Trudell Medical, London, Ontario, Canada) comprised of a jet nebulizer (Whisper Jet, Marquest Medical Products, Englewood, CO, USA), holding chamber and mask. Compressed air (25 psi) was used to jet nebulize inoculum directly into the holding chamber. Upon inspiration, the nebulized inoculum was inhaled and delivered through a one-way valve into the mask and directly into the nostrils during 8 min. The inoculums chamber was washed with 1 ml of PBS every time previous to the next delivery. Experimental infection was done in an isolated area to prevent contamination. Personnel used appropriate protective equipment, including full-face respirators with HEPA-filtered canisters and disposable protective clothing to prevent exposure to aerosolized *M. bovis*. The workplace and instruments were decontaminated with a 10% phenol solution and recipients used sterilized in autoclave.

## Blood Collection and IFN-g Release Assay

Blood samples were collected at day 0, 15, 40 and then about every two weeks until the end of the experiment, eight months after vaccination. These samples were shipped to the laboratory at ambient temperature and processed within 2 h of collection. To measure T cell response in whole-blood cultures after 16-h *in vitro* antigen stimulation, a commercial bovine IFN-g microplate enzyme-linked immunosorbent assay (ELISA, Bovigam; Commonwealth Serum Laboratories, Australia) kit was used. The kit was used according to the manufacturer instructions with a few modifications in reagent concentration. Briefly, samples were collected from the coccygeal vein and placed into heparin tubes. From there, 750 µl of whole blood were incubated in microplates, in duplicate with 50 µl for each antigen (bovine-PPD, avian-PPD). Negative control wells with phosphate-buffered saline (PBS) were included for each animal tested. Positive controls containing 50 µl of pokeweed mitogen of a 1 µg/ml concentration solution (Sigma–Aldrich, United Kingdom) were also included before incubating in a humidified 5% $CO_2$ incubator at 37°C for 20 h. Optical densities (OD) of PBS from the control wells were used to normalize individual readouts and to calculate optical density. Final OD readings were obtained by subtracting sample readings from PBS control readings.

**Table 3.** Number and proportion of positive tissue-samples to the presence of lesions TB-compatible detected in histopathological analysis in vaccinated (BCG and BCG+CFP) and unvaccinated (control) calves against tuberculosis and challenged with a field strain of *M. bovis*. At least one sample from each affected organ was included in the analysis.

| Group | Positive | Proportion of positive | Total |
|---|---|---|---|
| Control | 22 | 55 | 40 |
| BCG | 18 | 45 | 40 |
| BCG+CFP | 14 | 37 | 38 |

## Slaughter and Lesion Counting

Six months after challenge, experimental animals were sent to slaughter for carcass inspection, visible lesions counting and tissue sampling for histopathological and microbiological analysis. In order to perform a careful inspection of carcasses, animals were slaughtered in a three-week period; slaughtering three animals per day. Although all organs were reviewed, especial attention was place on lymph nodes: retropharyngeal, mediastinal and tracheobronquial, and lungs, liver and spleen. Lymph nodes were cut in half and all lesions counted. Liver, spleen and lungs were sliced in pieces of about 1 cm wide to count all visible lesions. Special care was focus in no counting the same lesion twice in both sides of the slice. Animals were slaughtered according to recommendations in the *NORMA Oficial Mexicana NOM-033-ZOO-1995, Sacrificio humanitario de los animales domésticos y silvestres* (Official norm for the humanitarian slaughter of domestic and wildlife animals), which is the legal document ruling this practice. Briefly, previous to slaughter, animals were kept in a quite environment, managed with care throughout the chute on the way to the final step and finally shut in the forehead with an air pistol.

## Statistical Analysis

The average IFN-g concentration (optical densities raw data) per group for every sampling period was compared with a one-way ANOVA test. The number of lesions per animal was first categorized into 5 groups according to the number of lesions in each experimental animal (1 = 0, 2 = 1 to 30, 3 = 31 to 100, 4 = 101 to 1000 and 5>1000) and the groups compared with the H statistic in the Kruskal-Wallis test. The association between the different experimental groups and the proportion of TB-compatible lesions was estimated throughout a chi-square test. Comparison of the number of bacilli between vaccinated and

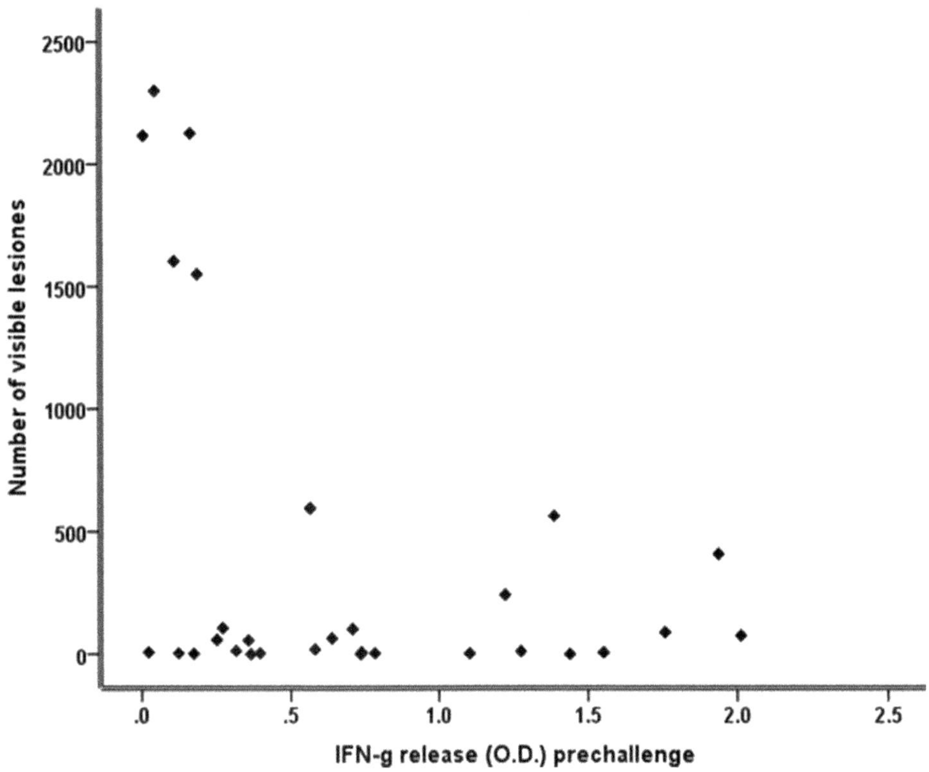


**Figure 3. Correlation of the total number of lesions with the average IFN-g response postvaccination-prechallenge (optical densities) of calves vaccinated and unvaccinated against tuberculosis.** Correlation was performed ignoring vaccination status.

control groups were performed throughout the nonparametric Kruskal-Wallis test. All analyses were performed in the SPSS software.

## Results

### IFN-g Release

IFN-g in plasma is used as a surrogate of vaccine efficacy against tuberculosis in cattle [22], [9], [10], [5], [22], [23], [13], [8], [24]. In this experiment, IFN-g response to bovine-PPD was measured at different time-periods after vaccination and after challenge (Figure 1). No significant difference was observed between groups at day 0 ($P = 0.70$). At day 30, even though the difference between the vaccinated and the control groups was not statistically significant ($P = 0.13$), the concentration of IFN-g was higher in the vaccinated than in the control group. From the second period on, the difference between the vaccinated and the control groups was statistically significant ($P<0.05$) until day 90, the challenge date. After challenge, no difference between groups was observed ($P>0.05$) until day 230, just previous to slaughter, where the control group had a significant higher concentration than the vaccinated groups ($P = 0.03$).

### Lesion Counts

All animals were sent to slaughter 6 months after challenge for carcass inspection and visible lesions counting; except for one animal in the BCG+CFP group and two animals in the BCG group, all animals showed at least two visible lesions in at least one organ. The average number of lesions per group (lesions from all organs) is shown in Figure 2. Since normality was difficult to accomplish given the data values, which ranked from 0 to more than 1000, a non-parametric rank test (Kruskal-Wallis) was used. No difference between the vaccinated and the control groups was observed (Chi-Square test = 2.402, 2 df and $P = 0.301$, with a mean rank of 18.75, 13.0 and 14.7 for the control, the BCG and the BCG+CFP+Polygen groups respectively); however, the average number of lesions was much lower in the vaccinated groups ($n = 322 \pm 720$ and $132 \pm 726$, for the BCG and the BCG+CFP groups, respectively) than in the control group ($n = 760 \pm 954.6$). It is important to mention that in the BCG group, a single animal with 2,300 lesions increased considerable the average of lesions for that group. In the control group, four animals had more than 1,000 lesions. Similar results were observed when counting lesions per organ; the higher number of lesions was observed in the mediastinal and the tracheobronchial lymph nodes (Table 1).

### Histopathology

Four samples from each experimental animal were sent for histopathological analysis. The average number of bacilli in 100 fields with a 40× lens per group is shown in Table 2. Comparison of groups was performed throughout the ranks Kruskal-Wallis test. No significant difference between groups was observed (Chi-Square = 3.164, 2 df, $P = 0.206$); however, the average number of bacilli observed in the control group (4.8) was considerable higher than that observed in the vaccinated groups (0.8 for the BCG and 2.0 for the BCG+CFP groups). In relation to the number of samples with TB-compatible lesions (Table 3), the proportion was higher in the control (55%) than in the vaccinated groups (45% for the BCG and 36% for the BCG+CFP groups); however, the association in a chi-square test was not significant ($P = 0.27$).

### Tuberculin Conversion

All experimental animals were tuberculin tested at day 0, 8 weeks post vaccination and 20 weeks post challenge (one week previous to slaughter). All animals were tuberculin negative in the first test. In the second test, 1, 9 and 9 animals from the control, BCG and BCG+CFP groups respectively, were positive. In the third test, all animals were positive.

## Discussion

Efficacy of the BCG vaccine in reducing dissemination of *M. bovis* in cattle has been variable [25], [26], [27], [9], [10], [28], [5], [23]. Different factors have been associated to this variability: doses, the inoculation route and the BCG strain used [29], age to vaccination and previous exposure to environmental *Mycobacterium* [12]. In our study we used a $10^6$ CFU doses, based on reports from previous studies [5], [9], [30].

In relation to the vaccination route, we used the SC route. Other studies have used the oral and the nasal routes; however, those are more impractical in the field [10], [23]. The most frequently used strains in experimental studies in cattle have been the Pasteur and the Danish strains [31], [32], [33], [30], there is general consent that all strains provide similar results. In this study we used the Phipps strain because in a previous study with ten substrains tested in a mouse model, it was the one that conferred the best protection against TB lesions in lung [20]. To prevent bias in the immune response from previous exposure to environmental *Mycobacterium* [34], [16], only calves negative to the tuberculin test and the IFN-g test were used.

Doses and Challenge routes play an important role in the pathogenesis of TB after vaccination [1]. In experimental conditions, different routes have been used: oral, intranasal, intratracheal and aerosol [35], [29], [36]. In our study the aerosol route was used because it better resembles natural conditions of infection; however, our doses was higher ($5\times10^5$ CFU) that that used by others. We were interested in a more aggressive dose, and a longer period post challenge-slaughter to have more notorious lesions and make a better contrast between vaccinated and unvaccinated animals.

Some studies have shown that boosting induces better protection against TB in cattle [37], [38], [39], [13], [40], [41], our results agree with those reports, the boosted group of animals had the highest concentration of IFN-g, the lowest number of lesions at slaughter and a the lowest number of bacilli in affected tissue.

IFN-g has been used as a surrogate for vaccine efficacy [9], [11], [42], [35], [43]; however, this asseveration has been questioned [44], [45], [46]. In our study, when the average of all IFN-g values postvaccination-prechallenge was correlated with the total number of lesions, it was observed that the five animals with the highest number of lesions (>1500) were from those with the lowest IFN-g concentration (<0.3 O.D.), clearly suggesting a negative correlation (Figure 3).

In experimental studies it is always difficult to come up with a single and definitive value reflecting vaccine efficacy; however, if we base efficacy in terms of the total number of lesions and consider the control group as the reference group (0% efficacy), the relative efficacy is 58% [efficacy = 1−(number of lesions in the vaccinated group/number of lesions in the control group) × 100] for the BCG group and 83% for the BCG+CFP group.

No significant differences were observed in the number of animals with visible lesions at slaughter between the vaccinated and the control groups, as reported in other studies [30], [47], [48]. This could be a consequence of the high doses of bacilli and the route of inoculation of the challenge inoculum in our study, we believed that both influenced the implantation of infection and, consequently, favor the development of lesions.

There are some reports about a low proportion of vaccinated animals responding to the tuberculin test, in our study, two months after vaccination 90% of the vaccinated animals were positive, a similar result to that reported for at least another study [49]. Therefore, there is an urgent need for tests that perform differential diagnosis if the vaccine is to be used in the field.

Even though no significant differences between vaccinated and unvaccinated groups of animals were observed in this study, our results support the hypothesis that the vaccine has the potential to reduce the dissemination of TB in cattle. We believe that no significant difference between groups was more a consequence of the high variability of lesions provided for one or two animals in the vaccinated groups, than for a real failure of the vaccine. In general, the average number of lesions in the vaccinated groups was considerable lower than that in the control group; however, the high standard deviation coming from one or two animals with a high number of lesions caused failure to detect significant difference between groups. In real life, it is highly probable that some animals are more susceptible to suffer infection and to develop disease for different reasons, some known (genetics for example) and some unknown. In our study, the high number of lesions in some vaccinated animals could have been due to factors such as natural susceptibility to infection, the high number of bacilli in the challenged doses and the method used to deliver the challenge inoculums. In natural conditions, hardly the doses of infection is going to be as high, and hardly the pressure of bacilli into the respiratory tract is going to be 25 psi. Therefore, in order to know how useful the vaccine is in control programs, field trials need to be performed.

## Conclusion

This study shows that the BCG vaccine, alone or in combination with a CFP boost, has the potential to reduce tuberculosis dissemination in cattle by reducing the number of lesions and the bacterial load per lesion. However, in order to make definitive conclusions about the usefulness of the vaccine in programs against TB in the field, it is necessary to perform long term field trials.

## Author Contributions

Conceived and designed the experiments: GJCA FMS. Performed the experiments: GJCA FMS OEPM LGC ANV YRV AMOR LBN. Analyzed the data: YRV FMS. Contributed reagents/materials/analysis tools: LGC SSG ANV LVN AMOR. Wrote the paper: GJCA FMS.

## References


1. Pollock JM, Rodgers JD, Welsh MD, McNair J (2006) Pathogenesis of bovine tuberculosis: the role of experimental models of infection. Vet Microbiol 112: 141–150.
2. Bawinec F, Taylor NM (1996) Assessment of the socio-economic importance of bovine tuberculosis in Turkey and possible strategies for control or eradication. Turkish-German Animal Health Information Project. General Dictatoriate of Protection and Control, Ankara TK.
3. Nader A, Husberg H (1988) Estimación de pérdidas por tuberculosis bovina en un rodeo lechero. Rev Med Vet 68: 46–49.
4. Kaneene BJ, Miller RA, Meyer RM (2006) Abattoir surveillance: the US experience. Vet Microbiol 112: 273–282.
5. Skinner MA, Wedlock DN, Buddle BM (2001) Vaccination of animals against *Mycobacterium bovis*. Rev Sci Tech 20112–20132.

6. Buddle BM, Wedlock DN, Denis M (2006) Progress in the development of tuberculosis vaccines for cattle and wildlife. Vet Microbiol 112 (2–4): 191–200.
7. Fine PEM (2001) BCG vaccines and vaccination. In: Reichman LB, Hershfield, ES, editors. Tuberculosis: A Comprehensive International Approach. New York: Marcel Dekker. pp. 503–524.
8. Hope JC, Villareal-Ramos B (2008) Bovine TB and the development of new vaccines. Comp Immunol Microbiol and Infect Dis 31: 77–100.
9. Buddle BM, de Lisle GW, Pfeiffer A, Aldwell FE (1995) Immunological responses and protection against *Mycobacterium bovis* in calves vaccinated with a low dose of BCG. Vaccine 13: 1123–1130.
10. Buddle BM, Parlane NA, Keen DL, Aldwell FE, Pollock JM, et al. (1999) Differentiation between *Mycobacterium bovis* BCG-vaccinated and *M. bovis* infected cattle by using recombinant mycobacterial antigens. Clin Diagn Lab Immunol 6 (1): 1–5.
11. Buddle BM, Pollock JM, Skinner MA, Wedlock DN (2003) Development of vaccines to control bovine tuberculosis in cattle and relationship to vaccine development for other intracellular pathogens. Int J Parasitol 33: 555–566.
12. Buddle BM, Wedlock DN, Parlane NA, Corner LA, de Lisle GW, et al. (2003) Revaccination of neonatal calves with *Mycobacterium bovis* BCG reduces the level of protection against bovine tuberculosis induced by a single vaccination. Infect Immunol 71: 6411–6419.
13. Wedlock DN, Denis M, Skinner MA, Koach J, de Lisle GW, et al. (2005) Vaccination of cattle with a CpG oligodeoxynucleotide-formulated mycobacterial protein vaccine and *Mycobacterium bovis* BCG induces levels of protection against bovine tuberculosis superior to those induced by vaccination with BCG alone. Infect Immun 73: 3540–3546.
14. Vordermeier HM, Huygen K, Singh M, Hewinson RG, Xing Z (2006) Immune responses induced in cattle by vaccination with a recombinant adenovirus expressing mycobacterial antigen 85A and *Mycobacterium bovis* BCG. Infect Immun 74: 1416–1418.
15. Buddle BM, Collins DM (2007) Development of tuberculosis vaccines for cattle. CAB Rev 73: 1–8.
16. Buddle BM, Wards BJ, Aldwell FE, Collins DM, de Lisle GW (2002) Influence of sensitization to environmental mycobacteria on subsequent vaccination against bovine tuberculosis. Vaccine 20 (7–8): 1126–1133.
17. Hope JC, Thom ML, Villareal-Ramos B, Vordermeier HM, Hewinson RG, et al. (2005) Vaccination of neonatal calves with *Mycobacterium bovis* BCG induces protection against intranasal challenge with virulent *M. bovis*. Clin Exp Immunol 139: 48–56.
18. Vordermeier HM, Gordon SV, Hewinson RG (2011) *Mycobacterium bovis* antigens for the differential diagnosis of vaccinated and infected cattle. Vet Microbiol 151: 8–13.
19. Milián F, Salman M, Ramirez C, Payeur J, Rhyan J, et al. (2000) Identification of tuberculosis in cattle slaughtered in México. Am J Vet Res 61(1): 86–89.
20. Castillo-Rodal A, Castañón-Arreola M, Hernández-Pando R, Calva-Mercado J, Sada DE, et al. (2006) BCG sub strains confers different protection against *Mycobacterium tuberculosis* infection in BALB/c model of progressive pulmonary tuberculosis. Infect Immunol 74: 1718–1724.
21. Palmer MV, Waters WR, Whipple DL (2002) Aerosol delivery of virulent *Mycobacterium bovis* to cattle. Tuberculosis (Edinb) 82(6): 275–282.
22. Vordermeier HM, Chambers MA, Cockle PJ, Whelan AO, Simmons J, et al. (2002) Correlation of ESAT-6-specific gamma interferon production with pathology in cattle following *Mycobacterium bovis* BCG vaccination against experimental bovine tuberculosis. Infect Immun 70: 3026–3032.
23. Corner LAL, Buddle BM, Pfeiffer DU, Morris RS (2002) Vaccination of the brushtail possum (*Trichosurus vulpecula*) against *Mycobacterium bovis* infection with bacille Calmette-Guerin: the response to multiple doses. Vet Microbiol 84: 327–336.
24. Milián-Suazo F, Pabello GJA, Bojorquez NL, Anaya EAM, Cantó AGJ, et al. (2011) IFN-g response to vaccination against tuberculosis in dairy heifers under commercial settings. Res Vet Scien 90: 419–424.
25. Francis J (1950) Control of infection with the bovine tubercle bacillus. Lancet 1: 34–39.
26. Waddington FG, Elwood DC (1972) An experiment to challenge the resistance to tuberculosis in BCG vaccinated cattle in Malawi. British Vet. J 128: 541–552.
27. Elwood DC, Waddington FG (1972) A second experiment to challenge the resistance to tuberculosis in BCG vaccinated cattle in Malawi. British Vet J 619–626.
28. Corner LAL, Buddle BM, Pfeiffer DU, Morris RS (2001) Aerosol vaccination of the brushtail possum (*Trichosurus vulpecula*) with bacille Calmette-Guerin: the duration of protection. Vet Microbiol 81: 181–191.
29. Vordermeier HM, Chambers MA, Buddle BM, Pollock JM, Hewinson RG (2006) Progress in the development of vaccines to control tuberculosis in cattle. *Vet. J.* 171: 229–244.
30. Wedlock DN, Denis M, Vordermeier HM, Hewinson RG, Buddle BM (2007) Vaccination of cattle with Danish and Pasteur strains of *Mycobacterium bovis* BCG induce different levels of IFN-gamma post-vaccination, but induce similar levels of protection against bovine tuberculosis. Vet Immunol Immunopathol 118: 50–58.
31. Ameni G, Tibbo M, (2002) Kinetics of interferon-gamma (IFNγ) release in the peripheral blood of calves vaccinated with BCG. J. Immun Immunochem 23: 245–253.
32. Skinner MA, Buddle BM, Wedlock DN, Keen D, de Lisle GW, et al. (2003). A DNA prime-*Mycobacterium bovis* BCG boost vaccination strategy for cattle induces protection against bovine tuberculosis. Infect Immun 71: 4901–4917.
33. Cai H, Yu DH, Hu XD, Li SX, Zhu YX (2006) A combined DNA vaccine-prime, BCG-boost strategy results in better protection against *Mycobacterium bovis* challenge. DNA Cell Biol 25: 438–447.
34. Brandt L, Feino CJ, Weinreich OA, Chilima B, Hirsch P, et al. (2002) Failure of the *Mycobacterium bovis* BCG vaccine: some species of environmental mycobacteria block multiplication of BCG and induction of protective immunity to tuberculosis. Infect Immun 70(2): 672–8.
35. Buddle BM, Wedlock DN, Denis M, Skinner MA (2005) Identification of immune response correlates for protection against bovine tuberculosis. Vet Immunol Immunopathol 10: 45–51.
36. Waters WR, Whelan AO, Lyashchenko KP (2010) Immune responses in cattle inoculated with *Mycobacterium bovis*, *Mycobacterium tuberculosis*, or *Mycobacterium kansasii*. Clin.Vaccine Immunol 7(2): 247–252.
37. Ferraz JC, Stavropoulos E, Yang M, Coade S, Espitia C, et al. (2004) A heterologous DNA priming *Mycobacterium bovis* BCG boosting immunization strategy using mycobacterial Hsp70, Hsp65, and Apa antigens improves protection against tuberculosis in mice. Infect Immun 72: 6945–6950.
38. Vordermeier HM, Rhodes SG, Dean G, Goonetilleke N, Huygen K, et al. (2004). Cellular immune responses induced in cattle by heterologous prime-boost vaccination using recombinant viruses and bacille Calmette-Guerin. Immunol 112: 461–470.
39. McShane H, Hill A (2005) Prime-boost immunization strategies for tuberculosis. Microb Infect 7: 962–967.
40. McShane H, Pathan AA., Sander CR, Goonetilleke NP, Fletcher HA, et al. (2005) Boosting BCG with MVA85A: the first candidate subunit vaccine for tuberculosis in clinical trials. Tuberculosis (Edinb.). 85, 47–52.
41. Vordermeier HM, Villareal-Ramos B, Cockle PJ, McAulay M, Rhodes SG, et al. (2009) Viral booster vaccines improve *Mycobacterium bovis* BCG-induced protection against bovine tuberculosis. Infect Immun 77: 3364–3373.
42. Logan KE, Chambers MA, Hewinson RG, Hogarth PJ (2005). Frequency of IFN-γ producing cells correlates with adjuvant enhancement of bacille Calmette-Guérin induced protection against *Mycobacterium bovis*. Vaccine 23: 5526–5532.
43. Wedlock DN, Skinner MA, Parlane NA, Vordermeier HM, Hewinson RG, et al. (2003) Vaccination with DNA vaccines encoding MPB70 or MPB83 or a MPB70 DNA prime-protein boost does not protect cattle against bovine tuberculosis. Tuberculosis (Edinb) 83: 339–349.
44. Kipnis A, Irwin S, Izzo AA, Basaraba RJ, Orme IM (2005) Memory T lymphocytes generated by *Mycobacterium bovis* BCG vaccination reside within a CD4 CD44lo CD62hi population. Infect Immun 73: 7759–7764.
45. Elias D, Akuffo H, Britton S (2005) PPD induced in vitro interferon gamma production is not a reliable correlate of protection against *Mycobacterium tuberculosis*. Trans R Soc Trop Med Hyg 99: 363–368.
46. Langermans JA, Andersen P, van Soolingen D, Vervenne RA, Frost PA, et al. (2001) Divergent effect of bacillus Calmette-Guerin (BCG) vaccination on *Mycobacterium tuberculosis* infection in highly related macaque species: implications for primate models in tuberculosis vaccine research. Proc Natl Acad Sci 98: 11497–11502.
47. Hope JC, Thom ML, McAulay M, Mead E, Vordermeier HM, et al. (2011) Identification of Surrogates and Correlates of Protection in Protective Immunity against *Mycobacterium bovis* Infection Induced in Neonatal Calves by Vaccination with *M. bovis* BCG Pasteur and *M. bovis* BCG Danish. Clin Vacc Immun 18: 373–379.
48. Vordermeier HM, Jones GJ, Whelan AO (2011) DIVA reagents for bovine tuberculosis vaccines in cattle. Expert Rev Vaccines 10(7): 1083–1091.
49. Ameni G, Vordermeier M, Aseffa A, Young DB, Hewinson G (2010) Evaluation of the Efficacy of *Mycobacterium bovis* Bacillus Calmette-Guerin against Bovine Tuberculosis in Neonatal Calves in Ethiopia. Clin Vaccine Immunol 17(10): 1533–1538.

# 16

# Estimation of the *Relative* Sensitivity of the Comparative Tuberculin Skin Test in Tuberculous Cattle Herds Subjected to Depopulation

**Katerina Karolemeas[1], Ricardo de la Rua-Domenech[2], Roderick Cooper[3], Anthony V. Goodchild[4], Richard S. Clifton-Hadley[4], Andrew J. K. Conlan[1], Andrew P. Mitchell[4], R. Glyn Hewinson[4], Christl A. Donnelly[5], James L. N. Wood[1], Trevelyan J. McKinley[1]***

**1** Disease Dynamics Unit, Department of Veterinary Medicine, University of Cambridge, Cambridge, United Kingdom, **2** Animal Health and Veterinary Laboratories Agency and Bovine Tuberculosis Programme of the Department for Environment, Food and Rural Affairs, London, United Kingdom, **3** Animal Health and Veterinary Laboratories Agency, West Midlands Regional Office, Stafford, United Kingdom, **4** Animal Health and Veterinary Laboratories Agency, Weybridge, United Kingdom, **5** MRC Centre for Outbreak Analysis and Modelling, Department of Infectious Disease Epidemiology, Imperial College London, London, United Kingdom

## Abstract

Bovine tuberculosis (bTB) is one of the most serious economic animal health problems affecting the cattle industry in Great Britain (GB), with incidence in cattle herds increasing since the mid-1980s. The single intradermal comparative cervical tuberculin (SICCT) test is the primary screening test in the bTB surveillance and control programme in GB and Ireland. The sensitivity (ability to detect infected cattle) of this test is central to the efficacy of the current testing regime, but most previous studies that have estimated test sensitivity (relative to the number of slaughtered cattle with visible lesions [VL] and/or positive culture results) lacked post-mortem data for SICCT test-negative cattle. The slaughter of *entire* herds ("whole herd slaughters" or "depopulations") that are infected by bTB are occasionally conducted in GB as a last-resort control measure to resolve intractable bTB herd breakdowns. These provide additional post-mortem data for SICCT test-negative cattle, allowing a rare opportunity to calculate the animal-level sensitivity of the test *relative* to the total number of SICCT test-positive and negative VL animals identified post-mortem (rSe). In this study, data were analysed from 16 whole herd slaughters (748 SICCT test-positive and 1031 SICCT test-negative cattle) conducted in GB between 1988 and 2010, using a Bayesian hierarchical model. The overall rSe estimate of the SICCT test at the *severe* interpretation was 85% (95% credible interval [CI]: 78–91%), and at *standard* interpretation was 81% (95% CI: 70–89%). These estimates are more robust than those previously reported in GB due to inclusion of post-mortem data from SICCT test-negative cattle.

**Editor:** Lionel G. Filion, University of Ottawa, Canada

**Funding:** KK was supported by a fellowship from the Veterinary Training and Research Initiative programme at Cambridge (Defra grant VT0105). TJM is supported by the Biotechnology and Biological Sciences Research Council grant (BB/I012192/1). JLNW is supported by the Alborada Trust and the Research and Policy for Infectious Disease Dynamics program of the Science and Technology Directorate, Department of Homeland Security, Fogarty International Center, National Institutes of Health. AJKC was funded by Defra grant SE3230 (Cambridge reference PU/T/WL/07/46), sponsored by the Veterinary Laboratories Agency. The work was also funded by Defra, United Kingdom, under contract SE3230. CAD thanks the Medical Research Council for Centre funding. The funders had no role in study design, data collection and analysis, decision to publish, or preparation of the manuscript.

**Competing Interests:** The authors have declared that no competing interests exist.

* E-mail: tjm44@cam.ac.uk

## Introduction

Bovine tuberculosis (bTB), caused by *Mycobacterium bovis*, is a zoonotic disease that is endemic in many countries worldwide, impacting on animal health, welfare, productivity and trade. In Great Britain (GB), despite a compulsory test-and-slaughter programme introduced in 1950, the disease is now endemic in West and South-west England and also in Wales (although Scotland was declared officially bTB free in 2009). Bovine TB has been described as GB's "*biggest endemic animal health issue*" [1] and in 2010–2011, the Department for the Environment, Food and Rural Affairs (Defra) spent over £91 million on bTB control in England [2]. The imperfect sensitivity (the probability that an infected animal is detected by the test) of the current diagnostic field test is one plausible contributor to the sustained increase in the incidence of bTB in GB over the last 25 years.

The intradermal tuberculin test is recognised by the World Organisation for Animal Health (OIE) and the European Commission as the primary screening test for detection of bTB in cattle [3,4]. The application of this test, supplemented with slaughterhouse surveillance and movement controls on infected and suspect herds, underpins many bTB control programmes worldwide and, in the absence of wildlife reservoirs for *M. bovis*, has led to the successful eradication of bTB in many countries [4–6].

Two variants of the intradermal tuberculin test are in use today: the single intradermal tuberculin test, which measures a cell-mediated delayed-type hypersensitivity response to injection of *M. bovis* tuberculin in the caudal skin fold (e.g. New Zealand, USA) or the skin of the mid-cervical region (e.g. most European Union Member States); and the single intradermal comparative cervical tuberculin (SICCT) test, which compares immune responses to

*M. bovis* (bovine) and *M. avium* (avian) tuberculins in the cervical region. In GB, Ireland and Portugal, the more specific, but less sensitive, SICCT test is used to minimise the likelihood of false positive results caused by exposure to cross-reacting mycobacteria [7,8].

Nearly half of all cattle herds in GB are tested for bTB every year, 41% are tested every four years and the remainder tested at 2- or 3-yearly intervals (Defra, unpublished data, 2012). Cattle with a positive SICCT test result (known as "reactors"), or those from which *M. bovis* is cultured from suspect bTB lesions detected during commercial slaughter, trigger a bTB herd "breakdown". All reactors are compulsorily slaughtered, whilst the affected herd is placed under movement restrictions and then re-tested at a minimum interval of 60 days (short-interval testing) until a series of tests with negative results is obtained. The number of short-interval tests, and their interpretation, is largely (but not exclusively) dictated by the detection of evidence of *M. bovis* infection at post-mortem examination (VLs) or on culture.

The sensitivity of the SICCT test is a measure of its ability to detect infected animals. Knowledge of SICCT test sensitivity is fundamental to understanding the potential success and limitations of bTB control programmes. Reduced test sensitivity increases the risk of within-herd persistence of infection, recurrent breakdowns, onward transmission through cattle movements and, potentially, infection spill-over into local wildlife reservoirs. Given the chronic nature of bTB and the relative insensitivity of post-mortem techniques used to confirm the presence of *M. bovis*, the true infection status of every animal cannot be determined and thus sensitivities have been estimated *relative* to post-mortem confirmation methods, namely detection of typical lesions of bTB (VLs) and/or laboratory culture of *M. bovis*.

SICCT test sensitivities in different countries and settings have been reported to range from 75.0–95.5% [8] and 77–95% [9], but for the majority of the studies from which these estimates were reported or derived, post-mortem data for a representative group of SICCT test-negative animals from the same population were lacking. Indeed, post-mortem data are rarely available for SICCT test-negatives, as normally only reactor animals are compulsorily slaughtered during herd breakdowns. Conducting experimental studies to collect such data would be expensive, labour-intensive and suffer from problems of extrapolation to the field situation. Without these data, the true proportion of infected animals not detectable by the test (false negatives) remains unknown, which could substantially impact on estimated values.

Whole herd slaughter ("stamping out" or depopulation) is a last-resort control option in GB, reserved only for severely or chronically infected herds in which repeated SICCT testing (occasionally supplemented since 2003 by interferon-gamma parallel testing [10]) has failed to resolve the breakdown. Depopulated herds provide a rare opportunity to improve estimates of relative SICCT test sensitivity (rSe), since they generate data on both SICCT test-positive *and* SICCT test-negative animals. In this study we analyse data from 16 such herds in GB, comprising a total of 748 SICCT test-positive and 1031 SICCT test-negative animals slaughtered between 1988 and 2010.

## Results

### Descriptive Results

Sixteen British cattle herds that underwent depopulation because of bTB infection between 1988 and 2010, and met the five inclusion criteria, were identified from VetNet (Table S1). A total of 748 reactors and 1031 direct contacts were slaughtered in those herds: the median number of reactors (per herd) was 34 (interquartile range [IQR] 23–45) and the median number of direct contacts removed was 53 (IQR 38–58.5). There were eleven beef and five dairy herds and the median number of animals slaughtered during the breakdown was 93 (IQR 58–128). The median herd size (of the maximum herd size recorded during the breakdown) was 82 (IQR 57–113).

Seven out of the 16 selected herds had at least one inconclusive reactor (IR: borderline test result) recorded during the breakdown. For five of these herds, all IRs were slaughtered (either immediately or after being retested later in the breakdown). Two herds had IRs recorded which were not all slaughtered but represented only a small percentage of the maximum herd size during the breakdown (1.3 and 5.4% respectively). Three breakdowns were triggered by *M. bovis* culture-positive lesions detected during routine slaughterhouse surveillance. These animals (infected but not SICCT-tested prior to slaughter), which represented between 1 and 8% of the number of animals slaughtered in each herd respectively, were excluded from the rSe calculations, as described in the Materials and Methods section.

### rSe Calculations

The overall rSe of the SICCT test at severe interpretation for the 16 slaughtered herds was estimated to be 85% (95% CI: 78–91%) from the model fitted with herd-level effects, and 84% (95% CI: 81–87%) when calculated assuming no herd-level heterogeneity (Table 1 and Figure 1A). Although data were available on skin test positivity or negativity for all 16 herds, data on exact skin test reaction sizes were only available for 11 of these (the five herds for which these data were not available had a breakdown date prior to September 2000, before which these data were not routinely recorded). When recalculated at the *standard* interpretation, the overall rSe was estimated to be 81% (95% CI: 70–89%) from the model fitted with herd-level effects, and 78% (95% CI: 74–81%) when calculated without (Table 1 and Figure 1B). This estimate is based on 11 herds only. The numbers of reactors at both interpretations are provided in Tables S2 and S3.

Figures 1A and 1B show the between-herd heterogeneity in the overall rSe estimates at severe and standard interpretation, obtained from the hierarchical model. Overall adjusted rSe estimates were slightly higher when the herd-level effects were included at both severe and standard interpretation, as a result of accounting for the herd-level heterogeneity. Although exceeding 90% in some herds, the estimated relative sensitivity of the test at severe interpretation was as low as 60% in others (Figure 1A). At the standard interpretation, one herd had a sensitivity estimate as low as 52% (Figure 1B).

Since we do not have SICCT test results for those animals that were detected via slaughterhouse surveillance, we conducted a sensitivity analysis assuming in the first instance that all of these animals would have been SICCT test positive, and in the second case assuming they would have been SICCT test negative. The results from these two extreme scenarios are shown in Table S4 and show negligible impacts on the estimates.

### Sensitivity of VL Detection Relative to Culture

Two of the slaughtered herds had culture data available for >90% of the animals slaughtered, in addition to post-mortem VL/NVL data. The rSe estimates for VL detection relative to culture for the two herds were 79% and 95% respectively, with an overall estimate of 90% (95% CI: 82–96%) [fitted without herd-level effects due to a lack of convergence when herd-level effects were included].

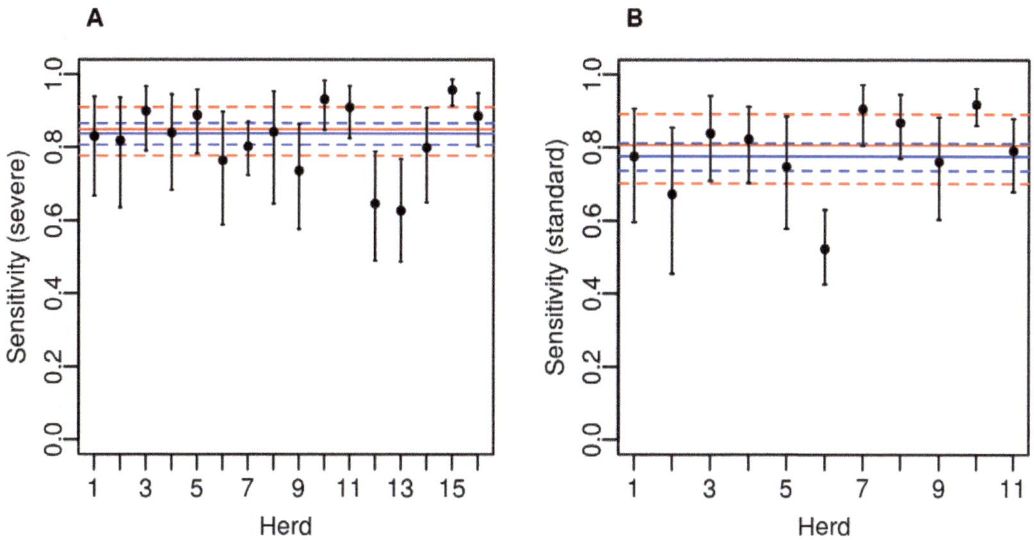


**Figure 1. Overall mean *relative* Single Intradermal Comparative Cervical Tuberculin test sensitivity (rSe) estimates for each herd.** Estimates are shown for severe and standard test interpretation, derived from logistic regression models fitted with and without herd-level effects. The solid lines represent the *mean* estimates, and the dashed lines represent the 95% credible intervals. The red lines are from the model *including* herd-level effects, and the blue lines are from the model *excluding* herd-level effects.

## Discussion

The SICCT test remains the primary bTB surveillance tool in several countries worldwide. Elsewhere, it is used as a second-line confirmatory (serial) test of caudal-fold and mid-cervical test reactors. The application of this test is central to restoring the officially TB-free status of infected herds in GB. A dearth of post-mortem data for SICCT test-negative animals has hindered the generation of reliable estimates for SICCT rSe. Depopulation, a radical bTB control method for severely affected herds, generates data for positive *and* negative-testing animals, providing a rare opportunity to improve current rSe estimates. Data for 16 cattle holdings, depopulated in GB between 1988 and 2010 due to bTB, were analysed calculating rSe relative to post-mortem detection of typical lesions of bTB (VLs) in slaughtered animals. The overall rSe was estimated to be 85% (95% CI: 78–91%) at the severe interpretation (16 herds) and 81% (95% CI: 70–89%) at the standard interpretation (11 herds) after accounting for herd-level heterogeneity.

Previously-reported median sensitivity estimates of the SICCT test at the standard (77.5%) and severe (92.2%) interpretation, were derived from a range of published studies that varied in location (some outside GB), tuberculin source and dosage [8] (these values were reported as 80% and 94% respectively; however, when the Costello *et al.* estimate [11] was re-examined, it was noted that this should have been included in the estimate for the *severe* interpretation, and thus the values have been recalculated to take account of this). For the five studies [11–15] where the test was performed as in GB, estimates ranged from 75.0–95.5% at the standard interpretation [8]. When the estimate from Costello *et al.* (1997) is excluded (since it was not interpreted at standard interpretation–as discussed previously), the range remains the same. SICCT test sensitivity ranged from 77–95% in an earlier review [9]: three of these studies [13–15] were included in the review by de la Rua-Domenech et al. (2006), and the remaining four studies [16–19] were not comparable, employing tuberculin extracted from *M. tuberculosis* that was replaced in GB by the current (*M. bovis*-derived) tuberculin in 1975 [9,13].

Critically, the majority of these previous studies have lacked post-mortem data for SICCT-negative animals, which could plausibly have inflated the estimated sensitivities by reducing the denominator. However, one study carried out in Ireland [11], did slaughter the entire SICCT-tested study population (2528 animals), including all test-negatives, and also calculated sensitivity relative to VL detection (rSe). The rSe reported was 90.9% (severe interpretation), which is higher than our overall estimate although just inside our upper credible interval limit (overall mean 85%, 95% CI: 78–91%). The avian and bovine tuberculins used in the Costello study, manufactured in Lelystad (Netherlands), have been shown to be associated with a slightly higher probability of confirmation (VLs and/or culture) when compared with tuberculin manufactured in Weybridge (UK) [20], the SICCT test antigen used in the majority (69%) of the herds in our study (for the remaining herds, 25% herds used Lelystad and 6% herds [1 herd]

**Table 1.** Overall mean *relative* Single Intradermal Comparative Cervical Tuberculin (SICCT) test sensitivity (rSe) estimates (%).

| | SICCT test interpretation[#] | Model *with* herd-level effects | Model *without* herd-level effects |
|---|---|---|---|
| **Overall mean *relative* sensitivity (95% CI)** | Severe | 85 (78–91) | 84 (81–87) |
| | Standard | 81 (70–89) | 78 (74–81) |

[#]At severe test interpretation, results are based on 16 herds. At standard interpretation, results are based on 11 herds. 95% CI represent the 95% credible intervals. Estimates are relative to visible lesion (VL) detection.

had the tuberculin type classified as undetermined). Costello *et al.* also calculated sensitivity relative to VL detection, and although their higher rSe could have been due to a reportedly more thorough post-mortem inspection, this was not reflected in the percentage of reactors that had VLs: 41% of reactors and IRs in their study compared with 40% of reactors nationally in GB [21], although inclusion of IRs here could have artificially lowered this figure due to the proportion of VL animals being lower in IRs than in reactors (in both GB and Ireland).

The true infection status of all individual animals is impossible to determine. Sensitivities calculated in this study from naturally-infected herds were *relative* to detection of gross bTB lesions post-mortem. While in some respects our estimates may provide a lower bound for the true sensitivity, the impact of well-developed disease on lesion visibility in some of these herds is hard to determine as there may be biases in the depopulation data when compared to other situations, which could influence the rSe estimate in an unpredictable manner.

Any culture data that were available from the selected herds were not combined with the post-mortem VL data for rSe calculations, as there may have been bias towards reactors being more likely to be cultured than direct contacts. Sensitivity of VL detection, *relative* to culture positivity, was explored for two depopulated herds that had VL/NVL *and* culture data for >90% of animals slaughtered. The overall rSe derived from a logistic regression model was 90%, suggesting that VL detection is a good measure of infection status, as defined by detection by culture. This is supported by the fact that, in GB, an estimated 95% of all VL reactors yield *M. bovis* on culture [21].

Although in our calculations we do not assume that visible lesions are synonymous with infection, the majority of suspect lesions do indeed culture *M. bovis*. In 2011, around 75% of suspect granulomas detected during routine slaughterhouse surveillance (i.e. not in reactor animals identified during a breakdown) cultured *M. bovis* (Defra, unpublished, 2012). However, this proportion of culture-positive lesions is likely to be much higher in severe bTB breakdowns with previously confirmed infection and a high prevalence of within-herd infection. VLs have been shown to develop as early as 17 days in experimentally-infected calves infected with doses to mimic naturally-occurring infection [22], (albeit in an experimental situation where postmortem inspection for visible lesions was conducted at a higher level of detail with serial slicing of target organs and lymph nodes), which is earlier than the minimum time taken to develop a detectable SICCT test response, reported to be 3–6 weeks after infection [8].

Mechanisms underlying imperfect skin test sensitivity include factors relating to the tuberculin source and potency, the operator's adherence to the prescribed test protocol [23–25], the stage of *M. bovis* infection in the tested animal (very early or too advanced to mount a detectable hypersensitivity response to tuberculin) [9], and factors that can mask or suppress the cell-mediated immune response to tuberculin, such as anti-inflammatory drugs and simultaneous infections with other pathogens. In addition, a period of transient and partial desensitisation has been shown to occur shortly after tuberculin injection [26] and, although EU protocols recommend a minimum interval of 60 days between sequential skin tests [27], recent evidence has shown that repeated SICCT testing at 60-day intervals can result in a progressive reduction in the comparative bovine-avian tuberculin response in naturally infected cattle [28].

Although the rSe at standard interpretation calculated in our study is in line with previously reported estimates, the mean rSe at severe interpretation calculated by Costello *et al.* (1997) only just fell within the upper credible interval limit of our estimate. Although the overall sensitivity generated from our study was lower at standard compared to severe interpretation (as one might expect), the 95% CIs overlapped considerably. This may have been due to the small numbers of animals reclassified during the change in interpretation (possibly limited by the number of herds in the study). Although this may suggest that in the field there may be little additional gain in applying the more severe interpretation, the small number of herds in this study should be considered when attempting to extrapolate the findings. The degree of between-herd heterogeneity for the rSe estimates also suggests that caution should be exercised in estimation from single herd breakdowns.

Veterinarians, farmers and government place rely on the SICCT test to detect bTB through surveillance testing and to clear infection from breakdown herds. It should be noted that our rSe estimates (and those from previous studies) are calculated at the *animal*-level, which provide a lower bound for the *herd*-level sensitivity (the ability to detect infected herds as part of a surveillance programme). The results suggest that even with the increased sensitivity at the more severe interpretation, undetected infected cattle may remain in some breakdown herds that are officially released from movement restrictions.

Almost a quarter of all breakdowns that are de-restricted after a programme of repeated skin testing at 60-day intervals with negative results recur in the same herd within 12 months, and 38% recur within 24 months [29]. Re-infection from badgers has been postulated to be a key mechanism behind persistence of infection in cattle herds in areas of endemic bTB incidence [30] and is anecdotally thought to be responsible for recurrent breakdowns in those areas. However, the failure of the SICCT test to detect every infected animal should also be recognised, as well as considering how the test is applied, both nationally as well at as the individual-herd level.

In many parts of GB the skin test has worked well as a bTB surveillance and eradication tool, with Scotland, for instance, gaining officially TB-free status in 2009 [31] in the apparent absence of a wildlife reservoir of *M. bovis*. As herds became attested and more counties became accredited TB free in the 1960s and 1970s, TB herd testing frequency in GB was gradually relaxed since the blanket enforcement of annual testing in the 1950s. It is only over the last 10–15 years that the herd testing coverage has been increasing year-on-year, with the whole of Wales and West and SW England now on yearly testing, reflecting the endemic bTB situation in those regions of GB [32]. The sensitivity of the overall bTB testing regime in GB has recently been boosted by the wider deployment of the interferon-gamma blood test in known infected herds and the introduction of compulsory pre-movement skin testing since 2006 [33], by the earlier removal of cattle with repeat inconclusive SICCT test results since 2010 [34] and by additional testing of high risk breakdown herds since 2011. Substantial efforts continue to be made to address gaps in the bTB eradication programme in GB and enhance the overall ability to detect and control infection (including training and auditing of operators conducting skin tests on farms).

## Conclusions

No ante-mortem diagnostic test for bTB is perfect and the SICCT test is no exception. Over 100 years ago, Bang (1892: cited in Monaghan, 1994) wrote "*the tuberculin test is no more perfect than most things in the world. Sometimes it fails, but it would be the greatest folly to reject this method because it is not able to give everything we desire*". We have robustly estimated the overall rSe of the SICCT test, as applied in GB, and found it to be consistent with previous estimates at both standard and severe interpretations, but with some noticeable variation between individual herds. This information should form a key component when developing control and eradication plans and in evaluating their potential success.

## Materials and Methods

A national animal disease surveillance database (VetNet) recorded information relating to bTB testing and herd breakdowns in GB [35]. (This has now been replaced by a new IT system [Sam] since September 2011.) For each herd, this includes the ante-mortem SICCT test results for animals classed as reactors (R: positive), or inconclusive (IR: borderline result) during a bTB breakdown, and the post-mortem results i.e. whether typical lesions were detected on post-mortem inspection (or during further laboratory examination) for all cattle slaughtered to control the infection. In addition, these data are also available for at-risk cattle compulsorily removed as SICCT test-negatives (direct contacts), as is the case in partial or whole-herd depopulations.

### Identification of Whole Herd Slaughters

VetNet data were analysed up until 29th October 2010. As bTB-infected holdings undergoing depopulation are not explicitly identifiable in this database, bTB breakdowns were selected for the study if they met *all five* of the following criteria:

1. **The presence of *M. bovis* infection had been confirmed** in the herd by detection of VLs and/or culture of the bacterium from at least one animal in the herd slaughtered during the breakdown.
2. **The percentage of the herd slaughtered was >90%.** The denominator used here is the *maximum* herd size recorded during the breakdown and all animals slaughtered over the entire duration of the breakdown were included in the numerator. Herd size may fluctuate between tests due to births/deaths or cattle moved (under licence) onto the farm/removed for bTB control, making it possible for the number of animals slaughtered to exceed the maximum herd size. Thus, those breakdowns in which the calculated percentage was >100% were not immediately excluded from the study.
3. **>90% of the herd size recorded at the last test during the breakdown was removed to slaughter.**
4. **The percentage of the animals slaughtered in the course of the breakdown that had a post-mortem (VL/NVL) result was >90%.** This cut-off was chosen to minimise the effect of missing VL/NVL data. This information should be recorded for all animals slaughtered, but it may be missing from the database on rare occasions.
5. **The number of animals slaughtered during the breakdown was >50.** This criterion was set to minimize spurious effects from small within-herd sample sizes.

Where the ancillary parallel interferon-gamma blood test had been performed in a herd (to accelerate the detection of infection in cattle that had tested negative to the SICCT test), any additional animals slaughtered as interferon-gamma positives were deemed to be SICCT test negative.

### Relative Test Sensitivity (rSe) Calculations

Culture of tissue samples is only conducted in GB in a representative sample of all the cattle slaughtered in a bTB breakdown or herd depopulation, thus preventing the inclusion of culture-positive cattle in the sensitivity calculations. In this study, SICCT test sensitivity for each herd was estimated *relative* (rSe) to the number of cattle presenting with typical tuberculous lesions during post-mortem carried out by veterinary and meat inspectors in approved slaughterhouses (the reference standard).

The SICCT test was conducted and its results interpreted according to the protocol set out at Annex B of EU Directive 64/432/EEC [27]. Slaughtered animals were recorded in VetNet as reactors, IRs or direct contacts. Any IRs slaughtered in the course of a breakdown as reactors/direct contacts/repeat IRs, were included in the rSe calculations if they had VLs (IRs with VLs were classified as SICCT test-positive). Untested cattle from eligible herds that had triggered a breakdown following detection of culture-positive bTB lesions during routine slaughterhouse surveillance were excluded, as no information on the SICCT test result was available for such animals.

In GB, SICCT test results can be interpreted using two different cut-offs for reactors, known as *standard* or *severe* interpretation [7]. For normal surveillance testing and in any bTB breakdowns without post-mortem evidence of *M. bovis* infection, so-called *unconfirmed* breakdowns, a standard interpretation is used whereby cattle with a positive *M. bovis* tuberculin reaction >4 mm greater than a positive or negative avian tuberculin reaction are classified as reactors, and those with a *M. bovis* tuberculin reaction exceeding the avian reaction by 1 to 4 mm are deemed IRs. In breakdowns with post-mortem or laboratory evidence of infection, so-called *confirmed* breakdowns, a severe interpretation is used whereby cattle are removed if they have a positive bovine reaction that exceeds the avian reaction by more than 2 mm (or where a positive bovine and a negative avian reaction are measured). Consequently, the standard interpretation IRs with the strongest bovine-avian tuberculin reaction differences (i.e. more than 2 mm and up to 4 mm) are culled as reactors under the severe interpretation of the SICCT test. Severe interpretation is therefore applied in an effort to accelerate the detection of infected cattle by increasing the sensitivity at the expense of the specificity.

On infected holdings that undergo depopulation due to bTB, the vast majority of SICCT tests will have been interpreted at severe interpretation. The overall rSe for this interpretation was calculated using a logistic regression model incorporating an individual herd-level effect to account for between-herd heterogeneity, such that:

$$\ln\left(\frac{\pi_i}{1-\pi_i}\right)=\beta_0+\gamma_i \quad (1)$$

Where $\pi_i$ is the true unknown relative sensitivity for herd $i$ where $i=1,\ldots,n_H$. We fit this model in a Bayesian framework, and in order to complete the specification we give the intercept parameter $\beta_0$ a vague N(0,100) prior distribution, and the herd-level effects, $\gamma$, $N(0,1/\tau)$ prior distributions with the precision $\tau \sim G(1,1)$. We note that this hierarchical model is analogous to a random intercepts model in a frequentist framework (though we avoid the use of this terminology since in the Bayesian framework all parameters are considered random). For comparison, we fit the model with and without the herd-level effects.

For each herd, the observed relative sensitivity for herd $i$, is $p_i=\frac{a}{a+c}$, where $a$ is the number of animals that were SICCT test-positive *and* had VLs post-mortem, and $a+c$ is the total number of animals that had VLs post-mortem. The overall mean rSe is then $\frac{e^{\beta_0}}{1+e^{\beta_0}}$. Fitting the model with the herd-level effects allows assessment of the degree of between-herd heterogeneity. For comparison, the sensitivity can then be recalculated without the herd-level effects. In addition, the rSe can be calculated using a *standard* SICCT test interpretation. IR cattle that were deemed reactors under the severe interpretation of the test were not classified as reactors under standard interpretation.

The models were fitted using Markov chain Monte Carlo (MCMC), and in each case a burn-in of 2500 iterations was used, followed by 10000 updates with the posterior distributions thinned to return 1000 samples. The overall (mean) estimate and 95% credible interval (CI) are reported to 2 significant figures (s.f.).

### Culture Data

Although culture is not routinely conducted for every animal in a breakdown, on those occasions where culture *and* VL/NVL data were available for >90% animals slaughtered in a herd, the rSe of VL detection was measured against a reference standard of culture positivity using equation (1) as before. This was possible in only two herds.

Analyses were carried out using the R statistical language [36], except the fitting of the Bayesian model which was conducted in WinBUGS [37] using the R2WinBUGS package [38].

## Supporting Information

**Table S1 Descriptive data of herds identified as whole herd slaughters.**

**Table S2 Data for calculating Single Intradermal Comparative Cervical Tuberculin (SICCT) test sensitivity at the *severe* interpretation.**

**Table S3 Data for calculating Single Intradermal Comparative Cervical Tuberculin (SICCT) test sensitivity at the *standard* interpretation.**

**Table S4 Sensitivity analysis assuming extreme cases for slaughterhouse animals that were not tested. Values are posterior means (95% credible intervals).**

## Acknowledgments

The authors are grateful to the Animal Health and Veterinary Laboratories Agency for compilation and provision of VetNet data and for systematic recording of bTB test and post-mortem examination data on VetNet.

## Author Contributions

Conceived and designed the experiments: KK RR-D RC AVG RSC-H AJCK APM RGH CAD JLNW TJM. Performed the experiments: KK TJM JLNW. Analyzed the data: KK TJM JLNW. Wrote the paper: KK RdlR-D RC AVG RSC-H AJKC APM RGH CAD JLNW TJM. Provided detailed whole-herd slaughter data: RC APM RdlR-D.

## References

1. Bovine TB Advisory Group (2009) Bovine Tuberculosis in England: Towards Eradication. Final Report of the Bovine TB Advisory Group. http://www.defra.gov.uk/foodfarm/farmanimal/diseases/atoz/tb/documents/tbag-finalreport.pdf [Accessed 15th October 2009].
2. Defra (2011) Bovine TB Eradication Programme for England. Available from: http://www.defra.gov.uk/publications/files/pb13601-bovinetb-eradication-programme-110719.pdf. [Accessed 23 June 2012].
3. Office International des Epizooties (2009) O.I.E. Terrestrial Manual. Chapter 2.4.7. Bovine tuberculosis. Available from: http://www.oie.int/fileadmin/Home/eng/Health_standards/tahm/2.04.07_BOVINE_TB.pdf. [Accessed 1st April 2011].
4. Reviriego Gordejo FJ, Vermeersch JP (2006) Towards eradication of bovine tuberculosis in the European Union. Vet Microbiol 112(2–4): 101–109.
5. Radunz B (2006) Surveillance and risk management during the latter stages of eradication: experiences from Australia. Vet Microbiol 112(2–4): 283–290.
6. Buddle BM, Livingstone PG, de Lisle GW (2009) Advances in ante-mortem diagnosis of tuberculosis in cattle. N Z Vet J 57(4): 173–180.
7. Defra (2008) Bovine TB: The tuberculin skin test. Available from: http://www.archive.defra.gov.uk/foodfarm/farmanimal/diseases/atoz/tb/control/tuberculin.htm. [Accessed 7th April 2011].
8. de la Rua-Domenech R, Goodchild AT, Vordermeier HM, Hewinson RG, Christiansen KH, et al. (2006) Ante mortem diagnosis of tuberculosis in cattle: a review of the tuberculin tests, gamma-interferon assay and other ancillary diagnostic techniques. Res Vet Sci 81(2): 190–210.
9. Monaghan ML, Doherty ML, Collins JD, Kazda JF, Quinn PJ (1994) The tuberculin test. Vet Microbiol 40(1–2): 111–124.
10. Defra (2009) Bovine TB: Gamma interferon. Available from: http://www.archive.defra.gov.uk/foodfarm/farmanimal/diseases/atoz/tb/control/gamma.htm. [Accessed 7th April 2011].
11. Costello E, Egan JW, Quigley FC, O'Reilly PF (1997) Performance of the single intradermal comparative tuberculin test in identifying cattle with tuberculous lesions in Irish herds. Vet Rec 141(9): 222–224.
12. Doherty ML, Monaghan ML, Bassett HF, Quinn PJ (1995) Effect of a recent injection of purified protein derivative on diagnostic tests for tuberculosis in cattle infected with *Mycobacterium bovis*. Res Vet Sci 58(3): 217–221.
13. Lesslie IW, Hebert CN, Burn KJ, MacClancy BN, Donnelly WJC (1975) Comparison of the specificity of human and *M. bovis tuberculin* PPD for testing cattle [Article 3: National trial in GB]. Vet Rec 96: 332–341.
14. O'Reilly LM (1986) Field trials to determine a suitable injection dose of bovine PPD tuberculin for the diagnosis of bovine tuberculosis in naturally infected cattle. Dev Biol Stand 58 (Pt B): 695–703.
15. O'Reilly LM, MacClancy BN (1975) A Comparison of the Accuracy of a Human and a *M. bovis tuberculin* PPD for Testing Cattle with a Comparative-Cervical Test. Irish Veterinary Journal 29: 63–70.
16. Kerr WR, Lamont HG, McGirr JL (1946) Studies on Tuberculin Sensitivity in the Bovine. Vet Rec 58: 443–446.
17. Kerr WR, Lamont HG, McGirr JL (1949) Further studies on Tuberculin Sensitivity in the Bovine. Vet Rec 61: 466–475.
18. Paterson AB, Stuart P, Leslie IW, Leech FB (1958) The use of tests on slaughterhouse cattle for estimating relative potencies of tuberculin and the calculation of discrimination tests. J Hyg 56: 1–18.
19. de Jong H, Ekdahl MO (1969) Evaluation of a number of ancillary tuberculin tests in cattle. N Z Vet J 17(11): 213–226.
20. Veterinary Laboratories Agency (2007) Tuberculin Usage. Report for 1 January 2005 to 31 March 2007.
21. Animal Health and Veterinary Laboratories Agency (2011) Project SB4500 (Bovine tuberculosis: Infection status in cattle annual surveillance report). Report for the period January to December 2010. Available from: http://www.defra.gov.uk/animal-diseases/files/Surveillance-annual-report-2010.pdf. [Accessed 23 June 2012].
22. Cassidy JP, Bryson DG, Pollock JM, Evans RT, Forster F, et al. (1999) Lesions in cattle exposed to *Mycobacterium bovis*-inoculated calves. J Comp Pathol 121(4): 321–337.
23. Enticott G (2010) Everyday ethics. Ethical Dilemmas in Practice. TB testing protocol. In Practice 32: 318–319.
24. DNV Consulting (2006) Review of TB testing procedures: Report for Defra and the Welsh Assembly Government. Report no.: 22115204–02.
25. Asselbergs M, Coulson N (2010) Comments on the dilemma in the July/August issue: 'TB testing protocol'. In Practice 32: 414–415.
26. Radunz BL, Lepper AW (1985) Suppression of skin reactivity to *M. bovis* tuberculin in repeat tests. Aust Vet J 62(6): 191–194.
27. EU Council (1964) EU Council Directive 64/432/EEC. Available from: http://eur-lex.europa.eu/LexUriServ/LexUriServ.do?uri=CELEX:31964L0432:EN:HTML [Accessed 15th October 2009].
28. Coad M, Clifford D, Rhodes SG, Hewinson RG, Vordermeier HM, et al. (2010) Repeat tuberculin skin testing leads to desensitisation in naturally infected tuberculous cattle which is associated with elevated interleukin-10 and decreased interleukin-1 beta responses. Vet Res 41(2): 14.
29. Karolemeas K, McKinley TJ, Clifton-Hadley RS, Goodchild AV, Mitchell A, et al. (2011) Recurrence of bovine tuberculosis breakdowns in Great Britain: Risk factors and prediction. Prev Vet Med 102: 22–29.
30. Defra (2005) Report by the Independent Scientific Review Group on TB in cattle and badgers. Available from: http://www.archive.defra.gov.uk/foodfarm/farmanimal/diseases/atoz/tb/publications/krebs.htm. [Accessed 7th April 2011].
31. EU Commission (2009) EU Commission Decision 2009/761/EC. Available from: http://eur-lex.europa.eu/LexUriServ/LexUriServ.do?uri=OJ:L:2009:271:0034:01:EN:HTML [Accessed 1st April 2011].

32. Animal Health and Veterinary Laboratories Agency (2012). TB Parish Testing Interval. Available from: http://animalhealth.defra.gov.uk/managing-disease/notifiable-disease/bovine-tb/tb-testing-intervals.asp. [Accessed 26 June 2012].
33. Defra (2009) Gamma Interferon diagnostic blood test for bovine tuberculosis. A Review of the Gamma Interferon testing policy for tuberculosis in cattle. Available from: http://archive.defra.gov.uk/foodfarm/farmanimal/diseases/atoz/tb/documents/gamma-review090710.pdf. [Accessed 26 June 2012.].
34. Defra (2010) Bovine TB Eradication Group for England. Available from: http://archive.defra.gov.uk/foodfarm/farmanimal/diseases/atoz/tb/partnership/eradication-group/index.htm. [Accessed 1st April 2011].
35. Defra (2008) Bovine TB: Data. Available from: http://www.archive.defra.gov.uk/foodfarm/farmanimal/diseases/atoz/tb/research/tbdata.htm. [Accessed 7th April 2011].
36. R Development Core Team (2012) R: A language and environment for statistical computing. Available from: http://www.R-project.org/. [Accessed 1st August 2012].
37. Lunn DJ, Thomas A, Best N, Spiegelhalter D (2000) WinBUGS. A Bayesian modelling framework: concepts, structure and extensibility. Statistics and Computing 10: 325–327. Available from: http://www.mrc-bsu.cam.ac.uk/bugs/winbugs/contents.shtml. [Accessed 1st August 2012].
38. Sturtz S, Ligges U, Gelman A (2005) R2WinBUGS: A Package for Running WinBUGS from R. Journal of Statistical Software 12(3): 1–16.

# 17

# Investigating the Role of State Permitting and Agriculture Agencies in Addressing Public Health Concerns Related to Industrial Food Animal Production

**Jillian P. Fry[1,2]*, Linnea I. Laestadius[1,3], Clare Grechis[1,4], Keeve E. Nachman[1,2,4], Roni A. Neff[1,2,4]**

**1** Center for a Livable Future, Johns Hopkins University, Baltimore, Maryland, United States of America, **2** Department of Environmental Health Sciences, Johns Hopkins Bloomberg School of Public Health, Baltimore, Maryland, United States of America, **3** Joseph J. Zilber School of Public Health, University of Wisconsin-Milwaukee, Milwaukee, Wisconsin, United States of America, **4** Department of Health Policy and Management, Johns Hopkins Bloomberg School of Public Health, Baltimore, Maryland, United States of America

## Abstract

***Objectives:*** Industrial food animal production (IFAP) operations adversely impact environmental public health through air, water, and soil contamination. We sought to determine how state permitting and agriculture agencies respond to these public health concerns.

***Methods:*** We conducted semi-structured qualitative interviews with staff at 12 state agencies in seven states, which were chosen based on high numbers or rapid increase of IFAP operations. The interviews served to gather information regarding agency involvement in regulating IFAP operations, the frequency and type of contacts received about public health concerns, how the agency responds to such contacts, and barriers to additional involvement.

***Results:*** Permitting and agriculture agencies' responses to health-based IFAP concerns are constrained by significant barriers including narrow regulations, a lack of public health expertise within the agencies, and limited resources.

***Conclusions:*** State agencies with jurisdiction over IFAP operations are unable to adequately address relevant public health concerns due to multiple factors. Combining these results with previously published findings on barriers facing local and state health departments in the same states reveals significant gaps between these agencies regarding public health and IFAP. There is a clear need for regulations to protect public health and for public health professionals to provide complementary expertise to agencies responsible for regulating IFAP operations.

**Editor:** Suminori Akiba, Kagoshima University Graduate School of Medical and Dental Sciences, Japan

**Funding:** Salary support for all authors was provided by a gift agreement from the GRACE Communications Foundation (www.gracelinks.org). The funders had no role in study design, data collection and analysis, decision to publish, or preparation of the manuscript.

**Competing Interests:** The authors have declared that no competing interests exist.

* E-mail: jfry@jhsph.edu

## Introduction

A dramatic series of changes in the landscape of animal agriculture have taken place over the past seventy years, accompanied by multiple public health concerns [1]. Small farms raising a diversity of crops and food animals have increasingly and steadily given way to a model of industrial food animal production (IFAP) that raises large numbers of animals in concentrated quarters, often on farms whose only crops are animal feed grown on spray fields. Hog production, for example, has shifted significantly since the 1980s. As Figure 1 illustrates, from 1987 to 2007 the number of hog operations in the US decreased from over 320,000 to about 75,000 [2]. Mid-aggregate enterprise size is a measurement of farm production that shows the point where half of a product comes from larger farms, and half from smaller farms. In 1987, the mid-aggregate enterprise size of US hog farms was 1,200 hogs, according to the USDA. By 2007, this measurement increased to 30,000, reflecting a 2,400% increase [3], as shown in Figure 1. The mid-aggregate enterprise size of dairy and broiler (chicken meat) production operations increased by 613% and 127%, respectively, during the same interval [3]). There has also been geographic concentration in production. In 2007, 75% of all U.S. hogs were raised in only 220 counties, down from 508 counties in 1987 [3]. These changes in the model, methods, and system within which animals are produced for food pose both a regulatory challenge and clear concerns for environmental health and public health more generally.

Research linking IFAP to public health concerns and impacts continues to increase. In addition to posing respiratory health risks to those residing near operations [4]–[8] due to emissions that include hydrogen sulfide [9], particulate matter [9], endotoxins [10], ammonia [11], allergens [12], and volatile organic compounds [13], [14], odor generated by IFAP operations and spray fields has been associated with a broad range of health problems. Public access to information regarding hazardous airborne releases from IFAP operations is hindered due to exemptions in federal laws that require disclosure of such releases [15], despite research linking chronic exposure to odors from IFAP

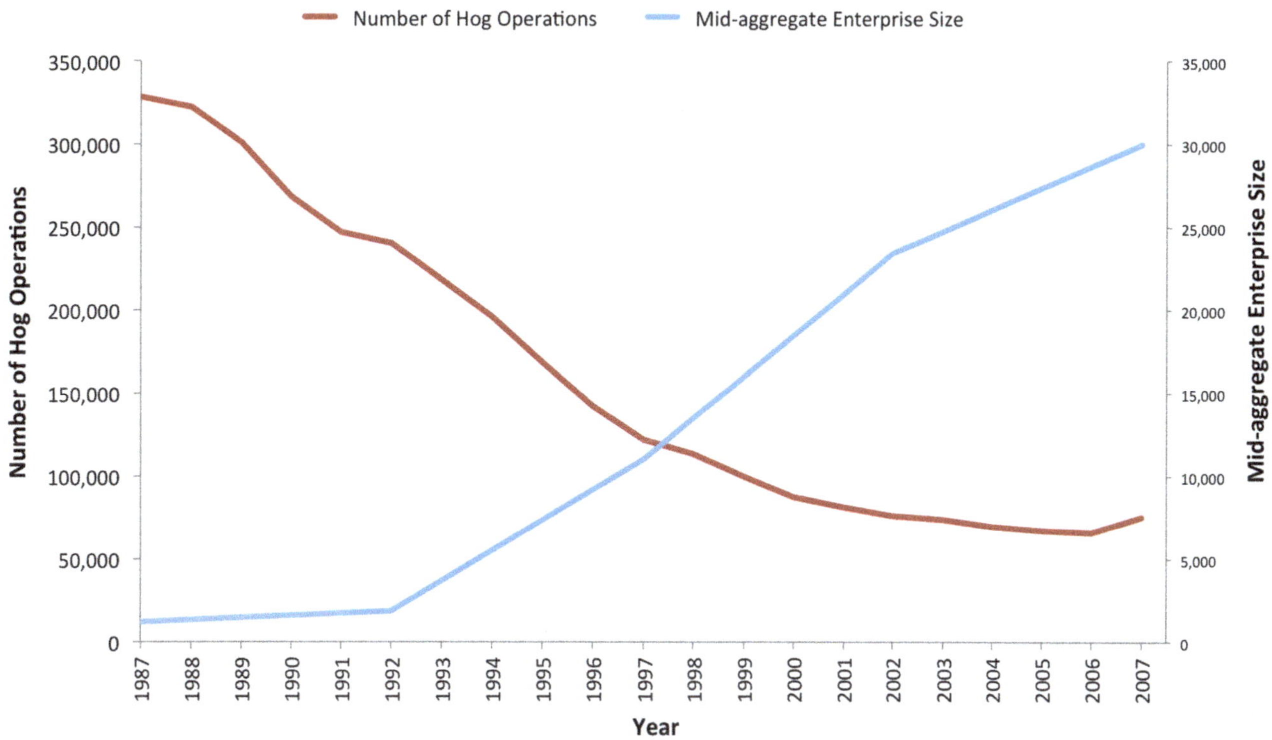


**Figure 1. Hog production in the US: Number of operations and mid-aggregate enterprise size; 1987–2007.** Data from the United States Department of Agriculture (USDA) shows the decline in number of hog operations and growth in operation size from 1987 to 2007. Sources: The Changing Organization of U.S. Farming, USDA Economic Research Service (http://www.ers.usda.gov/amber-waves/2011-december/changing-farming-practices.aspx#.UgqVWOu9xEo) and USDA Census of Agriculture QuickStats, various years (http://quickstats.nass.usda.gov/?source_desc=CENSUS).

to headaches, nausea, upset stomach, mood disorders, high blood pressure, and sleep problems [16]–[20]. Additionally, there is growing evidence that livestock can transmit methicillin-resistant *Staphylococcus aureus* (MRSA) to humans [21]–[23]. Rural water supplies are also at risk, as IFAP-generated animal waste contaminants, including nitrates, pathogens, pharmaceuticals, metals, and hormones, can leach into ground water [24], [25]. All of these concerns may be compounded by the fact that IFAP operations are disproportionately located in low-income communities with high-percentages of minority populations [26]–[28], which are more likely to experience limited political power [29] and barriers to healthcare access [30].

Despite health risks posed by these operations, regulation of IFAP is limited and characterized by a patchwork of different regulatory approaches from state to state. Under the Clean Water Act (CWA), IFAP operations that are designated as Concentrated Animal Feeding Operations (CAFOs) are required to obtain National Pollution Discharge Elimination System (NPDES) permits in order to discharge into U.S. waterways. Designation as a CAFO is determined by size and potential to pollute the surrounding environment. States are granted authority to determine some requirements, issue NPDES permits, monitor compliance, and impose penalties by the U.S. Environmental Protection Agency if the state adopts federal requirements into law. As of 2008, 44 states had permitting authority for CAFOs [31]. Animal operations that fail to meet the technical definition of a CAFO may still cause numerous concerns, thus the more inclusive term IFAP is utilized here.

The state agency responsible for permitting IFAP operations varies among states, with almost all delegating the responsibility to Departments of Environmental Protection/Management, Natural Resources, or Agriculture [31]. The organization and jurisdiction of state agencies responsible for regulations pertaining to the environment and natural resources varies significantly by state [32]; so defining the broader roles and responsibilities of these agencies is difficult. We focused on environmental agencies that have been given authority by the EPA to implement the NPDES permitting program for CAFOs, and we refer to these agencies as "permitting agencies". Responsibilities of Departments of Agriculture vary less by state, and in general they include promotion and regulation of agriculture in the state, as well as other common responsibilities such as conservation and farmland protection [33]. Regulatory authority is shared among agencies in some states, and the transfer of regulatory authority to Departments of Agriculture has been favored by industry [34]. Some states have developed additional regulations for smaller IFAP operations that do not meet the size threshold to be considered a CAFO. Additionally, states vary regarding resources available to monitor and enforce regulations, and potential penalties for violations [31], [35]. Common across most states, however, is delegating the permitting to an agency without a primary mandate to address public health [31], raising concerns that public health issues may not be adequately monitored or addressed by the agencies tasked with regulating IFAP operations.

A 2008 report by the National Conference on State Legislatures provided an overview of agencies in each state responsible for regulating IFAP [31], but the report did not address how state agencies respond to community health concerns arising from IFAP operations. We conducted interviews with state permitting agency and agriculture department staff members to determine the ability of these agencies to: 1) monitor IFAP operations and enforce current regulations, and 2) respond to citizen health concerns arising from IFAP operations. This manuscript represents the second part of a study examining how state and local agencies address health concerns associated with IFAP. Our previously published research [36] found that despite the evidence linking IFAP to public health concerns, state and local health departments play a limited role in addressing health issues linked to IFAP operations. Health departments report a lack of formal jurisdiction over IFAP operations as the primary barrier to regulatory response. As a result, we found that many health departments refer IFAP concerns, including community member concerns related to health, to departments of agriculture and permitting agencies.

Assessing the roles of the main agencies with jurisdiction over issues relevant to public health is essential to a full understanding of how health concerns, including those related to IFAP, are addressed. In fact, defining inter-organizational relationships, evaluating capacity as financial priorities shift, and identifying gaps in services or jurisdiction are key priorities in public health systems research [37], [38]. We aimed not only to characterize the current role, capacity, and barriers of the primary agencies potentially involved with IFAP and public health, but also to highlight gaps in agency responsibilities that would not be identified if only one type of agency were studied.

## Methods

Given the absence of research on the role of state agriculture and permitting agencies in responding to health concerns stemming from IFAP, an exploratory qualitative approach was chosen for data collection and analysis. See Fry (2013) for further details on the initial study, which examined state and county health department responses [36]. An inductive qualitative research approach was used, with a focus on flexibility rather than allegiance to any specific theoretical approach [39], [40]. Important aspects of study design, data collection, and data analysis are described below.

We used a purposive sampling strategy to select the states included in this study. First, all U.S. counties were ranked by USDA Census of Agriculture hog inventory data in two ways: 1) 2007 county hog inventory of operations with 1,000 or more hogs, and 2) increase in hog inventory of 1,000+ head operations between 2002 and 2007. It was anticipated that many hogs produced in intensive settings or large increases in hog numbers could lead community members or others to request that agencies take action to address health concerns. We also sought to examine settings where hog operations were near residential areas by ranking the top sixty counties from each list by population density using 2000 Census data. We then selected the top fifteen counties by population density from the two lists. These counties were located in eight states. Although states were chosen based on hog production and census data, the interviews asked about IFAP in general, and did not focus solely on hog operations.

We contacted state agriculture and permitting agency employees in the eight states to perform semi-structured interviews. Both types of agencies were included in order to understand their roles and to identify potential gaps in agency responses to health concerns. Permitting agencies and some agriculture departments were identified through the National Conference on State Legislatures' Survey of State Policies on CAFOs [31]. Through web-based searching and investigation, we confirmed the information from the report and collected contact information for additional agriculture departments. Most commonly, permitting responsibilities fell under the authority of departments of environmental protection or natural resources. We contacted staff members who worked on livestock permitting and/or livestock production, identified using agency websites. The specific roles and titles of interviewees varied across states and included, among others, livestock permitting and water quality specialists, state veterinarians, and community-relations liaisons. In some cases, staff we contacted referred us to others in their agencies who could better answer our questions.

All interviews were conducted over the telephone by JF or LL, with a note-taker present on each call to document responses. Interviews were not audio recorded. The questionnaires included mostly open-ended and some closed-ended questions (See Appendix S1 for survey instruments), and some follow-up or clarifying questions were asked based on participants' responses. We also specifically queried permitting agency staff members about their states' permitting regulations. Interview questions were developed based on our prior familiarity with the topic, which is drawn from extensive background reading, original research, and interactions with community members impacted by IFAP operations. We did not use the terms, IFAP or CAFO, in the interviews. Before beginning each interview, we read participants a confidentiality statement. The Johns Hopkins Bloomberg School of Public Health Institutional Review Board (IRB) determined the study was exempt from IRB oversight and did not require an informed consent process.

After data collection was complete, notes from three interviews were double coded through an inductive coding process using *HyperRESEARCH 3.0.3* (ResearchWare, Randolph, MA). Codes were jointly discussed to develop a uniform codebook, which was then applied to the remaining interviews. As this is a qualitative study with a small, purposive sample, we provide limited numeric information to avoid implying that the findings are generalizable to a larger population [41]. Instead, we describe important themes identified in the interviews that allow better understanding of the situation in states with significant industrial hog production.

## Results

We conducted telephone interviews with staff members from permitting agencies in seven states and departments of agriculture in five states. In one state, both the permitting agency and department of agriculture declined to participate, and agriculture department staff in two other states also declined. In total, twelve interviews in seven states were conducted between November 2010 and October 2011. Most of the interviews lasted between 35 and 45 minutes.

### Regulatory Oversight

In nearly all instances, state-level environment or natural resource departments managed NPDES permits. One state delegated permitting authority to counties. In two states, departments of agriculture shared regulatory authority with other departments, and in another state the permitting agency held only narrow authority over NPDES permitting with many activities transferred to the agriculture department in recent years.

As specified in the federal Clean Water Act, NPDES permitting requirements apply only to certain IFAP operations that match the defined criteria for CAFOs. States also reported additional non-NPDES regulatory measures. For example, some states require permits for IFAP operations that do not match the federal CAFO definition, although requiring less information for approval than for CAFOs. Some permitting agency staff members also said they were trying to develop regulations that would strengthen reporting requirements for small/medium size operations, thereby bringing them in line with CAFO permitting requirements. One agriculture department staff member noted that, "Issues are more frequently from smaller farms, but people are quick to blame bigger farms".

We asked interviewees if their agencies had any setback/zoning policies related to human health concerns. While several counties and states require that IFAP operations be a certain distance away from property lines, wells, waterways, homes, churches, schools, and/or parks, many interviewees said the requirements are not necessarily based on health standards. One staff member said,

> "There are setbacks, but those are more for odor rather than health. This can be from residences or populated areas, it's just a form of odor management. There are some setbacks for wells/private water wells too, that have more direct relevance to public health issues."

Setbacks can reduce public health impacts of odor and poor air quality, but interviewees indicated that most setbacks are not based on evidence-based health standards specifically designed to limit exposure to gases and particulate matter emitted from IFAP operations.

One permitting agency staff member said two counties in the state had established more stringent requirements for setbacks through health ordinances. Two states reported having air quality standards that apply to IFAP operations, but in one state the standard only applies to new operations.

## Inspections

States varied significantly in the frequency of inspections of permitted CAFOs for compliance with regulations. One permitting agency staff member said that they "can only afford to [inspect] on a complaint basis" because they "don't have staff or money." By contrast, a staff member from another state's agriculture department indicated that they inspected CAFOs every six months. The staff member also stated, "before reforms a few years ago, some facilities had gone 25 years without inspection."

## Contacts and Concerns Reported

All interviewees noted that their agencies had been contacted by people concerned about issues associated with living or spending time near animal production operations and/or manure (i.e., manure storage or spray fields). Estimates from permitting agency staff ranged from 25 to 120 contacts per year, and agriculture agency staff estimates ranged from fewer than 5 times per year to an average of 5 times per week. Some interviewees said that calls were generally more frequent when new or expanding operations were proposed and that they believed that calls had decreased due to the implementation of additional regulations. The absence of regulations pertaining to common community concerns with IFAP also appeared to reduce contacts as people began to learn what agencies were able to do in response to concerns. One permitting agency staff member said there "used to be more calls about odor, but there are no odor regulations, so there is nothing we can do about it; the public learned there's no point in calling about odor complaints."

We read to interviewees a list of types of people who may have contacted their agency about public health concerns (Appendix S1). All participants said they had been contacted by individuals describing their own concerns, and most also said they had been contacted by members of an organized campaign regarding CAFOs/IFAP. Few agencies were contacted by health care providers.

We also questioned staff members about topics related to animal agriculture that their agencies may have ever been contacted about. A list of topics was provided, and all interviewees answered affirmatively regarding odor. Table 1 presents a ranking based on the number of interviewees reporting their agency had been contacted about the issue as it relates to animal agriculture. Rankings varied little between permitting and agriculture agency staff.

## Response to Citizen Concerns

Permitting agency staff reported a variety of responses depending on the concern; they often mentioned gathering additional information and checking the validity of a concern by contacting the person reporting the issue or by investigating the issue. Inspection activities were frequently delegated to regional field staff or county conservation district staff. Permitting agencies also made referrals to other agencies including departments of agriculture and health, as well as state Farm Bureaus (agriculture trade group), when issues were beyond their jurisdictions. Staff indicated that at times they would contact the person who had raised the concern to let them know that they could not address it. One interviewee said: "If the problem is not covered under the agency, it might be a phone call or email to let people know why we can't address their concerns. Water issues are our primary jurisdiction. There are no state/federal regulations over air emissions." The person was referring to a lack of air quality regulations in their state. As mentioned above, only two states reported that they were able to address some air quality concerns under their current regulatory authority. A staff member in a state with an air quality standard that applies to all IFAP operations (i.e., sites that existed when the regulation was adopted and new sites) noted that if there were a "major complaint" they would use their equipment to take air samples.

Some agency staff reported reaching out to producers directly to try to resolve issues. One interviewee described an agreement with the state Farm Bureau in which the agriculture group would send someone to talk to the producer "farmer to farmer" in the event of a reported concern. They thought that producers are more willing to speak to another farmer than to an agency staff member, and that this process results in issues being resolved faster. Another staff member described talking to producers initially to see if rules are being followed. Then they explain the concern and determine if it can be resolved without further agency action. Some concerns could be related to a regulatory violation, but few agency personnel described a process for enforcement in response to community member concerns. Based on interviewees' responses,

**Table 1.** Topics of concern ranked in descending order according to number of agency staff stating people have ever contacted the agency about the issue.

| |
|---|
| Odor |
| Respiratory health |
| Ground water quality/contaminated well |
| Violations of regulations |
| Waste getting on property |
| General health |
| Traffic |

this phenomenon appeared to reflect factors including limited jurisdiction over common concerns and preference for voluntary and informal solutions to violations of regulations. All permitting agency interviewees stated that records were maintained of reported concerns, but it was not clear if those records always included concerns that were not formally investigated.

Agriculture department staff members' responses to concerns varied due to the state-by-state differences in agriculture agencies' jurisdiction over IFAP. In general, staff said that they investigate and/or refer to relevant agencies in response to concerns. Staff members from several states said they would refer concerns to the state permitting agency, either right away or after learning more about the situation. Most department of agriculture interviewees indicated that they had limited authority over many IFAP concerns and said the state permitting agencies would be more involved. Issues related to manure or dead animals, however, were more likely to be dealt with by agriculture departments. A small number of interviewees said they might contact the state health department in response to a call from a concerned citizen, although one said specifically that they would not. As with permitting agencies, several department of agriculture staff members said they try to work directly with producers to resolve issues. No interviewees in agriculture departments talked about enforcement or penalties that would be imposed by their department after investigating a reported concern. Most staff members said no records were kept of concerns; one said they kept records and another said records are kept only if they investigate the concern.

## Potential Role of Health Departments

The majority of permitting agency staff members and all agriculture department personnel interviewed thought that health departments should play a role in resolving citizen concerns related to IFAP. Interviewees indicated that they believed health departments should respond to health related concerns, work on air quality issues, provide technical assistance, collect and disseminate relevant data, and/or participate and provide input when regulations are under consideration. A small number of permitting agency and department of agriculture interviewees were not sure what the role of health departments should be, since health departments did not have regulatory mandates to enforce. Some staff stressed the importance of health department involvement due to permitting and agriculture agencies' lack of health expertise. One permitting agency staff member stated, "If anyone's going to [address health issues], it would have to be the health departments. From our perspective, we don't really have expertise in that area," and another said, "Obviously the health department has more expertise in the health area than us." Some agriculture agency staff members thought that involving health departments would provide additional assurance to the public that departments of agriculture are unable to provide on health issues.

## Additional Engagement

**Education.** We asked interviewees if they perform any health education activities related to "animal production farms," and the majority of permitting agency staff members said that they present information to producer/farmer groups in order to keep them informed on current regulations and requirements. A few talked about presenting to community groups when asked to do so, and that such presentations also addressed current regulatory requirements. Agriculture agency staff members also said that they provided information to farmers/producers and agriculture groups on current regulations, and at times presented information to community groups. No staff member, in permitting or agriculture agencies, said that they provided information regarding potential health issues related to IFAP.

**Data collection.** In response to a query about collection of environmental monitoring or health data, most permitting agency participants said that their agencies did little or no routine data collection. Data collection was usually performed in response to a concern, and generally involved surface water and well testing. One interviewee said, "If we have a well water complaint we'll do sampling case by case. We need to be sure there is good evidence the well is tainted or everyone would want their well water tested at our expense." One staff member said that some permits they issue require water monitoring and they do some monitoring at inspections, and that the data they collect is used for internal purposes only. One permitting staff member said they collected data on air quality in the past because they were directed to do so by the state legislature and that that monitoring led to a rulemaking process for farm emissions. The legislature, however, ultimately prevented the rule from being finalized and implemented. No agriculture department staff member reported that they collected environment or health data, but one interviewee said their department was working with a university to look into air quality issues.

**Contact with community/environment groups.** All permitting agency personnel reported some contact with organizations or groups of citizens that work to address local animal production farm issues. A few interviewees said their contact with community groups primarily consisted of receiving comments when new operations were proposed or receiving repeated requests for increased monitoring and stricter regulations. In one state, citizens petitioned the U.S. EPA to have the permitting agency's delegated authority over CAFOs revoked because they believed the agency was not doing enough to regulate CAFOs in the state. At the time of the interview, the action was still pending and six new staff had been hired in response to the petition. Another interviewee said that environment groups serve as "watch dogs" and are "good partners from that standpoint" because they look for problems, conduct their own monitoring, and quickly contact the agency about problems so they can investigate. One staff member spoke about serving on an advisory committee with community representatives and another said a group working to revise their state's manure manual, a guidance document for producers, includes citizen groups.

Most agriculture department staff said community/environment groups contact them, and some said that they have worked with them to address issues of concern. One interviewee said there are fewer groups now due to improved regulations in their state, and another said contacts from these groups increase in frequency when new operations are proposed.

## Barriers and Needs

While most interviewees appeared to consider their agency's current level of engagement with IFAP to be appropriate, many staff members did describe a number of barriers preventing them from implementing their current oversight efforts more fully and from expanding their efforts to monitor IFAP operations and respond to concerns.

**Barriers.** Permitting agency staff described the main barriers affecting their oversight of IFAP as limited budgets, staff size, and political factors. Some also mentioned that recent budget and staff reductions due to the economic downturn have affected their departments' ability to meet their regulatory obligations. Interviewees spoke about the effects of budget cuts on inspection frequency and farm visits, resources available for environmental monitoring, and ability of staff to conduct or attend in-person meetings and provide technical assistance. One staff member said "face to face meetings are much more effective when working with agriculture, and staff cutbacks are not good for that," and another said that they "would like to spend more time working with producers so they know the regulations and what causes water problems." Some interviewees said political factors have prevented them from more effectively regulating IFAP. For example, one permitting agency staff member explained that they had attempted to begin surface water monitoring at farms in response to concerns from environmental groups, but that financial and political issues had prevented this from moving forward. Only one agriculture agency staff member responded to an open-ended question regarding barriers with concerns about the agency's budget, but when agriculture departments were presented with a list of potential needs (Table 2), all staff members responded affirmatively regarding the need for increased funding.

The threat of legal action was an additional difficulty and barrier brought up by some permitting agency staff members and one agriculture department representative. The agriculture department interviewee said they "feel uncomfortable responding when a group brings in a lawyer" because they are not prepared to respond to legal action against the agency. The permitting agency participants who mentioned lawsuits said specifically that federal regulations are unclear about who needs a permit, and that they worry about lawsuits stemming from the lack of clarity.

**Needs.** We asked staff an open-ended question about items that could increase their agencies' effectiveness when addressing IFAP-related issues and further prompted them with a list of potential items (Appendix S1). The items most frequently identified as needs by interviewees are listed in Table 2. Some wording was adapted slightly in order to be more appropriate for and relevant to each agency. The most commonly expressed needs/resource improvements identified by permitting agency staff were: increased funding, more staff for environmental health, and educational materials for distribution to producers or the general public.

Even though most agriculture agency staff members did not mention funding as a barrier in an open-ended question about barriers/needs, it was the only item on the list that all agriculture interviewees said would be helpful to their department. Other needs identified were: updated information from researchers regarding the health effects of concern and more staff.

## Discussion

Our study reveals that sampled state permitting and agriculture agencies have taken limited actions to prevent and/or respond to public health concerns arising from IFAP operations. The main barriers identified that prevent further engagement include narrow or inadequate regulations, a lack of public health expertise within the agencies, and limited resources. There was widespread agreement among permitting and agriculture agency interviewees that health departments (HDs) should play a role in regulating IFAP operations, partly due to their own agencies' limited mandates and available expertise in public health. Yet previously published findings show limited involvement by local and state HDs due to political barriers and a lack of jurisdiction, expertise, and resources [36].

These results indicate a fragmented system to protect public health where no agency has ownership of monitoring or addressing the impact of IFAP on people's health. In short, HDs generally lack jurisdiction over IFAP operations [36] and permitting and agriculture agencies generally lack jurisdiction over and the capacity to address public health concerns. A

**Table 2.** Needs indicated by interviewees from list (Descending order).

| |
|---|
| **Permitting Agency Staff** |
| More staff dedicated to environmental health |
| Increased funding for environmental health |
| Educational materials for distribution |
| Funding specifically for animal production activities |
| Training for staff on public health issues relevant to animal production farms |
| Different political climate |
| Information on health effects of concern |
| Environmental quality tracking tools |
| Clearer federal regulations/guidelines (item added during study in response to data collected; not asked of all interviewees) |
| Connections to experts (i.e., university researchers) |
| **Agriculture Department Staff** |
| Increased funding |
| Updated information from researchers on health effects of concern |
| More staff |
| Funding specifically for animal production activities |
| Educational materials for distribution |
| Training for staff on issues relevant to animal production farms |
| Different political climate |
| Connections to experts |
| Environmental quality tracking tools |

growing divide between environmental and public health agencies was identified in the 1990's as a trend that threatens public health protections [42]. Research has found that the main foci of environment agencies have shifted to permitting, enforcement, record keeping, and standard setting, and away from public health evaluations [43]. Our findings are consistent with these trends.

Ideally, a new role for HDs in responding to community concerns over IFAP operations would be defined by legislation aimed at remedying the lack of explicitly health-focused protections in current IFAP regulations. Unfortunately, this seems unlikely at present due to economic and political factors at the local, state, and federal levels [44]. In the absence of needed legislative and regulatory reforms, we suggest that partnerships between HDs and agriculture and permitting agencies could begin to improve the situation by including public health considerations in local and state-level decisions regarding the permitting, monitoring, and enforcement of regulations pertaining to IFAP operations. For example, greater agriculture and permitting agency collaboration with HDs could result in more comprehensive setback requirements for IFAP operations or the implementation of air quality standards for IFAP emissions linked to health concerns. Given the aforementioned support that several study participants voiced for greater involvement of HDs in IFAP regulation, agencies appear to be receptive to these partnerships. The drawback of this approach, as compared to new regulations, is that these relationships would have to be created one-by-one and maintained over time in the absence of regulations providing HDs with a defined role. We also suggest a potential approach whereby panels of experts on IFAP and public health could develop training programs and provide technical support to agencies interested in such partnerships in order to help foster their development and improve effectiveness. Future work should seek to examine evidence-based strategies that can inform and serve as models for these types of collaborations.

We did not seek to fully characterize how IFAP regulations are implemented by sampled agencies, but we are concerned that there is significant variability in inspection frequency among states. The inconsistency reflects varying requirements by state with no federal requirement for inspection frequency [31]. Further investigation is also needed of the practice of state agencies contacting farmer and agriculture trade groups for assistance when called about a concern. It is possible that this is an effective way to deal with minor issues; however, if problems that could result in fines and other regulatory action are routinely handled in an informal manner, the deterrent function of penalties for poor practices may be weakened and/or lost [45]. In addition, concerns about vulnerability to legal action stemming from a lack of clarity in federal regulations could cause agencies to be less aggressive in their enforcement of regulations. Additional research is needed to determine if this more informal approach to enforcement puts the environment and public health at greater risk than a formal approach to the enforcement of IFAP regulations.

### Strengths and Limitations

To our knowledge, this is the most in-depth study to date investigating how state agencies with jurisdiction over IFAP operations respond to health concerns. The value of the findings is amplified when combined with previously published results on the engagement of HDs with this issue. Including multiple agencies that could be involved with IFAP and public health–health, permitting, and agriculture agencies–allowed us to compile a more comprehensive profile of a system to protect public health that has substantial gaps. This approach was especially critical since previously published results showed that HDs routinely refer concerned citizens to permitting and agriculture agencies, with unknown outcomes for those referrals [36]. These findings may hold relevance for other issues characterized by problematic public health system gaps.

This study provides new information on a relatively unexplored topic; but the sample size was small, and the results cannot be interpreted as representing all agriculture and permitting agencies in the US. Also, we interviewed only one staff member per agency, and other staff members at the sampled agencies might have provided different responses to our questions. Finally, as with all studies of this nature, our sample reflects only those agencies that agreed to participate in the study.

## Conclusion

In light of steadily increasing evidence regarding the multifaceted impact of IFAP on the public's health, it is crucial to examine how regulatory agencies respond to concerns and to understand what factors encourage and restrict agency staff from addressing health issues. A fragmented regulatory approach, narrow regulations, a lack of health expertise among agency staff, and limited resources are barriers preventing effective responses by sampled state permitting and agriculture agencies, despite the fact that jurisdiction over IFAP lies with these agencies. These findings are particularly troubling in light of prior research indicating that HDs also face multiple barriers to engagement with IFAP. The human health implications of IFAP operations have thus largely fallen by the wayside from a regulatory perspective, with rural communities suffering the consequences.

Given the near absence of explicit public health protections in current IFAP regulations, and the findings about responses from government agencies when issues are brought to their attention, there is a clear need for a more comprehensive public health response to IFAP. New regulations giving HDs a formal role in regulating IFAP and/or requiring public health experts on staff at regulatory agencies would go a long way toward addressing current gaps in the system. Short of new regulations, many actions could be taken to encourage capacity building and partnering among agencies. Future efforts should determine best practices for establishing these types of partnerships and evaluate their effectiveness in addressing the current disconnect between environmental regulations, public health, and IFAP. That said, voluntary efforts should not be seen as a replacement for much needed regulatory reforms. The public health implications of IFAP are increasingly clear, and regulations should ensure proper monitoring, oversight, and response by government agencies to protect public health.

## Supporting Information

**Appendix S1 Permitting Agency and Department of Agriculture Staff Member Questionnaires.**

## Acknowledgments

We thank the permitting and agriculture agency staff members who participated in the study.

## Author Contributions

Conceived and designed the experiments: JF LL CG KN RN. Performed the experiments: JF LL CG KN RN. Analyzed the data: JF LL. Wrote the paper: JF LL KN RN.

## References


1. The Pew Commission on Industrial Farm Animal Production (2008) *Putting Meat on the Table: Industrial Farm Animal Production in America*. Pew Charitable Trusts and Johns Hopkins Bloomberg School of Public Health. Available: http://www.ncifap.org/_images/PCIFAPFin.pdf. Accessed 2013 April 12.
2. US Department of Agriculture. Census of Agriculture Quick Stats. Available: http://quickstats.nass.usda.gov/?source_desc = CENSUS. Accessed 2013 April 12.
3. O'Donoghue EJ, Hoppe RA, Banker DE, Ebel R, Fuglie K, et al. (2011) The Changing Organization of U.S. Farming. EIB-88. U.S. Department of Agriculture, Economic Research Service. Accessed 2013 April 12.
4. Schinasi L, Horton RA, Guidry VT, Wing S, Marshall SW, et al. (2011) Air Pollution, Lung Function, and Physical Symptoms in Communities Near Concentrated Swine Feeding Operations. Epidemiology 22: 208–215.
5. Thu K, Donham K, Ziegenhorn R, Reynolds S, Thorne P, et al. (1997) A control study of the physical and mental health of residents living near a large-scale swine operation. J Agric Saf Health 3: 13–26.
6. Merchant JA, Naleway AL, Svendsen ER, Kelly KM, Burmeister LF, et al. (2005) Asthma and farm exposures in a cohort of rural Iowa children. Environ Health Perspect 113: 350–356.
7. Radon K, Schulze A, Ehrenstein V, van Strien RT, Praml G, et al. (2007) Environmental exposure to confined animal feeding operations and respiratory health of neighboring residents. Epidemiology 18: 300–308.
8. Pavilonis BT, Sanderson WT, Merchant JA (2013) Relative exposure to swine animal feeding operations and childhood asthma prevalence in an agricultural cohort. Environ Research 122: 74–80.
9. Schiffman SS, Bennett JL, Raymer JH (2001) Quantification of odors and odorants from swine operations in North Carolina. Agr Forest Meteorol 108: 213–240.
10. Schenker MB, Christiani D, Cormier Y, Dimich-Ward H, Doekes G, et al. (1998) American Thoracic Society: Respiratory Health Hazards in Agriculture. Am J Resp Crit Care Med 158: S1–S76.
11. Wilson SM, Serre ML (2007) Examination of atmospheric ammonia levels near hog CAFOs, homes, and schools in Eastern North Carolina. Atmos Environ 41: 4977–4987.
12. Williams DA, Breysse P, McCormack M, Diette G, McKenzie S, et al. (2011) Airborne cow allergen, ammonia and particulate matter at homes vary with distance to industrial scale dairy operations: an exposure assessment. Environ Health 10: 72. doi:10.1186/1476-069X-10-72.
13. Schinasi L, Horton RA, Guidry VT, Wing S, Marshall SW, et al. (2011) Air Pollution, Lung Function, and Physical Symptoms in Communities Near Concentrated Swine Feeding Operations. Epidemiology 22: 208–215.
14. Trabue S, Scoggin K, Li H, Burns R, Xin H, et al. (2010) Speciation of volatile organic compounds from poultry production. Atmos Environ 44: 3538–3546.
15. Smith TJS, Rubenstein LS, Nachman KE (2013) Availability of Information about Airborne Hazardous Releases from Animal Feeding Operations. PLoS ONE 8(12): e85342. doi:10.1371/journal.pone.0085342.
16. Schiffman SS, Miller EA, Suggs MS, Graham BG (1995) The effect of environmental odors emanating from commercial swine operations on the mood of nearby residents. Brain Res Bull 37: 369–375.
17. Avery RC, Wing S, Marshall SW, Schiffman SS (2004) Odor from industrial hog farming operations and mucosal immune function in neighbors. Arch Environ Health 59: 101–108.
18. Nimmermark S (2004) Odour influence on well-being and health with specific focus on animal production emissions. Ann Agr Environ Med 11: 163–173.
19. Horton RA, Wing S, Marshall SW, Brownley KA (2009) Malodor as a trigger of stress and negative mood in neighbors of industrial hog operations. Am J Public Health 99: S610–615.
20. Wing S, Horton RA, Rose KM (2013) Air Pollution from Industrial Swine Operations and Blood Pressure of Neighboring Residents. Environ Health Perspect 121(1): 92–96.
21. Voss A, Loeffen F, Bakker J, Klaassen C, Wulf M (2005) Methicillin-resistant *Staphylococcus aureus* in Pig Farming. Emerg Infect Dis 11(12): 1965–1966.
22. Rinsky JL, Nadimpalli M, Wing S, Hall D, Baron D, et al. (2013) Livestock-Associated Methicillin and Multidrug Resistant *Staphylococcus aureus* Is Present among Industrial, Not Antibiotic-Free Livestock Operation Workers in North Carolina. PLoS ONE 8(7): e67641. doi:10.1371/journal.pone.0067641.
23. Casey JA, Curriero FC, Cosgrove SE, Nachman KE, Schwartz BS (2013) High-Density Livestock Operations, Crop Field Application of Manure, and Risk of Community-Associated Methicillin-Resistant *Staphylococcus aureus* Infection in Pennsylvania. JAMA Intern Med, 173(21): 1980–90. doi:10.1001/jamainternmed.2013.10408.
24. Osterberg D, Wallinga D (2004) Addressing externalities from swine production to reduce public health and environmental impacts. Am J Public Health 94: 1703–1708.
25. Burkholder J, Libra B, Weyer P, Heathcote S, Kolpin D, et al. (2007) Impacts of waste from concentrated animal feeding operations on water quality. Environ Health Perspect 115: 308–312.
26. Wing S, Cole D, Grant G (2000) Environmental injustice in North Carolina's hog industry. Environ Health Perspect 108: 225–231.
27. Wilson SM, Howell F, Wing S, Sobsey M (2002) Environmental injustice and the Mississippi hog industry. Environ Health Perspect 110: 195–201.
28. Mirabelli MC, Wing S, Marshall SW, Wilcosky TC (2006) Race, poverty, and potential exposure of middle-school students to air emissions from confined swine feeding operations. Environ Health Perspect 114: 591–596.
29. Freudenberg N, Pastor M, Israel B (2011) Strengthening Community Capacity to Participate in Making Decisions to Reduce Disproportionate Environmental Exposures. Am J Public Health 101(S1): S123–S130.
30. US Census Bureau (2010) Income, Poverty, and Health Insurance Coverage in the United States: 2009. Available: http://www.census.gov/prod/2010pubs/p60-238.pdf. Accessed 2013 July 29.
31. Hendrick S, Farquhar D (2008) Concentrated Animal Feeding Operations A Survey of State Policies. National Conference of State Legislatures. Available: http://www.ncifap.org/bin/k/w/CAFO_State_Survey_Introduction.pdf. Accessed 2013 April 12.
32. Environmental Council of the States. Organization of Environmental Agencies. Available: http://www.ecos.org/section/states/natural_resources_org. Accessed 2014 Jan 6.
33. National Association of State Departments of Agriculture. NASDA & Members: A General Overview. Available: http://www.nasda.org/File.aspx?id = 4120. Accessed 2014 Jan 6.
34. Centner TJ, Newton GL (2011) Reducing concentrated animal feeding operations permitting requirements. J Anim Sci 89(12): 4364–4369.
35. Koski C (2007) Examining State Environmental Regulatory Policy Design. J Environ Plann Man 50(4): 483–502.
36. Fry JP, Laestadius LI, Grechis C, Nachman KE, Neff RA (2013) Investigating the Role of State and Local Health Departments in Addressing Public Health Concerns Related to Industrial Food Animal Production Sites. PLoS ONE 8(1): e54720. doi:10.1371/journal.pone.0054720.
37. Lenaway D, Halverson P, Sotnikov S, Tilson H, Corso L, et al. (2006) Public Health Systems Research: Setting a National Agenda. Am J Public Health 96: 410–413.
38. Gerding J, Sarisky J (2011) Environmental Public Health Systems and Services Research. J Environ Health 73: 24–25.
39. Smith J, Bekker H, Cheater F (2011) Theoretical versus pragmatic design in qualitative research. Nurse Researcher 18(2): 39–51.
40. Avis M (2003) Do we need methodological theory to do qualitative research? Qual Health Res 13(7): 995–1004.
41. Maxwell JA (2010) Using Numbers in Qualitative Research. Qual Inq 16: 475–482.
42. Burke TA, Shalauta NM, Tran NL (1995) Strengthening the Role of Public Health in Environmental Policy. Policy Stud J 23(1): 76–84. doi: 10.1111/j.1541-0072.1995.tb00507.x.
43. Burke TA, Tran NL, Shalauta NM (1995) Who's in Charge? 50-State Profile of Environmental Health and Protection Services. Available: http://www.asph.org/uploads/whos2.pdf. Accessed 2014 Jan 6.
44. Johns Hopkins Center for a Livable Future (2013) Industrial Food Animal Production in America: Examining the Impact of the Pew Commission's Priority Recommendations. Available: http://www.jhsph.edu/research/centers-and-institutes/johns-hopkins-center-for-a-livable-future/_pdf/research/clf_reports/CLF-PEW-for%20Web.pdf. Accessed 2014 Jan6.
45. Zinn MD (2002) Policing Environmental Regulatory Enforcement: Cooperation, Capture, and Citizen Suits. Stan Envtl L J 21: 81.

# Demographic Model of the Swiss Cattle Population for the Years 2009-2011 Stratified by Gender, Age and Production Type

**Sara Schärrer[1]*, Patrick Presi[1], Jan Hattendorf[2], Nakul Chitnis[2,3], Martin Reist[4], Jakob Zinsstag[2]**

**1** Veterinary Public Health Institute/University of Berne, Berne, Switzerland, **2** Swiss Tropical and Public Health Institute/University of Basel, Basel, Switzerland, **3** Fogarty International Center, National Institutes of Health, Bethesda, Maryland, United States of America, **4** Federal Food Safety and Veterinary Office, Bern, Switzerland

## Abstract

Demographic composition and dynamics of animal and human populations are important determinants for the transmission dynamics of infectious disease and for the effect of infectious disease or environmental disasters on productivity. In many circumstances, demographic data are not available or of poor quality. Since 1999 Switzerland has been recording cattle movements, births, deaths and slaughter in an animal movement database (AMD). The data present in the AMD offers the opportunity for analysing and understanding the dynamic of the Swiss cattle population. A dynamic population model can serve as a building block for future disease transmission models and help policy makers in developing strategies regarding animal health, animal welfare, livestock management and productivity. The Swiss cattle population was therefore modelled using a system of ordinary differential equations. The model was stratified by production type (dairy or beef), age and gender (male and female calves: 0–1 year, heifers and young bulls: 1–2 years, cows and bulls: older than 2 years). The simulation of the Swiss cattle population reflects the observed pattern accurately. Parameters were optimized on the basis of the goodness-of-fit (using the Powell algorithm). The fitted rates were compared with calculated rates from the AMD and differed only marginally. This gives confidence in the fitted rates of parameters that are not directly deductible from the AMD (e.g. the proportion of calves that are moved from the dairy system to fattening plants).

**Editor:** Edna Hillmann, ETH Zurich, Switzerland

**Funding:** The study was funded as part of a PhD project by the Swiss Federal Veterinary Office. The funders had no role in study design, data collection and analysis, decision to publish, or preparation of the manuscript.

**Competing Interests:** The authors have declared that no competing interests exist.

* Email: sara.schaerrer@vetsuisse.unibe.ch

## Introduction

Switzerland has been collecting data about cattle including date of birth, date of slaughter, date of death (other than slaughter for consumption) and information regarding movements on a mandatory basis since 1999. The purpose of a national database of animal movements was originally to restore consumer trust during the BSE crisis by assuring traceability and therefore a better food safety of beef products and to provide a tool for epizootic disease surveillance and control [1,2]. The AMD contains detailed and complete datasets about the Swiss cattle population for several years offering the opportunity to get an insight into the population dynamics. Understanding the demographic of the livestock population in turn provides accurate parameters needed to develop models of disease transmission and helps policy makers in developing strategies regarding animal health, animal welfare and livestock management [3].

Early detection of disease, monitoring of present agents and substantiation of freedom from disease are described as key tasks of modern public veterinary services in order to allow international trade with agricultural goods and to document a good sanitary status of domestic livestock [4–6].

To monitor the health status of the cattle population, the Swiss veterinary authorities invest substantial resources in yearly surveillance programmes that have to meet international standards. One way to maintain the standards while reducing the costs is the application of risk based targeted approaches (e.g. [7]). Other approaches comprise logistical improvements such as better exploiting infrastructures where already a lot of potential information carriers are available e.g. slaughterhouse or milk quality testing laboratories [8]. With the implementation of bulk milk testing in 2010 [9,10] the production type became an important criterion for shaping the sampling strategy of national surveillance programs. As beef and fattening cattle, correspond to one third of the population, they have to be handled separately. The two production types (dairy and beef) do not only differ with respect to purpose but also with respect to management practices. The resulting differences in age distribution and slaughter rates in the two sub populations are of interest for the planning of stratified surveillance programmes to assure the representativeness of the sample (e.g. for sampling at the slaughterhouse level).

The objective of this study was therefore to create an AMD data driven demographic model that simulates the age and gender specific dynamics of the Swiss cattle population according to the production type. The derived rates describing population dynamics can be used for livestock development planning and associated economic analyses, as a backbone for disease transmission models

**Table 1.** The Swiss cattle population 2009–2011.

| Year | No of farms | No of cattle (January 1th) | No of dairy cows (January 1th) | No of slaughtered animals | No of births |
|---|---|---|---|---|---|
| 2009 | 42'966 | 1'608'062 | 675'285 | 647'715 | 721'810 |
| 2010 | 42'233 | 1'610'277 | 671'874 | 648'313 | 719'004 |
| 2011 | 41'465 | 1'612'230 | 676'253 | 653'754 | 718'697 |

Numbers are extracted from the Swiss animal movement database (AMD).

or for the design of cost-effective disease control and monitoring programmes.

Here we present the first dynamic demographic model of the Swiss cattle population. It is based on over 30 million data points collected in the Swiss animal movement database (AMD) between 2009 and 2011.

## Materials and Methods

### 2.1 The Swiss cattle population

The major livestock species in Switzerland is cattle. Although the number of farms decreases, for the years 2009–2011 the number of cattle in Switzerland is stable at roughly 1.6 million animals (table 1). Two thirds of the Swiss cattle industry is dedicated to dairy production. As a consequence, adult dairy cows (older than two years) make the largest demographic segment (figure 1). The average lifespan of a dairy cow in Switzerland is 6.2 years and the average number of calves in a lifetime is 3.7. The oldest cow that died between 2009 and 2011 was 25 years old.

Due to subsidies for ecological and behaviourally sound husbandry and strict animal protection legislation, small holdings with less than hundred animals are still the most common farm type. Over the summer month (May–October) one fourth of the livestock is moved to alpine pastures.

### 2.2 Data management

The Swiss animal movement database (AMD) contains information on farm level (e.g. location, production type), animal level (e.g. birthdate, gender, and breed), movement records (date, movement type) and stays (i.e. for every animal the start and end date of a stay on any holding is recorded). The data used for the models was an extract from the AMD, containing all recorded movements (25.5 million entries) and stays (15.8 million entries) from January 1999 until January 2012.

Birthdate, date of death (slaughter or natural) and gender are recorded on individual animal level, while the production type is available on farm level. The production type for each animal was consequently determined by the farm it stayed on at the given time step. Calf mortality consisted of notified stillbirths and mortality. As stays on alpine pastures are recorded only since 2008 and the quality of those recordings improved notably in 2009, only data from 2009 to 2011 was used for fitting of the population model.

### 2.3 The model

The Swiss cattle population was simulated using a system dynamic software [11]. The model is composed of a series of coupled difference equations. Compartments were defined by production type (dairy or beef), age class and gender. Calves were defined as animal being less than one year old, heifer and young bulls as one to two years old and cows and bulls as older than two years. We assumed that cows calve for the first time at the age of two and therefore the category "heifer" doesn't contribute to births. The beef and dairy system are connected through the transfer of calves from dairy farms to fattening plants, which is represented in the model as "fattening". The model is represented in figure 2.

The dynamic of the cattle population is simulated by month as time unit. Equations (1)–(12) show the number of animals per compartment (for parameter notation see table 2 and 3).

To represent the seasonal fluctuations in the number of births and death calves, we used a sinusodial-function with amplitude (a), phase ($\varphi$) and average ($\mu$) as parameters to fit (equations (13)–(20)). The frequency ($\omega$) was set to $\frac{2\pi}{12}$.

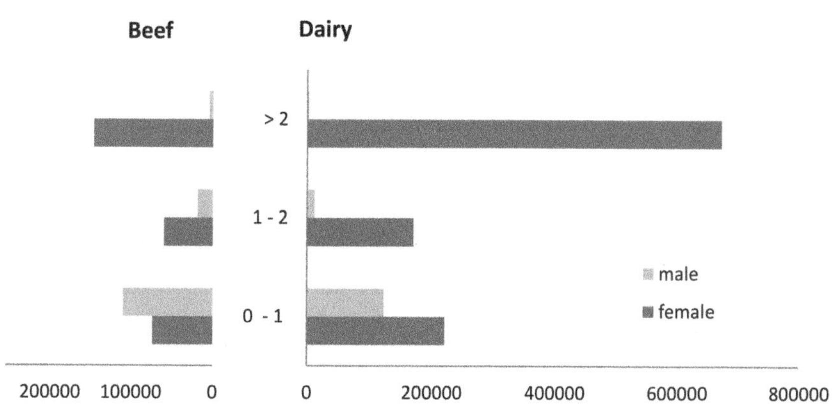


**Figure 1. Demographic of the Swiss cattle population per age class and sex in number of animals.**

$$\frac{dX_{DF}(t)}{dt} = b_{XDF}(t) * Z_{DF} - (m_{XDF}(t) + s_{XDF} + f_{XDF} + tr_{XDF}) * X_{DF} \quad (1)$$

$$\frac{dX_{DM}(t)}{dt} = b_{XDM}(t) * Z_{DF} - (m_{XDM}(t) + s_{XDM} + f_{XDM} + tr_{XDM}) * X_{DM} \quad (2)$$

$$\frac{dX_{BF}(t)}{dt} = b_{XBF}(t) * Z_{BF} + f_{XDF} * X_{DF} - (m_{XBF}(t) + s_{XBF} + tr_{XBF}) * X_{BF} \quad (3)$$

$$\frac{dX_{BM}(t)}{dt} = b_{XBM}(t) * Z_{BF} + f_{XDM} * X_{DM} - (m_{XBM}(t) + s_{XBM} + tr_{XBM}) * X_{BM} \quad (4)$$

$$\frac{dY_{DF}(t)}{dt} = tr_{XDF} * X_{DF} - (m_{YDF} + s_{YDF} + tr_{YDF}) * Y_{DF} \quad (5)$$

$$\frac{dY_{DM}(t)}{dt} = tr_{XDM} * X_{DM} - (m_{YDM} + s_{YDM} + tr_{YDM}) * Y_{DM} \quad (6)$$

$$\frac{dY_{BF}(t)}{dt} = tr_{XBF} * X_{BF} - (m_{YBF} + s_{YBF} + tr_{YBF}) * Y_{BF} \quad (7)$$

$$\frac{dY_{BM}(t)}{dt} = tr_{XBM} * X_{DF} - (m_{YBM} + s_{YBM} + tr_{YBM}) * Y_{BM} \quad (8)$$

$$\frac{dZ_{DF}(t)}{dt} = tr_{YDF} * Y_{DF} - (m_{ZDF} + s_{ZDF}) * Z_{DF} \quad (9)$$

$$\frac{dZ_{DM}(t)}{dt} = tr_{YDM} * Y_{DM} - (m_{ZDM} + s_{ZDM}) * Z_{DM} \quad (10)$$

$$\frac{dZ_{BF}(t)}{dt} = tr_{YBF} * Y_{BF} - (m_{ZBF} + s_{ZBF}) * Z_{BF} \quad (11)$$

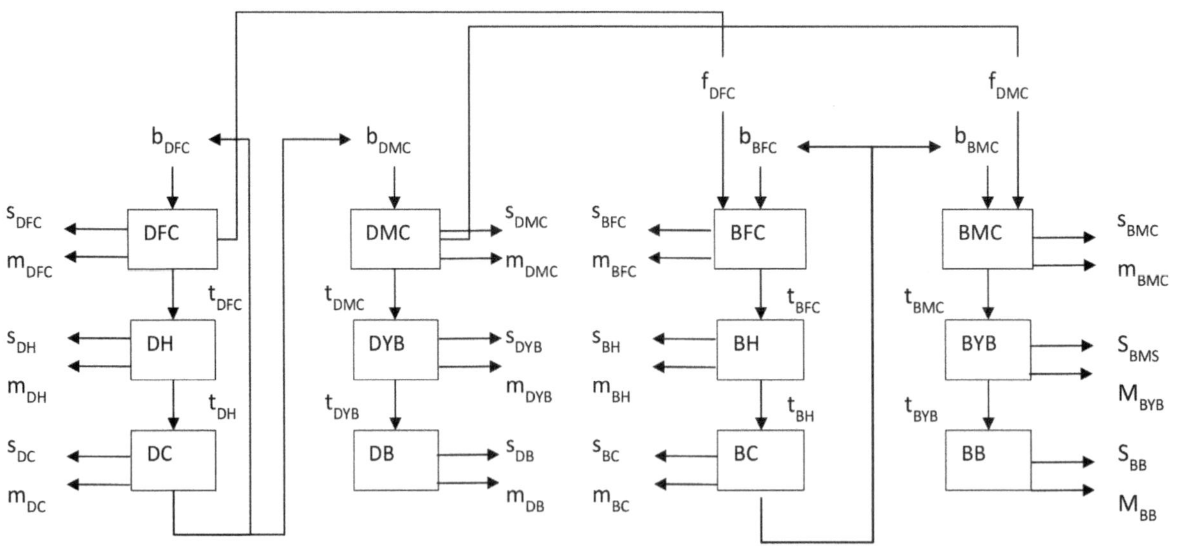


**Figure 2. Schematic representation of the Vensim model.** Arrows represent flows of animals into or out of a compound, boxes represents numbers of animals at a given time point in a category. s: slaughter; m: mortality; b: birth; tr: transition; f: fattening; D: dairy; B: beef; F: female; M: male; X: calves; Y: subadults; Z: adults.

**Table 2.** Nomenclature for subscripts in Equations 1–12.

| | Description | type |
|---|---|---|
| X | Calves | age class |
| Y | Subadults | age class |
| Z | Adults | age class |
| D | Dairy | production type |
| B | Beef | production type |
| F | Female | gender |
| M | Male | gender |

$$\frac{dZ_{BM}(t)}{dt} = tr_{YBM} * Y_{BM} - (m_{ZBM} + s_{ZBM}) * Z_{BM} \quad (12)$$

$$b_{XDF}(t) = \mu_{1XDF} + a_{1XDF} * \sin(t * \omega + \varphi_{1XDF}) \quad (13)$$

$$m_{XDF}(t) = \mu_{2XDF} + a_{2XDF} * \sin(t * \omega + \varphi_{2XDF}) \quad (14)$$

$$b_{XDM}(t) = \mu_{1XDM} + a_{1XDM} * \sin(t * \omega + \varphi_{1XDM}) \quad (15)$$

$$m_{XDM}(t) = \mu_{2XDM} + a_{2XDM} * \sin(t * \omega + \varphi_{2XDM}) \quad (16)$$

$$b_{XBF}(t) = \mu_{1XBF} + a_{1XBF} * \sin(t * \omega + \varphi_{1XBF}) \quad (17)$$

$$m_{XBF}(t) = \mu_{2XBF} + a_{2XBF} * \sin(t * \omega + \varphi_{2XBF}) \quad (18)$$

$$b_{XBM}(t) = \mu_{1XBM} + a_{1XBM} * \sin(t * \omega + \varphi_{1XBM}) \quad (19)$$

$$m_{XBM}(t) = \mu_{2XBM} + a_{2XBM} * \sin(t * \omega + \varphi_{2XBM}) \quad (20)$$

**Table 3.** Compartments and parameters in Equations 1–12.

| | Description | Unit |
|---|---|---|
| X | No of calves | Animals |
| Y | No of subadults | Animals |
| Z | No of adults | Animals |
| s | slaughter rate | $month^{-1}$ |
| m | mortality rate | $month^{-1}$ |
| b | birth rate | $month^{-1}$ |
| tr | transition rate | $month^{-1}$ |
| f | fattening rate | $month^{-1}$ |
| μ | Average | $month^{-1}$ |
| a | Amplitude | $month^{-1}$ |
| ω | Frequency | $month^{-1}$ |
| φ | Phase | Dimensionless |

**Table 4.** Monthly population parameters for the Swiss cattle population.

| | Dairy | | | | Beef | | |
|---|---|---|---|---|---|---|---|
| | | | **Month$^{-1}$** | **95%-CI** | | **Month$^{-1}$** | **95%-CI** |
| slaughter rates | Female calf | $s_{XDF}$ | 0.0197 | [0.0192, 0.0201] | $s_{XBF}$ | 0.0396 | [0.0389, 0.0403] |
| | Heifer | $s_{YDF}$ | 0.0065 | [0.0062, 0.0069] | $s_{YBF}$ | 0.0261 | [0.0253, 0.0269] |
| | Cow | $s_{ZDF}$ | 0.0190 | [0.0189, 0.0191] | $s_{ZBF}$ | 0.0233 | [0.0231, 0.0235] |
| | Male calf | $s_{XDM}$ | 0.1123 | [0.1103, 0.1144] | $s_{XBM}$ | 0.0638 | [0.0631, 0.0645] |
| | Young bull | $s_{YDM}$ | 0.1702 | [0.1658, 0.1748] | $s_{YBM}$ | 0.2834 | [0.2768, 0.2902] |
| | Bull | $s_{ZDM}$ | 0.1113 | [0.1072, 0.1156] | $s_{ZBM}$ | 0.0606 | [0.0590, 0.0623] |
| mortality rates | Female calf | $\mu_{2XDF}$ | 0.0094 | [0.0089, 0.0098] | $\mu_{2XBF}$ | 0.0059 | [0.0055, 0.0062] |
| | Heifer | $m_{YDF}$ | 0.0007 | [0.0006, 0.0007] | $m_{YBF}$ | 0.0008 | [0.0007, 0.0009] |
| | Cow | $m_{ZDF}$ | 0.0013 | [0.0012, 0.0013] | $m_{ZBF}$ | 0.0013 | [0.0013, 0.0014] |
| | Male calf | $\mu_{2XDM}$ | 0.0255 | [0.0241, 0.0269] | $\mu_{2XBM}$ | 0.0074 | [0.0071, 0.0078] |
| | Young bull | $m_{YDM}$ | 0.0017 | [0.0015, 0.0019] | $m_{YBM}$ | 0.0017 | [0.0015, 0.0019] |
| | Bull | $m_{ZDM}$ | 0.0022 | [0.0016, 0.0028] | $m_{ZBM}$ | 0.0026 | [0.0021, 0.0031] |
| transition rates | Female calf | $tr_{XDF}$ | 0.0684 | [0.0678, 0.0689] | $tr_{XBF}$ | 0.0718 | [0.0710, 0.0725] |
| | Heifer | $tr_{YDF}$ | 0.0804 | [0.0797, 0.0812] | $tr_{YBF}$ | 0.0615 | [0.0607, 0.0624] |
| | Male calf | $tr_{XDM}$ | 0.0207 | [0.0203, 0.0212] | $tr_{XBM}$ | 0.0511 | [0.0505, 0.0518] |
| | Young bull | $tr_{YDM}$ | 0.0234 | [0.0226, 0.0243] | $tr_{YBM}$ | 0.0161 | [0.0157, 0.0165] |
| fattening rates | Female calf | $f_{XDF}$ | 0.0172 | [0.0170, 0.0175] | | | |
| | Male calf | $f_{XDM}$ | 0.0731 | [0.0722, 0.0740] | | | |
| birth rates | Female calf | $\mu_{1XDF}$ | 0.0374 | [0.0373, 0.0376] | $\mu_{1XBF}$ | 0.0335 | [0.0332, 0.0339] |
| | Male calf | $\mu_{1XDM}$ | 0.0392 | [0.0389, 0.0396] | $\mu_{1XBM}$ | 0.0352 | [0.0347, 0.0357] |

D: dairy; B: beef; F: female; M: male; X: calf, Y: subadult, Z: adult. Small letters indicate rates (s: slaughter, m: mortality, f: fattening, tr: transition to next age class). $\mu$1: average birth rate; $\mu$2: average mortality rate;

**Table 5.** Values for the amplitudes and phases in the trigonometric functions of the presented Swiss cattle population model.

| | Dairy | | | Beef | |
|---|---|---|---|---|---|
| | | 95%-CI | | | 95%-CI |
| $a_{1XDF}$ | 0.0031 | [0.0022, 0.0041] | $a_{1XBF}$ | 0.0009 | [0, 0.0023] |
| $a_{1XDM}$ | 0.0091 | [0.0073, 0.0109] | $a_{1XBM}$ | 0.0040 | [0.0024, 0.0056] |
| $a_{2XDF}$ | 0.0029 | [0.0020, 0.0038] | $a_{2XBF}$ | 0.0013 | [0.0008, 0.0018] |
| $a_{2XDM}$ | 0.0063 | [0.0037, 0.0088] | $a_{2XBM}$ | 0.0016 | [0.0010, 0.0022] |
| $\varphi_{1XDF}$ | 1.6799 | [1.4046, 1.9574] | $\varphi_{1XBF}$ | 2.9510 | [1.1437, 4.7768] |
| $\varphi_{1XDM}$ | 1.9245 | [1.7428, 2.1096] | $\varphi_{1XBM}$ | 2.4772 | [2.0699, 2.8935] |
| $\varphi_{2XDF}$ | 1.6576 | [1.3443, 1.9727] | $\varphi_{2XBF}$ | 1.0713 | [0.6834, 1.4582] |
| $\varphi_{2XDM}$ | 1.7900 | [1.3820, 2.1969] | $\varphi_{2XBM}$ | 0.9218 | [0.5575, 1.2856] |

D: dairy; B: beef; F: female; M: male; X: calf, Y: subadult, Z: adult. a 1: amplitude for birth rate; a 2: amplitude for mortality rate; $\varphi_1$: phase for birth rate; $\varphi_2$: phase for mortality rate;

**2.3.1 Model fitting.** The number of living animals was extracted at the beginning of each month, number of birth, slaughter and death from the AMD per month, age class, production type and gender from January 2009 to December 2011. This data-set served to optimize the model parameters on the basis of the goodness-of-fit of the nonlinear maximum-likelihood optimization using the Powell algorithm [12]. Parameters were fitted stepwise, adding a variable at every step to the payoff values, using the outcome rates from the previous step as

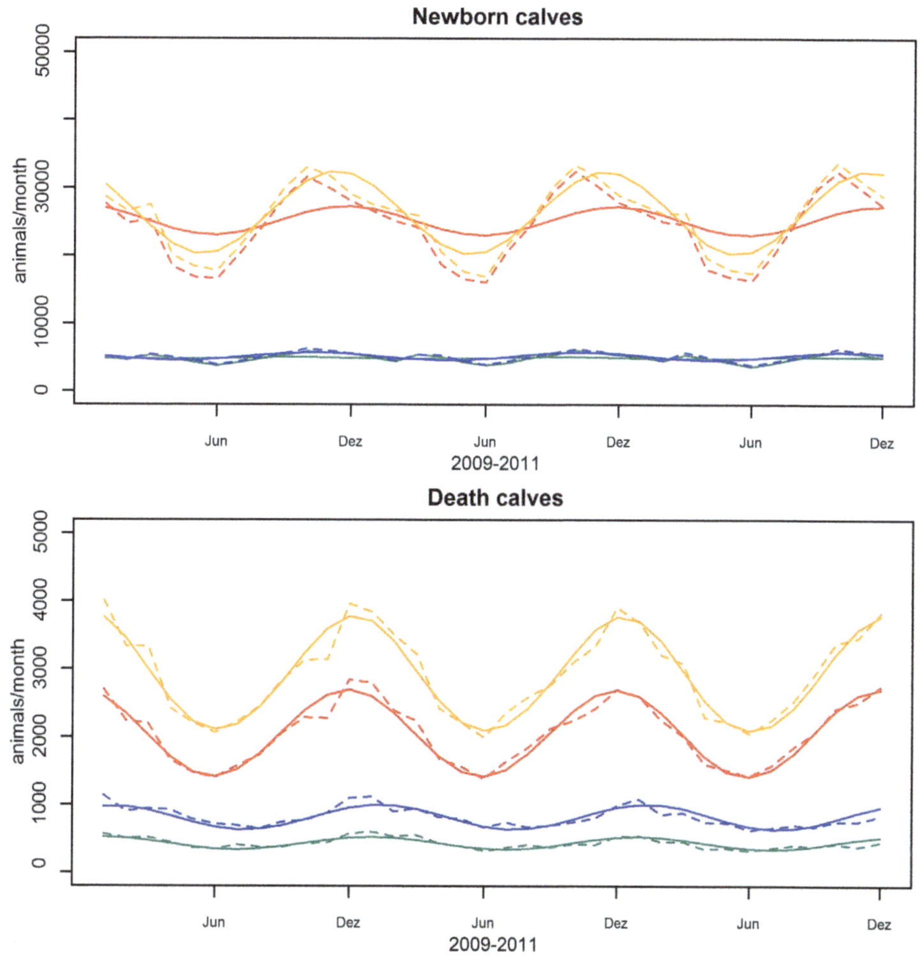


**Figure 3. Seasonal pattern of birth and mortality in calves.** Solid line: model data, dashed lines: AMD data. Orange: dairy male calf, red: dairy female calf, blue: beef male calf, green: beef female calf.

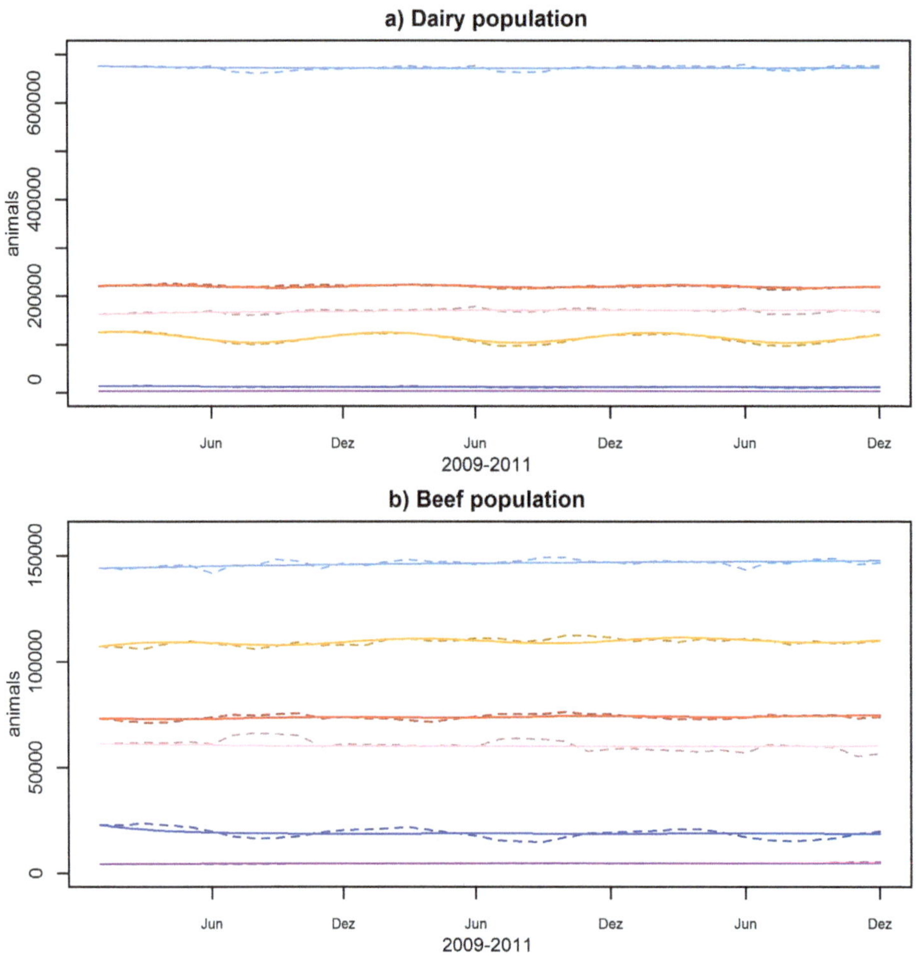


**Figure 4. Animal numbers per age category.** a) Dairy population. b) Beef population. Solid line: model data, dashed lines: AMD data. Light blue: cow, orange: male calf, red: female calf, pink: heifer, blue: young bull, purple: bull.

initial search point (maximum and minimum values set to +/− 10%).

**2.3.2 Comparison of calculated and fitted rates.** Birth, slaughter and mortality rates were calculated from the AMD data and compared to the fitted values from the model. Average birth rates were calculated as number of calves per month and category divided by the number of cows on the first of the months of the according production type and averaged over the 3 years period. Mortality and slaughter rates were calculated as number of death or slaughtered animals per month divided by the number of animals of the same age category and production type on the first of the month and averaged over the 3 years period. Model and empirical estimates were correlated in R [13].

**2.3.3 Sensitivity analysis.** The model was rebuilt with the statistical software R. To assess the sensitivity of the model, each parameter was varied separately using a range from −10% to +10% of the fitted value from the Vensim model (baseline), divided in 100 steps. For each value, the resulting absolute change in total numbers of animals compared to the baseline was represented graphically (Figures S1–S10, supplementary material).

## Results

In table 4 the fitted parameter values from the demographic model are shown. The model allowed the calculation of parameters that are not directly deductible from the AMD (transition rates and fattening rates).

By introducing parameters (amplitude and phase, table 5) to describe calf mortality and birth rates as trigonometric functions, the seasonal dynamic of changes in the population can be described more accurately than with the corresponding linear parameters deducted from the monthly extracts of the AMD (figure 3).

The correlation of the empirical parameters from the AMD and the fitted values gives a correlation coefficient of 0.994. The good fit of the model to the empirical data is also illustrated in figure 4.

As expected, the beef and dairy sector show differences in the demographic composition. While the proportions of young female animals are comparable (18.5% dairy female calves, 17.8% beef female calves 14.2% dairy heifers and 14.6% beef heifers), dairy cows account for around 56.7% of the dairy population while beef cows account for 35.5% of the beef population. For male animals the differences are even more noticeable: beef male calves, young bulls and bulls make 26.5%, 4.6% and 1.1% of the beef population compared to 9.6%, 1.0% and 0.2% for dairy male calves, dairy young bulls and dairy bulls respectively (all proportions are means over the 36 month of data analysis).

As import and export of live cattle are negligible for Switzerland (6'787 imported animals from 2009 to 2011 and 3'318 exported animals over the same period), the beef population is maintained

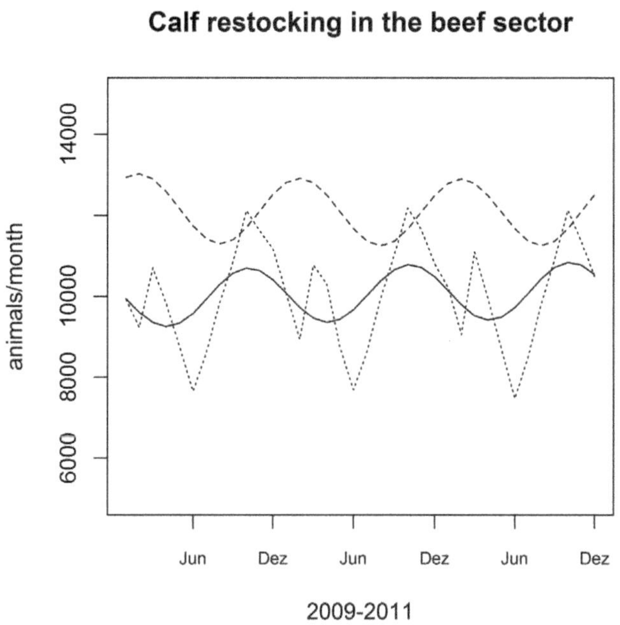


**Figure 5. Restocking of calves in the beef sector.** Dashed line: dairy calves transferred to fattening plants (VENSIM), solid line:born beef calves (VENSIM), dotted line: born beef calves (AMD).

to a considerable extend by calves from the dairy industry. Almost every month more dairy calves are transferred to fattening plants (i.e. to from the dairy to the beef industry) than were born within the beef industry (figure 5).

The number of slaughtered animals does not show a clear seasonal pattern (AMD data, figure 6) and the slaughter rate in the model is linear.

The sensitivity analysis shows, that the dairy female calf birth average and the dairy cow slaughter rate have the biggest influence on the total population with a change in animal numbers bigger than 50′000 after 3 years of simulation (figures S1–S10, supplementary material).

## Discussion

### 4.1 The Swiss cattle population

The composition of the Swiss cattle population accentuates that the milk industry dominates the domestic production and shapes the population dynamic. Adult dairy cows account for over 40% of all animals (figure 4). The importance of dairy female animals for the total population is reflected in the high sensitivity of the beef population to changes in the dairy cow slaughter rate and the dairy female birth rate (figures S2, S6, supplementary material). The irregular slaughter pattern indicates that the farmers keep the population constant by management decisions.

The higher monthly average mortality of dairy male calves compared to their contemporaries (0.0255 compared to 0.0094 (XDF), 0.0074 (XBM) and 0.0059 (XBF)) is in line with findings of other authors. [14] and [15] found higher mortality rates in dairy breeds than in beef breeds and higher mortality rates in male calves than female calves. As they all defined calves as maximum 180 days of age, the broader categories in our model might explain why dairy male calves differ as much from the others as the effect of early perinatal mortality with higher risk of dystocia for male calves [14] is combined with management decisions, i.e. less care for the economically relatively uninteresting male dairy calves [15]. As we also determined the production type on farm level and not according to the breed as in the above mentioned studies, effects of management decisions on the calve mortality might be even more manifest.

When deducting yearly rates roughly by multiplying the monthly age transition rates by 12, the difference in the management of beef and dairy animals becomes more obvious: while 82% of female dairy calves reach the next age class, only 25% of dairy male calves live through their first year. For beef calves 86% of the females and 61% of the males reach the next age class which reflects the interest of fattening beef breeds for more than 12 month. The most valued group of animals, dairy heifers, reach adulthood in 96% of the cases while more beef heifers are slaughtered and only 74% get two years old.

### 4.2 Model assumptions

In high productive agriculture systems of the developed world the population dynamics of livestock is controlled by the farmer and depends on policy and economics rather than on resource limitation or other external factors e.g. [16]. Bleul [14] states, that 80% of Swiss cows are inseminated artificially. For this reason we did not consider a resource constraint i.e. a carrying capacity in our model. The results may be of use for countries in similar economic situation but with less complete records but are to be applied carefully to cattle population that live under more resource dependent natural conditions.

The difference in the birth rates of dairy female and male calves in the model is an artefact presumably due to the difference in the dynamic of the two compartments. Dairy female calves are the most important segment to maintain the population which makes the model sensitive to any change in dairy female calf births. A conservative simulation gives a more stable overall result.

As alpine pastures usually use the gained milk directly for cheese production and it enters therefore not in commerce or they have young stock not yet lactating, they are mostly in the beef category regardless the provenience of the cattle. Therefore the data was corrected over the summer months, using the production type of the farm of origin from the movement records to alpine pastures. The visible seasonal bumps in beef heifers in figure 4 show, that the correction is imperfect due to an incomplete registration of the movements from and to alpine pastures. Since 2012 these are mandatory and improvement of the data quality can be expected.

To integrate the seasonality of birth and mortality in calves, we assumed a sinusoidal pattern and did not investigate other functions.

### 4.3 Future applications of the model

This is the first dynamic population model for Swiss cattle. As the data source is the complete record of the cattle population, a very good fit could be expected. Nonetheless the fitted population parameters allow a close to reality simulation of the population for future development planning scenario analysis, serve as a backbone to disease transmission models and for the simulation of disease surveillance and control (e.g. [17]).

The fitted population parameters allow building age and sex structured transmission models to simulate disease dynamics with different prevalences in different age classes (e.g. infectious bovine rhinotracheitis IBR, Brucellosis).

Furthermore the transmission rates of different age and production type categories to the slaughterhouse give precise information, which proportions of populations and subpopulations would be basically available for testing at the slaughterhouse in which time period. The slaughterhouse is a very convenient spot for sampling, because it allows taking samples from many animals

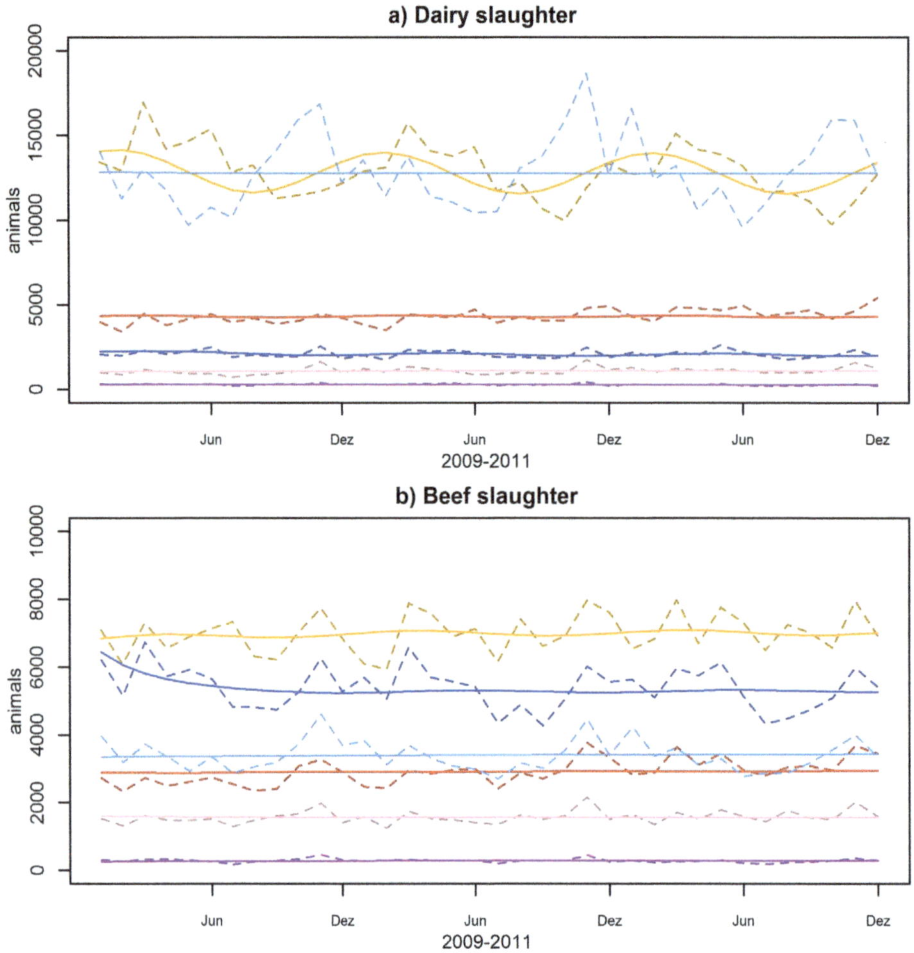


**Figure 6. Slaughter numbers per age category.** a) Dairy population. b) Beef population. Solid line: model data, dashed lines: AMD data. Light blue: cow, orange: male calf, red: female calf, pink: heifer, blue: young bull, purple: bull.

from different farms of origin within a short time period. Furthermore, there are diseases such as bovine spongiform encephalopathy (BSE) that can only be diagnosed in tissue matrices accessible at slaughter, e.g. brainstem.

As the outcome parameters in the model are calculated for the dairy and beef sector separately, surveillance systems with different components for the different production types can be simulated (e.g. IBR, Brucellosis). For example the efficacy of combining bulk tank milk sampling with slaughterhouse or on farm sampling can be evaluated. As the transfer from calves from the dairy sector to the beef sector is included, the model allows a realistic simulation of disease transmission in the overall population and of the effect of different surveillance strategies on the system sensitivity for different production types.

The fitted population parameters can also be interpreted as baseline parameters for the healthy Swiss cattle population. As seasonal effects are included in the parameter fitting, they can be used to search for aberrations in present data (e.g. increased mortality) to detect health events in an early stage.

In the healthy population most female calves are kept to restock the dairy population, as can be inferred from the relatively low transmission rates of female dairy calves to slaughter. If that segment is affected by an epidemic leading to increased abortions, calf mortality or decreased fertility, consequences on population structure and management are to be expected. Achievement of breeding objectives might be delayed or even out of reach. Impacts on the milk and meat markets are to be expected. The impact on population structure such as decrease of adult dairy cows in the slaughter population can be estimated by model derived transmission factors.

## Conclusions

The Swiss animal movement database is a reliable source of information about the Swiss cattle population and can provide stakeholders and decision makers with important knowledge without expensive and laborious field work. The presented demographic model allows a simulation of Swiss cattle production and economics under different policy scenarios and can be used as the demographic backbone for disease transmission models.

## Supporting Information

**Figure S1 Influence of varying slaughter rates on the number of animals in the dairy population.**

**Figure S2 Influence of varying slaughter rates on the number of animals in the beef population.**

**Figure S3 Influence of varying mortality rates on the number of animals in the dairy population.**

**Figure S4 Influence of varying mortality rates on the number of animals in the beef population.**

**Figure S5 Influence of varying average birth rates on the number of animals in the dairy population.**

**Figure S6 Influence of varying average birth rates on the number of animals in the beef population.**

**Figure S7 Influence of varying fattening rates (calves transferring from the dairy to the beef sector) on the number of animals in the dairy population.**

**Figure S8 Influence of varying fattening rates (calves transferring from the dairy to the beef sector) on the number of animals in the beef population.**

**Figure S9 legends for the colour scales for the dairy population.**

**Figure S10 legends for the colour scales for the beef population.**

**Text S1**

## Author Contributions

Analyzed the data: SS JZ JH NC. Wrote the paper: SS JH NC MR JZ PP. Model design: SS JZ JH NC PP MR.

## References


1. Golan E, Krissoff B, Kuchler F, Calvin L, Nelson K, et al. (2004) Traceability in the U. S. Food Supply: Economic Theory and Industry Studies. Agric Econ Rep 830.
2. Lüdi F (2004) Demographie, räumliche Verteilung und Dynamik der schweizerischen Rindviehpopulation in den Jahren 2002 und 2003.
3. O'Connor J, More S, Griffin J, O'Leary E (2009) Modelling the demographics of the Irish cattle population. Prev Vet Med 89: 249–254. Available: http://www.ncbi.nlm.nih.gov/pubmed/19327855. Accessed 13 June 2012.
4. Anonymous (2002) Agreement between the European Community and the Swiss Confederation on trade in agricultural products.
5. Anonymous (1995) Tierseuchenverordnung, SR 916.401. Bern. Available: http://www.admin.ch/opc/de/classified-compilation/19950206/index.html.
6. WTO (n.d.) The WTO Agreement on the Application of Sanitary and Phytosanitary Measures (SPS Agreement). Available: http://www.wto.org/english/tratop_e/sps_e/spsagr_e.htm.
7. Stärk K, Regula G, Hernandez J, Knopf L, Fuchs K, et al. (2006) Concepts for risk-based surveillance in the field of veterinary medicine and veterinary public health: Review of current approaches. BMC Health Serv Res 6: 20. doi:10.1186/1472-6963-6-20.
8. Hadorn D, Racloz V, Schwermer H, Stärk K (2009) Establishing a cost-effective national surveillance system for Bluetongue using scenario tree modelling. Vet Res 40: 57. doi:10.1051/vetres/2009040.
9. Reber A, Reist M, Schwermer H (2012) Cost-effectiveness of bulk-tank milk testing for surveys to demonstrate freedom from infectious bovine rhinotracheitis and bovine enzootic leucosis in Switzerland. SAT 154: 189–197. doi:10.1024/0036-7281/.
10. Reist M, Jemmi T, Stärk K (2012) Policy-driven development of cost-effective, risk-based surveillance strategies. Prev Vet Med 105: 176–184. Available: http://www.ncbi.nlm.nih.gov/pubmed/22265642. Accessed 4 April 2013.
11. Ventana Systems I (2008) Vensim: Simulator for Business.
12. Press W, Flannery B, Teukolsky S, Viatour P (1991) Numerical Recipies in C. Cambridge.
13. R Development Core Team (2008) R: A language and environment for statistical computing. Available: http://www.r-project.org.
14. Bleul U (2011) Risk factors and rates of perinatal and postnatal mortality in cattle in Switzerland. Livest Sci 135: 257–264. Available: http://linkinghub.elsevier.com/retrieve/pii/S1871141310004142. Accessed 21 August 2014.
15. Perrin J, Ducrot C, Vinard J, Hendrikx P, Calavas D (2011) Analyse de la mortalité bovine en France de 2003 à 2009. INRA Prod Anim 24: 235–244.
16. Rosen S, Murphy K, Scheinkman J (2012) Cattle Cycles. J Polit Econ 102: 468–492.
17. Riley S (2007) Large-scale spatial-transmission models of infectious disease. Science 316: 1298–1301. Available: http://www.ncbi.nlm.nih.gov/pubmed/17540894. Accessed 11 July 2014.

# 19

# Pesticide Residues and Bees – A Risk Assessment

**Francisco Sanchez-Bayo[1]*, Koichi Goka[2]**

1 Faculty of Agriculture and Environment, The University of Sydney, Eveleigh, New South Wales, Australia, 2 National Institute for Environmental Sciences, Tsukuba, Ibaraki, Japan

## Abstract

Bees are essential pollinators of many plants in natural ecosystems and agricultural crops alike. In recent years the decline and disappearance of bee species in the wild and the collapse of honey bee colonies have concerned ecologists and apiculturalists, who search for causes and solutions to this problem. Whilst biological factors such as viral diseases, mite and parasite infections are undoubtedly involved, it is also evident that pesticides applied to agricultural crops have a negative impact on bees. Most risk assessments have focused on direct acute exposure of bees to agrochemicals from spray drift. However, the large number of pesticide residues found in pollen and honey demand a thorough evaluation of all residual compounds so as to identify those of highest risk to bees. Using data from recent residue surveys and toxicity of pesticides to honey and bumble bees, a comprehensive evaluation of risks under current exposure conditions is presented here. Standard risk assessments are complemented with new approaches that take into account time-cumulative effects over time, especially with dietary exposures. Whilst overall risks appear to be low, our analysis indicates that residues of pyrethroid and neonicotinoid insecticides pose the highest risk by contact exposure of bees with contaminated pollen. However, the synergism of ergosterol inhibiting fungicides with those two classes of insecticides results in much higher risks in spite of the low prevalence of their combined residues. Risks by ingestion of contaminated pollen and honey are of some concern for systemic insecticides, particularly imidacloprid and thiamethoxam, chlorpyrifos and the mixtures of cyhalothrin and ergosterol inhibiting fungicides. More attention should be paid to specific residue mixtures that may result in synergistic toxicity to bees.

**Editor:** Raul Narciso Carvalho Guedes, Federal University of Viçosa, Brazil

**Funding:** These authors have no support or funding to report.

**Competing Interests:** The authors have declared that no competing interests exist.

* E-mail: sanchezbayo@mac.com

## Introduction

Growing concern about the impact of pesticides on pollinators is reflected in the enormous literature on the topic in the past few years [1]. In response to this concern, considerable amounts of new data on toxic effects of pesticides on wild bees, in particular bumble bees, have been obtained from laboratory and semi-field experiments [2,3].

A number of reviews on the topic have highlighted the importance of bees as natural pollinators not only for our crops but also for wildflowers and plants of forests and tropical ecosystems [4,5]. That is why the current declining trend of pollinators is worrying [6]. For example, it has been estimated that without bees, some 60 species of crop plants would fail to produce fruit [7]; the economic consequences of this impact are obvious. Importation of bumble bees to make up for the losses of pollinators in the areas affected not only does not solve the issue but also creates more problems by exporting parasites to other regions or countries [8,9] or competing with native species [10].

Of particular importance is the collapse of honey bee (*Apis mellifera*) colonies (CCD) in America and other developed countries, because they provide honey and wax commodities to our society. Attempts to explain the CCD have focussed on two main fronts: i) biological diseases, which includes virus [11] *Nosema* infections [12], parasites such as mites [13,14] and hive beetles [15]; and ii) pesticides, including not only insecticides and acaricides but also fungicides and herbicides [16,17]. Naturally, low levels of pesticides may act as stressors that make bees more prone to biological infections [3,18,19]. Among the pesticides, newly developed systemic insecticides such as fipronil and neonicotinoids have been targeted as the main culprits involved in the collapses since they were launched to the market in the mid-1990s [20,21,22,23].

Biological factors have been responsible for many of the problems that beekeepers have with their bee hives [24], but they are unlikely to be the main cause of disappearance of a number of wild bee species, or the decline of bumble bees in North America and Europe in recent years [12,25]. Although there are scant data on bee populations from other parts of the world to make a proper evaluation, the fact that bee declines have been observed in countries that have a long history of using pesticides in agriculture points to these agrochemicals as one of the important factors underlying wild bee and honey bee colony losses. To resolve this issue, several surveys have been carried out in recent years in North America [26,27,28], France [29,30], Spain [31] and India [32] among others, to find out the amounts and prevalence of pesticide residues present in pollen, honey, wax and other matrices of the bee hives (e.g. combs). They constitute a useful dataset to evaluate the impact that current pesticide residue levels have on honey bees and, possibly, wild bees as well; this risk is different to the risk of being affected by spray drift of these plant-protection products [33,34].

Typical risk assessments consider only acute toxicity of the chemicals either by topical or oral exposure in 24 or 48 hours, ignoring thus the negative effects derived from constant exposure to pesticide residues over longer periods. Some assessments have focused on environmental fate of pesticides and their application rates to estimate Toxicity Exposure Ratios (TERs) that were then used as indicators of the risk for honeybees due to particular exposure routes, e.g. ingestion of pollen or contact with it [35]. Recently, an individual study on pollen residues evaluates the possible risk of such residues to honey bees by both contact with and ingestion of contaminated pollen [28]. Neither study, however, includes the frequency of contaminated pollen among the risk parameters, while they also ignore the residues in honey or nectar. This we consider a serious flaw, as risk assessments should be based on the probability of exposure to actual residue levels. Indeed, none of the frequency data from the surveys mentioned above have been used to assess the impact that individual chemical residues and their combinations may or may not have on bees.

Some authors have tried to link the residue levels to the CCD in America [36], but by and large no risk assessment that includes residue levels, their prevalence and toxicity has been carried out to date. The handicap here is not insufficient residue data or acute toxicity data, but rather a lack of understanding as to how chronic toxicity by constant dietary exposure to residues found in pollen and honey affect the mortality of individual bees and the growth and reproduction of their colonies. Such effects include not only sublethal impairments but also delayed mortality [37]. Experiments with bumble bees have demonstrated that the lethal effects of new insecticidal compounds, including insect growth regulators and neonicotinoids, cannot be assessed based on acute toxicity data alone [22]. To understand the impact of small but constant doses of toxic residues on bee colonies it is necessary to apply different approaches where the time of exposure is taken into account [38].

Here, we attempt to provide a comprehensive risk assessment for all pesticide residues found in pollen and honey, or nectar, to bumble and honey bees using all residue and toxicity information available to date in the open literature and databases. Residue data originate from application of pesticides in accordance with standard agricultural practices in the countries surveyed, not from worst case, theoretical scenarios. Bees rely on nectar and pollen to meet the majority of their nutritional requirements, and therefore our risk assessment is focused on these two plant materials; honey is just concentrated nectar. Residues in wax are not included in this assessment since their availability to the bees was considered to be negligible compared to the direct exposure by contact with or dietary intake of pollen and honey [39]. However, recent research indicates that wax residues may also have an impact higher than expected until now [40], so available residue data in wax is presented for comparison only. Inhalation of volatile pesticides near treated crops is also excluded, since this is considered a minor route of exposure for most pesticides [41]. Traditional as well as novel methods of risk assessment will be used and compared in their predictions.

This assessment differs from those intended for regulatory purposes in several aspects: i) our focus is on the actual exposure of bees to the current pesticide residues found in the environment of developed and developing countries, not on the predicted exposure levels determined by models used in the tiered process of pesticide registration; ii) our assessment does not consider the particular application method of individual chemicals to their specific crops (e.g. foliar spray, granular, seed treatment, etc.) as it is based on the residue levels that are actually found in pollen and honey, regardless of the way they get there; iii) our assessment considers bee larvae and two castes of worker bees with different food requirements: nurses that feed on pollen, and nectar foragers. While the viability of the bee colonies depends largely on the queen's health and her reproductive output, at present there is insufficient knowledge to assess the impact that pesticides have on the queen's performance – the exception being recent studies with honey bees [42] and bumble bees [43].

The aim of this risk assessment is to identify the main chemicals that may pose a threat to the life of bees in their natural environment, which is currently contaminated with a large array of pesticides and other chemicals. By highlighting the compounds with higher risk to bees, we hope that apiculturists, beekeepers and policy makers involved in agricultural production will be able to screen the products most harmful to bees and find the appropriate remedies to avoid further damage.

## Materials and Methods

This assessment is restricted to honey bees (*Apis mellifera*) and bumble bees (*Bombus* spp.), which are very important pollinators and have been well studied. Information on ecotoxicity of a few pesticides to other wild bees exists [44,45,46], and their assessment can be inferred from the risk to the most common bee species presented here.

### Residues Data

Data on pesticide residues in pollen, honey and wax from bee hives were taken from several sources, including recent pesticide surveys in the USA [26,27], France [47] and Spain [31] as well as a survey of neonicotinoids in Poland [48]. The review by Johnson et al. [17] provided further data on maximum residues in all these matrices. Residues in honey include additional data from surveys in Greece [49], Spain [50], Brazil [51] and India [52], complemented with sparse data from other sources as well as with residues in nectar from treated plants [44,47,53,54,55,56]. Residues in wax also include other data from Spain [57] and the USA [26,58]. The data were compiled to obtain average and maximum residue loads for each compound, and their frequency, in pollen, wax, honey or nectar (see Table S1).

### Toxicity Data

Acute oral and contact toxicity of pesticides to honey bees are available for the majority of pesticides as either median lethal doses per bee (LD50) or median lethal concentrations (LC50) in the tested media. Median values are preferred to no-observed effect level (NOEL) or the lowest-observed effect level (LOEL) values, which are only available for a small number of compounds and which relevance for risk assessment has been questioned on statistical grounds [59] and inaccuracy [60].

Toxicity data for honey bees were obtained from the Pesticide Manual [61], the ECOTOX database of the U.S. Environment Protection Agency (http://cfpub.epa.gov/ecotox/) and the Agri-Tox Database of the Agence Nationale de Sécurité Sanitaire de l'Alimentation, de l'Environnement et du Travail in France (http://www.agritox.anses.fr/index.php). Toxicity of 29 insecticides to bumble bees was obtained from ECOTOX and the open literature [2].

Agreement between the toxicity data sources was remarkably high (>95% of all compounds), with only a handful of compounds (8) showing obvious discrepancies. It is concerning, however, that LD50 values for 30% of the most highly toxic compounds to bees are not reported in the Pesticide Manual, since this is the database most commonly used by consultants in the agricultural business. Notable among these omissions are imidacloprid, emamectin benzoate, etofenprox, flumethrin, prallethrin and several organophosphorus compounds: dicrotophos, parathion (ethyl), omethoate and acephate. Surprisingly, toxicity data for coumaphos – which is widely used in apiaries for mite control – were absent from the Pesticide Manual and Agritox databases, as noted also by other researchers [28].

Oral toxicities were available for 221 compounds of the 322 pesticides compiled (69%), whereas contact toxicity (topical) covered 96% of the pesticides (see Table S2). Included in the data are 76% of existing insecticides and fungicides and 83% of acaricides registered for use in agriculture. Herbicides were excluded since they are non-toxic to bees, i.e. LD50 values above 100 or 200 μg bee$^{-1}$. When more than one oral or contact LD50 value was available, a geometric mean was calculated. For data reported as "more than a given value", that value was used in the calculations. Oral toxicities were referred in almost all cases to 48-h exposures, whereas contact exposures varied between less than a day and 96-h, with a median of 48-h, so the average LD50s or LC50s used here can be considered representative of acute exposures to bees in about 2 days.

Unfortunately, no toxicity data for larvae are available (but see [40]), so here we assume the same LD50 values for larvae as for adult bees. Chronic data for bees are extremely rare and only reported for 1 systemic insecticide [62,63], and 6 insecticide growth regulators [22], as indicated in Table S2.

A regression of insecticides' LD50s (μg bee$^{-1}$) between honey and bumble bees reveals that the sensitivity of honey bees by oral exposure is similar to that of bumble bees (slope = 0.34, $r^2 = 0.94$, $p<0.001$, n = 13), whereas bumble bees are 28 times less sensitive than honey bees in regard to contact exposure with insecticides (slope = 28.3, $r^2 = 0.93$, $p<0.001$, n = 16) (Fig. 1). Even after correcting for weight between species, bumble bees are about 7 times less sensitive to insecticides by contact than honey bees. Because such difference varies from chemical to chemical, extrapolations of toxicity from honey bee to bumble bee have been avoided in this study, even if they may be useful in some situations [33,64].

## Data Analysis

**Standard risk approach.** It should be recognised that the standard hazard quotient (i.e. HQ = PEC/LD50, where PEC is the predicted environmental concentration) is not a measure of risk because it does not indicate the probability of a hazard to occur. And yet, previous studies on pesticides and bees used HQs in their evaluations [28,35]. To estimate the risk of bees being affected by contaminated pollen or nectar it is necessary to consider the frequency of detection of pesticides residues in such matrices, because prevalence indicates the probability of exposure to the contaminants. Therefore, a simple risk assessment should incorporate this probability as follows

$$Risk = \frac{frequency\ [\%] \times residue\ dose\ [\mu g]}{LD50\ [\mu g\ bee^{-1}]} \quad (1)$$

This expression indicates that a given pesticide residue has a certain probability of causing 50% mortality among the bees that come into contact with or ingest the contaminated pollen or nectar. Since we only use here LD50 values to estimate risks, our assessment should be considered very conservative. For estimation of risks at the lowest effect level, approximate estimates of LD10 can be calculated as 0.1 × LD50. Such approximation is based on previous field studies that determined the lowest effect levels of many pesticides on aquatic organisms [65].

For expression (1) to represent the actual probability of risk, residue loads should be first converted to the actual doses of residue that come in contact with (topical exposure) or are ingested (oral exposure) by the bees. Having data on average and maximum residue loads allows us establish a range of possible risks to bees. For average loads, the frequency of detection shown in Table S1 was used in the calculations; for maximum loads it should be noted that their frequency of detection is 1/total number of samples analysed in each survey. As the number of samples per survey varies between 99 and 845, the frequency of appearance of maximum residues is in the range 0.1–1.0%, i.e. much lower than the average prevalence of residues.

In the case of exposure by contact with pollen, topical LD50s shown in Table S2 were used to calculate the risk for a worker bee that comes in contact with 1 g of contaminated pollen per day. For oral exposure we focus on nurses, which feed exclusively on pollen for 10 days, and nectar foragers, which feed on nectar/honey for another 20 days during the summer season (Table 1). These types are considered representative of the bee colonies during the summer, when most pesticides are applied to crops, although winter bees can be exposed to residues in nectar for up to 100 days or more [66]. Lack of specific data on intake by bumble bees obliged us to scale the same rates as honey bees multiplied by a factor of 5, based on average intake of syrup by workers of *Bombus terrestris* and *Apis mellifera* - see File SI for estimation of contact doses and daily intake of residues.

**Risk of synergistic mixtures.** Because of the known synergism between ergosterol inhibiting fungicides with pyrethroids and cyano-substituted neonicotinoids (i.e. thiacloprid and acetamiprid), risks of residue mixtures were also included in this assessment for both contact and dietary exposures. These fungicides disable the monooxygenase detoxification system in honey bees, thus increasing the lethal effects several fold [67,68,69]. Synergistic factors vary for each combination and are reported only for topical exposure, but here we assume the same factor applies to oral exposures. In any case, the LD50 of the mixture was estimated as

$$LD50_{mixture} = \frac{LD50_{insecticide}}{synergistic\ factor} \quad (2)$$

Risks of mixtures are estimated using equation (1), with residue loads of the insecticide and the combined frequency of the compounds. Since the probability of finding residues of both insecticide and synergist in the same pollen or honey cannot be estimated here we considered the lowest frequency of either compound only.

**New approaches to risk.** The above expressions of risk indicate probabilities of causing serious effects (e.g. 20% risk of resulting in 50% mortality) within short periods of exposure, i.e. about 2 days. They suit well the assessment of risks by contact exposures. However, they may not be appropriate to assess risks by chronic, dietary exposure because the bees constantly consume pollen, nectar and honey. Assuming the residues ingested remain in the body, the median lethal dose may be reached after some time; in practice, there is some elimination and metabolism for most compounds [70], so the cumulative residue amounts estimated this way represent a worse case scenario. As the residue loads in pollen and honey are already known, the only limitation is the life-span of the individual bees, which varies from 5 days in worker larvae to 100 or more days in winter worker bees (Table 1).

Consequently, a simple way to assess the dietary risk of pesticide residues is by estimating the time to reach their corresponding LD50s, and compare those times with the actual life-span of each stage of development. Only times which are shorter than the life-

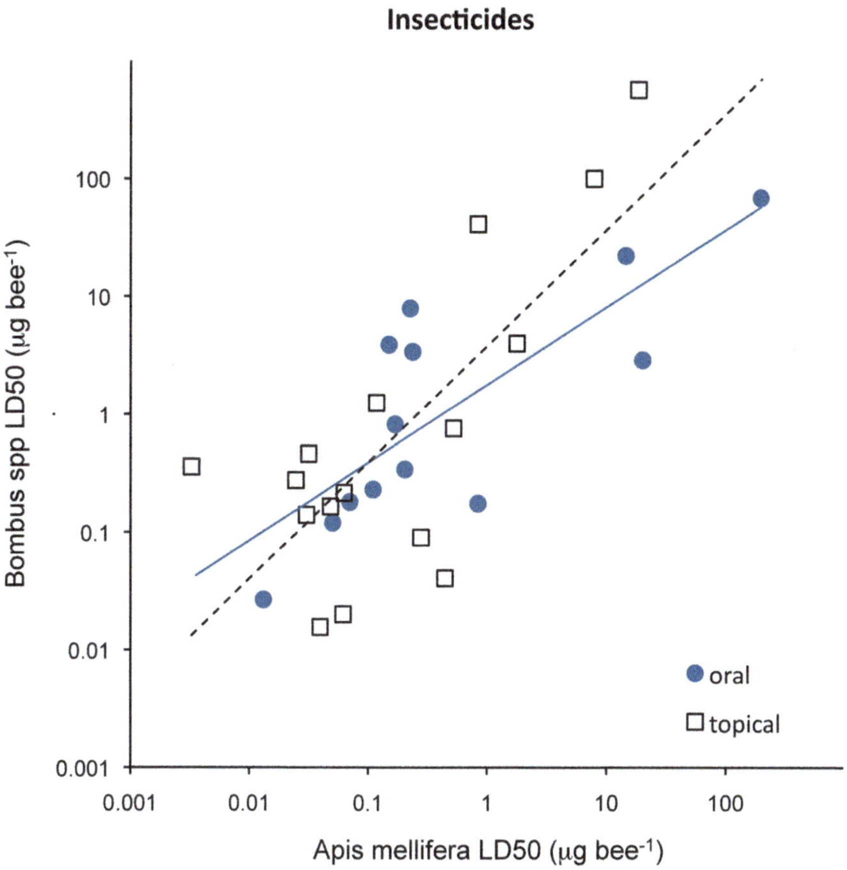


**Figure 1. Comparison of the sensitivity of honey bees (*Apis mellifera*) and bumble bees (*Bombus* spp.) to 29 insecticides, as expressed by their contact or oral LD50s (μg bee$^{-1}$).** Susceptibility of both species by oral exposure is similar (line, slope = 0.34, $p < 0.001$), whereas on average bumble bees are 7 times less sensitive than honey bees by contact exposure, after correcting for weight (stippled line, slope = 28.3, $p < 0.001$).

span would represent a serious risk, as they indicate that surely more than 50% of the bees exposed would die. We use two distinct new approaches for assessing dietary exposure:

i) Fixed dose approach. Assuming that acute LD50 values are constant for each pesticide, estimates of the time to reach the dietary LD50 (henceforth T50) of each pesticide were calculated as follows

$$T50\ [days] = \frac{LD50\ [\mu g\ bee^{-1}]}{daily\ dose\ [\mu g]} \qquad (3)$$

**Table 1.** Life-span of larvae and worker bees and their consumption rates of pollen and honey (After [76]).

| | Apis mellifera | | | Bombus spp.[1] | |
|---|---|---|---|---|---|
| | Daily rate (mg/day) | | Life-span (days) | Daily rate (mg/day) | |
| | Honey | Pollen | | Honey | Pollen |
| Drone larvae | 15.1 | 1.1 | 6.5 | 75.5 | 5.5 |
| Worker larvae | 28.9 | 1.1 | 5 | 144.5 | 5.5 |
| Brood attendant | 34–50 | | 8 | 170–250 | |
| Nectar forager | 80.2 | | 30 | 401 | |
| Nurse worker | | 6.5 | 10 | | 32.5 |
| Pollen forager | 13 | | 30 | 65 | |
| Wax-bees | 18 | | 6 | 90 | |
| Winter bees | 8.8 | | 91+ | 44 | |

[1]Assuming 5 times the consumption of *Apis mellifera* in the same proportion.

As with the standard risk assessment, T50s were estimated for intake of average residues as well as maximum residues, so as to provide a range rather than an exact number of days. This approach may be valid for most pesticides, but there are some exceptions that justify another way of assessing risks.

ii) Time-cumulative effects. This approach is based on the experimental observation that dietary LD50s for certain compounds decrease with exposure time [37]. Consequently, the estimated T50s will be reached earlier than expected. The rate of change of LD50 with time can be estimated experimentally by a simple log-to-log regression of the LD50s on the exposure times

$$Ln\,T50\,[days] = a + b \bullet Ln\,LD50\,[\mu g\ bee^{-1}] \qquad (4)$$

where a (intercept) and b (slope) are empirical parameters specific to each chemical and species tested [71]. Slope values <1 result in an exponential increase of effects over time, according to the Druckrey-Küpfmüller equation (D Tn = constant, where the exponent n = 1/slope, D = dose and T = time) [37]. To date, there is empirical evidence of time-cumulative toxicity for some carcinogenic substances, neonicotinoid insecticides, rodenticides and methylmercury, and its underlying mechanism is thought to be the irreversible binding of the toxicant to specific receptors [72]. In the case of bees, the only data available are for the neonicotinoids imidacloprid [62,63] and thiamethoxam [73], so this new approach will be used here only for these two compounds.

## Results

### Residue Data

A total of 161 pesticides have been found so far in bee hives, of which 124 appeared in pollen, 95 in wax and 77 in honey or nectar. The majority were insecticides (83), with fungicides (40), herbicides (27) and acaricides (10) making up the remainder; only one insecticide synergist (piperonyl butoxide) was found [17]. Among the survey's data,15 metabolites were reported due to their toxicity and persistence (e.g. aldicarb sulfoxide, endosulfan sulfate, fipronil sulfone, 5-hydroxy-imidacloprid, DDE, etc.). Whenever metabolites are reported, total loads for a compound were calculated as the sum of loads of the individual metabolites and their respective parent compound, weighted for the frequency of appearance in the respective surveys. Some persistent compounds are no longer used in one or several of the countries surveyed (e.g. dieldrin, DDT, HCB) but their residues are still present in their environment and need to be taken into account for risk assessment of bees. Obviously not all chemicals appeared in all surveys, and their concentrations and frequency differed markedly among surveys, reflecting the usage pattern of agrochemicals in each country or region.

The highest residue concentrations were found in wax and pollen (average 126 and 66 μg kg$^{-1}$ respectively), whereas the highest frequency of detection corresponds to wax (over 50% for chlorfenvinphos, tau-fluvalinate, bromopropylate, coumaphos and chlorothalonil) and honey (over 50% for thiacloprid, thiamethoxam and acetamiprid [48]). It is unclear whether the residues detected in pollen collected from apiaries (Fig. 2a) originated from sprayed fields or from hives treated with pesticides; they are probably a mixture of both. Whatever their source, such pollen feeds the nurse bees and larvae. Among the residues in honey, systemic insecticides stand out for their high prevalence: neonicotinoids are the most commonly found, while phorate, dimethoate and carbofuran are typically present in more than 5% of nectar collected from treated plants (Fig. 2b). The presence of hydrophobic pesticides such as coumaphos or vinclozolin, and to a lesser extent tau-fluvalinate, in honey suggests contamination from the comb, since honey bee colonies are commonly treated with these pesticides for mite and fungal control [74].

### Risk by Contact Exposure

A total of 92 individual compounds could be assessed for risk to contaminated pollen by contact exposure after matching residue and toxicity data. To these were added the synergistic combinations of cyhalothrin, thiacloprid and acetamiprid with three ergosterol inhibiting fungicides: propiconazole, penconazole and myclobutanil. Table 2 shows the risk for honey bees and bumble bees exposed to average and maximum residue levels of each compound, after taking into account their average prevalence in Europe, America and Asia. Only 33 compounds and 5 mixtures that have some relevance (i.e. risk >0.1) are shown, since 65% of the compounds have negligible or no risk to the bees. Risks above 5% can be considered high, as they correspond to T50s of 2 days or less; between 1 and 5% the risk is moderate, usually corresponding to T50s between 2 and 7 days, which are within the life-span of larvae and adult workers; risk below 1% can be regarded as low, for which T50s range from 7 to 60 days and more, covering the life-span of nectar foragers in summer and most of the life-span of winter bees.

Not surprisingly, the bulk of chemicals posing contact risk to bees are insecticides (20) or insecticide-acaricides (10), with only 2 acaricides, 1 fungicide and 5 fungicide mixtures appearing in that list. The risk of being seriously affected by contact with pollen residues is generally low, with only 5 compounds showing high risks (>5%): thiamethoxam (3.7–29.6% for honey bees), phosmet (14.6–23.9% for honey bees), chlorpyrifos (8.3–12.9% for both bees), imidacloprid (10.3–16% for honey bees but 31.8–49% for bumble bees) and clothianidin (1.0–5.3% for honey bees and 2.5–13.3% for bumble bees). It should be noted that the risk of these neonicotinoids to bumble bees is about two to three times as high as for honey bees, due to the different sensitivity among the species (Fig. 1). These compounds pose high risk to bees on account of their extremely high toxicity to both honey and bumble bees, with topical LD50s in the range 0.02–0.09 μg bee$^{-1}$, and also because their average residues (12–35 ppb) were present in 11 to 16% of the pollen surveys worldwide. By contrast, the high risk of phosmet is mainly due to average residues of 339 ppb (highest 16.5 ppm) in spite of its moderate toxicity to honey bees (topical LD50 = 0.62 μg bee$^{-1}$). While six other compounds were more common among the residues (coumaphos, tau-fluvalinate, chlorothalonil, acetamiprid, amitraz and thiacloprid), their toxicity to bees is 100–5000 times lower than that of thiamethoxam or chlorpyrifos.

Mixtures of fungicides with cyhalothrin or thiacloprid pose also high risks for honey bees (3.7–8.8%) and a moderate risk to bumble bees (1.1–2.6%), even if the prevalence of the three fungicides is relatively low (1.8–5.5%). Attention should be paid to the synergism of propiconazole with such insecticides, as it changes markedly the risk of the individual compounds from being moderate (0.2–1.8% cyhalothrin) or negligible (<0.1% thiacloprid) to a high risk. The synergistic factor of propiconazole for thiacloprid is 560 [67,75] and for cyhalothrin 16.2 [67,68]. Only the mixtures acetamiprid with propiconazole and fenuconazole showed low risk for honey bees (0.1–0.7%) and negligible risk for bumble bees (0.01–0.07%) based on synergistic factors of 100-fold (propiconazole) or 4.5-fold (fenuconazole) and the low frequency of such fungicides in pollen (1.8–3.3%).

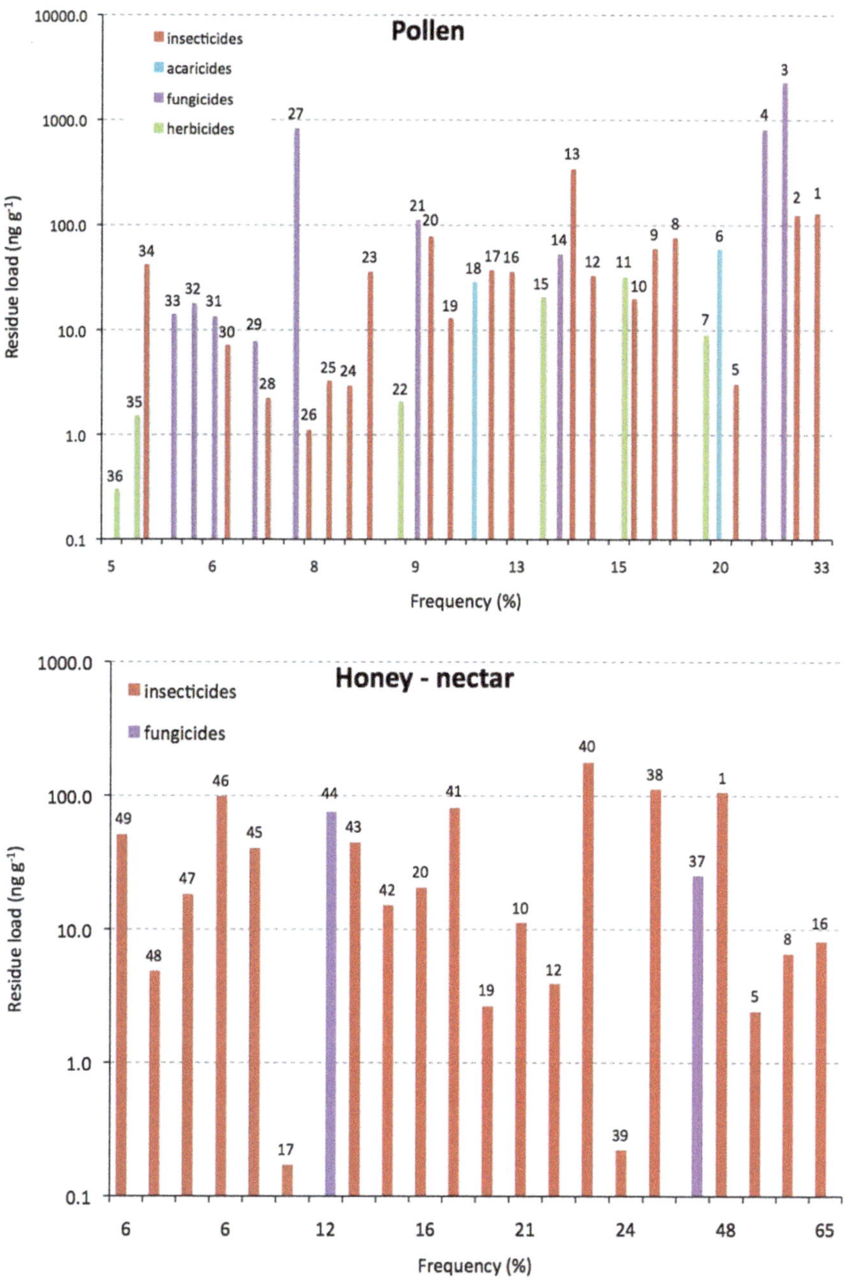


**Figure 2. Residue loads of the most common pesticides plotted against their prevalence (frequency) in: A) pollen; B) honey or nectar.** Key: 1 coumaphos (total); 2 tau-fluvalinate; 3 thymol; 4 chlorothalonil; 5 acetamiprid; 6 amitraz (total); 7 dithiopyr; 8 thiacloprid; 9 carbaryl; 10 imidacloprid (total); 11 pendimethalin; 12 chlorpyrifos; 13 phosmet; 14 carbendazim; 15 atrazine; 16 thiamethoxam; 17 chlorfenvinphos; 18 fenpyroximate; 19 clothianidin; 20 endosulfan (total); 21 thiophanate-methyl; 22 metolachlor; 23 fenpropathrin; 24 methoxyfenozide; 25 esfenvalerate; 26 tebufenozide; 27 captan (total); 28 bifenthrin; 29 azoxystrobin; 30 lambda-cyhalothrin; 31 diphenylamine; 32 penconazole; 33 trifloxystrobin; 34 fenthion; 35 norflurazon; 36 metribuzin; 37 hexachlorobenzene; 38 HCH (alpha and beta); 39 phorate; 40 gamma-HCH (lindane); 41 heptenofos; 42 methiocarb; 43 DDT (total); 44 vinclozolin; 45 methidathion; 46 malathion; 47 cypermethrin; 48 dimethoate; 49 carbofuran (total).

Moderate risk by contact exposure (i.e.1–5%) includes 6 pyrethroids (acrinathrin, deltamethrin, cypermethrin, bifenthrin, esfenvalerate and lambda-cyhalothrin, in that order), the carbamate carbaryl, the organophosphorus fenthion, the neonicotinoid dinetofuran, the pyrazole fipronil and the acaricide pyridaben. However, risks of these compounds to bumble bees are below 1%, because they are less toxic to the large pollinators (Table S2). Notice that, despite fipronil and bifenthrin being among the most toxic insecticides to honey bees (topical LD50 0.007 and 0.015 μg bee$^{-1}$ respectively), their risk by contact exposure is reduced because of their low residue loads (1.6–29 ppb and 2.2–13 ppb respectively) and low prevalence in pollen residues (average 2.8 and 6.6% respectively).

The remaining 17 compounds pose a low risk (<1%) for being less toxic (chlorothalonil, coumaphos, tau-fluvalinate, endosulfan, chlorfenvinphos), appearing rarely (phenothrin, prallethrin, ace-

**Table 2.** Risk (% probability) and time to reach topical LD50 (T50 in days) for worker bees, estimated as contact exposure with 1 g of contaminated pollen, at average or maximum levels, during 2 days.

| Use[1] | Chemical | Honey bee | | | | | Bumble bee | | | | |
|---|---|---|---|---|---|---|---|---|---|---|---|
| | | Topical LD50 (μg bee$^{-1}$) | Risk (%) | | TD50 (days) | | Topical LD50 (μg bee$^{-1}$) | Risk (%) | | TD50 (days) | |
| | | | Average | Max | Average | Max | | Average | Max | Average | Max |
| I | thiamethoxam | 0.02 | 29.58 | 3.66 | 1 | 0.2 | | | | | |
| I | phosmet | 0.62 | 14.56 | 23.89 | 2 | 0.04 | | | | | |
| I | chlorpyrifos | 0.07 | 12.92 | 10.33 | 2 | 0.1 | 0.09 | 10.32 | 8.26 | 3 | 0.1 |
| I | imidacloprid (total) | 0.06 | 10.34 | 16.00 | 3 | 0.1 | 0.02 | 31.77 | 49.15 | 1 | 0.02 |
| I+F | cyhalothrin+propiconazole | 0.003[2] | 8.79 | 5.90 | 0.4 | 0.1 | 0.01[2] | 2.56 | 1.71 | 1.4 | 0.3 |
| I+F | cyhalothrin+myclobutanil | 0.004[2] | 7.86 | 5.52 | 0.6 | 0.1 | 0.02[2] | 2.28 | 1.60 | 2.1 | 0.4 |
| I+F | cyhalothrin+penconazole | 0.011[2] | 7.23 | 3.68 | 2 | 0.3 | 0.04[2] | 2.10 | 1.07 | 5 | 1.0 |
| I | clothianidin | 0.04 | 5.28 | 0.99 | 4 | 1 | 0.02 | 13.26 | 2.49 | 2 | 0.4 |
| I+F | thiacloprid+propiconazole | 0.065[2] | 4.21 | 7.43 | 1 | 0.1 | | | | | |
| A | acrinathrin (total) | 0.17 | 3.44 | 1.21 | 1 | 0.2 | | | | | |
| I | deltamethrin | 0.02 | 3.27 | 2.26 | 1 | 0.3 | 0.28 | 0.29 | 0.20 | 11 | 3 |
| I-A | cypermethrin | 0.03 | 2.40 | 1.75 | 2 | 1 | | | | | |
| I | carbaryl | 0.84 | 2.32 | 0.95 | 14 | 1 | 41.16 | 0.05 | 0.02 | 699 | 39 |
| I | bifenthrin | 0.01 | 1.97 | 1.14 | 7 | 1 | | | | | |
| I | esfenvalerate | 0.03 | 1.96 | 2.99 | 8 | 0.4 | | | | | |
| I | fenthion | 0.22 | 1.89 | 0.57 | 5 | 1 | | | | | |
| I | dinotefuran | 0.05 | 1.84 | 2.18 | 1 | 0.3 | | | | | |
| I | lambda-cyhalothrin | 0.05 | 1.83 | 0.71 | 7 | 1 | 0.17 | 0.53 | 0.21 | 23 | 5 |
| I | fipronil (total) | 0.007 | 1.19 | 2.49 | 5 | 0.3 | | | | | |
| A | pyridaben | 0.05 | 1.00 | 0.29 | 3 | 2 | | | | | |
| I-A | tau-fluvalinate | 8.66 | 0.92 | 0.28 | 70 | 3 | | | | | |
| I | indoxacarb | 0.59 | 0.81 | 0.43 | 5 | 1 | | | | | |
| I | permethrin | 0.06 | 0.60 | 1.90 | 6 | 1 | 0.22 | 0.18 | 0.55 | 20 | 2 |
| I | beta-cyfluthrin | 0.03 | 0.56 | 1.41 | 14 | 1 | 0.46 | 0.04 | 0.10 | 209 | 14 |
| I | prallethrin | 0.03 | 0.42 | 0.16 | 4 | 4 | | | | | |
| I-A | coumaphos (total) | 20.29 | 0.41 | 0.28 | 158 | 3 | | | | | |
| I | phenothrin | 0.13 | 0.38 | 0.37 | 2 | 2 | | | | | |
| F | chlorothalonil | 135.32 | 0.32 | 0.35 | 169 | 1 | | | | | |
| I-A | endosulfan (total) | 6.35 | 0.26 | 0.48 | 82 | 2 | | | | | |
| I-A | carbofuran (total) | 0.16 | 0.22 | 0.55 | 13 | 1 | | | | | |
| I-A | chlorfenvinphos | 4.10 | 0.22 | 0.05 | 112 | 10 | | | | | |
| I-A | aldicarb (total) | 0.38 | 0.17 | 1.61 | 29 | 0.3 | | | | | |

**Table 2.** Cont.

| Use[1] | Chemical | Honey bee | | | | | Bumble bee | | | | |
|---|---|---|---|---|---|---|---|---|---|---|---|
| | | Topical LD50 (µg bee$^{-1}$) | Risk (%) | | TD50 (days) | | Topical LD50 (µg bee$^{-1}$) | Risk (%) | | TD50 (days) | |
| | | | Average | Max | Average | Max | | Average | Max | Average | Max |
| I-A | methomyl | 0.49 | 0.16 | 0.03 | 48 | 21 | | | | | |
| I+F | acetamiprid+propiconazole | 0.076[2] | 0.14 | 0.85 | 25 | 0.6 | 0.95[2] | 0.01 | 0.07 | 318 | 7.1 |
| I-A | diazinon | 0.38 | 0.14 | 0.05 | 45 | 9 | | | | | |
| I | acephate | 1.78 | 0.11 | 0.06 | 20 | 11 | 3.99 | 0.05 | 0.02 | 44 | 24 |
| I-A | malathion | 0.47 | 0.11 | 0.07 | 28 | 7 | | | | | |
| I+F | acetamiprid+fenbuconazole | 1.76[2] | 0.01 | 0.09 | 587 | 13 | 22.2[2] | 0.00 | 0.01 | 7407 | 166 |
| IGR | diflubenzuron | 114.80 | 0.00 | 0.00 | 1441 | 897 | 0.10[3] | 1.68 | 1.66 | 1 | 1 |

[1]A = acaricide; F = fungicide; I = insecticide; IGR = insect growth regulator; IS = insecticide synergist.
[2]Mixture LD50 estimated in accordance with known synergistic ratios [67,68,69].
[3]Chronic LD50 for 77 days exposure.

phate, carbofuran, malathion and permethrin) or both (diazinon, methonyl, aldicarb, beta-cyfluthrin and indoxacarb). Except for pyrethroid residues, which can have almost immediate effects by contact exposure, the average T50 for all other pesticides in this group is above 60 days for honey bees, denoting a very low risk by contact with pollen. Obviously, maximum residues of these pesticides would result in serious effects in very few days or even less (Table 2). Presumably, a similar risk would apply to contact with residues in wax.

## Risk by Dietary Exposure

Information on oral toxicity to bees is less comprehensive than that of topical toxicity (see Table S2), so only 77 compounds could be evaluated here. Average and maximum daily doses of residues ingested (Table S3) were calculated first to assess the risk of worker larvae, nurses and nectar foragers when exposed to the array of pesticides found in pollen and nectar (Table S1). Considering the life spans of each type of bee, the risk of consuming contaminated food during their lifetime and the T50 were assessed using the standard risk method. Results for 25 pesticides and 1 mixture that pose some risks (i.e. >0.1%) to honey bees are shown in Table 3; the remaining 67% of pesticides pose a negligible or no dietary risks to these bees.

**Dietary risk to honey bees.** Extremely high risks were found for thiamethoxam and lindane residues in honey, which affect primarily nectar foragers and secondarily the larvae. Daily consumption of nectar or honey contaminated with these compounds at the average residue levels found worldwide would cause nectar forager mortalities of 50% or above within 3 days in the case of lindane, or a week for thiamethoxam (Table 3). The risk of these two insecticides to larvae is moderate (0.6–4.0% lindane, 0.2–2.8% thiamethoxam), since larvae consume less amounts and their exposure is only during 5 days. In addition, two other neonicotinoid insecticides found in honey pose high risks to foragers (3–22% clothianidin, 6–23% imidacloprid) and moderate risks to larvae (0.2–1.2% for either compound).

Residues of the pyrethroid cypermethrin in honey pose a moderate risk to nectar foragers (4.0–6.8%) but a low risk to larvae (0.1%). Moderate risks (1–5%) are also found for the organophosphorus coumaphos and quinalphos, the neonicotinoid dinetofuran and the carbamate methiocarb, but only coumaphos and dinetofuran present some risk to larvae. Nectar foragers are at low risk (0.1–1%) when feeding on honey contaminated with 9 more insecticides: the organophosphorus chlorpyrifos, dimethoate, pirimiphos ethyl, diazinon and malathion, the carbamates carbaryl and pirimicarb, the pyrethroid beta-cyfluthrin and total residues of DDT (i.e. DDT and its metabolites). Among these, only carbaryl seems to pose a minor risk to bee larvae (0.03–0.4%) and foragers (0.5–0.8%) alike, but daily consumption of its residues would only inflict some mortality among the adult foragers (T50 of 45–80 days). Residues of the synergistic fungicides, myclobutanil, penconazole and propiconazole have so far not been detected in honey, and therefore nectar foragers are exempt of higher risks in this regard.

The residual composition of pollen is different from that of honey, with 70 out of the 124 pesticides found only in pollen (Table S1). Among the worker bees, only nurses depend entirely on this kind of food, but the queen and larvae are fed substantial amounts of pollen as well ([76], Table 1). Moderate risks of pollen residues (1–5%) to both nurses and larvae were found for thiamethoxam, clothianidin, imidacloprid and phosmet. Estimated T50s for thiomethoxam are 6–27 days for nurses and 8–23 days for larvae, depending on the residue load. Obviously, the high toxicity of this insecticide to honey bees (oral LD50 0.005 µg

**Table 3.** Risk (% probability) and time to reach oral LD50 (T50 in days) for larvae and workers of honey bees feeding on contaminated pollen and/or nectar at average or maximum residue levels (see Table S3).

| Chemical | Oral LD50 (μg bee$^{-1}$) | Risk (%) | | | | | | T50 (days) | | | | | |
|---|---|---|---|---|---|---|---|---|---|---|---|---|---|
| | | Worker larvae[1] | | Nurses[2] | | Nectar forager[3] | | Worker larvae | | Nurses | | Nectar forager | |
| | | Average | Max | Average | Max | Average | Max | Average | Max | Average | Max | Average | Max |
| thiamethoxam | 0.005 | 2.77 | 0.23 | 4.80 | 0.59 | 200.18 | 3.86 | 23 | 8 | 27 | 6 | 10 | 4 |
| gamma-HCH (lindane) | 0.05 | 0.62 | 4.02 | 0.01 | 0.01 | 200.40 | 313.09 | 9 | 0.4 | 979 | 323 | 3 | 0.1 |
| imidacloprid (total) | 0.013 | 1.19 | 0.64 | 1.57 | 2.43 | 23.33 | 5.93 | 68 | 4 | 103 | 2 | 28 | 2 |
| clothianidin | 0.004 | 1.02 | 0.23 | 1.91 | 0.36 | 22.04 | 3.25 | 54 | 10 | 58 | 13 | 23 | 4 |
| cypermethrin | 0.06 | 0.13 | 0.10 | 0.04 | 0.03 | 4.00 | 6.77 | 119 | 23 | 711 | 154 | 44 | 9 |
| coumaphos (total) | 4.61 | 0.11 | 0.03 | 0.06 | 0.04 | 2.62 | 4.37 | 1444 | 71 | 5524 | 120 | 545 | 28 |
| dinotefuran | 0.02 | 0.10 | 0.06 | 0.13 | 0.16 | 1.50 | 2.37 | 49 | 27 | 74 | 20 | 20 | 13 |
| quinalphos | 0.07 | 0.00 | 0.00 | | | 1.29 | 0.69 | 253 | 236 | | | 91 | 85 |
| methiocarb | 0.47 | 0.00 | 0.00 | 0.00 | 0.00 | 1.08 | 0.28 | 1080 | 601 | 51648 | 51648 | 391 | 217 |
| chlorpyrifos | 0.24 | 0.04 | 0.01 | 0.13 | 0.10 | 0.86 | 0.30 | 1605 | 176 | 1118 | 44 | 764 | 197 |
| carbaryl | 0.15 | 0.41 | 0.03 | 0.42 | 0.17 | 0.54 | 0.82 | 202 | 63 | 392 | 22 | 80 | 45 |
| beta-cyfluthrin | 0.05 | 0.10 | 0.03 | 0.01 | 0.03 | 0.43 | 0.62 | 190 | 123 | 3497 | 226 | 69 | 49 |
| dimethoate | 0.17 | 0.01 | 0.00 | 0.00 | 0.00 | 0.40 | 0.24 | 1198 | 662 | 11303 | 6190 | 440 | 243 |
| DDT (total) | 5.08 | 0.00 | 0.00 | 0.00 | 0.00 | 0.29 | 0.62 | 3871 | 266 | 25061 | 7108 | 1432 | 96 |
| pirimiphos ethyl | 0.22 | 0.00 | 0.00 | | | 0.21 | 0.24 | 401 | 346 | | | 144 | 125 |
| diazinon | 0.21 | 0.04 | 0.01 | 0.01 | 0.00 | 0.19 | 0.39 | 426 | 202 | 3869 | 781 | 156 | 76 |
| malathion | 9.17 | 0.00 | 0.00 | 0.00 | 0.00 | 0.15 | 0.12 | 3218 | 1292 | 82696 | 20162 | 1167 | 471 |
| pirimicarb | 3.84 | 0.00 | 0.00 | | | 0.10 | 0.09 | 3500 | 1873 | | | 1261 | 675 |
| thiacloprid+propiconazole | 0.03[4] | 0.08 | 0.29 | 0.30 | 0.53 | 0.00 | 0.00 | 109 | 4 | 61 | 5 | 57 | 2 |
| phosmet | 0.37 | 0.07 | 0.11 | 0.79 | 1.29 | | | 991 | 20 | 168 | 3 | | |
| fipronil (total) | 0.001 | 0.02 | 0.05 | 0.27 | 0.57 | | | 596 | 33 | 101 | 6 | | |
| acrinathrin (total) | 0.12 | 0.01 | 0.36 | 0.17 | 0.06 | | 49.74 | 719 | 2 | 122 | 20 | | 1 |
| acephate | 0.23 | 0.00 | 0.01 | 0.03 | 0.01 | | 0.55 | 2264 | 135 | 383 | 214 | | 54 |
| permethrin | 0.13 | 0.00 | 0.02 | 0.01 | 0.03 | | 0.52 | 10842 | 142 | 1835 | 209 | | 58 |
| methoxychlor | 5.02 | 0.00 | 0.00 | | | | 0.28 | | 293 | | | | 106 |
| dichlorvos | 0.29 | 0.00 | 0.00 | 0.00 | 0.00 | 0.07 | 0.11 | 1215 | 735 | 6561 | 4746 | 452 | 272 |

[1]Exposure period of 5 days.
[2]Exposure period of 10 days.
[3]Exposure period of 30 days.
[4]Mixture LD50 estimated in accordance with known synergistic ratios [67].

**Table 4.** Risk (% probability) and time to reach oral LD50 (T50 in days) for larvae and workers of bumble bees feeding on contaminated pollen and/or nectar at average or maximum residue levels (see Table S3).

| Chemical | Oral LD50 (μg bee$^{-1}$) | Risk (%) | | | | | | T50 (days) | | | | | |
|---|---|---|---|---|---|---|---|---|---|---|---|---|---|
| | | Worker larvae[1] | | Nurses[2] | | Nectar forager[3] | | Worker larvae | | Nurses | | Nectar forager | |
| | | Average | Max | Average | Max | Average | Max | Average | Max | Average | Max | Average | Max |
| imidacloprid (total) | 0.03 | 2.93 | 1.57 | 3.86 | 5.98 | 57.39 | 14.58 | 28 | 2 | 42 | 1 | 11 | 1 |
| heptenophos | 0.53 | 0.00 | 0.00 | | | 28.98 | 10.42 | 46 | 16 | | | 17 | 6 |
| chlorpyrifos | 0.23 | 0.23 | 0.07 | 0.65 | 0.52 | 4.44 | 1.57 | 313 | 34 | 218 | 9 | 149 | 38 |
| quinalphos | 0.18 | 0.00 | 0.00 | | | 2.51 | 1.34 | 130 | 122 | | | 47 | 44 |
| beta-cyfluthrin | 0.12 | 0.22 | 0.05 | 0.02 | 0.06 | 0.90 | 1.28 | 92 | 59 | 1683 | 109 | 33 | 23 |
| dimethoate | 0.82 | 0.01 | 0.00 | 0.00 | 0.00 | 0.41 | 0.25 | 1159 | 641 | 10930 | 5986 | 425 | 235 |
| lambda-cyhalothrin | 0.18 | 0.02 | 0.00 | 0.08 | 0.03 | 0.19 | 0.10 | 1248 | 567 | 757 | 149 | 624 | 572 |
| carbaryl | 3.88 | 0.08 | 0.01 | 0.08 | 0.03 | 0.10 | 0.16 | 1050 | 328 | 2034 | 114 | 415 | 231 |
| cyhalothrin+propiconazole | 0.01[4] | 0.12 | 0.03 | 0.11 | 0.13 | 0.19 | 0.10 | 77 | 35 | 47 | 9 | 624 | 572 |
| cyhalothrin+myclobutanil | 0.02[4] | 0.11 | 0.03 | 0.07 | 0.12 | 0.19 | 0.10 | 114 | 52 | 69 | 14 | 624 | 572 |
| cyhalothrin+penconazole | 0.04[4] | 0.10 | 0.02 | 0.03 | 0.08 | 0.19 | 0.10 | 282 | 128 | 171 | 34 | 624 | 572 |
| acetamiprid+propiconazole | 0.21[4] | 0.02 | 0.02 | 0.00 | 0.02 | 0.07 | 0.00 | 585 | 80 | 2179 | 49 | 23125 | 4173 |
| acetamiprid | 22.20 | 0.00 | 0.00 | 0.00 | 0.00 | 0.07 | 0.00 | 61259 | 8370 | 228262 | 5110 | 23125 | 4173 |
| diflubenzuron | 1.46 | 0.00 | 0.00 | 0.02 | 0.02 | | | 3349 | 2084 | 567 | 353 | | |
| bifenthrin | 0.34 | 0.00 | 0.00 | 0.01 | 0.01 | | 0.11 | 28186 | 675 | 4770 | 807 | | 284 |
| rotenone | 0.83 | 0.00 | 0.00 | 0.00 | 0.00 | | | 6538 | 6538 | 1106 | 1106 | | |
| acephate | 7.87 | 0.00 | 0.00 | 0.00 | 0.00 | | 0.08 | 15762 | 937 | 2667 | 1488 | | 378 |
| phosalone | 3.98 | 0.00 | 0.00 | 0.00 | 0.00 | | | 23177 | 23177 | 3922 | 3922 | | |
| methomyl | 3.38 | 0.00 | 0.00 | 0.00 | 0.00 | | | 59814 | 25670 | 10122 | 4344 | | |

[1]Exposure period of 5 days.
[2]Exposure period of 10 days.
[3]Exposure period of 30 days.
[4]Mixture LD50 estimated in accordance with known synergistic ratios for honeybees [67,68].

bee$^{-1}$), together with its relatively high residue loads (29 ppb) and worldwide prevalence (12.8%) are the main reasons behind this risk. Although clothianidin is 4 times more toxic than imidacloprid, average residues of the latter insecticide are slightly higher and more frequently found in pollen than those of the former, so their overall risk is very similar. Nevertheless, only the highest residues of imidacloprid would seriously affect nurses and larvae alike, with T50s of 2 and 4 days respectively, whereas the highest clothianidin residues would have a smaller impact because the T50s are longer than the life spans of the bees (Table 3). In addition to neonicotinoids, the highest residues of phosmet and fipronil in pollen can result in 50% mortality of nurses in 3 and 6 days, respectively; but the risk can be considered low due to their low average residues (0.8% phosmet and 0.3% fipronil).

Residues in pollen of 4 other insecticides (carbaryl, acrinathrin, dinotefuran and chlorpyrifos) have low risk to honey bees (0.1–1%), as their T50s exceed by a long margin the life spans of nurses and larvae (Table 3). Also, the mixture of thiacloprid+propiconazole may pose some risk to larvae and nurses (T50s of 4 and 5 days respectively) only when residues of thiacloprid in pollen are at the highest recorded levels (1 ppm); otherwise, under normal circumstances the average residues of this neonicotiniod in pollen (75 ppb) wouldn't be of concern to either forager bees (0.5%, and T50 57 days) or larvae (0.08%, and T50 109 days). Risk of the remaining compounds found in pollen is considered negligible.

**Dietary risk to bumble bees.** In the case of bumble bees, estimates of risks for 15 compounds and 4 mixtures for which toxicity data are available are shown in Table 4. Contrasting with honey bees, the dietary risk of imidacloprid to bumble bees is very high: 14.5–57.4% for nectar foragers that consume honey or nectar and 3.8–6% for nurses that feed on pollen, while a moderate risk (1.6–2.9%) was found for worker larvae that consume both types of food (Table 4). Moreover, the maximum imidacloprid residues ingested by the different types of bumble bee would reach the oral LD50 within their respective life spans, while average residues in honey result in T50 of 11 days for nectar foragers, indicating than half of them would probably die before reaching the end of their lives. In addition, residues of heptenophos in honey present high risk to forager bumble bees (10.4–29% and T50 6–17 days) but not to their larvae. Lack of toxicity data for thiamethoxam on bumble bees prevents us from a further assessment of this toxic compound using standard methods, even if some researchers have proven its negative effects on experimental bee colonies [42,43].

Moderate risks to nectar foragers (1–5%) were determined for the organophosphorus chlorpyrifos and quinalphos as well as the pyrethroid beta-cyfluthrin, but once again the risk of theses insecticides to larvae are low or negligible (Table 4). Based on the estimated T50s, only the highest residues of chlorpyrifos in pollen and beta-cyfluthrin in honey may represent a considerable risk to nurses and foragers respectively.

Residues of dimethoate, lambda-cyhalothrin, bifenthrin and carbaryl in honey pose a low risk to forager bumble bees. The risk of fungicides mixtures with cyhalothrin and acetamiprid is estimated as low as well. Since the three fungicides are only present in pollen, their synergism only affects the larvae and nurses, and even then the resulting risks are low: <0.12 for larvae and <0.13% for nurses, with T50s above their life span (Table 4). The only exception is with the highest residues of cyhalothrin in pollen (36 ppb), for which a T50 of 9 days was estimated.

## Risks by Cumulative Toxicity

Estimates of T50s for imidacloprid and thiamethoxam in honey bees were also carried out using the time-cumulative approach (Table 5). Indeed, the time to reach their oral LD50 is shorter than the T50s estimated by the standard method. For example, nurse bees feeding on pollen contaminated with imidacloprid would reach their LD50 in 7 to 9 days (within their life-span), and those ingesting thiamethoxam would die in large numbers after one day, no matter what the residue loads are. The same applies to nectar foragers, which would be at serious risk when feeding on nectar or honey contaminated with either chemical, and so would be the larvae consuming thiamethoxam.

This approach should apply to all insecticides that exhibit time-cumulative toxicity, which requires the binding to specific receptors to be persistent [72]. However, it would not apply to the majority of chemical residues found in pollen or nectar. Although fipronil and its toxic metabolites have systemic properties and high toxicity to bees [77], so far there is no evidence of time-cumulative effects of fipronil on bees or other organisms. Also, residues of pyrethroid insecticides have little effect when ingested by bees, as they are either metabolised or quickly

**Table 5.** Comparison of estimated times to LD50 (T50 range in days) for dietary exposure of honey bees to two neonicotinoid insecticides, using standard and cumulative risk approaches.

| Risk approach | Imidacloprid | | | Thiamethoxam | | |
|---|---|---|---|---|---|---|
| | T50 (days) | | | T50 (days) | | |
| | Worker larvae | Nurse | Nectar forager | Worker larvae | Nurse | Nectar forager |
| Cumulative | 6–10 | 7-9 | 4-8 | <0.1-1 | 0.2-1 | <0.1-0.5 |
| Standard | 4-68 | 2-103 | 2-28 | 8-23 | 6-27 | 4-10 |
| | Experimental data[1] | | | Experimental data[2] | | |
| Exposure (days) | 2 | 4 | 31 | 1 | 5.2 | 8.04 |
| Oral LD50 (ng bee$^{-1}$) | 28.5 | 10.79 | 0.18 | 0.109 | 0.057 | 0.009 |
| Equation | Ln T50 = 2.55−0.53 *Ln LD50 | | | Ln T50 = −1.03−0.70 *Ln LD50 | | |
| Power exponent (n = 1/slope) | 1.89 | | | 1.44 | | |
| r2 | 1.00 | | | 0.68 | | |

[1]Sources: [62,63].
[2]Source : [73].

eliminated; consequently, their oral toxicity is on average 3 times lower than their topical toxicity (Table S2). Organophosphorus and carbamate compounds undergo a similar fate, so only a handful of persistent compounds (i.e chlorpyrifos, coumaphos and chlorfenvinphos) may last long enough to cause time-dependent toxicity, if any, during chronic ingestion of their residues.

## Discussion

Bees can be exposed to plant protection products in two ways:

i) by direct exposure to either drift droplets, which are scattered during the foliar spraying of crops [33], dust from seed drilling at planting [78], or inhalation of volatile pesticides during or after application to the crops; and
ii) by exposure to residues present in pollen, wax, nectar, honey and guttation drops, which may result either from direct spray contamination of flowers, translocation through the treated plants or soil [20,54], or direct contamination during treatment of the combs (for honey bees only). Bees also drink water [79], and we have observed them drinking from paddy field waters contaminated with pesticides.

Our risk assessment in this paper deals only with residues in pollen and honey or nectar, because these constitute the essential food of bees [39]. Exposure to residues in these matrices may be by contact, while gathering pollen in the field and in the storages of the comb, or most likely by dietary and chronic ingestion of contaminated nectar, honey and pollen. Foragers presumably feed on nectar from flowers, rather than consuming the honey stores. Nectar is carried in the insect's honey stomach, and then processed by the bees before it becomes honey. Foragers carry and process far more nectar than they consume. We do not know how much of the active ingredients enter the insects during these processes, but it can be assumed that the exposure of foragers could thus be much higher than estimated here. Exposure to guttation drops was not considered here, as it is unlikely to affect most bees since such drops only appear in the early hours of the day [80]. For risk during agricultural operations the reader can consult [81,82].

The large number of agricultural chemicals found in pollen demands a rigorous evaluation of their risk to bee pollinators. Of the 124 parent compounds found in pollen from honey bee apiaries, about half of them appear with a frequency of 2% or more, 20 are present in more than 10%, and two insecticide-acaricides (coumaphos and tau-fluvalinate) appear regularly in more than 30%, particularly in North America [26,27]. It is also worrying that residues of the four most common compounds (tau-fluvalinate, coumpahos, thymol and chlorothalonil) are present at average concentrations above 100 ng $g^{-1}$ of pollen (Fig. 2a). Highest residues can be up to 20 times higher (see Table S1), although they only appear occasionally.

Some 77 compounds have also been found in honey, with 23 of them being exclusive to this matrix. Residues in honey include mainly systemic compounds, among which the most commonly found are neonicotinoid insecticides – up to 65% prevalence (Fig. 2b). Systemic insecticides can move from the soil, where they are applied as granules or seed-coatings, through the sap of the plants and reach the nectar glands at the time of pollination, when the bees are attracted to the flowers [83]. It is no surprise, therefore, that many residues found in honey are of hydrophilic herbicides (5) and fungicides (15), as they are known to translocate within the various parts of the treated plants [84,85]. The highest residue loads in honey, however, correspond to hydrophobic compounds such as lindane and coumaphos, the latter being used to treat the combs for mite control [86].

Traditional risk assessments have considered only the residue loads in pollen and the acute oral or contact toxicity of the compounds [28,33]. We draw attention here to this important distinction, as the toxicity of hydrophobic insecticides and acaricides is mostly by contact exposure whereas the toxicity of hydrophilic fungicides and systemic insecticides is mainly by oral ingestion of residues in pollen and honey. It should be noted that pyrethroids, which are highly hydrophobic compounds, are on average 3 times more toxic to bees by contact than by oral exposure. By contrast, 60% of the systemic (hydrophilic) pesticides have oral toxicities higher than their contact toxicities, up to 11 and 13 times higher in the case of clothianidin and phorate, respectively (Table S2). It follows that regulators should pay more attention to dietary toxicities of any hydrophilic compound suspected of getting into the food chain.

Also essential to any risk assessment are not just the actual residue loads, but the frequency with which they appear in pollen and/or nectar. This is because the risk of bees being affected by pesticide residues is directly proportional to the prevalence of such residues in the environment (see equation (1)). For example, chlorpyrifos and methomyl have equal oral toxicities to honey bees (0.24 ng $bee^{-1}$), so assuming equal residue loads in pollen their hazard quotient is the same. However, because chlorpyrifos is present in 14.3% of pollen and methomyl only in 3.8%, a nurse bee is more likely to be intoxicated with the former compound while feeding on pollen, and hence the risk of chlorpyrifos to the bees is greater than that of methomyl.

When considering the relative weight of these three factors, residue loads, prevalence and toxicity, in the estimation of risks, it is evident that toxicity is the most important factor. Thus, risks above 1% by contact exposure are obtained for compounds with topical LD50s of 1 μg $bee^{-1}$ or below, and this includes fipronil, four neonicotinoids (clothianidin, dinetofuran, imidacloprid and thiamethoxam), all synthetic pyrethroid insecticides (except tau-fluvalinate), pyridaben and six cholinesterase inhibitors: phosmet, chlorpyrifos, carbaryl, carbofuran, fenthion and aldicarb (Table 2). Risks by contact with residues in pollen or wax are likely overestimated here because not all residues are bioavailable by this exposure route [39]. High acute toxicity also determines the risk through dietary exposure, with neonicotinoids, cypermethrin, lindane and three cholinesterase inhibitors (coumaphos, quinalphos and methiocarb) posing the highest risks for honey bees (Table 3). In stark contrast, low risks were determined for three acaricides used in apiaries to control mites, tau-fluvalinate (0.3–1%), coumaphos (0.3–0.4%) and chlorfenvinphos (0.05–0.2%), even if their residues loads (36–128 ppb) are above the 60 ppb average and appear in pollen with a frequency of 12 to 32% ([26], Table S1). All of them present little risk to the bees because their toxicities by contact are low (4, 8 and 20 ng $bee^{-1}$). They may be of concern, however, when present in high concentrations, and it is only then that they can reach the topical LD50 in 2 to 4 days (Table 2).

The second factor used in the estimation of risk is a combination of both residue loads and prevalence. Indeed, the risk of 10 ppb of residues of a compound appearing in 10% of the pollen is equivalent to the risk of 100 ppb of the same compound appearing only in 1% of pollen. For example, among the highest average residue loads found in pollen are the fungicides captan (821 ppb) and chlorothalonil (802 ppb); chlorothalonil is of greater concern (risk 0.32%) not only because is twice as toxic as captan (135 μg $bee^{-1}$ vs 215 μg $bee^{-1}$) but also its residues are found in 27% of the pollen, whereas captan is present only in 7% of the pollen.

Since toxicity is the main factor affecting risk, the synergistic combinations of ergosterol inhibiting fungicides with pyrethroids and cyano-substituted neonicotinoids are of great concern for one reason: the intrinsic toxicity of the individual compounds is already very high in the case of pyrethroids, and it is boosted up to 16-fold when propiconazole is present among the residues [68]. For thiacloprid the synergistic factor is as high as 560-fold, and for acetamiprid 100-fold [67], so their safety features [87] become all of a sudden hazardous. However, combinations of anilinopyrimidine fungicides with the same neonicotinoids do not show synergism in bees [75], perhaps because they do not interfere with the P-450 detoxification system. The relatively low prevalence of these fungicides among pollen residues (1.8–5.5%) could be an ameliorating factor for nurses, queens and larvae, while nectar foragers would not be affected by the synergism as honey appears to be free of these fungicides. Not considered here is the synergism of chlorothalonil with fluvalinate and coumpahos (which only occurs at high concentrations of fluvalinate), because the presence of coumaphos significantly reduces the toxicity of the fluvalinate and chlorothalonil mixture [40]. The risks of fungicide-insecticide mixtures calculated here are based on the lowest prevalence among the fungicide-insecticide pairs, but even then they may be overestimated: it is obvious that not all pollen contaminated with the insecticides (e.g. 6.2% lambda-cyhalothrin and 17.7% thiacloprid) contains at the same time residues of the synergistic fungicides. Also, risks of some residue mixtures are high for contact exposures (Table 2), but low or negligible for dietary exposures (Tables 3 and 4). Experimental evidence has shown that mixtures of imidacloprid and lambda-cyhalothrin increase mortality of bumble bees (*B. terrestris*) and reduce brood production in their colonies more than when fed on pollen contaminated with only one insecticide [88]. However, the effects of insecticide mixtures are additive, not synergistic.

The risk of dietary exposure was estimated for representatives of three different types of bees (larvae, nurses and nectar foragers) in the assumption that ingested residues and/or toxic metabolites [89] remain in their bodies. As mentioned above, this is a worst-case scenario, since elimination and metabolism over time are not taken into account; therefore, dietary risks may be overestimated for some compounds in this assessment. Although only a handful of individual pesticides appear to pose a serious threat to the bees (Tables 3 and 4), we should not forget that our evaluation considered oral LD50 values, not NOEL or LOEL values. For example, nurse honey bees feeding on pollen contaminated with imidacloprid may never reach the oral LD50 for that insecticide during their short life-time of 10 days, but toxic effects will be felt among those bees well before reaching the median dose, including some mortality. There is ample evidence that honey bees and bumble bees exposed to relevant sublethal doses of imidacloprid suffer motor and learning difficulties [90,91,92] and may even die in small proportions [88]. More meaningful risk assessments can be done using 1/10 of the oral LD50 values, which represent the lowest doses required for causing toxic effects. In this way, the probabilities of risk shown in Tables 3 and 4 would increase 10 times, and the T50s will be reduced correspondingly.

What is clear from the dietary assessment shown here is that systemic insecticides rank at the top of the list of risky chemicals: thiamethoxam, clothianidin, imidacloprid, dinetofuran, and to a lesser extent methiocarb, dimethoate and carbaryl [93]. These are more likely than any other pesticide to produce long-term toxic effects in workers and larvae of bumble and honey bees. However, while the systemic aldicarb is known to translocate to nectar and affect bees in the first four weeks after treatment of plants [94] its residues have not been found in honey from apiaries in recent years. In view of these findings, banning of some neonicotinoids by the European Community seems to be justified alone on the grounds of residues in the food of bees, apart from other considerations [20] and side-effects that these compounds may have [95]. Surely, the high prevalence of neonicotinoids in honey (17–65%) is of great concern not only for worker bees but also for larvae (Fig. 2b and Table 3). Presumably, queens would be affected in a similar way as larvae, because both consume royal jelly and pollen, with the queens consuming larger quantities. Some experimental evidence indicates that the reproductive output of bumble bee queens is seriously curtailed when fed on pollen contaminated with imidacloprid [88] or thiamethoxam [43].

Moreover, the risk of neonicotinoids by dietary exposure above appears to be underestimated because it is known that these insecticides have chronic toxicities that exceed the known acute toxicities [62,73,96]. Time-cumulative effects justify a new approach to calculate T50s based on the exponential effects with time during dietary exposure. Indeed, mortality of bees increases by a power factor of 1.5 to 2, so the LD50s are reached sooner than expected. Consequently, average residues of thiamethoxam found in honey and pollen would approach the oral LD50 within the life span of larvae and worker honey bees, while average residues of imidacloprid would cause more than 50% mortality among nectar foragers and nurses and substantial mortality among larvae (Table 5). The latter predictions are deemed more realistic than the standard risks as they are in agreement with the negative effects of these insecticides observed in laboratory and semi-field experiments [63,97]. They contrast, however, with the conclusions of a recent report, funded by the chemical industry, suggesting that residues of thiamethoxam in pollen and nectar of oilseed rape and maize do not reduce the performance of honey bee colonies [98]. It should also be noted that sublethal and side-effects of neonicotinoids, such as immune suppression [99], have not been taken into account in our assessment.

Among the hydrophobic pesticides, the highest risks by dietary exposure correspond to four organophosphorus compounds (coumaphos, chlorpyrifos, heptenophos and quinalphos) on account of their high toxicity, residue loads and average prevalence in pollen (14–32%) and/or honey (4–47%). Other highly toxic insecticides such as fipronil, and pyrethroids could also have some impact on larvae and nurse bees, but low prevalence of their residues in pollen (usually <5%) and their absence in nectar or honey ensures their risks are low compared to that of neonicotinoids and cholinesterase inhibitors. Despite being designed to stop moulting in insects, average residues of diflubenzuron in pollen (80 ppb) pose little risk to bumble bees under chronic exposure because they are rarely found in that matrix (1% prevalence).

## Conclusions

The large number and frequency of pesticide residues found in pollen and nectar of crop plants pose a clear risk to bee pollinators. Based solely on contact exposure, some 18 compounds (mostly pyrethroids and neonicotinoids) pose a threat to worker bees, but only five insecticides, namely thiamethoxam, phosmet, imidacloprid, chlorpyrifos and clothianidin, and four insecticide-fungicide mixtures pose risks with probabilities above 5%.

Those three neonicotinoids plus the organochlorine lindane pose the highest risk to worker bees and larvae when feeding on contaminated honey or nectar, but only thiamethoxam is of great concern when they feed on contaminated pollen, honey or nectar. In addition, risks of systemic neonicotinoids are probably underestimated because of their time-cumulative toxicity, synergistic effects with ergosterol inhibiting fungicides, and additive effects in combination with pyrethroids. Further research on the combined effects of such mixtures is needed to fully understand the reasons behind the collapse of honey bee and bumble bee colonies.

## Supporting Information

**Table S1 Pesticide residues (μg $kg^{-1}$ or ppb) found in pollen, honey or nectar and wax together with their average prevalence (%) in Europe, the Americas and Asia.**

**Table S2 Acute toxicity (LD50 μg $bee^{-1}$) of pesticides to honey bees and bumble bees.**

**Table S3 Estimated average and maximum daily doses (ng $bee^{-1}$) of pesticide residues ingested by bees – herbicides excluded.**

**File SI Estimation of doses and risks by contact and dietary exposure.**

## Acknowledgments

We thank P. Mineau for making useful comments to earlier drafts of this paper, S. Kegley for providing data on neonicotinoids and bees, and J. Chapman for editing the manuscript.

## Author Contributions

Conceived and designed the experiments: FSB. Performed the experiments: FSB. Analyzed the data: FSB. Contributed reagents/materials/analysis tools: FSB KG. Wrote the paper: FSB KG.

## References

1. Osborne JL (2012) Ecology: Bumblebees and pesticides. Nature 491: 43–45.
2. Marletto F, Patetta A, Manino A (2003) Laboratory assessment of pesticide toxicity to bumble bees. Bulletin of Insectology 56: 155–158.
3. Mommaerts V, Smagghe G (2011) Side-effects of pesticides on the pollinator *Bombus*: An overview. In: Stoytcheva M, editor. Pesticides in the Modern World – Pests Control and Pesticides Exposure and Toxicity Assessment. Rijeka, Croatia: InTech. pp. 508–552.
4. Klein A-M, Vaissile BE, Cane JH, Steffan-Dewenter I, Cunningham SA, et al. (2007) Importance of pollinators in changing landscapes for world crops. Proceedings of the Royal Society B: Biological Sciences 274: 303–313.
5. Allen-Wardell G, Bernhardt P, Bitner R, Burquez A, Buchmann S, et al. (1998) The potential consequences of pollinator declines on the conservation of biodiversity and stability of food crop yields. Conservation Biology 12: 8–17.
6. Potts SG, Biesmeijer JC, Kremen C, Neumann P, Schweiger O, et al. (2010) Global pollinator declines: trends, impacts and drivers. Trends in Ecology & Evolution 25: 345–353.
7. Heard TA (1999) The role of stingless bees in crop pollination. Annual Review of Entomology 44: 183–206.
8. Graystock P, Yates K, Evison SEF, Darvill B, Goulson D, et al. (2013) The Trojan hives: pollinator pathogens, imported and distributed in bumblebee colonies. Journal of Applied Ecology 50: 1207–1215.
9. Goka K, Okabe K, Yoneda M (2006) Worldwide migration of parasitic mites as a result of bumblebee comercialization. Population Ecology 48: 285–291.
10. Kenta T, Inari N, Nagamitsu T, Goka K, Hiura T (2007) Commercialized European bumblebee can cause pollination disturbance: an experiment on seven native plant species in Japan. Biological Conservation 134: 298–309.
11. Cox-Foster DL, Conlan S, Holmes EC, Palacios G, Evans JD, et al. (2007) A metagenomic survey of microbes in honey bee Colony Collapse Disorder. Science 318: 283–287.
12. Cameron SA, Lozier JD, Strange JP, Koch JB, Cordes N, et al. (2011) Patterns of widespread decline in North American bumble bees. Proceedings of the National Academy of Science USA 108: 662–667.
13. Thompson H, Ball R, Brown M, Bew M (2003) *Varroa destructor* resistance to pyrethroid treatments in the United Kingdom. Bulletin of Insectology 56: 175–181.
14. Underwood RM, vanEngelsdorp D (2007) Colony Collapse Disorder: have we seen this before? Bee Culture 135: 13–15.
15. Buczek K (2009) Honey bee colony collapse disorder (CCD) [Zespo masowego giniecia pszczoy miodnej (CCD)]. Annales Universitatis Mariae Curie-Skodowska Sectio DD, Medicina Veterinaria 64: 1–6.
16. Maini S, Medrzycki P, Porrini C (2010) The puzzle of honey bee losses: a brief review. Bulletin of Insectology 63: 153–160.
17. Johnson RM, Ellis MD, Mullin CA, Frazier M (2010) Pesticides and honey bee toxicity - USA. Apidologie 41: 312–331.
18. Pettis J, vanEngelsdorp D, Johnson J, Dively G (2012) Pesticide exposure in honey bees results in increased levels of the gut pathogen *Nosema*. Naturwissenschaften 99: 153–158.
19. Vidau C, Diogon M, Aufauvre J, Fontbonne R, Vigues B, et al. (2011) Exposure to sublethal doses of fipronil and thiacloprid highly increases mortality of honeybees previously infected by *Nosema ceranae*. PloS One: e21550.
20. Goulson D (2013) An overview of the environmental risks posed by neonicotinoid insecticides. Journal of Applied Ecology 50: 977–987.
21. Blacquière T, Smagghe G, van Gestel C, Mommaerts V (2012) Neonicotinoids in bees: a review on concentrations, side-effects and risk assessment. Ecotoxicology 21: 973–992.
22. Mommaerts V, Reynders S, Boulet J, Besard L, Sterk G, et al. (2010) Risk assessment for side-effects of neonicotinoids against bumblebees with and without impairing foraging behavior. Ecotoxicology 19: 207–215.
23. Aliouane Y, El-Hassani AK, Gary V, Armengaud C, Lambin M, et al. (2009) Subchronic exposure of honeybees to sublethal doses of pesticides: effects on behavior. Environmental Toxicology and Chemistry 28: 113–122.
24. Williams GR, Tarpy DR, vanEngelsdorp D, Chauzat M-P, Cox-Foster DL, et al. (2010) Colony Collapse Disorder in context. BioEssays 32: 845–846.
25. Goulson D, Lye GC, Darvill B (2008) Decline and conservation of bumble bees. Annual Review of Entomology 53: 191–208.
26. Mullin CA, Frazier M, Frazier JL, Ashcraft S, Simonds R, et al. (2010) High levels of miticides and agrochemicals in North American apiaries: implications for honey bee health. PloS One 5: e9754.
27. Rennich K, Pettis J, vanEngelsdorp D, Bozarth R, Eversole H, et al. (2012) 2011–2012 National Honey Bee Pests and Diseases Survey Report. USDA. 17 p.
28. Stoner KA, Eitzer BD (2013) Using a hazard quotient to evaluate pesticide residues detected in pollen trapped from honey bees (*Apis mellifera*) in Connecticut. PLoS One 8: e77550.
29. Chauzat MP, Carpentier P, Martel AC, Bougeard S, Cougoule N, et al. (2009) Influence of pesticide residues on honey bee (Hymenoptera: Apidae) colony health in France. Environmental Entomology 38: 514–523.
30. Chauzat M-P, Faucon J-P, Martel A-C, Lachaize J, Cougoule N, et al. (2006) A survey of pesticide residues in pollen loads collected by honey bees in France. Journal of Economic Entomology 99: 253–262.
31. Bernal J, Garrido-Bailon E, Nozal MJd, Gonzalez-Porto AV, Martin-Hernandez R, et al. (2010) Overview of pesticide residues in stored pollen and their potential effect on bee colony (*Apis mellifera*) losses in Spain. Journal of Economic Entomology 103: 1964–1971.
32. Choudhary A, Sharma DC (2008) Dynamics of pesticide residues in nectar and pollen of mustard (*Brassica juncea* (L.) Czern.) grown in Himachal Pradesh (India). Environmental Monitoring and Assessment 144: 143–150.
33. Thompson HM (2001) Assessing the exposure and toxicity of pesticides to bumblebees (*Bombus* sp.). Apidologie 32: 305–321.
34. Greig-Smith PW, Thompson HM, Hardy AR, Bew MH, Findlay E, et al. (1994) Incidents of poisoning of honeybees (*Apis mellifera*) by agricultural pesticides in Great Britain 1981–1991. Crop Protection 13: 567–581.
35. Villa S, Vighi M, Finizio A, Bolchi Serini G (2000) Risk assessment for honeybees from pesticide-exposed pollen. Ecotoxicology 9: 287–297.
36. VanEngelsdorp D, Speybroeck N, Evans JD, Nguyen BK, Mullin C, et al. (2010) Weighing risk factors associated with Bee Colony Collapse Disorder by classification and regression tree analysis. Journal of Economic Entomology 103: 1517–1523.
37. Tennekes HA, Sánchez-Bayo F (2012) Time-dependent toxicity of neonicotinoids and other toxicants: implications for a new approach to risk assessment. Journal of Environmental & Analytical Toxicology S4: S4–001.
38. Halm M-P, Rortais A, Arnold G, Taséi JN, Rault S (2006) New risk assessment approach for systemic insecticides: the case of honey bees and imidacloprid (Gaucho). Environmental Science & Technology 40: 2448–2454.
39. EPA (2012) White Paper in Support of the Proposed Risk Assessment Process for Bees. In: Office of Chemical Safety and Pollution Prevention OoPP, Environmental Fate and Effects Division, editor. Washington, D. C. pp. 275.
40. Zhu W, Schmehl DR, Mullin CA, Frazier JL (2014) Four common pesticides, their mixtures and a formulation solvent in the hive environment have high oral toxicity to honey bee larvae. PLoS One 9: e77547.
41. Geoghegan TS, Kimberly JH, Scheringer M (2013) Predicting honeybee exposure to pesticides from vapour drift using a combined pesticide emission and

atmospheric transport model. SETAC Australasia - Multidisciplinary approaches to managing environmental pollution. Melbourne 1–3 October. pp. 174.
42. Henry Ml, Beguin M, Requier F, Rollin O, Odoux JFo, et al. (2012) A common pesticide decreases foraging success and survival in honey bees. Science 336: 348–350.
43. Whitehorn PR, O'Connor S, Wackers FL, Goulson D (2012) Neonicotinoid pesticide reduces bumble bee colony growth and queen production. Science 336: 351–352.
44. Scott-Dupree CD, Conroy L, Harris CR (2009) Impact of currently used or potentially useful insecticides for canola agroecosystems on *Bombus impatiens* (Hymenoptera: Apidae), *Megachile rotundata* (Hymentoptera: Megachilidae), and *Osmia lignaria* (Hymenoptera: Megachilidae). Journal of Economic Entomology 102: 177–182.
45. Helson BV, Barber KN, Kingsbury PD (1994) Laboratory toxicology of six forestry insecticides to four species of bee (Hymenoptera: Apoidea). Archives of Environmental Contamination and Toxicology 27: 107–114.
46. Tasei JN (2002) Impact of agrochemicals on non-*Apis* bees. In: Devillers J, Pham-Delegue MH, editors. Honey bees: estimating the environmental impact of chemicals. pp. 101–131.
47. Chauzat M-P, Martel A-C, Cougoule N, Porta P, Lachaize J, et al. (2011) An assessment of honeybee colony matrices, *Apis mellifera* (Hymenoptera: Apidae) to monitor pesticide presence in continental France. Environmental Toxicology and Chemistry 30: 103–111.
48. Pohorecka K, Skubida P, Miszczak A, Semkiw P, Sikorski P, et al. (2012) Residues of neonicotinoid insecticides in bee collected plant materials from oilseed rape crops and their effect on bee colonies. Journal of Apicultural Science 56: 115–134.
49. Balayiannis G, Balayiannis P (2008) Bee honey as an environmental bioindicator of pesticides' occurrence in six agricultural areas of Greece. Archives of Environmental Contamination and Toxicology 55: 462–470.
50. Blasco C, Fernandez M, Pena A, Lino C, Silveira MI, et al. (2003) Assessment of pesticide residues in honey samples from Portugal and Spain. Journal of Agricultural and Food Chemistry 51: 8132–8138.
51. Rissato SR, Galhiane MS, de Almeida MV, Gerenutti M, Apon BM (2007) Multiresidue determination of pesticides in honey samples by gas chromatography-mass spectrometry and application in environmental contamination. Food Chemistry 101: 1719–1726.
52. Amit C, Sharma DC (2008) Pesticide residues in honey samples from Himachal Pradesh (India). Bulletin of Environmental Contamination and Toxicology 80: 417–422.
53. Schmuck R, Schöning R, Stork A, Schramel O (2001) Risk posed to honeybees (*Apis mellifera* L, Hymenoptera) by an imidacloprid seed dressing of sunflowers. Pest Management Science 57: 225–238.
54. Stoner KA, Eitzer BD (2012) Movement of soil-applied imidacloprid and thiamethoxam into nectar and pollen of squash (*Cucurbita pepo*). PLoS One 7: e39114.
55. Dively GP, Kamel A (2012) Insecticide residues in pollen and nectar of a cucurbit crop and their potential exposure to pollinators. Journal of Agricultural and Food Chemistry 60: 4449–4456.
56. Byrne FJ, Visscher PK, Leimkuehler B, Fischer D, Grafton-Cardwell EE, et al. (2014) Determination of exposure levels of honey bees foraging on flowers of mature citrus trees previously treated with imidacloprid. Pest Management Science 70: 470–482.
57. Serra-Bonvehí J, Orantes-Bermejo J (2010) Acaricides and their residues in Spanish commercial beeswax. Pest Management Science 66: 1230–1235.
58. Wu JY, Anelli CM, Sheppard WS (2011) Sub-lethal effects of pesticide residues in brood comb on worker honey bee (*Apis mellifera*) development and longevity. PLoS One 6: e14720.
59. Landis WG, Chapman PM (2011) Well past time to stop using NOELs and LOELs. Integrated Environmental Assessment and Management 7: vi–viii.
60. Fox DR, Billoir E, Charles S, Delignette-Muller ML, Lopes C (2012) What to do with NOECS/NOELS–prohibition or innovation? Integrated Environmental Assessment and Management 8: 764–766.
61. Tomlin CDS (2009) The e-Pesticide Manual. In: Tomlin CDS, editor. 12 ed. Surrey, U.K.: British Crop Protection Council.
62. Suchail S, Guez D, Belzunces LP (2001) Discrepancy between acute and chronic toxicity induced by imidacloprid and its metabolites in *Apis mellifera*. Environmental Toxicology and Chemistry 20: 2482–2486.
63. Dechaume-Moncharmont F-X, Decourtye A, Hennequet-Hantier C, Pons O, Pham-Delegue M-H (2003) Statistical analysis of honeybee survival after chronic exposure to insecticides. Environmental Toxicology and Chemistry 22: 3088–3094.
64. Thompson HM, Hunt LV (1999) Extrapolating from honeybees to bumblebees in pesticide risk assessment. Ecotoxicology 8: 147–166.
65. Wijngaarden RPAV, Brock TCM, van den Brink PJ (2005) Threshold levels for effects of insecticides in freshwater ecosystems: a review. Ecotoxicology 14: 355–380.
66. Mattila HR, Harris JL, Otis GW (2001) Timing of production of winter bees in honey bee (*Apis mellifera*) colonies. Insectes Sociaux 48: 88–93.
67. Iwasa T, Motoyama N, Ambrose JT, Roe RM (2004) Mechanism for the differential toxicity of neonicotinoid insecticides in the honey bee, *Apis mellifera*. Crop Protection 23: 371–378.
68. Pilling ED, Jepson PC (1993) Synergism between EBI fungicides and a pyrethroid insecticide in the honeybee (*Apis mellifera*). Pesticide Science 39: 293–297.
69. Biddinger DJ, Robertson JL, Mullin C, Frazier J, Ashcraft SA, et al. (2013) Comparative toxicities and synergism of apple orchard pesticides to *Apis mellifera* (L.) and *Osmia cornifrons* (Radoszkowski). PLoS One 8: e72587.
70. Suchail S, Debrauwer L, Belzunces LP (2004) Metabolism of imidacloprid in *Apis mellifera*. Pest Management Science 60: 291–296.
71. Sánchez-Bayo F (2009) From simple toxicological models to prediction of toxic effects in time. Ecotoxicology 18: 343–354.
72. Tennekes HA, Sánchez-Bayo F (2013) The molecular basis of simple relationships between exposure concentration and toxic effects with time. Toxicology 309: 39–51.
73. Oliveira RA, Roat TC, Carvalho SM, Malaspina O (2013) Side-effects of thiamethoxam on the brain and midgut of the africanized honeybee *Apis mellifera* (Hymenopptera: Apidae). Environmental Toxicology 28: (in press) DOI: 10.1002/tox.21842.
74. Chauzat M-P, Faucon J-P (2007) Pesticide residues in beeswax samples collected from honey bee colonies (*Apis mellifera* L.) in France. Pest Management Science 63: 1100–1106.
75. Schmuck R, Stadler T, Schmidt HW (2003) Field relevance of a synergistic effect observed in the laboratory between an EBI fungicide and a chloronicotinyl insecticide in the honeybee (*Apis mellifera* L, Hymenoptera). Pest Management Science 59: 279–286.
76. Rortais A, Arnold G, Halm MP, Touffet-Briens F (2005) Modes of honeybees exposure to systemic insecticides: estimated amounts of contaminated pollen and nectar consumed by different categories of bees. Apidologie 36: 71–83.
77. Jacob CRO, Soares HM, Carvalho SM, Nocelli RCF, Malaspina O (2013) Acute toxicity of fipronil to the stingless bee *Scaptotrigona postica* Latreille. Bulletin of Environmental Contamination and Toxicology 90: 69–72.
78. Girolami V, Marzaro M, Vivan L, Mazzon L, Greatti M, et al. (2012) Fatal powdering of bees in flight with particulates of neonicotinoids seed coating and humidity implication. Journal of Applied Entomology 136: 17–26.
79. Kovac H, Stabentheiner A, Schmaranzer S (2010) Thermoregulation of water foraging honeybees - Balancing of endothermic activity with radiative heat gain and functional requirements. Journal of Insect Physiology 56: 1834–1845.
80. Thompson HM (2010) Risk assessment for honey bees and pesticides–recent developments and 'new issues'. Pest Management Science 66: 1157–1162.
81. Tapparo A, Marton D, Giorio C, Zanella A, Solda L, et al. (2012) Assessment of the environmental exposure of honeybees to particulate matter containing neonicotinoid insecticides coming from corn coated seeds. Environmental Science & Technology 46: 2592–2599.
82. Mineau P, Harding KM, Whiteside M, Fletcher MR, Garthwaite D, et al. (2008) Using reports of bee mortality in the field to calibrate laboratory-derived pesticide risk indices. Environmental Entomology 37: 546–554.
83. Tasei JN, Ripault G, Rivault E (2001) Hazards of imidacloprid seed coating to *Bombus terrestris* (Hymenoptera: Apidae) when applied to sunflower. Journal of Economic Entomology 94: 623–627.
84. Hsieh Y-N, Liu L-F, Wang Y-S (1998) Uptake, translocation and metabolism of the herbicide molinate in tobacco and rice. Pesticide Science 53: 149–154.
85. Vieira RF, Sumner DR (1999) Application of fungicides to foliage through overhead sprinkler irrigation – a review. Pesticide Science 55: 412–422.
86. Higes M, Martín-Hernández R, Martínez-Salvador A, Garrido-Bailón E, González-Porto AV, et al. (2010) A preliminary study of the epidemiological factors related to honey bee colony loss in Spain. Environmental Microbiology Reports 2: 243–250.
87. Schmuck R (2001) Ecotoxicological profile of the insecticide thiacloprid. Pflanzenschutz-Nachrichten Bayer 54: 161–184.
88. Gill RJ, Ramos-Rodriguez O, Raine NE (2012) Combined pesticide exposure severely affects individual- and colony-level traits in bees. Nature 491: 105–108.
89. Suchail S, Sousa Gd, Rahmani R, Belzunces LP (2004) *In vivo* distribution and metabolisation of 14C-imidacloprid in different compartments of *Apis mellifera* L. Pest Management Science 60: 1056–1062.
90. Decourtye A, Armengaud C, Renou M, Devillers J, Cluzeau S, et al. (2004) Imidacloprid impairs memory and brain metabolism in the honeybee (*Apis mellifera* L.). Pesticide Biochemistry and Physiology 78: 83–92.
91. Desneux N, Decourtye A, Delpuech J-M (2007) The sublethal effects of pesticides on beneficial arthropods. Annual Review of Entomology 52: 81–106.
92. Teeters BS, Johnson RM, Ellis MD, Siegfried BD (2012) Using video-tracking to assess sublethal effects of pesticides on honey bees (*Apis mellifera* L.). Environmental Toxicology and Chemistry 31: 1349–1354.
93. Moffett JO, Macdonald RH, Levin MD (1970) Toxicity of carbaryl-contaminated pollen to adult honey bees. Journal of Economic Entomology 63: 475–476.
94. Johansen CA, Rincker CM, George DA, Mayer DF, Kious CW (1984) Effects of aldicarb and its biologically active metabolites on bees. Environmental Entomology 13: 1386–1398.
95. van der Sluijs JP, Simon-Delso N, Goulson D, Maxim L, Bonmatin J-M, et al. (2013) Neonicotinoids, bee disorders and the sustainability of pollinator services. Current Opinion in Environmental Sustainability 5: 293–305.
96. Laurino D, Manino A, Patetta A, Porporato M (2013) Toxicity of neonicotinoid insecticides on different honey bee genotypes. Bulletin of Insectology 66: 119–126.

97. Cresswell J (2011) A meta-analysis of experiments testing the effects of a neonicotinoid insecticide (imidacloprid) on honey bees. Ecotoxicology 20: 149–157.
98. Pilling E, Campbell P, Coulson M, Ruddle N, Tornier I (2014) A four-year field program investigating long-term effects of repeated exposure of honey bee colonies to flowering crops treated with thiamethoxam. PLoS One 8: e77193.
99. Di Prisco G, Cavaliere V, Annoscia D, Varricchio P, Caprio E, et al. (2013) Neonicotinoid clothianidin adversely affects insect immunity and promotes replication of a viral pathogen in honey bees. Proceedings of the National Academy of Sciences 110: 18466–18471.

# Prospects of Apicultural Entrepreneurship in Coastal Districts of Eastern India: A Melissopalynological Evaluation

**Debasis Upadhyay[1,4], Swapan Bhattacharya[2], David K. Ferguson[3], Subir Bera[4]***

1 Department of Botany, Budge Budge College, Kolkata, West Bengal, India, 2 Department of Microbiology, Moulana Azad College, Kolkata, West Bengal, India, 3 Department of Paleontology, University of Vienna, Vienna, Austria, 4 Centre of Advanced Studies, Department of Botany, University of Calcutta, West Bengal, India

## Abstract

A melissopalynological analysis of fifty-one natural honey samples (twenty four spring, fifteen summer and twelve winter) collected during 2010–2011 from two east-coastal districts ($20^0 20'$ to $22^0 11'$ N, $82^0 39'$ to $87^0 01'$ E) of Orissa, India was performed. Out of 37 unifloral samples found 25 were contributed by *Apis cerana indica*, seven by *A. dorsata* and the remaining five by *A. florea*. Out of 14 multifloral samples five were contributed by *A. cerana indica*, five by *A. dorsata* and the remaining four by *A. florea*. Principal component analysis confirmed the palynological classification of the unifloral honey samples. Eighty-two bee-plant taxa belonging to forty four families were recovered. The predominant nectariferous taxa of the spring season were *Acanthus ilicifolius, Avicennia marina, Bruguiera gymnorrhiza, Cocos nucifera, Eucalyptus globulus, Phoenix paludosa, Pongamia pinnata, Prosopis juliflora, Sonneratia apetala* and *Syzygium cumini*. In the summer the predominant nectariferous taxa were *Borassus flabellifer, C. nucifera, E. globulus, Syzygium cumini, Terminalia arjuna, Aegiceras corniculatum, P. paludosa* and *Sonneratia apetala* while those of the winter were *Brassica nigra, Coriandrum sativum, Zizyphus jujuba, Alstonia scholaris, E. globulus and Bruguiera gymnorrhiza*. Very low ($<0.09$) HDE/P for 98% of the samples and absence of toxic palynotaxa assure that these honeys are suitable for human consumption. Quite extended honey flow period with spring and summer as best forage seasons for the honeybees and occurrence of 82% of these honeys with APC Group II, III and IV justify the sustainability of the present study area for establishing moderate to large-scale apicultural entrepreneurship. This should improve the socio-economic status of the people of this region.

**Editor:** Fabio S. Nascimento, Universidade de São Paulo, Faculdade de Filosofia Ciências e Letras de Ribeirão Preto, Brazil

**Funding:** Fund from any funding agency was not procured. The basic requirements of the work have been supported by the laboratory. The first author (research scholar) worked on honorary basis.

**Competing Interests:** The authors have declared that no competing interests exist.

* E-mail: berasubir@yahoo.co.in

## Introduction

Honey is produced by mutual interactions between bees and nectariferous plants. Microscopic analysis of pollen grains of natural honeys known as melissopalynology, allows identification of the different nectar sources over the season. Bees use these sources for the production of honey in a region, classifying the honey botanically and geographically according to its origin [1], [2], [3]. Melissopalynology is therefore extremely useful for hive management and allows the identification of likely periods of production of unifloral honey (honey from only one floral source), which has high commercial value [4], [5]. Usually the beekeepers do not have information of all the important nectar plants contributing to honey production and thus pollen analysis proves to be a useful guide to beekeeping in a region [6].

The per capita income of people in the coastal state of Orissa, is among those of the seven poorest states of India during 2010–2011 which amounts to $ 600 as against the Indian national figure of $ 1000 [7]. Moreover during 2009–2010 the rural poverty line of this state was as low as approximately $ 0.30 per day, against the urban poverty line $ 0.40 per day [8]. Tribal populations and forest dwellers in the coastal districts of Orissa viz. Kolha, Munda, Santal, Panpano, Shabar, Gokha, Dhoba, Kondhs [9] often depend on honey collection from wild hives as their traditional profession. It provides valuable nutrition in the form of honey, protein-rich pollen and brood for their family. In addition bee products contribute important ingredients of folk and traditional medicine. Melissopalynological analysis and its application may thus improve the economic status of the local people of an area [10], [11]. Approximately 112 families, 370 genera and 632 species of angiosperms represent high floristic diversity in Orissa [12] that can be exploited by the beekeeping industry. As there are only a few small-scale beekeepers in Orissa producing honey for the local market and personal consumption and are hardly sponsored by the Government, only 10 kg of honey on an average is produced per hive-box per year [13].

Bhadrak and Kendraparha ($20^0 20'$ to $22^0 11'$ N, $82^0 39'$ to $87^0 01'$ E) are two coastal districts of Orissa with dry deciduous and mangrove forests and a vast stretch of coast. These are bounded on the north by the Baleshwar district, Kendujhar, Jajpur and Cuttack districts on the west, Jagatsinghapur district on the south and Bay of Bengal in the east (Fig. 1). The climate of the two coastal districts under study is tropical - moist sub humid to dry sub humid type. It belongs to the mega thermal type with very high temperature during the months of April and May. The

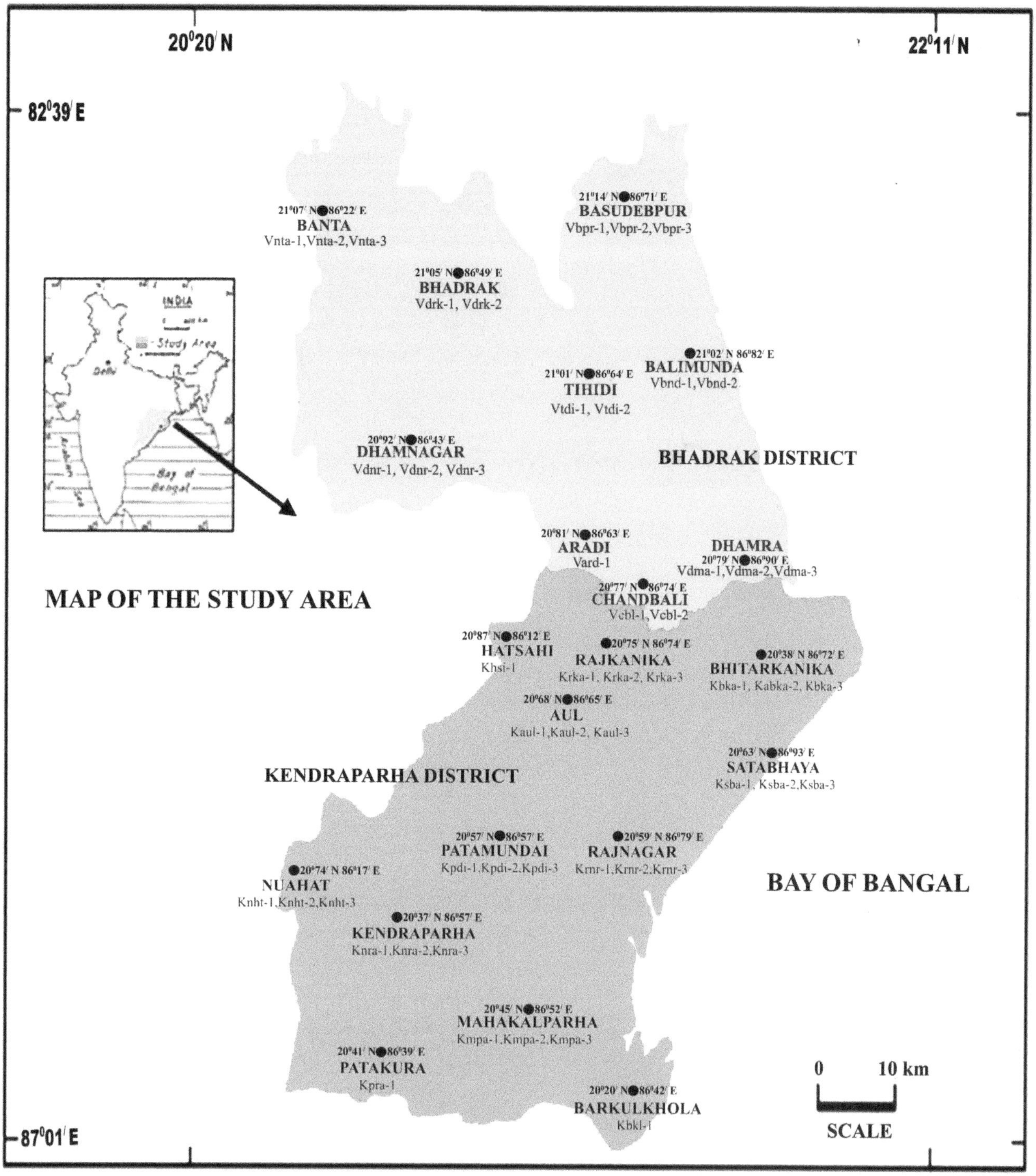


**Figure 1. Location Map of study areas in the coastal districts of Orissa, India.** Latitudes and longitudes of each study area and honey sample numbers have been mentioned.

rainfall in this area is mostly contributed by the southwest monsoon during the months of June to September [12], [28]. The predominant plants of this area include *Acacia nilotica, Acanthus illicifolius, Aegle marmelos, Aegialitis rotundifolia, Aegiceras corniculatum, Aglaia cucullata, Amaranthus viridis, Antigonon leptopus, Avicennia officinalis, Avicennia marina, Barringtonia acutangula, Bauhinia purpurea, Borassus flabellifer, Bruguiera gymnorrhiza, Capparis indica, Ceiba protandra, Ceriops decandra, Clerodendron inerne, Cocos nucifera, Crataeva religiosa, Dalbergia spinous, Derris scandens, Diospyros ebenum, Eucalyptus globulus, Eupatorium odoratum, Euphoria longa, Excoecaria agallocha, Fimbristylis falcata, Grewia tiliaefolia, Guettarda speciosa, Heritiera fomes, H. littoralis, Hibiscus tiliaceus, Hydrophylax maritime, Lagerstroemia parviflora, Madhuca indica, Mangifera indica, Merope angulata, Mikania scandens, Mimosa pudica, Myriostachya wightiana, Ocimum sanctum,*

**Table 1.** Melissopalynological data of natural honey samples from two east-coastal districts of India.

| Honey samples, extraction source area and month of collection | pH & Color of the honey samples | Floristic nature, Bee species and the season to which honey samples belong | APC (/10 g) & Group of honey | HDE/P and toxic pollen grains, if any | Bee-plant Species/Melisso palynotaxa (and their Families) recovered | No. of pollen grains & frequency of their occurrence (in %) |
|---|---|---|---|---|---|---|
| 1.Vnta-1 (Apiary, Banta, July) | 5.9, Yellowish brown | Multifloral, *A. cerana indica*, Summer | 18500, I | 14/227 = 0.06 | *Syzygium cumini* (Myrtaceae) | 96 (42.29) |
| | | | | | *Barringtonia acutangula* (Lecythidaceae) | 49 (21.59) |
| | | | | | *Saraca asoka* (Fabaceae) | 21 (9.25) |
| | | | | | *Ocimum sanctum* (Lamiaceae) | 03 (1.32) |
| | | | | | *Alternanthera sessilis* (Amarananthaceae) | 12 (5.29) |
| | | | | | *Mimosa pudica* (Fabaceae) | 17 (7.49) |
| | | | | | *Borassus flabellifer* (Arecaceae) | 29 (12.78) |
| | | | | | Total Number of Pollen grains counted | 227 |
| 2.Vnta-2 (Squeezed, Banta, April) | 6.2, Yellowish brown | Multifloral, *A. dorsata*, Spring | 95500, II | 29/662 = 0.04 | *Tecoma stans* (Bignoniaceae) | 114 (17.22) |
| | | | | | *Sesamum indicum* (Pedaliaceae) | 27 (4.08) |
| | | | | | *Eucalyptus globulus* (Myrtaceae) | 106 (16.01) |
| | | | | | *Syzygium cumini* (Myrtaceae) | 235 (35.50) |
| | | | | | *Punica granatum* (Punicaceae) | 19 (2.87) |
| | | | | | *Borassus flabellifer* (Arecaceae) | 21(3.17) |
| | | | | | *Barringtonia acutangula* (Lecythidaceae) | 08 (1.21) |
| | | | | | *Jasminum oleoides* (Oleaceae) | 29 (4.38) |
| | | | | | *Prosopis juliflora* (Fabaceae) | 16 (2.42) |
| | | | | | *Heliotropium indicum* (Boraginaceae) | 25 (3.78) |
| | | | | | *Aegle marmelos* (Rutaceae) | 62 (9.37) |
| | | | | | Total Number of Pollen grains counted | 662 |
| 3.Vnta-3 (Apiary, Banta, January) | 6.3, Amber | Unifloral, *A. florea*, Winter | 71200, II | 36/696 = 0.05 | *Tagetes patula* (Asteraceae) | 36 (5.17) |
| | | | | | *Mimosa pudica* (Fabaceae) | 473 (67.96) |
| | | | | | *Eupatorium odoratum* (Asteraceae) | 102 (14.66) |
| | | | | | *Crataeva religiosa* (Capparidaceae) | 72 (10.34) |
| | | | | | *Zizyphus jujuba* (Rhamnaceae) | 13 (1.87) |
| | | | | | Total Number of Pollen grains counted | 696 |
| 4.Vdma-2 (Squeezed, Dhamra, May) | 6.1, Light Amber | Unifloral, *A. cerana indica*, Spring | 63900, II | 21/598 = 0.03 | *Cocos nucifera* (Arecaceae) | 23 (3.85) |
| | | | | | *Borassus flabellifer* (Arecaceae) | 26 (4.35) |
| | | | | | *Anacardium occidentale* (Anacardiaceae) | 77 (12.88) |
| | | | | | *Avicennia marina* (Acanthaceae) | 429 (71.74) |
| | | | | | *Ceriops decandra* (Rhizophoraceae) | 43 (7.19) |
| | | | | | Total Number of Pollen grains counted | 598 |
| 5.Vbpr-2 (Squeezed, Basudebpur, March) | 6.3, Dark brown | Unifloral, *A. cerana indica*, Winter | 26800, II | 67/188 = 0.35 | *Anacardium occidentale* (Anacardiaceae) | 58 (30.85) |
| | | | | | *Coriandrum sativum* (Apiaceae) | 91 (48.40) |
| | | | | | *Phoenix sylvestris* (Arecaceae) | 39 (20.74) |
| | | | | | Total Number of Pollen grains counted | 188 |
| 6.Vbpr-3 (Squeezed, Basudebpur, May) | 5.5, Light yellow | Multifloral, *A. florea*, Spring | 25600, II | 16/587 = 0.02 | *Brassica nigra* (Brassicaceae) | 191 (32.54) |
| | | | | | *Scleichera oleosa* (Sapindaceae) | 95 (16.18) |
| | | | | | *Cocos nucifera* (Arecaceae) | 11 (1.87) |
| | | | | | *Mangifera indica* (Anacardiaceae) | 30 (5.11) |
| | | | | | *Eucalyptus globulus* (Myrtaceae) | 209 (35.60) |

**Table 1.** Cont.

| Honey samples, extraction source area and month of collection | pH & Color of the honey samples | Floristic nature, Bee species and the season to which honey samples belong | APC (/10 g) & Group of honey | HDE/P and toxic pollen grains, if any | Bee-plant Species/Melisso palynotaxa (and their Families) recovered | No. of pollen grains & frequency of their occurrence (in %) |
|---|---|---|---|---|---|---|
| | | | | | *Flacourtia indica* (Flacourtiaceae) | 41(6.98) |
| | | | | | *Tagetes patula* (Asteraceae) | 10 (1.70) |
| | | | | | Total Number of Pollen grains counted | 587 |
| 7.Vdrk-1 (Squeezed, Bhadrak, April) | 5.7, Amber | Multifloral, *A. florea*, Spring | 54400, II | 48/559 = 0.08 | *Phoenix sylvestris* (Arecaceae) | 135 (24.15) |
| | | | | | *Cocos nucifera* (Arecaceae) | 7 (1.25) |
| | | | | | *Peristrophe bicalyculata* (Acanthaceae) | 75 (13.42) |
| | | | | | *Capparis indica* (Capparidaceae) | 92 (16.46) |
| | | | | | *Coriandrum sativum* (Apiaceae) | 37 (6.62) |
| | | | | | *Salmalia malabarica* (Bombacaceae) | 18 (3.22) |
| | | | | | *Mimosa pudica* (Fabaceae) | 174 (31.13) |
| | | | | | Poaceae | 8 (1.43) |
| | | | | | *Carum copticum* (Apiaceae) | 13 (2.33) |
| | | | | | Total Number of Pollen grains counted | 559 |
| 8.Vcbl-1 (Squeezed Chandbali, July) | 6.1, Brown | Unifloral, *A. cerana indica*, Summer | 50800, II | 41/683 = 0.06 | *Citrus grandis* (Rutaceae) | 35 (5.20) |
| | | | | | *Papaver somnifera* (Papaveraceae) | 28 (4.16) |
| | | | | | *Sonneratia apetala* (Lythraceae) | 34 (5.05) |
| | | | | | *Aegiceras corniculatum* (Myrsinaceae) | 575 (85.56) |
| | | | | | Total Number of Pollen grains counted | 672 |
| 9.Vtdi-1 (Apiary, Tihidi, February) | 6.0, Reddish brown | Unifloral, *A. cerana indica*, Winter | 380000, III | 7/512 = 0.01 | *Mimosa pudica* (Fabaceae) | 468 (91.41) |
| | | | | | *Zizyphus jujuba* (Rhamnaceae) | 44 (8.59) |
| | | | | | Total Number of Pollen grains counted | 512 |
| 10.Vtdi-2 (Squeezed, Tihidi, April) | 5.7, Yellow | Multifloral, *A. dorsata*, Spring | 23000, II | 8/519 = 0.01 | *Phoenix sylvestris* (Arecaceae) | 82 (15.80) |
| | | | | | *Mimosa pudica* (Fabaceae) | 22 (4.24) |
| | | | | | *Cocos nucifera* (Arecaceae) | 11 (2.12) |
| | | | | | *Psidium guajava* (Myrtaceae) | 123 (23.70) |
| | | | | | *Coriandrum sativum* (Apiaceae) | 18 (3.47) |
| | | | | | *Eupatorium odoratum* (Asteraceae) | 12 (2.31) |
| | | | | | *Spondias dulcis* (Anacardiaceae) | 87 (16.76) |
| | | | | | *Euphoria longa* (Sapindaceae) | 26 (5.01) |
| | | | | | *Prosopis juliflora* (Fabaceae) | 18 (3.47) |
| | | | | | *Jasminum oleoides* (Oleaceae) | 30 (5.78) |
| | | | | | *Murraya paniculata* (Rutaceae) | 39 (7.51) |
| | | | | | *Zizyphus jujuba* (Rhamnaceae) | 51 (9.83) |
| | | | | | Total Number of Pollen grains counted | 519 |
| 11.Vbnd-1 (Apiary, Balimunda, May) | 6.2, Brown | Multifloral, *A. cerana indica*, Spring | 7400, I | 6/69 = 0.08 | *Brassica nigra* (Brassicaceae) | 8 (11.59) |
| | | | | | *Portulaca oleracea* (Portulaccaceae) | 13 (18.84) |
| | | | | | *Madhuca indica* (Sapotaceae) | 29 (42.03) |
| | | | | | *Phoenix sylvestris* (Arecaceae) | 5 (7.25) |
| | | | | | *Jasminum oleoides* (Oleaceae) | 12 (17.39) |
| | | | | | Poaceae | 2 (2.90) |

**Table 1.** Cont.

| Honey samples, extraction source area and month of collection | pH & Color of the honey samples | Floristic nature, Bee species and the season to which honey samples belong | APC (/10 g) & Group of honey | HDE/P and toxic pollen grains, if any | Bee-plant Species/Melisso palynotaxa (and their Families) recovered | No. of pollen grains & frequency of their occurrence (in %) |
|---|---|---|---|---|---|---|
| | | | | | Total Number of Pollen grains counted | 69 |
| 12.Vbnd-2 (Squeezed, Balimunda, February) | 5.9, Amber | Unifloral, *A. cerana indica*, Winter | 8100, I | 2/61 = 0.03 | *Mimosa pudica* (Fabaceae) | 11 (18.03) |
| | | | | | *Madhuca indica* (Sapotaceae) | 16 (26.23) |
| | | | | | *Tecoma stans* (Bignoniaceae) | 3 (4.92) |
| | | | | | *Zizyphus jujuba* (Rhamnaceae) | 31 (50.82) |
| | | | | | Total Number of Pollen grains counted | 61 |
| 13.Vdnr-1 (Squeezed, Dhamnagar, May) | 5.5, Light yellow | Unifloral, *A. dorsata*, Spring | 75900, II | 14/579 = 0.02 | *Syzygium cumini* (Myrtaceae) | 487 (84.11) |
| | | | | | *Cocos nucifera* (Arecaceae) | 9 (1.55) |
| | | | | | *Coriandrum sativum* (Apiaceae) | 13 (2.25) |
| | | | | | *Jasminum oleoides* (Oleaceae) | 15 (2.59) |
| | | | | | *Callistemon citrinus* ((Myrtaceae) | 41 (7.08) |
| | | | | | *Tridax procumbens* (Asteraceae) | 14 (2.42) |
| | | | | | Total Number of Pollen grains counted | 579 |
| 14.Vdnr-2 (Apiary, Dhamnagar, July) | 5.9, Yellowish amber | Unifloral, *A. cerana indica*, Summer | 16200, I | 12/440 = 0.02 | *Syzygium cumini* (Myrtaceae) | 412 (93.64) |
| | | | | | *Bauhinia purpurea* (Fabaceae) | 14 (3.18) |
| | | | | | *Barringtonia acutangula* (Lecythidaceae) | 14 (3.18) |
| | | | | | Total Number of Pollen grains counted | 440 |
| 15.Vdnr-3 (Squeezed, Dhamnagar, June) | 5.5, Yellowish brown | Multifloral, *A. dorsata*, Winter | 56800, II | 14/648 = 0.02 | *Eucalyptus gobulus* (Myrtaceae) | 104 (16.05) |
| | | | | | *Cocos nucifera* (Arecaceae) | 34 (5.25) |
| | | | | | *Borassus flabellifer* (Arecaceae) | 34 (5.25) |
| | | | | | *Coriandrum sativum* (Apiaceae) | 128 (19.75) |
| | | | | | *Rosa chinensis* (Rosaceae) | 6 (0.93) |
| | | | | | *Salmalia malabarica* (Bombacaceae) | 49 (7.56) |
| | | | | | *Carum copticum* (Apiaceae) | 55 (8.49) |
| | | | | | *Lagerstroemia parviflora* (Lythraceae) | 142 (21.91) |
| | | | | | *Spondias dulcis* (Anacardiaceae) | 20 (3.09) |
| | | | | | *Sterculia foetida* (Sterculiaceae) | 76 (11.73) |
| | | | | | Total Number of Pollen grains counted | 648 |
| 16.Vard-1 (Apiary, Aradi, May) | 5.3, Light brown | Unifloral, *A. cerana indica*, Sprlng | 35500, II | 18/586 = 0.03 | *Syzygium cumini* (Myrtaceae) | 579 (98.81) |
| | | | | | *Coriandrum sativum* (Apiaceae) | 7 (1.19) |
| | | | | | Total Number of Pollen grains counted | 586 |
| 17.Vdrk-2 (Squeezed, Bhadrak, December) | 5.6, Yellow | Unifloral, *A. dorsata*, Winter | 81000, II | 10/674 = 0.01 | *Cleome viscosa* (Cappaidaceae) | 167 (24.77) |
| | | | | | *Grewia tiliaefolia* (Tiliaceae) | 35 (5.19) |
| | | | | | *Anacardium occidentale* (Anacardiaceae) | 15 (2.22) |
| | | | | | *Carum copticum* (Apiaceae) | 10 (1.48) |
| | | | | | *Salmalia malabarica* (Malvaceae) | 04 (0.59) |
| | | | | | *Alstonia scholaris* (Apocynaceae) | 396 (58.75) |
| | | | | | *Schleichera oleosa* (Sapindaceae) | 47 (6.97) |
| | | | | | Total Number of Pollen grains counted | 674 |

**Table 1.** Cont.

| Honey samples, extraction source area and month of collection | pH & Color of the honey samples | Floristic nature, Bee species and the season to which honey samples belong | APC (/10 g) & Group of honey | HDE/P and toxic pollen grains, if any | Bee-plant Species/Melisso palynotaxa (and their Families) recovered | No. of pollen grains & frequency of their occurrence (in %) |
|---|---|---|---|---|---|---|
| 18.Vcbl-2 (Squeezed, Chandbali, April) | 4.9, Amber | Unifloral, *A. cerana indica*, Summer | 166000, III | 5/608 = 0.01 | *Anacardium occidentale* (Anacardiaceae) | 32 (5.36) |
| | | | | | *Aegiceras corniculatum* (Myrsinaceae) | 457 (76.67) |
| | | | | | *Sonneratia apetala* (Lythraceae) | 59 (9.89) |
| | | | | | *Ceiba protandra* (Bombacaceae) | 27 (4.53) |
| | | | | | *Pongamia pinnata* (Fabaceae) | 21 (3.52) |
| | | | | | Total Number of Pollen grains counted | 596 |
| 19.Vdma-1 (Squeezed, Dhamra, February) | 5.2, Light brown | Unifloral, *A. cerana indica*, Winter | 526500I, V | 4/665 = 0.01 | *Eucalyptus gobulus* (Myrtaceae) | 05 (0.75) |
| | | | | | *Cocos nucifera* (Arecaceae) | 03 (0.45) |
| | | | | | *Bruguiera gymnorrhiza* (Rhizophoraceae) | 649 (97.59) |
| | | | | | *Sonneratia apetala* (Lythraceae) | 8 (1.20) |
| | | | | | Total Number of Pollen grains counted | 665 |
| 20.Vbpr-1 (Apiary, Basudebpur, July) | 5.7, Brown | Unifloral, *A. cerana indica*, Spring | 64900, II | 48/530 = 0.09 | *Borassus flabellifer* (Arecaceae) | 6 (1.13) |
| | | | | | *Pongamia pinnata* (Fabaceae) | 429 (80.94) |
| | | | | | *Coriandrum sativum* (Apiaceae) | 21 (3.96) |
| | | | | | *Phoenix sylvestris* (Arecaceae) | 7 (1.32) |
| | | | | | *Anacardium occidentale* (Anacardiaceae) | 24 (4.52) |
| | | | | | *Mikania scandens* (Asteraceae) | 13 (2.45) |
| | | | | | *Phlogacanthus thyrsiflorus* (Acanthaceae) | 15 (2.83) |
| | | | | | Total Number of Pollen grains counted | 530 |
| 21.Vdma-3 (Squeezed, Dhamra, July) | 5.6, Light brown | Unifloral, *A. cerana indica*, Spring | 93000, II | 0/573 = 0.00 | Fabaceae | 36 (6.28) |
| | | | | | *Acanthus ilicifolius* (Acanthaceae) | 537 (93.71) |
| | | | | | Total Number of Pollen grains counted | 573 |
| 22.Khsi-1 (Squeezed, Hatsahi, June) | 5.5, Light amber | Unifloral, *A. florea*, Spring | 150100, III | 6/568 = 0.01 | *Eucalyptus globulus* (Myrtaceae) | 486 (85.56) |
| | | | | | *Antigonon leptopus* (Polygonaceae) | 36 (6.33) |
| | | | | | *Amaranthus viridis* (Amaranthaceae) | 46 (8.09) |
| | | | | | Total Number of Pollen grains counted | 568 |
| 23.Krka-1 (Apiary, Rajkanika, July) | 4.8, Yellowish brown | Unifloral, *A. cerana indica*, Spring | 98000, II | 7/536 = 0.01 | *Syzygium cumini* (Myrtaceae) | 289 (53.91) |
| | | | | | *Grewia tiliaefolia* (Tiliaceae) | 231 (43.09) |
| | | | | | *Ocimum sanctum* (Lamiaceae) | 16 (2.98) |
| | | | | | Total Number of Pollen grains counted | 536 |
| 24.Krka-2 (Squeezed, Rajkanika, May) | 5.0, Yellowish amber | Unifloral, *A. dorsata*, Spring | 89500, II | 4/482 = 0.01 | *Syzygium cumini* (Myrtaceae) | 367 (76.14) |
| | | | | | *Ocimum sanctum* (Lamiaceae) | 83 (17.21) |
| | | | | | *Barringtonia acutangula* (Lecythidaceae) | 32 (6.63) |
| | | | | | Total Number of Pollen grains counted | 482 |
| 25.Krka-3 (Squeezed, Rajkanika, July) | 5.5, Brown | Unifloral, *A. cerana indica*, Summer | 95000, II | 4/545 = 0.01 | *Sonneratia apetala* (Lythraceae) | 138 (25.32) |
| | | | | | *Aegiceras corniculatum* (Myrsinaceae) | 366 (67.15) |

**Table 1.** Cont.

| Honey samples, extraction source area and month of collection | pH & Color of the honey samples | Floristic nature, Bee species and the season to which honey samples belong | APC (/10 g) & Group of honey | HDE/P and toxic pollen grains, if any | Bee-plant Species/Melisso palynotaxa (and their Families) recovered | No. of pollen grains & frequency of their occurrence (in %) |
|---|---|---|---|---|---|---|
| | | | | | *Ceiba protandra* (Bombacaceae) | 41 (7.52) |
| | | | | | Total Number of Pollen grains counted | 545 |
| 26.Kaul-1 (Apiary, Aul, July) | 5.4, Light brown | Unifloral, *A. cerana indica*, Summer | 17600, I | 9/158 = 0.05 | *Eucalyptus globulus* (Myrtaceae) | 35 (22.15) |
| | | | | | *Borassus flabellifer* (Arecaceae) | 123 (77.84) |
| | | | | | Total Number of Pollen grains counted | 158 |
| 27.Kaul-2 (Squeezed, Aul, March) | 4.7, Light yellow | Unifloral, *A. florea*, Winter | 60400, II | 39/379 = 0.10 | *Brassica nigra* (Brassicaceae) | 243 (64.11) |
| | | | | | *Mimosa pudica* (Fabaceae) | 42 (11.08) |
| | | | | | *Eupatorium odoratum* (Asteraceae) | 94 (24.80) |
| | | | | | Total Number of Pollen grains counted | 379 |
| 28.Kaul-3 (Apiary, Aul, July) | 5.5, Amber | Unifloral, *A. cerana indica*, Spring | 196000, III | 14/576 = 0.02 | *Mimosa pudica* (Fabaceae) | 465 (80.72) |
| | | | | | *Cocos nucifera* (Arecaceae) | 74 (12.84) |
| | | | | | *Phyllanthus emblica* (Euphorbiaceae) | 37 (6.42) |
| | | | | | Total Number of Pollen grains counted | 576 |
| 29.Krnr-1 (Squeezed, Rajnagar, February) | 5.3, Light brown | Multifloral, *A.dorsata*, Spring | 210000, III | 6/545 = 0.01 | *Aegialitis rotundifolia* (Plumbaginaceae) | 19 (3.48) |
| | | | | | *Bruguiera gymnorrhiza* (Rhizophoraceae) | 203 (37.24) |
| | | | | | *Phoenix paludosa* (Arecaceae) | 238 (43.66) |
| | | | | | *Casuarina equisetifolia* (Casuarinaceae) | 85 (15.59) |
| | | | | | Total Number of Pollen grains counted | 545 |
| 30.Krnr-2 (Squeezed, Rajnagar, April) | 4.6, Yellowish brown | Unifloral, *A. dorsata*, Spring | 240000, III | 11/595 = 0.01 | *Bruguiera gymnorrhiza* (Rhizophoraceae) | 522 (87.73) |
| | | | | | *Solanum melongena* (Solanaceae) | 56 (9.41) |
| | | | | | Unidentified | 17 (2.85) |
| | | | | | Total Number of Pollen grains counted | 595 |
| 31.Krnr-3 (Squeezed, Rajnagar, June) | 5.4, Yellowish amber | Unifloral, *A. cerana indica*, Summer | 109000, III | 9/472 = 0.01 | *Syzygium cumini* (Myrtaceae) | 90 (19.06) |
| | | | | | *Aegiceras corniculatum* (Myrsinaceae) | 142 (30.08) |
| | | | | | *Phoenix paludosa* (Arecaceae) | 240 (50.84) |
| | | | | | Total Number of Pollen grains counted | 472 |
| 32.Kbka-1 (Squeezed, Bhitrkanika, July) | 5.1, Amber | Unifloral, *A. cerana indica*, Summer | 515000, IV | 0/565 = 0.00 | *Aegiceras corniculatum* (Myrsinaceae) | 516 (91.32) |
| | | | | | Unidentified | 49 (8.67) |
| | | | | | Total Number of Pollen grains counted | 565 |
| 33.Kbka-2 (Squeezed, Bhitrkanika, February | 5.5, Yellowish brown | Unifloral, *A. cerana indica*, Winter | 296000, III | 18/395 = 0.04 | *Bruguiera gymnorrhiza* (Rhizophoraceae) | 199 (50.37) |
| | | | | | *Derris scandens* (Fabaceae) | 87 (22.02) |
| | | | | | *Phoenix paludosa* (Arecaceae) | 66 (16.70) |
| | | | | | *Amaranthus viridis* (Amaranthaceae) | 43 (10.88) |
| | | | | | Total Number of Pollen grains counted | 395 |
| 34.Kbka-3 (Squeezed, Bhitrkanika, June | 4.9, Amber | Multifloral *A. cerana indica* Spring | 96500, II | 0/461 = 0.00 | *Bruguiera gymnorrhiza* (Rhizophoraceae) | 193 (41.86) |

**Table 1.** Cont.

| Honey samples, extraction source area and month of collection | pH & Color of the honey samples | Floristic nature, Bee species and the season to which honey samples belong | APC (/10 g) & Group of honey | HDE/P and toxic pollen grains, if any | Bee-plant Species/Melisso palynotaxa (and their Families) recovered | No. of pollen grains & frequency of their occurrence (in %) |
|---|---|---|---|---|---|---|
| | | | | | *Derris scandens* (Fabaceae) | 100 (21.69) |
| | | | | | *Aegialitis rotundifolia* (Plumbaginaceae) | 168 (36.44) |
| | | | | | Total Number of Pollen grains counted | 461 |
| 35.Ksba-1 (Squeezed, Satabhaya, April | 5.6, Yellowish brown | Unifloral, *A. dorsata*, Spring | 195000, III | 3/573 = 0.005 | *Derris scandens* (Fabaceae) | 63 (10.99) |
| | | | | | *Bruguiera gymnorrhiza* (Rhizophoraceae) | 212 (36.99) |
| | | | | | *Sonneratia apetala* (Lythraceae) | 298 (52.00) |
| | | | | | Total Number of pollen grains counted | 573 |
| 36.Ksba-2 (Squeezed, Satabhaya, February) | 4.7, Light brown | Unifloral, *A. cerana indica*, Spring | 190000, III | 0/551 = 0.00 | *Bruguiera gymnorrhiza* (Rhizophoraceae) | 202 (36.66) |
| | | | | | *Phoenix paludosa* (Arecaceae) | 302 (54.80) |
| | | | | | *Amaranthus viridis* (Amaranthaceae) | 47(8.52) |
| | | | | | Total Number of Pollen grains counted | 551 |
| 37.Ksba-3 (Squeezed, Satabhaya, June) | 5.6, Light amber | Unifloral, *A. cerana indica*, Summer | 78000, II | 2/402 = 0.004 | *Aegiceras corniculatum* (Myrsiniaceae) | 129 (32.08) |
| | | | | | *Sonneratia apetala* (Lythraceae) | 246 (61.19) |
| | | | | | *Amaranthus viridis* (Amaranthaceae) | 27 (6.71) |
| | | | | | Total Number of Pollen grains counted | 402 |
| 38.Kpdi-1 (Squeezed, Patamundai, June) | 4.8, Yellowish brown | Unifloral, *A. dorsata*, Summer | 13000, I | 05/433 = 0.02 | *Eucalyptus globulus* (Myrtaceae) | 83 (19.16) |
| | | | | | *Cocos nucifera* (Arecaceae) | 34 (7.85) |
| | | | | | *Borassus flabellifer* (Arecaceae) | 316 (72.97) |
| | | | | | Total Number of Pollen grains counted | 433 |
| 39.Kpdi-2 (Squeezed, Patamundai, April) | 6.4, Light yellow | Multifloral, *A. dorsata*, Spring | 17500, I | 09/352 = 0.02 | *Eucalyptus globulus* (Myrtaceae) | 26 (7.38) |
| | | | | | *Coriandrum sativum* (Apiaceae) | 102 (28.97) |
| | | | | | *Madhuca indica* (Sapotaceae) | 46 (13.06) |
| | | | | | *Rosa chinensis* (Rosaceae) | 38 (10.79) |
| | | | | | *Phoenix sylvestris* (Arecaceae) | 31 (8.80) |
| | | | | | *Moringa oleifera* (Moringaceae) | 43 (12.21) |
| | | | | | *Butea monosperma* (Fabaceae) | 66 (18.75) |
| | | | | | Total Number of Pollen grains counted | 352 |
| 40.Kpdi-3 (Squeezed, Patamundai, March) | 5.7, Yellowish brown | Unifloral, *A. florea*, Spring | 74000, II | 03/531 = 0.01 | *Eucalyptus globulus* (Myrtaceae) | 268 (50.47) |
| | | | | | *Mimosa pudica* (Fabaceae) | 79 (14.87) |
| | | | | | *Borassus flabellifer* (Arecaceae) | 33 (6.21) |
| | | | | | *Antigonon leptopus* (Polygonaceae) | 87 (16.38) |
| | | | | | *Amaranthus viridis* (Amaranthaceae) | 40 (7.53) |
| | | | | | *Phoenix sylvestris* (Arecaceae) | 24 (4.51) |
| | | | | | Total Number of Pollen grains counted | 531 |
| 41.Knht-1 (Apiary, Nuahat, June) | 6.1, Light brown | Multifloral, *A. cerana indica*, Summer | 81500, II | 6/567 = 0.01 | *Pongamia pinnata* (Fabaceae) | 194 (34.21) |
| | | | | | *Grewia tiliaefolia* (Tiliaceae) | 108 (19.04) |
| | | | | | *Amaranthus viridis* (Amaranthaceae) | 28 (4.93) |
| | | | | | *Madhuca indica* (Sapotaceae) | 55 (9.70) |
| | | | | | *Borassus flabellifer* (Arecaceae) | 36 (6.34) |

**Table 1.** Cont.

| Honey samples, extraction source area and month of collection | pH & Color of the honey samples | Floristic nature, Bee species and the season to which honey samples belong | APC (/10 g) & Group of honey | HDE/P and toxic pollen grains, if any | Bee-plant Species/Melisso palynotaxa (and their Families) recovered | No. of pollen grains & frequency of their occurrence (in %) |
|---|---|---|---|---|---|---|
| | | | | | *Punica granum* (Punicaceae) | 64 (11.28) |
| | | | | | *Capparis indica* (Capparidaceae) | 82 (14.46) |
| | | | | | Total Number of Pollen grains counted | 567 |
| 42.Knht-2 (Apiary, Nuahat, March) | 6.0, Light yellow | Unifloral, *A. cerana indica*, Winter | 96000, II | 2/511 = 0.003 | *Eucalyptus globulus* (Myrtaceae) | 357 (72.26) |
| | | | | | *Capparis indica* (Capparidaceae) | 64 (12.95) |
| | | | | | Poaceae | 73 (14.77) |
| | | | | | Total Number of Pollen grains counted | 494 |
| 43.Knht-3 (Apiary, Nuahat, May) | 5.6, Amber | Multifloral, *A. cerana indica*, Summer | 155000, III | 1/526 = 0.002 | *Eucalyptus globulus* (Myrtaceae) | 219 (41.63) |
| | | | | | *Grewia tiliaefolia* (Tiliaceae) | 218 (41.44) |
| | | | | | *Amaranthus viridis* (Amaranthaceae) | 21 (3.99) |
| | | | | | *Cleome viscosa* (Capparidaceae) | 68 (12.92) |
| | | | | | Total Number of Pollen grains counted | 526 |
| 44.Knra-1 (Squeezed, Kendrapara, February) | 4.7, Light yellow | Multifloral, *A. florea*, Winter | 14500, I | 17/382 = 0.04 | *Mimosa pudica* (Fabaceae) | 102 (26.70) |
| | | | | | *Brassica nigra* (Brassicaceae) | 105 (27.48) |
| | | | | | *Phoenix sylvestris* (Arecaceae) | 97 (25.39) |
| | | | | | *Mikania scandens* (Asteraceae) | 36 (9.42) |
| | | | | | *Capsicum frutescence* (Solanaceae) | 42 (10.99) |
| | | | | | Total Number of Pollen grains counted | 382 |
| 45.Knra-2 (Squeezed, Kendrapara, April) | 5.5, Amber | Unifloral, *A. cerana indica*, Spring | 533000, IV | 0/547 = 0.00 | *Phoenix paludosa* (Arecaceae) | 452 (82.63) |
| | | | | | *Amaranthus viridis* (Amaranthaceae) | 27 (4.93) |
| | | | | | *Lycopersicon esculentum* (Solanaceae) | 68 (12.43) |
| | | | | | Total Number of Pollen grains counted | 547 |
| 46.Knra-3 (Apiary, Kendrapara, July) | 4.9, Light amber | Unifloral, *A. cerana indica*, Summer | 503000, IV | 3/526 = 0.01 | *Terminalia arjuna* (Combretaceae) | 319 (60.64) |
| | | | | | *Capsicum frutescence* (Solanaceae) | 32 (6.08) |
| | | | | | *Acacia nilotica* (Fabaceae) | 14 (2.66) |
| | | | | | *Tamarindus indica* (Fabaceae) | 27 (5.13) |
| | | | | | *Mimusops elengi* (Sapotaceae) | 33 (6.27) |
| | | | | | *Lagerstroemia parviflora* (Lythraceae) | 101 (19.20) |
| | | | | | Total Number of Pollen grains counted | 526 |
| 47.Kmpa-1 (Apiary, Mahakalpara, June) | 5.4, Amber | Unifloral, *A. cerana indica*, Summer | 45000, II | 4/515 = 0.01 | *Eucalyptus globulus* (Myrtaceae) | 423 (82.13) |
| | | | | | *Amaranthus viridis* (Amaranthaceae) | 19 (3.68) |
| | | | | | *Phoenix sylvestris* (Arecaceae) | 73 (14.17) |
| | | | | | Total Number of Pollen grains counted | 515 |
| 48.Kmpa-2 (Squeezed, Mahaklpara, May) | 5.2, Light brown | Unifloral, *A. florea*, Spring | 105000, II | 7/424 = 0.01 | *Borassus flabellifer* (Arecaceae) | 92 (21.69) |
| | | | | | *Antigonon leptopus* (Polygonaceae) | 103 (24.29) |
| | | | | | *Coriandrum sativum* (Apiaceae) | 24 (5.66) |
| | | | | | *Prosopis juliflora* (Fabaceae) | 205 (48.34) |
| | | | | | Total Number of Pollen grains counted | 424 |

**Table 1.** Cont.

| Honey samples, extraction source area and month of collection | pH & Color of the honey samples | Floristic nature, Bee species and the season to which honey samples belong | APC (/10 g) & Group of honey | HDE/P and toxic pollen grains, if any | Bee-plant Species/Melisso palynotaxa (and their Families) recovered | No. of pollen grains & frequency of their occurrence (in %) |
|---|---|---|---|---|---|---|
| 49.Kmpa-3 (Squeezed, Mahaklpara, February) | 6.0, Light yellow | Multifloral, *A. florea*, Winter | 59500, II | 21/426 = 0.04 | *Eucalyptus globulus* (Myrtaceae) | 19 (4.46) |
| | | | | | *Coriandrum sativum* (Apiaceae) | 183 (42.95) |
| | | | | | *Brassica nigra* (Brassicaceae) | 103 (24.17) |
| | | | | | *Cleome viscosa* (Capparidaceae) | 27 (6.33) |
| | | | | | *Zizyphus jujuba* (Rhamnaceae) | 74 (17.37) |
| | | | | | *Ailanthus excelsa* (Simaroubaceae) | 20 (4.69) |
| | | | | | Total Number of Pollen grains counted | 426 |
| 50.Kpra-I (Squeezed, Patakura, May) | 5.6, Light yellow | Unifloral, *A. dorsata*, Summer | 64000, II | 22/574 = 0.03 | *Cocos nucifera* (Arecaceae) | 574 (100) |
| | | | | | Total Number of Pollen grains counted | 574 |
| 51.Kbkl-1 (Apiary, Barkulkhla, April) | 5.5, Light brown | Unifloral, *A. cerana indica*, Spring | 16000, I | 06/509 = 0.01 | *Cocos nucifera* (Arecaceae) | 461 (90.56) |
| | | | | | *Madhuca indica* (Sapotaceae) | 48 (9.43) |
| | | | | | Total Number of Pollen grains counted | 509 |

*Pandanas tectorius, Phoenix paludosa, Phoenix sylvestris, Phyllanthus rotundifolius, Pongamia pinnata, Prosopis juliflora, Psidium guajava, Rhizophora apiculata, R. mucronata, R. stylosa, Saraca asoka, Salmalia malabarica, Scleichera oleosa, Sonneratia apetala, Spinifex littoreus, Syzygium cumini, Spondias dulcis, Tamarix indica, Terminalia arjuna, Thespesia acutifolia, Tridax procumbens, Xylocarpus mekongensis* and *Zizyphus jujuba* [12], [28]. Honey pollen analytical data of this region is lacking and little is known from the whole east coastal India [10], [11], [14], [15]. In order to rectify this insufficiency of melissopalynological database the present study of qualitative and quantitative pollen analysis of natural honey samples from these two districts of Orissa has been undertaken.

## Materials and Methods

### Ethics statement

No specific permits were required for the described field studies. The sampling sites are not protected in any way and the field studies did not involve endangered or protected species.

### Materials

Fifty-one natural honey samples (out of which thirty five were squeezed and sixteen were from apiary) were collected from twenty-one different areas of Bhadrak and Kendraparha districts during different seasons in 2010–2011 (Fig. 1). The honey samples were obtained directly from the honey-bearing portion of the combs of *Apis cerana indica* (31 honey samples), *A. dorsata*, (12 samples) *and A. florea* (8 samples) as far as possible. The honey samples were mostly analyzed immediately after collection. The pH of each sample was measured with the help of a digital pH meter (GLOBAL DPH 500, India). The color of the honey samples was determined by a simple color grading method [16]. An inventory of the honey samples is presented in Table 1.

### Methods

For qualitative melissopalynogical analysis, honey samples were processed by acetolysis method adopted by Erdtman [17] with some modifications recommended by the International Commission for Bee Botany [1], [2]. Ten grams of each honey sample were transferred to a pointed glass centrifuge tube (50 ml capacity), stirred with a 1 ml pipette and dissolved in 20 ml of distilled water (20–40°C) and centrifuged for 10 min at 2000 *g*. The supernatant was decanted off. The process was repeated twice. The supernatant liquid was decanted off and all but the last drop is removed by placing the tube upside down at a 45° angle to allow the remaining excess liquid to be taken up on absorbent paper. The sediment obtained was treated with acetolysis mixture (acetic anhydride: conc. sulphuric acid = 9:1 v/v). The centrifuge tube with acetolysed pollen sediment was placed in a 70°C water bath for 10 min. After incubation the tube was centrifuged for 5 min and decanted carefully into a dry beaker. The tube was filled with distilled water and a drop of a strong detergent added and shaken vigorously and centrifuged again for 5 min. The supernatant liquid was drawn off. The entire sediment was distributed more or less equally on three slides and spread out over an area of about 20×20 mm. A drop of glycerin jelly was applied to the sediment in each slide. Another drop was poured on a cover glass, which was placed on the sediment-glycerin jelly mixture. To analyze the pollen content three pollen slides were prepared in this way from each honey sample studied and photographed under a compound microscope (Zeiss, Axioskop 2). Dried acetolysed pollen sediments were also photographed using a Scanning Electron Microscope (Carl Zeiss, Evo 40). The pollen grains were identified with the help of reference slides prepared from a collection of the local flora (Repository of spore-pollen slides of modern taxa, CUH, Palaeobotany and Palynology section, Center of Advanced Studies, Department of Botany, and University of Calcutta) and published literature [18], [19], [20], [14], [15], [21], [22]. The majority of the pollen types could be identified at the species level and those types that could not be identified were mentioned as unidentified.

For quantification of pollen grain types 500 pollen grains were randomly counted from each sample. Percentage frequency of the pollen taxa in all the samples was calculated. Unacetolysed honey samples were examined for computing the (HDE/P) ratio of Honey Dew Elements (HDE) and total number of pollen grains (P) of bee-plant taxa in each sample. Recommendations of the International Commission for Bee Botany [1] were followed for determination of the frequency classes. Pollen frequency classes are determined as predominant pollen (represented by >45% of the pollen grains counted), secondary pollen (16–45%), important minor pollen (3–15%) and minor pollen (less than 3%). A honey sample with one predominant pollen type was regarded as unifloral (and multifloral otherwise).

The absolute pollen count (APC) of the honey samples (i.e. the number of pollen grains per 10 g of honey) was calculated using a haemocytometer [23]. The samples were categorized under various groups in conformity with the universally followed grading parameters [1] viz., Group I (<20,000 pollen grains), Group II (20,000–100,000), Group III (100,000–500,000), Group IV (500,000–1,000,000) and Group V (>1,000,000).

A nectar calendar of the study area was constructed from the palynotaxa frequency data by considering all predominant pollen types in the unifloral honeys as chief nectar sources and all pollen types represented by at least 10% in the honey samples as alternative nectar sources. The seasons when a large amount of nectar from flowers is available are Honey Flow Period while the seasons when nectar source is limited and/or not available to the honey bees (for example due to heavy rain during the rainy season bee forage is affected seriously) are Honey Dearth Period.

We have attempted a multivariate analysis through Principal Component Analysis (PCA) of the melissopalynological data to evaluate the classification of these natural honey samples according to their botanical origin. Quality control methods, in conjunction with multivariate statistical analysis, have proved to be potent for classifying honey from different geographic regions [24], [25]. The matrix comprised the absolute values of all the parameters measured for the taxa recorded in each sample. PAST 2.17 version 3.0.9 [26] is used for principal component analysis (PCA) from the pollen data that are explored to derive the first principal components which are further analyzed by Microcal Origin version 6.0 [27] to examine the grouping of the samples in order to understand the relative distribution of unifloral honey samples according to their botanical origin.

## Results

Melissopalynological study assesses the status of native flora [4]. On the basis of microscopic analysis of fifty-one honey samples eighty-two angiospermic taxa belonging to forty-four families are demonstrated as bee plants in these coastal districts of India (Table 1). Three species of honeybees viz., *Apis cerana indica*, *A. dorsata* and *A. florea* contribute to these honeys. Thirty-seven samples are proved to be unifloral, twenty-five of which are contributed by *Apis cerana indica*, seven by *A. dorsata* and the remaining five by *A. florea*. Fourteen samples are multifloral, five of which contributed by *A. cerana indica*, five by *A. dorsata* and the remaining four by *A. florea*. The extent to which a given honey is derived from different plant sources can be deduced from the frequencies of the different pollen in it [28]. The predominant nectariferous taxa are *Acanthus ilicifolius, Aegiceras corniculatum, Alstonia scholaris, Avicennia marina, Borassus flabellifer, Brassica nigra, Bruguiera gymnorrhiza, Cocos nucifera, Coriandrum sativum, Eucalyptus globulus, Phoenix paludosa, Pongamia pinnata, Prosopis juliflora, Sonneratia apetala, Syzygium cumini, Terminalia arjuna*, and *Zizyphus jujuba*. Thirty

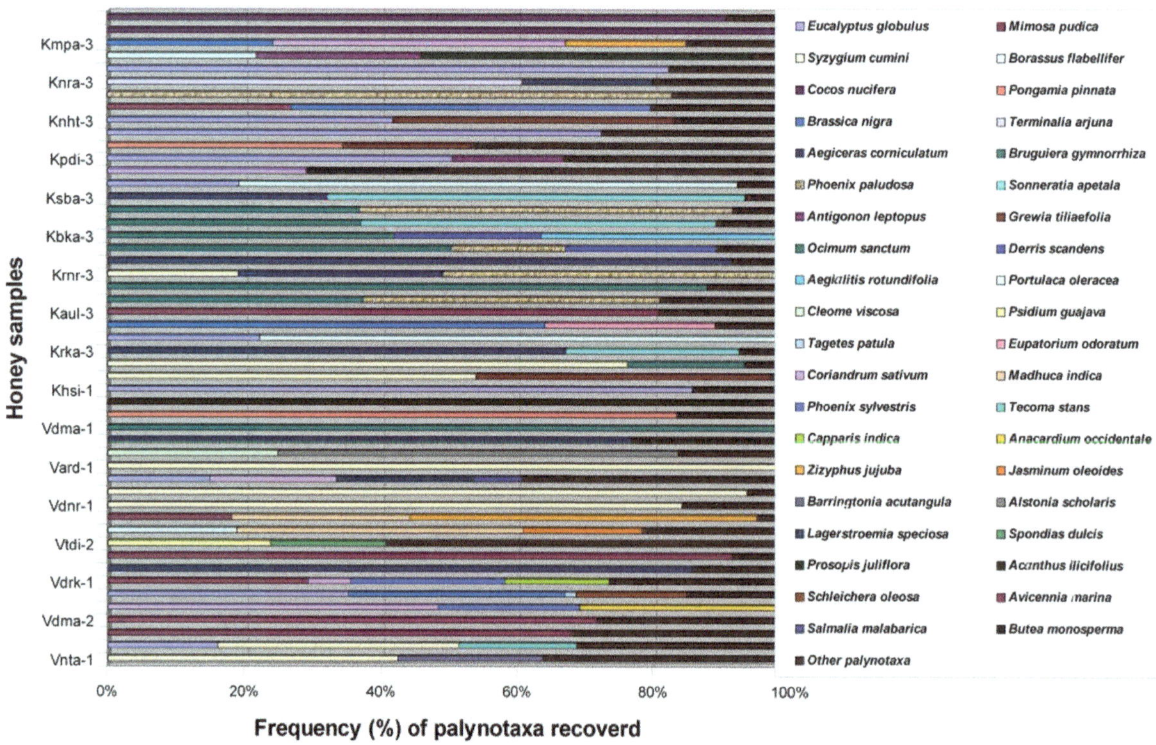


**Figure 2. Frequency (%) of palynotaxa recovered.**

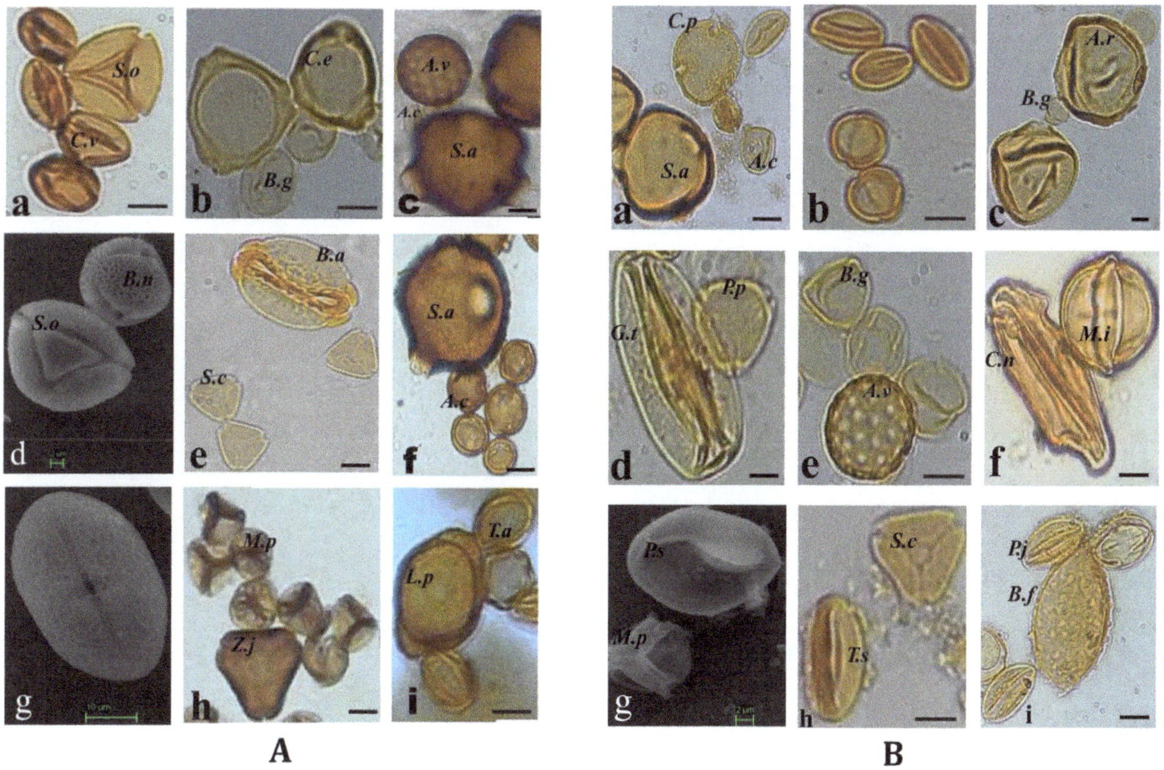

**Figure 3. A) Photomicrographs and scanning electron micrographs of some of the significant and characteristic pollen assemblage recovered from the natural honey samples of coastal district of Bhadrak and Kendraparha, India.** a. *Schleichera oleosa* (*S.o*) and *Cleome viscosa* (*C.v*) found in sample Vdrk-2; b. *Casuarina equisetifolia* (*C.e*) and *Bruguiera gymnorrhiza* (*B.g*) found in sample Krnr-1; c. *Sonneratia apetala* (*S.a*), *Aegiceras corniculatum* (*A.c*) and *Amaranthus viridis* (*A.v*) found in sample Ksba-3; d. *Brassica nigra* (*B.n*) and *Scleichera oleosa* (*S.o*) in sample Vbpr-3; e. *Syzygium cumini* (*S.c*) and *Barringtonia acutangula* (*B.a*) in sample Krka-2; f. *Aegiceras corniculatum* (*A.c*) and *Sonneratia apetala* (*S.a*) in sample Vcbl-2; g. *Acanthus ilicifolius* (*A.i*) in sample Vdma-3; h. *Mimosa pudica* (*M.p*) and *Zizyphus jujuba* (*Z.j*) in sample Vtdi-1; i. *Terminalia arjuna* (*T.a*) and *Lagerstroemia parviflora* (*L.p*) in sample Knra-3; *B.* a. *Aegiceras corniculatum* (*A.c*), *Ceiba protandra* (*C.p*) and *Sonneratia apetala* (*S.a*) in sample Krka-3; b. *Aegiceras corniculatum* (*A.c*) in sample Kbka-1; c. *Bruguiera gymnorrhiza* (*B.g*) and *Aegialitis rotundifolia* (*A.r*) in sample Kbka-3; d. *Pongamia pinnata* (*P.p*) and *Grewia tiliaefolia* (G.*t*) in sample Knht-1; e. *Bruguiera gymnorrhiza* (*B.g*) and *Amaranthus viridis* (*A.v*) in sample Ksba-2; f. *Cocos nucifera* (*C.n*) and *Madhuca indica* (*M.i*) in sample Kbkl-1; g. *Phoenix sylvestris* (*P.s*) and *Mimosa pudica* (*M.p*) in sample Vdrk-1; h. *Tecoma stans* (*T.s*) and *Syzygium cumini* (*S.c.*) in sample Vnta-2; i. *Prosopis juliflora* (*P.j*) and *Borassus flabellifer* (*B.f*) in sample Kmpa-2. Italicized initials within parenthesis after each palynotaxon above is given (as suggested by the referee) to indicate species level identification of the taxon in the respective figures of *Fig.3.A* and *B* Bar = 10 μm unless otherwise mentioned (original magnification: 400× and 1000×; d and g of Plate 1 and g of Plate 2 are scanning electron micrographs).

two secondary pollen taxa are recorded among which the significant are *Anacardium occidentale*, *Antigonon leptopus*, *Barringtonia acutangula*, *Borassus flabellifer*, *Cleome viscosa*, *Derris scandens*, *Eupatorium odoratum*, *Grewia tiliaefolia*, *Lagerstroemia parviflora*, *Madhuca indica*, *Phoenix sylvestris* and *Psidium guajava* (Table 1). *Mimosa pudica*, *Amaranthus viridis* and Poaceae types do not have nectar, so probably the bees forage these plants as a pollen source or are common probably because they become attached to the honeybees due to the close proximity of the non-nectariferous taxa to nectariferous ones. The absolute Pollen Count (APC) per 10 g of honey samples indicates that most of the samples (approximately 55%) belong to Group II (27 samples) and III (11 samples) 9 samples to Group I while 4 samples belong to Group IV. Absolute count and percentage frequency in all the samples are graphically represented in Fig. 2. HDE/P of the samples ranges from 0.0 to 0.1 with the lowest value (0.0) shown by the samples Vdma-III, Knra-II, Ksba-II, Kbka-I and Kbka-III and highest value (0.1) by Kaul-II (Table 1). The pH of the samples ranges from 4.6 to 6.4 with lowest pH (4.6) shown by the sample Krnr-II and highest pH (6.4) by Kpdi-II. The color range of the samples varies from light yellow (in eight samples) to dark brown (in the sample Vbpr-II). Photomicrographs and scanning electron micrographs of the significant and predominant palynotaxa recovered from the honey samples are recorded and presented in Fig. 3.A and B. Three taxa could not be identified (and are referred to as unidentified).

There are remarkable differences in the overall pollen contents of the honeys of Bhadrak and Kendraparha despite their geographical proximity. Fifty nine percent of the total honey samples are from Kendraparha while 41% of the samples are from Bhadrak. *Aegiceras corniculatum*, *Bruguiera gymnorrhiza*, *Mimosa pudica* and *Syzygium cumini* represent the predominant palynotaxa in both districts but *Acanthus ilicifolius*, *Alstonia scholaris*, *Avicennia marina*, *Coriandrum sativum*, *Pongamia pinnata* and *Zizyphus jujuba* are found to be predominant palynotaxa in Bhadrak. Whereas *Borassus flabellifer*, *Brassica nigra*, *Cocos nucifera*, *Eucalyptus globulus*, *Phoenix paludosa*, *Prosopis juliflora*, *Sonneratia apetala* and *Terminalia arjuna* are found to be predominant in Kendraparha. All three honeybees *A. cerana indica*, *A. dorsata and A. florea* prefer arboreal families especially Arecaceae, Myrtaceae and Fabaceae and in the mangrove areas Rhizophoraceae, Myrsinaceae and Lythraceae. *A. dorsata* never

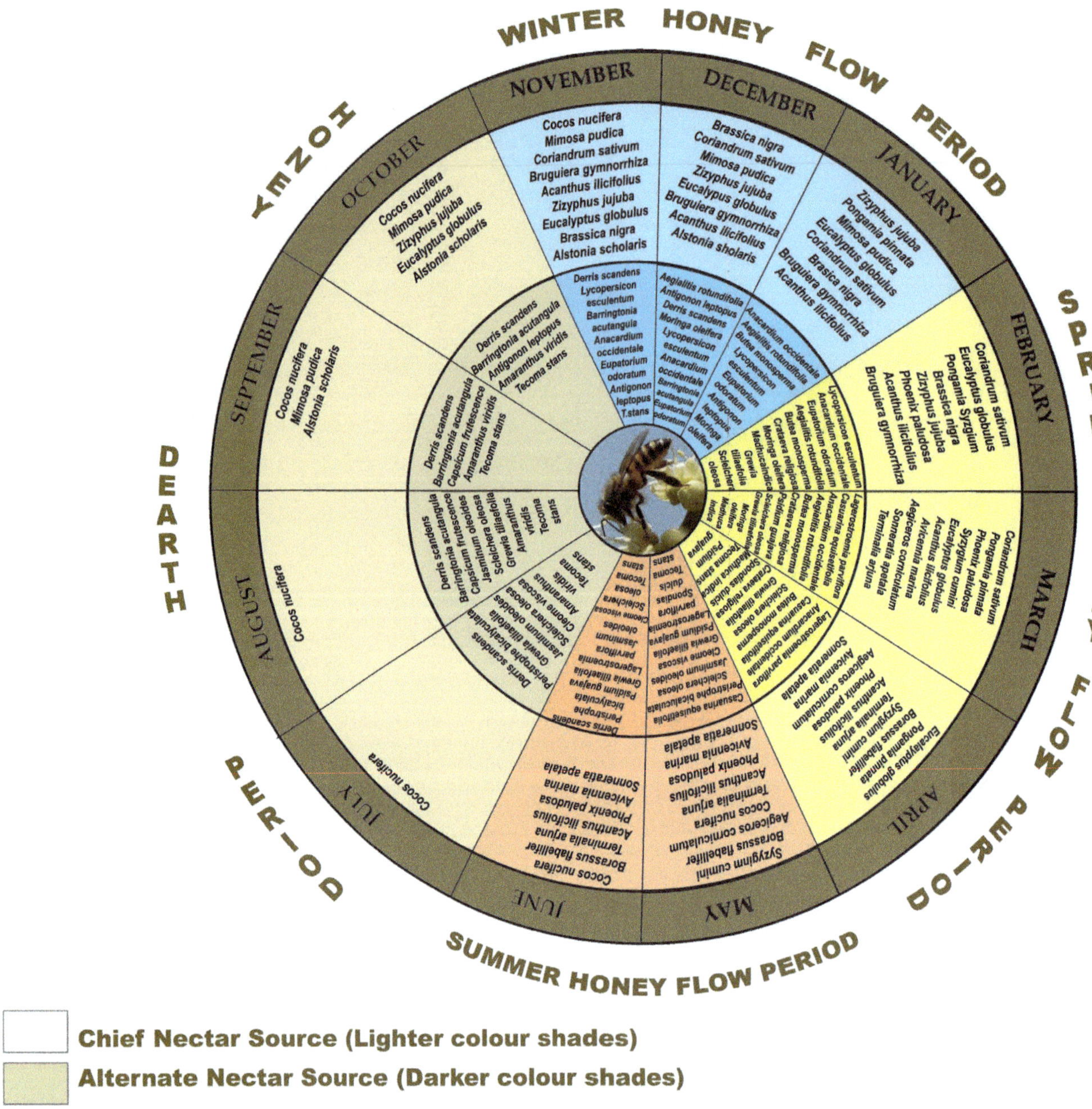


**Figure 4. Nectar calendar of Bhadrak and Kendraparha districts, Orissa, India.**

forage on *Alternanthera sessilis, Brassica nigra, Cleome gynandra, Cleome viscosa, Mikania scandens* probably due to their feeble and slender peduncles and the small size of these flowers. *A. florea* often forages on herbs near the agricultural fields. HDE/P values (Mean$\pm$SD) are lower ($0.01\pm0.02$) in Kendraparha samples than those in Bhadrak ($0.49\pm0.07$).

The nectar calendar of the study area shows that the overall honey flow period is from November to June which is divided into a Spring Honey Flow Period (Spring HFP) during February-April, a Summer Honey Flow Period (Summer HFP) from May to June and a Winter Honey Flow Period (Winter HFP) during November- January (Fig. 4). July to October is recognized as the Honey Dearth Period (HDP) because few plants serve as nectar sources possibly due to the heavy monsoon in the area. The score plot of the two principal components (PC1 and PC2) for the classification of honey samples according to their botanical origin is determined (Fig. 5). A distinct separation of honey samples is observed according to their botanical origin. The samples that are grouped in the clusters have pollen grains in common and the samples in this work are clustered into five groups. The left uppermost ellipse is a cluster in which the honey samples with *Syzygium* as predominant pollen type are grouped together. The right uppermost cluster includes the samples with *Aegiceras* as the predominant pollen type, the middle horizontal cluster represents the honey samples with *Aegiceras* as the secondary pollen type and the lowermost cluster includes honey samples with *Bruguiera* as the predominant pollen type while the middle vertical cluster includes all the remaining honey samples without *Syzygium*, *Aegiceras* and *Bruguiera* pollen grains. Principal Component Analysis has

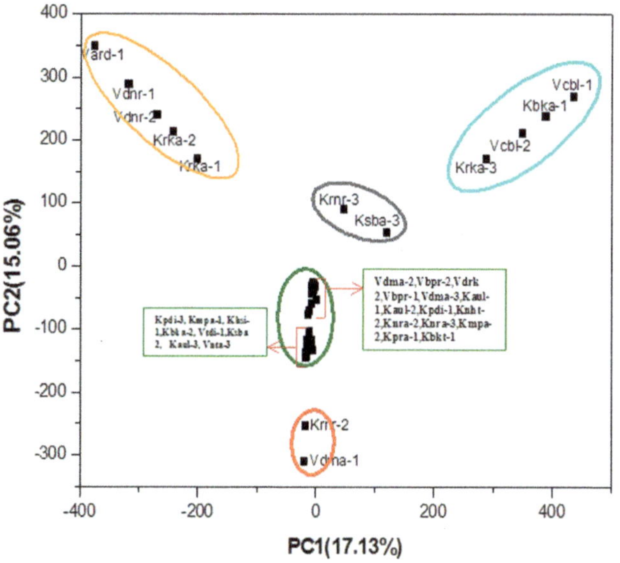


**Figure 5. Principal Component Analysis (PCA) of the honey samples using the absolute value variables per sample.**

confirmed that all of the thirty-seven unifloral honey samples have been classified correctly by the melissopalynological analysis.

## Discussion and Conclusion

Most of the indigenous people are tribes who have collected honey as a livelihood for generations. Due to an absence of organized bee keeping, since 1975 these people collect and sell honey \$ 0.40 per kg. Since cultivable land is scanty in this coastal area, 25% of the local tribes depend on fishing, 21% on farming and the remaining 20% of the population are wage-earning laborers in fishing, farming and forest product-related occupations while, nearly 35% people do not have any occupation [29]. Thus there is an urgent need for socioeconomic improvement of this eastern coastal region of India where three-fourths of the population lives under the poverty line [8].

Results of this work reveal that 70% of the honeys are unifloral among which 30% are from the mangrove forests. Principal Component Analysis has confirmed that all the thirty-seven unifloral honey samples have been classified correctly by the melissopalynological analysis. There is a high demand for unifloral honey in the local as well as in the global market [30]. Unifloral nature, very low HDE/P (<0.09 for most of the samples) and absence of toxic palynotaxa guarantee the suitability of these honey samples for human consumption.

Honeybees are effective and potential pollinators in agricultural and natural ecosystems in different seasons; this can enhance the crop productivity improving the overall socio-economic status of the local tribal communities. A monthly nectar calendar of all the honeybee species would educate rural beekeepers to produce unifloral honey in this region. This calendar of the study area and extensive field observations of bee pasturing in the present study reveal that during the Winter HFP, *Alstonia scholaris, Brassica nigra, Bruguiera gymnorrhiza, Coriandrum sativum, Eucalyptus globulus Pongamia pinnata, Syzyzium cumini* and *Zizyphus jujuba* are found to be nectar rewarding to the honeybees. During the Spring HFP the nectariferous taxa are *Acanthus ilicifolius, Avicennia marina, Bruguiera gymnorrhiza, Cocos nucifera, Eucalyptus globulus, Lagerstroemia parviflora, Moringa oleifera, Phoenix paludosa, Pongamia pinnata, Prosopis juliflora, Sonneratia apetala* and *Syzygium cumini* while during the Summer HFP the bees collect nectar from *Aegiceras corniculatum, Barringtonia acutangula, Borassus flabellifer, C. nucifera, E. globulus, Phoenix paludosa, Sonneratia apetala, Syzygium cumini, Tamarindus indica*, and *Terminalia arjuna*. Thus the honey flow period in this area is quite extended (November to June) with a relatively short honey dearth period (July to October), which is congenial for sustainable apicultural entrepreneurship.

Among the 82 nectariferous taxa Coriandrum sativum, Eucalyptus globulus, Prosopis juliflora and Terminalia arjuna are medicinally important plants; Borassus flabellifer, Syzygium cumini, and Zizyphus jujuba are fruit yielding plants; Brassica nigra and Cocos nucifera are oil yielding plants while Acanthus ilicifolius, Aegiceras corniculatum, Avicennia marina, Bruguiera gymnorrhiza, Phoenix paludosa, Pongamia pinnata, and Sonneratia apetala are mangrove plants. Arecaceae are represented in 29 samples followed by Myrtaceae and Fabaceae each in 22 and Apiaceae in 12 samples, Anacardiaceae, Rhizophoraceae and Lythraceae each in 8, Myrsinaceae 7 and Asteraceae and Sapotaceae each in 6 samples. Highest number of palynotaxa (12) is recorded in the sample Vtdi-II collected from Tihidi in Bhadrak and lowest number of palynotaxa (1) in Kpra-I collected from Patakura in Kendraparha. The vegetation cover in this region is thus diverse enough to sustain bee colonies. Local vegetation plays a significant role as to which pollens are collected by bees [31] and it is possible to produce several monofloral honeys in a region with great plant diversity [32].

This is the first case study of melissopalynological evaluation of apicultural entrepreneurship in India taking the coastal districts as a model. The findings of melissopalynology point to the potential commercial value of the honey market [4]. Qualitative and quantitative melissopalynological analyses in the present study demonstrate that these coastal regions are rich in bee plants with a potential of producing adequate unifloral honey, have an extended honey flow period and thus can be utilized commercially for a moderate to large-scale apicultural venture. We suggest that the best possible regions for apicultural entrepreneurship may be the areas near cultivated vegetation in Basudevpur, Bhadrak, Tihidi, Dhamnagar in Bhadrak district and Aul, Rajkanika, Patamundai, Kendraparha, Mahakalparha in Kendraparha district; the adjacent dry deciduous forests of Banta and Tihidi in Bhadrak, Nuahat, Patamundai, and Mahakalparha in Kendraparha and the mangrove forests of Dhamra and Chandbali in Bhadrak and Rajkanika, Bhitarkanika, Satavaya and Rajnagar in Kendraparha are also suitable for this venture. The competent authority, on the basis of the pollen analytical data presented in this work, should initiate apicultural enterprises in these two eastern coastal districts of India to improve the socio-economy of the local inhabitants of this region.


## Acknowledgments

We are sincerely grateful to Dr. Anup Sahoo, Department of Physics, Budge Budge College, Kolkata −700 137, India for analyzing the melissopalynological data to illustrate for its Principal components and Mr. Manoj Kumar Pradhan without whose active cooperation the honey samples could not have been collected from the interior rural and forested areas of Bhadrak and Kendrapara districts.


## Author Contributions

Conceived and designed the experiments: DU S. Bhattacharya DKF S. Bera. Performed the experiments: DU. Analyzed the data: DU DKF S. Bera. Contributed reagents/materials/analysis tools: DU S. Bera. Wrote the paper: DU DKF S. Bera.

## References


1. Louveaux J, Maurizio A, Vorwohl G (1978) Methods of Melissopalynology. Bee World 59(4): 139–157.
2. Von Der OW, Persano OL, Piana L, Morlot M, Martin P (2004) Harmonized methods of melissopalynology. Apidologie 35: S18–S25.
3. Barth OM (2004) Melissopalynology in Brazil: A review of pollen analysis of honeys, propolis and pollen loads of bees. Scientia Agricola 61(3): 342–350.
4. Costa MC, Vergara-Roig VA, Kivatinitz SC (2013) A melissopalynological study of artisanal honey produced in Catamarca (Argentina). Grana 52: (3), 229-237. doi:10.1080/00173134.2013.819525
5. Oliveira PP, van den Berg C, Santos FDeARD (2010) Pollen analysis of honeys from Caatinga vegetation of the state of Bahia, Brazil. Grana 49(1), 66–75 doi:10.1080/00173130903485122
6. Song XY, Yao YF, Yang WD (2012) Pollen Analysis of Natural Honeys from the Central Region of Shanxi, North China. PLoS ONE 7(11): e49545. doi:10.1371/journal.pone.0049545
7. Government of India (2011–2012) Economic survey.
8. Government of India (2009–2010) Planning Commission data.
9. Government of India (2011) Census report.
10. Upadhyay D, Bera S (2008) Pollen analysis of natural honey from Puri district, Orissa, India. Journal of Botanical Society of Bengal 61 (2): 13–17.
11. Upadhyay D, Bera S (2012) Pollen spectra of natural honey samples from a coastal district of Orissa, India. Journal of Apicultural Research 51 (1): 10–22. doi:10.3896/IBRA.1.51.1.03
12. Banerjee L K, Rao A, Shastri T, Ghosh D (2002) Diversity of coastal plant communities in India. ENVIS Centre, Botanical Survey of India; Kolkata, India.
13. Government of Orissa, India, (2010) Small scale industries in Orissa,
14. Ramanujam CGK, Kalpana TP, Fatima K (1992) Melittopalynology and recognition of major nectar and pollen sources for honey bees in some districts of Andhra Pradesh. Geophytology 22: 261–271.
15. Jana D, Bera S (2004) *Sonneratia apetala* Buch Ham as major nectar source for honey bees during mid-summer days in Sunderbans, West Bengal, India. Phytomorphology (International Society of Plant Morphologists, New Delhi) 54(1&2): 51–57.
16. Lipp J (1994) Der Honig. Verlag Eugen Ulmer Stuttgart, 205 p.
17. Erdtman G (1960) The acetolysis method: A revised description. Svensk Botanisk Tidskrift 54: 561–564.
18. Nayar T S (1990) Pollen flora of Maharashtra State, India. Today and Tomorrow's printers & publishers New Delhi, India.
19. Hari PG, Sharma C (1986) Pollen flora of North-West Himalaya. Indian Association of Palynostratigraphers, Janki Bhawan, Narhi, Lucknow- 226 001, India.
20. Chaya PK, Varma YNR (2004) Pollen diversity in the honey samples from Medak district of Andhra Pradesh. Geophytology 34(1&2): 95–100.
21. Jana D, Bandyopadhyay A, Bera S (2002) Pollen analysis of winter honey samples from Murshidabad district, West Bengal. Geophytology 30(1&2): 91–97.
22. Saha R, Bera S, Datta BK (2007) Melittopalynological Investigation of *Apis cerana indica* Fabr. Honeys from West Tripura, India. Advances in Plant Sciences 20 II: 571–578.
23. Suryanarayana MC, Seethalakshmi TS, Phadkae RP (1981) Pollen analysis of Indian honeys-1. Honeys from Litchi (*Nephelium litchi*) and Jamun (*Syzygium cumini*). Proc. IV Int. Palynol. Conf. Lucknow (1976–77) 3: 491–498.
24. Marini F, Magri AL, Balestrieri F, Fabretti F, Marini D (2004) Supervised pattern recognition applied to the discrimination of the floral origin of six types of Italian honey samples. Analytica Chimica Acta 515(1): 117–125.
25. Devillers J, Morlot M, Pham-Delegue MH, Dore JC (2004) Classification of monofloral honeys based on their quality control data. Food Chemistry 86(2): 305–312.
26. Hammer O, Harper DAT, Ryan PD (2012) PAST 2.17, version 3.0.9: Paleontological Statisticscs Software package for Education and data analysis. University of Oslo, Norway.
27. Microcal Origin version 6.0 (1991–1999) Microcal Software Inc. 1, Round House Plaza, Northampton, MA 01060 USA.
28. Dobre I, Alexe P, Escuredo O, Maria-Seijo C (2013) Palynological evaluation of selected honeys from Romania. Grana 52(2): 113–121 doi:10.1080/00173134.2012.72444319
29. Ravishankar T, Navamuniyammal M, Gnanappazham Nayak SS, Mahapatra GC, Selvam VM (2004) S. Research Foundation, Chennai, India. Atlas of Mangrove Wetlands of India, Part-3, Orissa.
30. Guemes-Ricalde FJ, Villanueva GR, Echazarretagonzalez C, Gomez-Alvarez R, Pat-Fernandez JM (2006) Production costs of conventional and organic honey in the Yucatán Peninsula of Mexico. Journal of Apicultural Research 45(3): 106–111. doi:10.3896/IBRA.1.45.3.02
31. Webby R (2004) Floral origin and seasonal variation of bee-collected pollens from individual colonies in New Zealand. Journal of Apicultural Research 43(3): 83–92.
32. Villanueva-Gutiérrez R, Moguel-Ordóñez YB, Echazarreta-González CM, Arana-López G (2009) Monofloral honeys in the Yucatán Peninsula, Mexico Grana 48(3): 214–223. doi:10.1080/00173130902929203

# 21

# Modeling Tuberculosis Dynamics, Detection and Control in Cattle Herds

**Mohammed El Amine Bekara[1], Aurélie Courcoul[1], Jean-Jacques Bénet[2], Benoit Durand[1]***

**1** University Paris Est, Anses, Laboratory of Animal Health, Epidemiology Unit, Maisons-Alfort, France, **2** University Paris Est, National Veterinary School of Alfort (ENVA), EpiMAI Unit, Maisons-Alfort, France

## Abstract

Epidemiological models are key tools for designing and evaluating detection and control strategies against animal infectious diseases. In France, after decades of decrease of bovine tuberculosis (bTB) incidence, the disease keeps circulating. Increasing prevalence levels are observed in several areas, where the detection and control strategy could be adapted. The objective of this work was to design and calibrate a model of the within-herd transmission of bTB. The proposed model is a stochastic model operating in discrete-time. Three health states were distinguished: susceptible, latent and infected. Dairy and beef herd dynamics and bTB detection and control programs were explicitly represented. Approximate Bayesian computation was used to estimate three model parameters from field data: the transmission parameter when animals are inside ($\beta_{inside}$) and outside ($\beta_{outside}$) buildings, and the duration of the latent phase. An independent dataset was used for model validation. The estimated median was 0.43 [0.16–0.84] month$^{-1}$ for $\beta_{inside}$ and 0.08 [0.01–0.32] month$^{-1}$ for $\beta_{outside}$. The median duration of the latent period was estimated 3.5 [2–8] months. The sensitivity analysis showed only minor influences of fixed parameter values on these posterior estimates. Validation based on an independent dataset showed that in more than 80% of herds, the observed proportion of animals with detected lesions was between the 2.5% and 97.5% percentiles of the simulated distribution. In the absence of control program and once bTB has become enzootic within a herd, the median effective reproductive ratio was estimated to be 2.2 in beef herds and 1.7 in dairy herds. These low estimates are consistent with field observations of a low prevalence level in French bTB-infected herds.

**Editor:** Caroline Colijn, Imperial College London, United Kingdom

**Funding:** This work was supported by the French Ministry of Agriculture. The funder had no role in study design, data collection and analysis, decision to publish, or preparation of the manuscript.

**Competing Interests:** The authors have declared that no competing interests exist.

* Email: benoit.durand@anses.fr

## Introduction

Bovine tuberculosis (bTB) is a chronic animal disease most often caused by *Mycobacterium bovis*, that mainly affects the respiratory system [1]. The main route of transmission of the infection between cattle is the respiratory route [2]. In France, a control program for bTB became mandatory in 1965. Detection of infected herds was based on an annual screening of animals using skin tests, and on routine inspection of carcasses at slaughter for bTB-like lesions (with subsequent isolation of *M. bovis* at the laboratory). Since 1990, the control program was reinforced by the compulsory screening (using skin tests) of animals introduced in bTB-free herds, these animals always originating from other bTB-free herds. In 1999, herd prevalence fell below 0.1% and the slaughter of all cattle in infected herds (termed below "total slaughter") was introduced in the control program. In 2001, France was declared officially free of bTB by the European Union. However, in recent years, an increase in bTB incidence has been observed in some departments. Similar trends have been observed in other European countries. In Great Britain and Ireland, the control programs became mandatory in 1950 and 1957, respectively. They were also based on an annual screening by skin test, the slaughter of positive animals and the inspection of carcasses at the slaughterhouse [3]. bTB incidence decreased in both countries, but the epidemiological situation began to deteriorate gradually from the 1980s, partly because of the existence of a wildlife reservoir in badger populations [4]. In France, such a wildlife reservoir has not been identified, although wild boars, red deer and badgers with *M. bovis* lesions have been found in several departments. Unlike France, Great Britain or Ireland, there are countries where "test and slaughter" control programs were successful in eradicating bTB: Sweden was considered bTB-free in 1958, after 40 years of this type of control program [5] (sporadic cases have been reported afterwards, the last one in 1978 [6]).

In France, data collected over 35 years, between 1965 and 2000, have shown parallel evolutions of herd management practices (with a disappearance of family farms and a professionalization of breeders) and of herd structures (with changes in herd types with a switch from dairy to beef and increasing herd sizes). Besides the effectiveness of control programs, the decrease of bTB incidence between 1965 and 2000 could be partly attributable to these changes in herd management practices and in herd structures [7].

**Table 1.** Description of the four datasets.

| Id | Data type | Number of herds (animals) | | Period (geographic origin) | Control program |
|---|---|---|---|---|---|
| | | Dairy | Beef | | |
| A | Aggregated | 5 (134) | | 1981–1983 (Nord) | Annual screening by SITT, slaughter of all cattle if>40% of SITT-positive animals, otherwise slaughter of SITT-positive only |
| B | Individual | 1 (60) | 12 (625) | 2004–2006 (Dordogne) | Biennial screening by SITT confirmed by SICCT, slaughter of SICCT-positive animals, slaughter of all cattle in case of lesions and *M. bovis* isolation |
| | Aggregated | 9 (684) | 16 (1,011) | | |
| C | Individual | 3 (142) | 11 (596) | 2007–2010 (Dordogne) | Annual screening by SITT with second control by γIFN[1] and confirmation by SICCT, slaughter of SICCT-positive, slaughter of all cattle in case of lesions and *M. bovis* isolation |
| | Aggregated | 7 (530) | 22 (1,344) | | |
| D | Aggregated | | 29 (6,317) | 2005–2009 (Côte d'Or) | Biennial screening by SICCT, slaughter of SICCT-positive animals, slaughter of all cattle in case of lesions and *M. bovis* isolation |

[1]To improve the specificity of the control program C, the γIFN assay used in this control program is a combination of two tests: the PPD γIFN and the ESAT-6 γIFN. The modified γIFN test can be positive (both tests are positive), negative (both tests are negative) or divergent (the results of the two tests are discordant).

Neither ante-mortem tests [8] nor post-mortem tests [6,9] used in bTB control programs have perfect sensitivity and specificity. With the decrease of infection prevalence and because of the increase of herd sizes (multiplied on average by 3.5 between 1965 and 2000 in France [7]), ante-mortem false-positive reactions became a major problem for the surveillance and control program and its acceptability by breeders. This led to an increasing complexity of testing procedures over the past fifteen years, with combinations of several tests: single intradermal tuberculin test (SITT), single intradermal comparative cervical tuberculin test (SICCT) and gamma interferon test (γIFN). Additionally, total slaughter has been replaced in some areas by the slaughter of reactive animals only. All these adaptations allowed reducing the number of animals slaughtered due to false-positive tests, the counterpart being the burdensomeness of control programs due to prolonged restriction of cattle movements.

The failure of the eradication of bTB in France and other European countries seems therefore the result of a complex interaction between the evolution of herd management practices and herd structures, the evolution of control programs, and the implication of wildlife as a reservoir for infection in cattle.

Mathematical modelling is used for the study of complex phenomena such as the dynamics of infectious agents [10]. Diseases with a long incubation period, such as bTB, are difficult to study experimentally or in the field because of prolonged waiting times for obtaining results and the high costs of conducting experimental studies [2]. Mathematical simulation models offer the possibility to test a range of control programs in a short time and to identify the most effective one [2].

A number of mathematical models have been developed to represent the spread of bTB and assess the effectiveness of control measures, especially for wildlife (badgers, opossums and white-tailed deer) [11]. Several models have been built to simulate the dynamics of bTB within cattle herds, to quantify the importance of within-farm bTB transmission by estimating the within-herd transmission coefficient of infection [2,12–14] or to evaluate detection and control strategies against bTB [2,13,15,16].

Each of these models incorporates three processes that shape the within-herd bTB infection dynamics: (i) the natural history of the disease, (ii) husbandry practices and (iii) the surveillance and control program [15]. Because points (ii) and (iii) are specific to a particular context (geographical area and time period), the models mentioned above are difficult to extrapolate to other countries.

The aim of this study was to design, calibrate and validate a model of spread of *M. bovis* within a cattle herd. Parameter definition was designed to allow simulating various herd management practices as well as control programs of arbitrary complexity.

## Materials and Methods

### Ethical statement

bTB is a notifiable disease for which there are control and surveillance campaigns in France. Official methods for diagnosis of this disease are culture, PCR and histopathology. Therefore, all the datasets included in this study are issued from animals analyzed within an official context. No purpose killing of animals was performed for this study. All datasets were in complete agreement with national and European regulations. No ethical approval was necessary.

### Data

Four datasets denoted A, B, C and D were used, which had been collected at different periods (between 1980 and 2010) in 3 French departments (Nord, Dordogne, and Côte d'Or). At that time and in these departments, four specific control programs were applied, also denoted A, B, C and D (Table 1).

The total number of herds in the four datasets was 115. All were submitted to a total slaughter. Out of all these 115 farms, detailed data for each animal (individual data) were available for 27 herds, namely the year of birth, the screening tests used, the date and results of these tests, the date of slaughter and the result of carcass inspection at the slaughterhouse (presence or absence of lesions). For the 88 remaining farms, we obtained a single data point aggregated at the herd level: the within-herd proportion of animals with detected lesions at total slaughter (Table 1). Dataset A was obtained from a case study conducted by Carnon [17]. The median percentage of animals with lesions per herd was 34.6% (min: 20.8%, max: 86.4%). Dataset B and C were obtained from the official veterinary services of Dordogne department. In dataset B, the median percentage of animals with lesions per herd was 6.4% (min: 0.1%, max 52%). In dataset C, it was 3.1% (min: 0.6%, max 43%). Data set D was obtained from the Alfort

National Veterinary School. The median percentage of animals with lesions per herd was 1% (min: 0.3%, max: 5%).

## Model

The within-herd spread of bTB was modelled using a compartmental stochastic model operating in discrete time (Figure 1). The time step was one month. Hereafter, $t$ represents the current time step and $m$ the current month. Each cattle was represented by its age in years $i \in A = \{0,\ldots,A\_max\}$ and its health state $j \in H = \{S,E,I\}$, with: $S$ (susceptible): the animal is not infected, $E$ (latent): the animal is infected but does not have lesions and does not excrete bacteria, and $I$ (infectious): the animal is infected, has lesions and excretes bacteria. Animals were assumed to be grouped into batches according to their age class. To each age class $i$ ($i \in A$) was assigned a batch denoted $L(i)$ (with $L(i) \subset A$). A given batch thus contained animals of one or several age classes. In practice, these batches are often placed by farmers on distant pastures and sometimes in separate buildings, animals from distinct batches having little contact with each other.

The control program was represented by a succession of steps numbered from 1 to $D_{max}$. During each of these steps, biological tests were performed on the entire or part of the herd; the results of these tests determined the following step. For example, in the control program B which was applied between 2004 and 2006 in the department of Dordogne (Figure 2), the first step is the yearly screening of all animals by SITT. If one animal is positive, the second step of control program is performed two months later: positive animals of step 1 are tested by SICCT, as well as a group of randomly selected negative animals. If one animal is positive in step 2, step 3 is performed 3 months later: positive animals of step 2 are slaughtered and, when lesions are observed, bacterial cultures are performed. If one animal is positive to the bacterial culture, the total slaughter (step 4) is performed 1 month later. The status of each animal with respect to the step of the control program was represented by an integer $k \in D = \{0,\ldots,D\_max\}$, with: $k = 0$ if the animal has never been tested or always had negative test results, and $k>0$ if the animal had positive results to tests performed in step $k$ of the control program, and negative tests thereafter.

The state of a herd at time step $t$ was represented by a triplet$\langle X_t,y_t,z_t \rangle$, where $y_t$ is the number of the current step of the control program ($1 \leq y_t \leq D_{max}$), $z_t$ is the date on which the biological tests associated with this step must be performed ($z_t \geq t$), and $X_t$ is a state variable structured on age class, health status and step of the control program, which represents the state of the animals at time step $t$: $X_t$ (i,j,k) represents the number of animals in age class $i$ ($i \in A$), health state $j$ ($j \in H$), which have the status $k$ ($k \in D$) with respect to the control program.

The dynamic of infection in the herd resulted from the implementation of three sequential processes (Figure 1): the demographic process (ageing and renewal of animals), the infectious process (transmission of infection and evolution of infected animals) and the process of detection and control. A full description of the model is given in Appendix S1.

The size of the herd was assumed constant over time and the herd was assumed closed. Slaughtered animals (routine slaughter or implementation of the control program) being replaced by susceptible young animals (0–1 years old) born in the same herd. The culling rate $k \in D$ was assumed to vary according to the month and to the age class.

We made the simplifying assumption that transmission can occur only between animals of the same batch. However, because of the ageing of animals, the composition of the batch changes every year. Animal transfers between batches then allowed the between-batch infection spread. The within-herd transmission of bTB was modelled by a frequency-dependent function. Because *M. bovis* transmission intensity was assumed to vary according to whether the animals are housed inside a stable (where the intensity of within-batch transmission is assumed to be high) or allowed to graze (where the intensity of within-batch transmission is assumed to be low), the transmission parameter ($\beta_m$) was assumed to vary according to the month. The transition from the latent ($E$) to the infectious ($I$) state was based on the transition rate $\alpha$, where $1/\alpha$ is the duration of the latency period in months.

A step of the control program represented the implementation of one or more biological tests, some of which potentially requiring the slaughter of the animal (for detecting lesions). Tests implemented in the context of a control program could be performed on the entire or part of the herd; the sample of tested animals depending on the results of previous steps of the control program. A step $k$ of the control program was then represented by the quadruplet$\langle Se_k,Sp_k,M_k,Q_k \rangle$, where:

- $Se_k$ and $Sp_k$ are the joint sensitivity and specificity of tests (assumed independent of age and identical for animals in health states $E$ and $I$)
- $m_k$ equals 1 if the tests require the slaughter of the animal and 0 otherwise
- $Q_k$ is a parameter structured according to age class and to the status of animals with respect to the step of the control program, which defines the sampling plan for testing animals: $Q_k$ $(i,k)$ is the probability that an animal is tested if it has age $i$ and status $k$ for the control program. This parameter verifies: $\sum_{((i,k) \in A \times D)} Q_k\ (i,k) = 1$.

The step $k = 1$ of the control program corresponded to a screening in a bTB-free herd. The steps $k>1$ corresponded to detection and control measures conducted in a herd known or suspected to be infected.

At each time step, it was further assumed that slaughterhouse surveillance allowed the detection of lesions in culled animals. This slaughterhouse surveillance combined the visual inspection of carcasses with the isolation of *M. bovis* from lesions (Figures 2 and 3).

## Parameterisation

**Fixed parameters.** Consistent estimates of the sensitivity and specificity of diagnostic tests are found in the literature, and herd management practices were assumed homogeneous in a given region. The values of the corresponding parameters were thus fixed according to literature or based on expert opinion.

Three batches were distinguished in beef herds: young heifers (1–2 years), heifers of reproductive age (2–3 years), and cows (after the first calving at 4 years) with calves and heifers fed by their mothers (<1year). In dairy herds, animals were also separated into three batches: young heifers (<1 year), heifers of reproductive age (1–2 years), and cows (after the first calving, at 3 years).

For both types of herds, the stabling period was between November and March. Renewal and routine cull of animals was assumed to occur between January and March, animals being culled from the age of 3 years in dairy herds and from the age of 4 years in beef herds. The maximum age of the cattle ($A_{max}$) was set to 15 years. Other fixed parameters (culling rate, sensitivity and specificity of screening and diagnostic tests) are given in Table 2.

Month-specific transmission parameters allowed representing the difference between the stabling and grazing period, with $\beta_m = \beta_{inside}$ during the stabling period (between November and March) and $\beta_m = \beta_{outside}$ for the remainder of the year. Because of

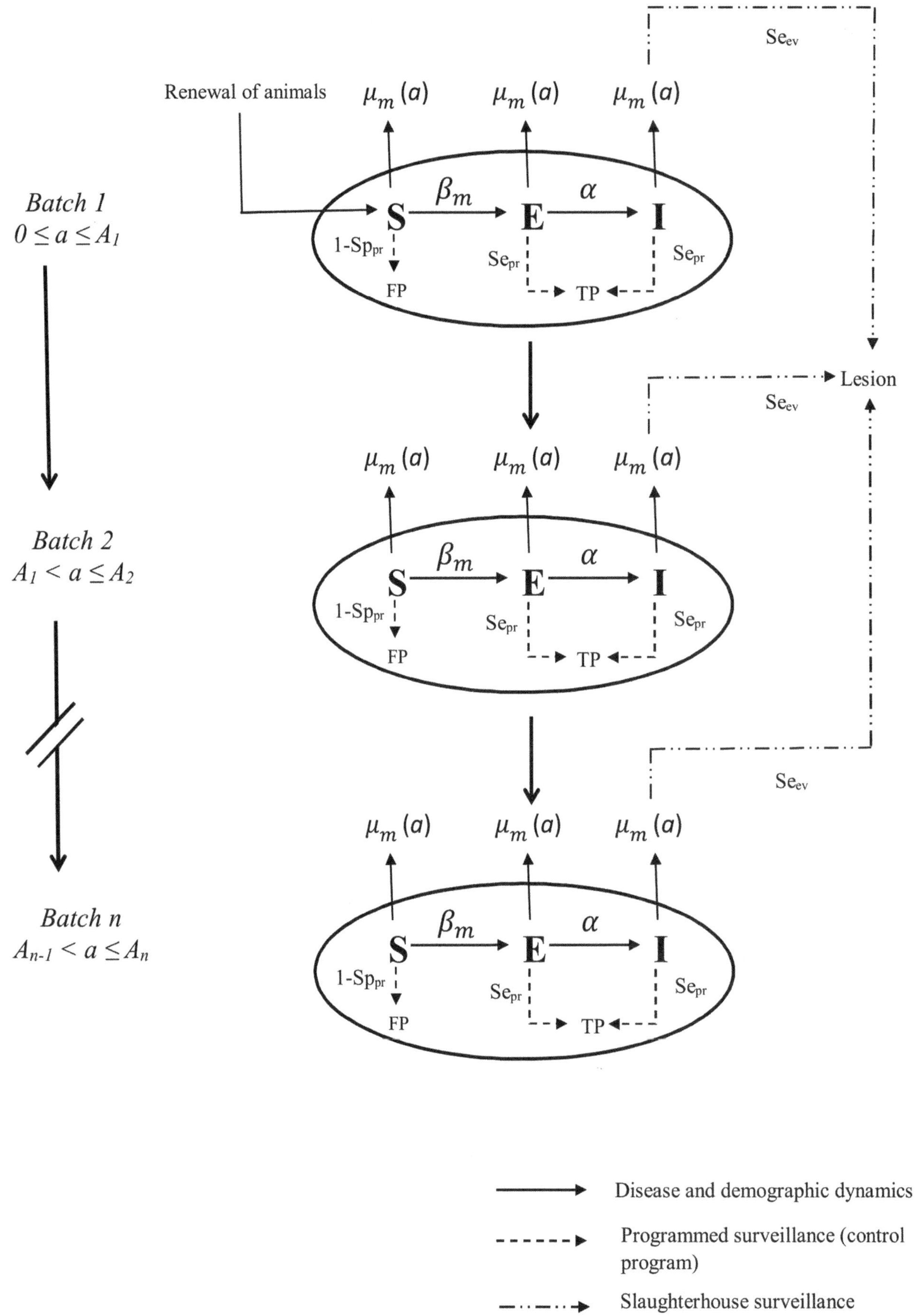
Renewal of animals
$\mu_m(a)$
$\beta_m$
$\alpha$
S
E
I
1-$Sp_{pr}$
$Se_{pr}$
FP
TP
$Se_{ev}$
Lesion
Batch 1
$0 \leq a \leq A_1$
Batch 2
$A_1 < a \leq A_2$
Batch n
$A_{n-1} < a \leq A_n$
Disease and demographic dynamics
Programmed surveillance (control program)
Slaughterhouse surveillance

**Figure 1. Schematic representation of the model of within-herd transmission of bTB by three processes: the demographic process (renewal of cattle, reform and change of batches), the infectious process (transitions between health states S: susceptible, E: latent, I: infectious) and the process of detection and control.** a: age class in years; Batch: group of animals; m: month; $\mu_m$ (a): culling rate for month m and age class a; $\beta_m$: month-specific transmission coefficient; $Se_{pr}$: test sensitivity of step k of programmed surveillance (control program); $Sp_{pr}$: test specificity of step k of programmed surveillance (control program); FP: false positive; TP: true positive; $Se_{ev}$: sensitivity of passive surveillance at the slaughterhouse.

the close contact between dairy cows that occurs twice per day in the milking parlour, $\beta_{inside}$ was used throughout the year in dairy herds for the cows' batch.

Initial conditions assumed the presence of a single infected animal in the state $E$. To represent the introduction of infection through contact during grazing, this animal was randomly selected among animals that had access to pasture (all batches for beef herds and the first two batches in dairy herds). Simulations were stopped when there were no more infected animals (thanks to routine cull and/or to the control program), the current step of the control program being 1 (i.e. screening in a bTB-free herd).

**Parameter estimation.** As estimates available in the literature are divergent or uncertain, and because many aspects of bTB pathogenesis are yet to be elucidated, three parameters were estimated from field data: $\alpha$ (transition rate between the health states $E$ and $I$), $\beta_{inside}$ (transmission parameter when animals are inside the stable) and $\beta_{outside}$ (transmission parameter when animals are grazing). Individual-level data from datasets B and C were used, corresponding to 27 herds and 1423 animals (Table 1). Control programs B (Figure 2) and C (Figure 3) were modelled according to memos of the French Ministry of Agriculture, Food and Forestry.

The ABC (Approximate Bayesian Computation) method was used to estimate the parameters ($\alpha$, $\beta_{inside}$ and $\beta_{outside}$). This method aims at estimating the posterior distributions of the parameters where the calculation of the likelihood is difficult or burdensome [18].

The generic form of the ABC method's "rejection algorithm" is [18]:

1- Sample parameter values $\theta^*$ from the prior distribution $\pi(\theta)$.

2- Simulate a dataset $\chi^*$ using parameter values $\theta^*$.

3- Calculate the summary statistics for the observed data $S(\chi o)$ and simulated data $S(\chi^*)$.

4- Compare the simulated data, $S(\chi^*)$, and the observed data, $S(\chi o)$, using the distance function, $d$, and the tolerance $\varepsilon$; if $d(S(\chi_o), S(\chi^*)) \leq \varepsilon$, the value of $\theta^*$ is accepted. When $\varepsilon = 0$, the posterior distribution is exactly $\pi(\theta/S(\chi o))$, whereas when $\varepsilon \rightarrow +\infty$, the posterior distribution is equal to the prior distribution [19].

Other algorithms have been proposed for the ABC method (e.g. Markov Chain Monte Carlo or Sequential Monte Carlo). We used the rejection algorithm followed by a step of local linear regression as proposed by Beaumont et *al.* [19]. This step aims at reducing the variance of the posterior estimation in order to try to correct errors that are due to a non-zero value of the tolerance $\varepsilon$. A total of 100,000 simulations were performed for parameter estimation.

Prior distributions were uniform distributions $U[0.027–0.50]month^{-1}$ for $\alpha$ and $U[0–2]month^{-1}$ for $\beta_{inside}$ and $\beta_{outside}$. For $\alpha$, these values correspond to a period of latency between 2 and 36 months. The lower bound (2 months) is the rounded value of the lower bound (87 days) of the estimation obtained by Neill et *al.* in an experimental study [20], while the upper bound (36 months) is a rounded value of the upper bound (34 months) of the estimated duration of latency in Perez et *al.* [2]. For $\beta_{inside}$ and $\beta_{outside}$, the upper bound of the prior distribution was chosen arbitrarily after having verified by simulation that, using this value, the median simulated percentage of animals with lesions at total slaughter (22% and 17% for dairy and beef farms, respectively) was significantly higher than the observed value in dataset B (6.4%).

The choice of summary statistics is a crucial step in the ABC method. This choice involves a compromise between information loss and size of the statistic [21]. Using a poorly chosen set of summary statistics can often lead to an overestimation of credible

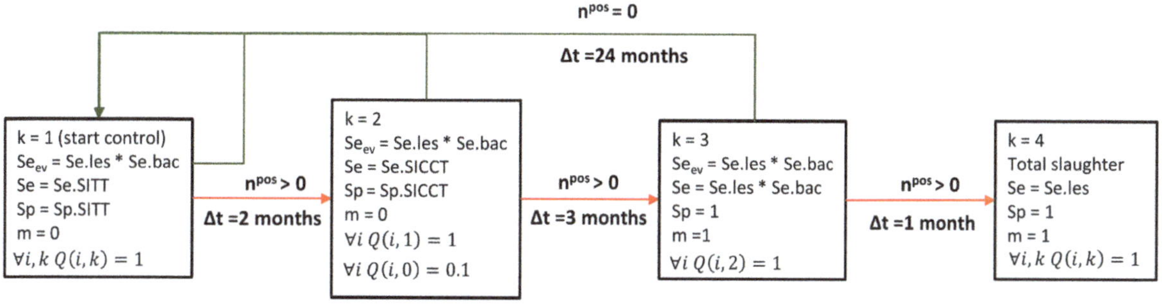


**Figure 2. Control program B applied in the Dordogne department between 2004 and 2006. Step 1: biennial bTB screening using SITT (the herd being considered disease-free).** All animals are tested ($\forall i,kQ(i,k) = 1$). Sensitivity (Se) and specificity (Sp): those of SITT. Transition to step 2 if an animal is positive ($n^{pos}>0$), two months later ($\Delta t = 2$ months. Step 2: confirmation of SITT positive results using SICCT. All the positive animals of step 1 are tested ($\mu$ i $Q(i,1) = 1$), as well as 10% of the negative animals ($\mu$ i $Q(i,1) = 1$). Sensitivity (Se) and specificity (Sp): those of SICCT. Transition to step 3 if an animal is positive, three months later. Step 3: slaughter of SICCT-positive animals and isolation of *M. bovis* from lesions. All the positive animals of step 2 ($\mu$ i $Q(i,2) = 1$) are slaughtered ($m = 1$) and bacterial culture is performed from observed lesions. Sensitivity (Se): sequential combination of a visual inspection at the slaughterhouse and of a bacterial culture. Transition to step 4 if an animal is positive, 1 month later. Step 4: total slaughter. All the animals ($\forall i,kQ(i,k) = 1$) are slaughtered ($m = 1$). Routine detection of lesions at slaughterhouse. Sensitivity ($Se_{ev}$): sequential combination of a visual inspection at the slaughterhouse and of a bacterial culture. If a positive animal is thus detected, regardless of the current stage of control program, transition to step 4. See table 2 for the definition of the other parameters.

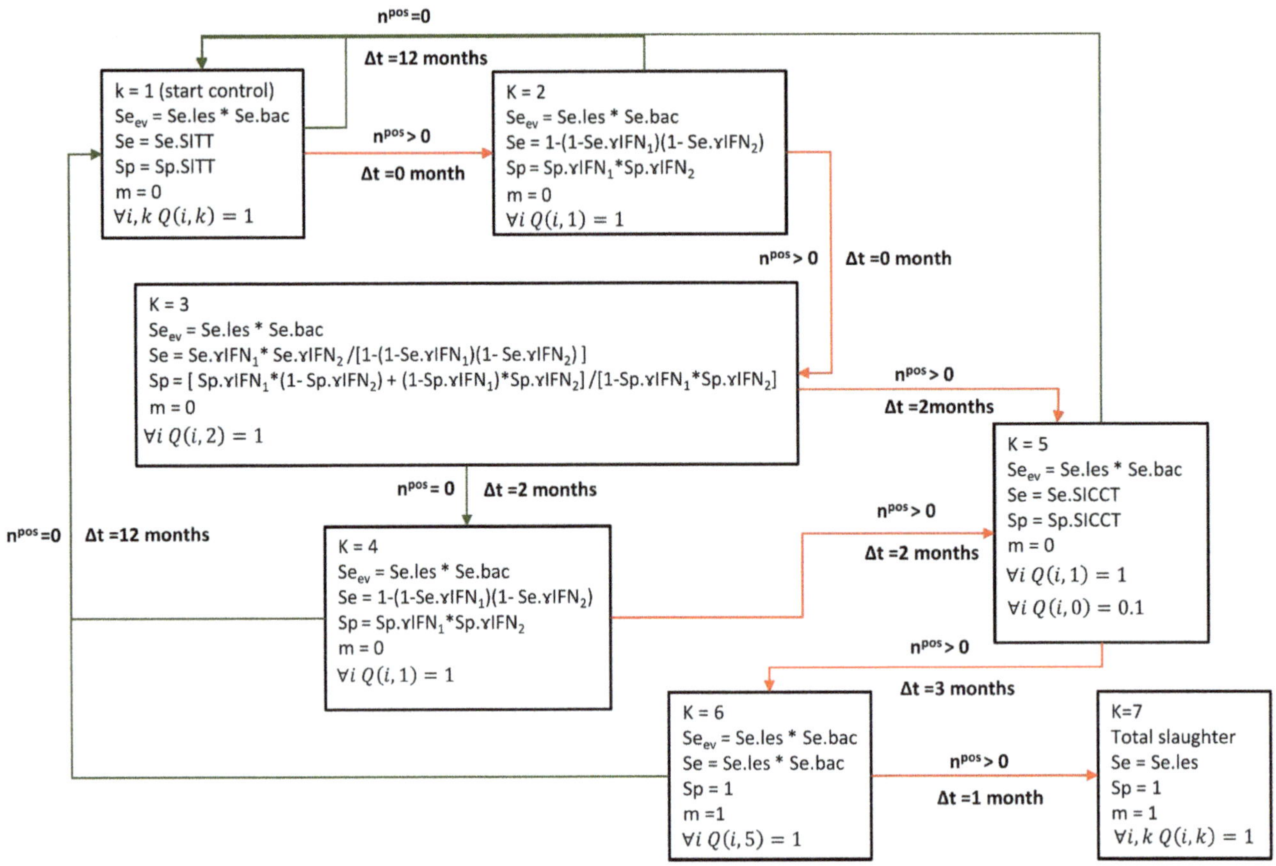


**Figure 3. Control program C applied in the Dordogne department between 2007 and 2010.** Step 1: yearly bTB screening using SITT (the herd being considered disease-free). All animals are tested (µ i,k Q(i,k) = 1). Sensitivity (Se) and specificity (Sp): those of SITT. Immediate transition to step 2 if an animal is positive ($n^{pos}>0$). Step 2: confirmation of SITT positive results using PDD γIFN (γIFN1) and ESAT-6 γIFN (γIFN2). All the positive animals of step 1 are tested (µ i Q(i,1) = 1. Sensitivity (Se) and specificity (Sp): parallel combination of both tests (positivity to any of both tests). Immediate transition to step 3 if an animal is positive; otherwise: transition to step 1. Step 3: interpretation of the positive results of step 2 as convergent (both γIFN tests are positive) or not (only one of both tests is positive). All the positive animals of step 2 (µ i Q(i,2) = 1) are concerned. Sensitivity (Se) and Specificity (Sp): sequential combination of both tests in positive animals of step 2. Transition to step 5 if an animal is positive, and to step 4 otherwise, two months later. Step 4: second confirmation of SITT positive results using PDD γIFN (γIFN1) and ESAT-6 γIFN (γIFN2). All the positive animals of step 1 are tested (µ i Q(i,1) = 1). Sensitivity (Se) and specificity (Sp): same as for step 2. Transition to step 5 if an animal is positive, two months later; otherwise: transition to step 1, 12 months later. Step 5: confirmation of SITT positive results using SICCT. All the positive animals of step 1 are tested (µ i Q(i,1) = 1), as well as 10% of the negative animals (µ i Q(i,0) = 0.1). Sensitivity (Se) and specificity (Sp): those of SICCT. Transition to step 6 if an animal is positive, three months later. Step 6: slaughter of SICCT-positive animals and isolation of *M. bovis* from lesions. All the positive animals of step 5 (µ i Q(i,5) = 1) are slaughtered (m = 1) and bacterial culture is performed from observed lesions. Sensitivity (Se): sequential combination of a visual inspection at the slaughterhouse and of a bacterial culture. Transition to step 7 if an animal is positive, 1 month later ($\Delta t = 1$ month). Step 7: total slaughter. All the animals (µ i,k Q(i,k) = 1) are slaughtered (m = 1). Routine detection of lesions at slaughterhouse. Sensitivity ($Se_{ev}$): sequential combination of a visual inspection at the slaughterhouse and of a bacterial culture. If a positive animal is thus detected, regardless of the current stage of control program, transition to step 7. See table 2 for the definition of the other parameters.

intervals due to the loss of information [22]. One way to capture most of the information present in the data would be to use many statistics, but the accuracy and stability of the ABC method decrease rapidly with an increasing number of summary statistics [22,23]. In order to choose the best combination of summary statistics, eight combinations (Table 3) were tested on simulated data obtained using arbitrarily fixed parameter values. The tested summary statistics combined two types of variables, the time between the last negative control and the detection of infection and the percentage of animals with detected lesions, taking into account or not the batch, the type of herd and the year. The procedure used to select the best summary statistics among the eight tested combinations is described in Appendix S2. Briefly, parameter values were set to arbitrarily chosen values and used to generate 100 simulated datasets. In a first step, one of these datasets was chosen and considered a pseudo-observation. Each set of summary statistics was then successively used to estimate the posterior distribution of the parameters, and we kept for further analysis the set of summary statistics for which the arbitrarily chosen parameter values were inside the corresponding 95% credibility interval. In a second step, the preceding procedure was successively applied to the 100 simulated datasets and we selected the set of summary statistics that maximized the proportion of cases for which the arbitrarily chosen parameter values were inside the 95% credibility interval of the posterior distribution of parameters.

The tolerance threshold ε was determined by choosing the proportion of retained simulations $p_\varepsilon$. A leave-one-out procedure was applied to a set of 100 datasets simulated using parameters ($\alpha$, $\beta_{inside}$, $\beta_{outside}$) sampled from the prior distributions, to evaluate the

**Table 2.** Definitions and values of fixed parameters in the model.

| Parameters | Description | Value | Range[1] | Reference |
|---|---|---|---|---|
| $\mu_m$ (*i*) | Culling rate | | | (Prof. Coureau, Alfort National Veterinary School, pers. Comm.) |
| | Dairy herd, 2000's | 0.35 | 0.25; 0.45 | |
| | Beef herd, 2000's | 0.25 | 0.15; 0.35 | |
| | All herds, 1980's | 0.25 | | |
| Se.SITT | Sensitivity of single intradermal tuberculin test | | 0.635; 1 | [8] |
| | 1980's | 0.72 | | |
| | 2000's | 0.839 | | |
| Sp.SITT | Specificity of single intradermal tuberculin test | | 0.755; 0.99 | [8] |
| | 1980's | 0.988 | | |
| | 2000's | 0.968 | | |
| Se.SICCT | Sensitivity of single intra dermal comparative cervical tuberculin test | 0.80 | 0.52; 1 | [8] |
| Sp.SICCT | Specificity of single intra dermal comparative cervical tuberculin test | 0.995 | 0.788; 1 | [8] |
| Se.PPD $\gamma$IFN | Sensitivity of gamma interferon test (PPD) | 0.876 | 0.73; 1 | [8] |
| Sp.PPD $\gamma$IFN | Specificity of gamma interferon test (PPD) | 0.966 | 0.850; 0.996 | [8] |
| Se.ESAT-6 $\gamma$IFN | Sensitivity of gamma interferon test (ESAT-6) | 0.763 | 0.690; 0.836 | [51] |
| Sp.ESAT-6 $\gamma$IFN | Specificity of gamma interferon test (ESAT-6) | 0.992 | 0.976; 1 | [51] |
| qSICCT | Proportion of animals non-reactors to SITT tested six weeks later by SICCT | 0.10 | 0.05; 0.15 | (French Ministry of agriculture, Food and Forestry) |
| Se.bac | Sensitivity of bacterial culture | 0.78 | 0.729; 0.828 | [52] |
| Se.les | Sensitivity of visual inspection of carcass | 0.50 | 0.255; 0.755 | [6] |

[1]Values used in the sensitivity analysis.

**Table 3.** Selection of the set of summary statistics based on simulated datasets generated using arbitrarily fixed parameter values: $\alpha = 0.083$ months, $\beta_{inside} = 0.5$, and $\beta_{inside} = 0.1$.

| Id | Summary statistics | Dimension | Median [95% credible interval] | | | p[1] |
|---|---|---|---|---|---|---|
| | | | $\alpha$ | $\beta_{inside}$ | $\beta_{outside}$ | |
| A | Average percentage of animals with detected lesions[2] | 1 | 0.72 [0.60–0.78] | 0.79 [0.44–1.27] | 0.61 [0,03–1,92] | – |
| B | Average percentage of animals with detected lesions in each batch | 3 | 0.42 [0.00–2.43] | 0.33 [0.28–0.70] | 0.29 [0.00–3.9] | 73% |
| C | Percentage of animals with detected lesions in each batch (n = 3) of each herd (n = 27) | 81 | 0.13 [0.06–0.67] | 0.35 [0.02–1.97] | 0.44 [0.01–2.21] | 49% |
| D | Same as C+percentage of herds submitted to total slaughter | 82 | 0.97 [0.48–2.04] | 0.05 [0.01–0.15] | 0.03 [0.00–0.12] | – |
| E | Same as C+time between the last negative test and the detection of infection for each herd | 108 | 0.29 [0.07–0.69] | 0.35 [0.00–0.77] | 0.07 [0.00–0.84] | 32% |
| F | Same as E+percentage of herds submitted to total slaughter | 109 | 0.33 [0.12–0.65] | 0.29 [0.08–0.57] | 0.08 [0.00–0.79] | – |
| G | Average percentage of animals with detected lesion for each batch, period[3] and herd type[4]; average time between the last negative test and the detection of infection for each period and herd type | 16 | 0.26 [0.07–0.74] | 0.27 [0.04–0.75] | 0.07 [0.00–0.52] | 71% |
| H | Same as G+percentage of herds submitted to total slaughter for each period and herd type | 20 | 0.073 [0.05–0.54] | 0.48 [0.03–0.76] | 0.12 [0.00–0.59] | 88% |

[1]Proportion of cases for which each of the three arbitrarily fixed parameter values were inside the 95% credible interval of the posterior distributions (100 repetitions).
[2]For herds submitted to a total slaughter.
[3]Before 2007 (control program B) and after 2007 (control program C).
[4]Dairy of beef.

robustness of the estimates to the tolerance rate. Each of the 100 simulated datasets was successively considered a pseudo-observation and parameter values were estimated using the above procedure. A prediction error coefficient was calculated for each estimated parameter, for values of $p_\varepsilon$ ranging from 0.1% to 5% [24]. The smallest value of $p_\varepsilon$ for which the prediction error coefficient was stable for the three parameters ($\alpha$, $\beta_{inside}$, $\beta_{outside}$) was thus selected.

We used the "abc" package of software R [25]. The function cv4abc was used to select the value for the proportion of retained simulations $p_\varepsilon$.

A sensitivity analysis was performed to evaluate the effect of changing the fixed parameter values (Table 2) on the estimated values of the transition rate ($\alpha$) and of the transmission parameters ($\beta_{inside}$ and $\beta_{outside}$). Two different values were selected for each fixed parameter (Table 2) according to literature (SITT, SICCT, γIFN and bacterial culture) or using fixed deviations (±10% or ±5%). Only first-order effects were analyzed and, for each scenario, $\alpha$, $\beta_{inside}$ and $\beta_{outside}$ were estimated by the ABC method, as described above. This operation was repeated 24 times (12 fixed parameters and 2 points per parameter) resulting in 24 posterior distributions for $\alpha$, $\beta_{inside}$ and $\beta_{outside}$. These were analysed using three generalised linear model (GLMs) (one per estimated parameter). The explanatory variables were the fixed parameters values (their deviation from their default values).

The GLMs were used to predict the effect of an increase of 5% of each fixed parameter on the average values of $\alpha$, $\beta_{inside}$ and $\beta_{outside}$. To identify the most influential fixed parameters, we compared the coefficient of variation of the posterior distributions (obtained using the default values of fixed parameters) with the relative error induced by a 5% change of fixed parameters, as predicted by the GLMs.

## Validation

An internal validation was first performed using a leave-one-out cross-validation procedure. One hundred triples of parameter values ($\alpha$, $\beta_{inside}$ and $\beta_{outside}$) were randomly chosen among the 100,000 used for parameter estimation. Each was successively used to generate a dataset considered a pseudo-observation, and posterior distributions were estimated using the above procedure (without using the pseudo-observation). The reliability of the estimation procedure was quantified by the proportion of cases for which the triple ($\alpha$, $\beta_{inside}$ and $\beta_{outside}$) used to generate the dataset was within the 95% credible interval.

An external validation was performed to test the ability of the model to reproduce the observational data of bTB in France between 1980 and 2010, using the data that were not used for parameter estimation: dataset A, aggregated data from datasets B and C, and dataset D (Table 1). Each herd of these datasets was successively used to parameterize the model (herd size and type, control program - Figures 2, 3, S1 and S2), and the posterior distributions of parameters ($\alpha$, $\beta_{inside}$ and $\beta_{outside}$) were sampled to perform 1000 simulations. Those ending by a total slaughter were kept to compute the distribution of the proportion of animals with detected lesions. For each herd of the above datasets, the observed proportion was compared with the simulated distribution thus obtained. Furthermore, for dataset A, the predicted proportion of simulations ending by a total slaughter was compared with the percentage observed in the corresponding department in the 1980s: 5%.

## Model exploitation

The basic reproduction ratio (R0) is used to measure the transmission potential of a disease. It is the number of secondary infections produced by an infectious animal in a fully susceptible population that is totally susceptible. This indictor is of great importance in infectious disease modelling as, when $R0<1$, the disease tends to vanish, whereas when $R0>1$, the disease tends to spread and a large epidemic may occur [26]. However, in real situations, the population is rarely entirely susceptible, mainly because of the spread of infection (and the corresponding decrease in the number of susceptible individuals). We computed the effective reproductive ratio R(t): the average number of cases secondary to an infectious case in a population consisting of susceptible and infected individuals. Two average herds, respectively representative of a French dairy herd (81 cattle) and a French beef herd (70 cattle) in the 2000s, were considered. Disease dynamics was simulated in these herds without a control program, and the distribution of the effective reproductive ratio R(t) was calculated each month after the introduction of *M. bovis*, during 360 months (twice the maximal age of cows in the model: 15 years). Details of R(t) calculation are given in Appendix S3.

# Results

## Parameter estimation

The procedure used to choose the set of summary statistics first led to select B, C, E, G and H, for which, when using one simulated dataset as a pseudo-observation, the arbitrarily chosen parameter values were all inside the 95% credible interval of posterior distributions of parameters (Table 3). Their dimensions varied between 3 for B and 109 for F (Table 3). Among these five sets of summary statistics, H maximized the proportion of cases (88%) where the arbitrarily fixed parameters ($\alpha$, $\beta_{inside}$ and $\beta_{outside}$) were all within the credible interval of 95% of posterior distributions when using the 100 simulated datasets as pseudo-observations (Table 3). This set of summary statistics was thus chosen: a combination of statistics describing the percentage of animals with detected lesions per batch and the percentage of herds with total slaughter per herd type (dairy/beef) and per period (periods where control program B and C were applied). The evolution of the values of the prediction error coefficient according to the proportion of retained simulations led to select $p_\varepsilon = 2\%$.

The median of the posterior distribution of $\alpha$ was 0.28 (95% credible interval [CI]: [0.13–0.56]), which corresponds to a median latency period for bTB of 3.5 months (95% CI: [2–8] months) (Figure 4). The median value of bTB transmission coefficients were 0.43 $month^{-1}$ (95% CI [0.16–0.84] $month^{-1}$) inside the stable ($\beta_{inside}$) and 0.08 $month^{-1}$ (95% CI [0.01–0.32] $month^{-1}$) outside the stable ($\beta_{outside}$) (Figure 4). The median of the ratio $\beta_{inside}/\beta_{outside}$ was 5 (95% CI [0.8–100]). It was>1 in 95.6% of cases (Figure 4). Negative values of Spearman's rank correlation coefficient were observed between $\alpha$ and $\beta_{inside}$ (−0.47, $p<0.0001$), between $\alpha$ and $\beta_{outside}$ (−0.31, $p<0.0001$), and between $\beta_{inside}$ and $\beta_{outside}$ (−0.35, $p<0.0001$). These negative correlations were expected, as the summary statistics (such as the proportion of animals showing lesions) may remain approximately the same if a higher value of $\alpha$ (i.e. a shorter latency period) is compensated by smaller values of $\beta_{inside}$ and $\beta_{outside}$, or if a higher value of $\beta_{inside}$ is compensated by a lower value of $\beta_{outside}$. The results of the sensitivity analysis of parameter estimates to variations of fixed parameter values are summarised in a tornado chart (Figure 5). A 5% increase in the values of the fixed parameters had limited effects on the estimation of $\alpha$ and $\beta_{outside}$ (Figure 5). For $\beta_{inside}$, the relative difference of the average posterior estimate greater than the coefficient of variation computed from the posterior distribution of for a single parameter: the specificity of the ESAT-6 γIFN

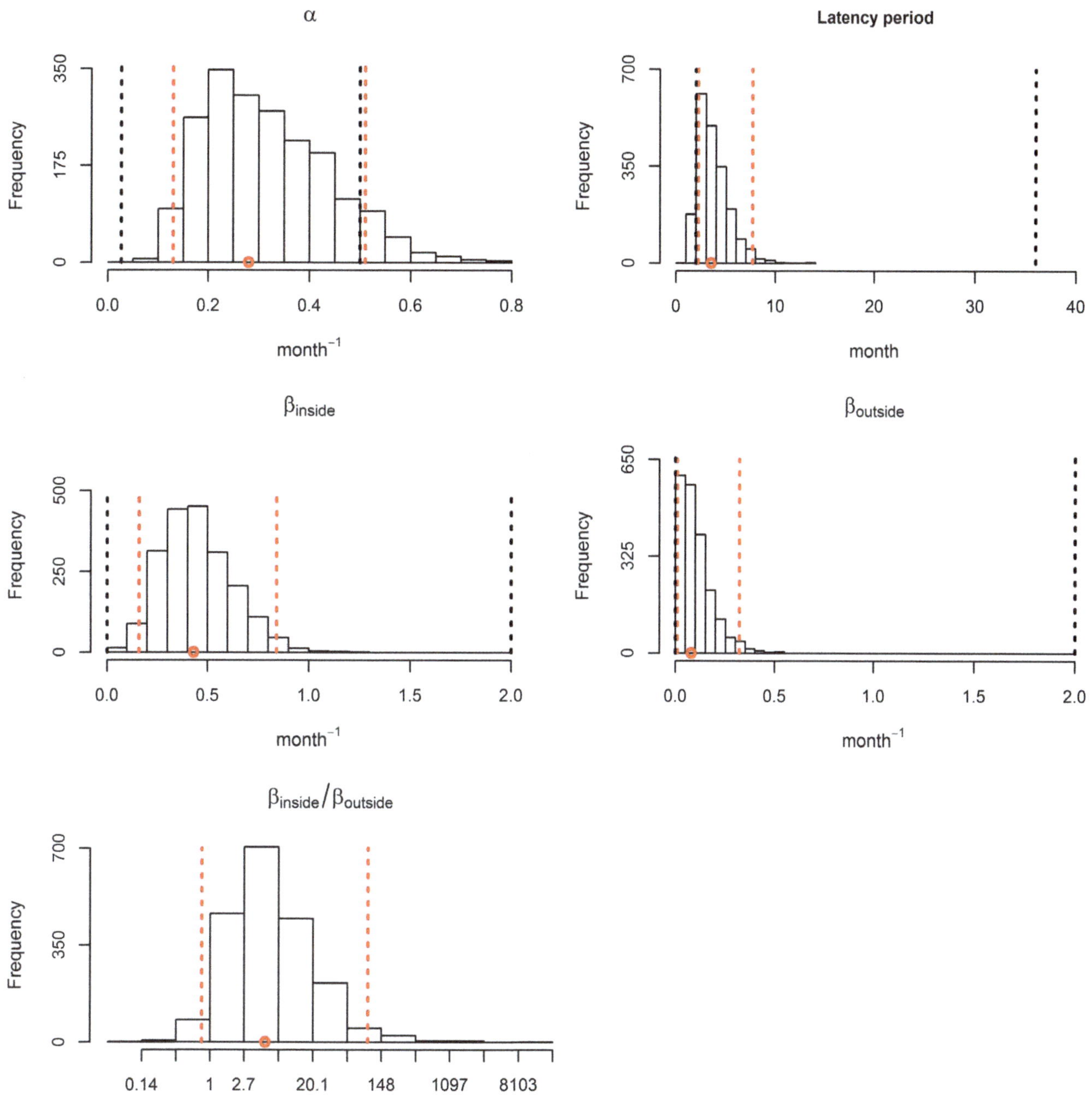


**Figure 4. Posterior distribution of parameters $\alpha$, $\beta_{inside}$ and $\beta_{outside}$.** Red dots: median value of parameter estimates; black dotted lines: lower and higher bounds of parameter prior distributions; red dotted lines: 95% credible intervals of the parameter posterior distributions. Histogram $\beta_{inside}/\beta_{outside}$ represents the value of log ratio $\beta_{inside}/\beta_{outside}$ but the x-axis values are not expressed in log.

test. A 5% increase of the specificity of the ESAT-6 γIFN test led to a 48.5% decrease of the average posterior estimate of $\beta_{inside}$.

## Predicted bTB dynamics

Figure 6 and Table 4 represents the predicted bTB dynamics in two herds representing dairy farms (average size in 2000:81 animals) and beef herds (average size in 2000:70 animals) in France, with and without a control program (control program B, Table 1 and Figure 2). When the control program was simulated, after its introduction in a susceptible herd, bTB was predicted to disappear in 22% of cases in beef herds and in 17% of cases in dairy farms, thanks to routine cull: the median time from disease introduction and disease extinction was less than 10 months, in both types of herds, with or without a control program (Table 4). In most cases, *M. bovis* did not disappear from the simulated herds and the disease was predicted to be detected through passive surveillance (at slaughterhouse) (13% for beef herds and 21% for dairy farms) or through active surveillance (screening using skin tests) (65% for beef herds and 62% for dairy farms). In both cases, the median time period between the introduction of infection and total slaughter was less than 3 years in the two types of herds (Table 4). In the absence of a control program (Figure 6a and 6b), the median proportion of infected animals was predicted to increase until reaching a plateau at 61 months in beef herds and

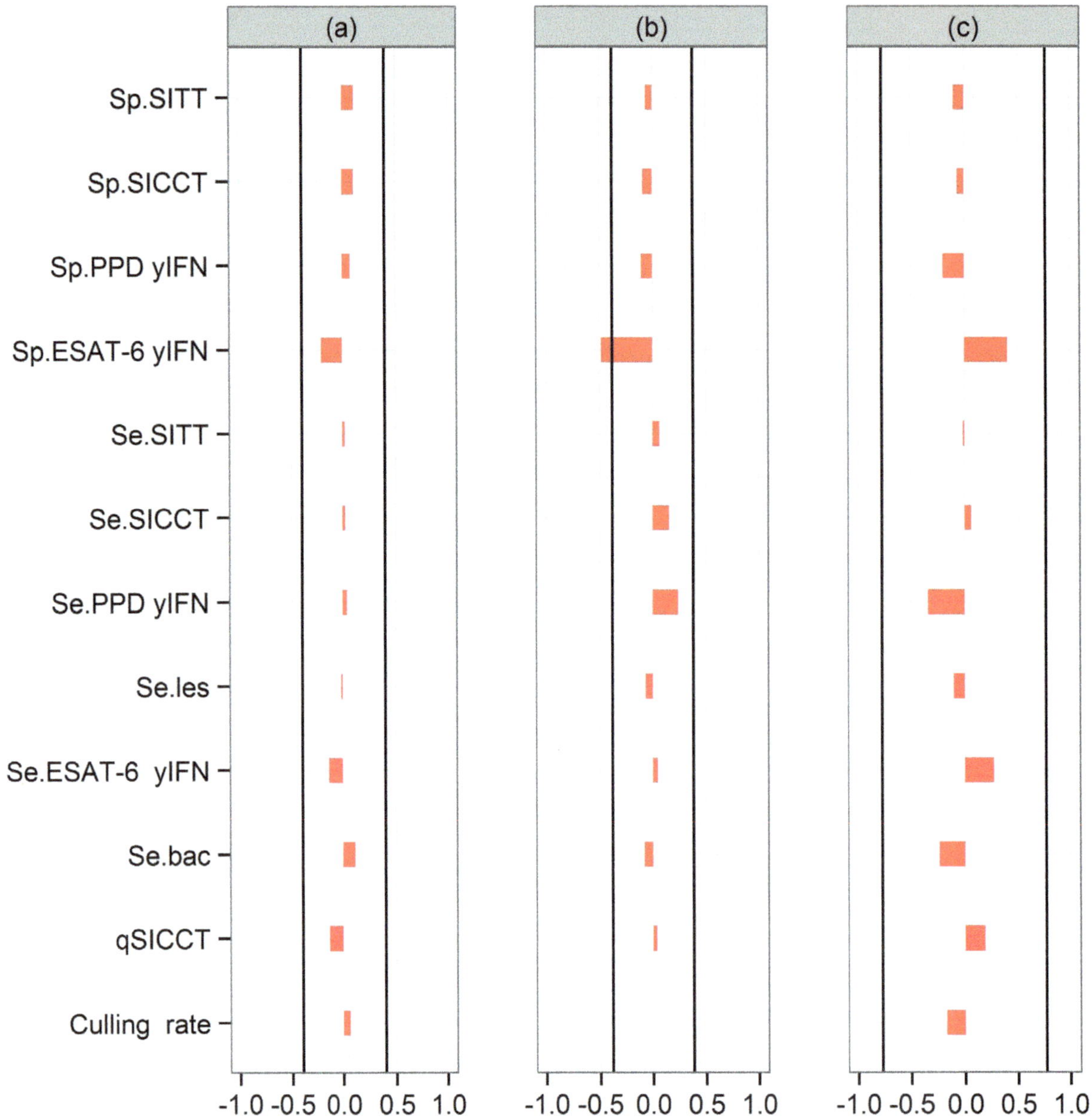


**Figure 5. Impact of changes of fixed parameter values on the estimates of $\alpha$, $\beta_{inside}$ and $\beta_{outside}$.** (a): relative error for $\alpha$; (b): relative error for $\beta_{inside}$; (c): relative error for $\beta_{outside}$; black line: value of the coefficient of variation of the posterior distributions obtained when we using the default values of fixed parameters; if the relative error value is within the two black lines, the effect of fixed parameters values on parameter estimates in the model is considered limited; qSICCT: percentage of SITT-negative animals tested 6 weeks after by SICCT; see table 2 for the definition of the other parameters.

40 months in dairy herds (Figure 6a and 6b). The corresponding level of infection prevalence was predicted to be higher in beef herds (Figure 6a) than in dairy herds (Figure 6b): once reached the enzootic infection level the median prevalence of infected animals was 93% in beef farms and 53% in dairy herds. As expected, when a control program was used, the maximum percentage of infected animals was predicted to be lower in both types of herds (Figure 6c and 7d).

## Validation

Internal validation showed that, for 83 of the 100 randomly selected simulations, the fixed values of the three parameters $\alpha$, $\beta_{inside}$ and $\beta_{outside}$, were inside the 95% credible interval of the posterior distributions.

Using dataset A, external validation showed that, in four of five herds (farms 1, 2, 3 and 4), the observed percentage of cattle with lesions was between the 2.5% and 97.5% percentiles of the simulated proportion of animals with lesions at total slaughter (Figure 7). In the fifth herd, the observed percentage of animals with lesions was very high (86.6%) and was above the 97.5% percentile of the simulated distribution (Figure 7). The proportion of simulations ending with a total slaughter varied between 3% and 7%, depending on the herd. These values are close to the proportion of 5% reported by the French Ministry of Agriculture, Food and Forestry for that department at that time.

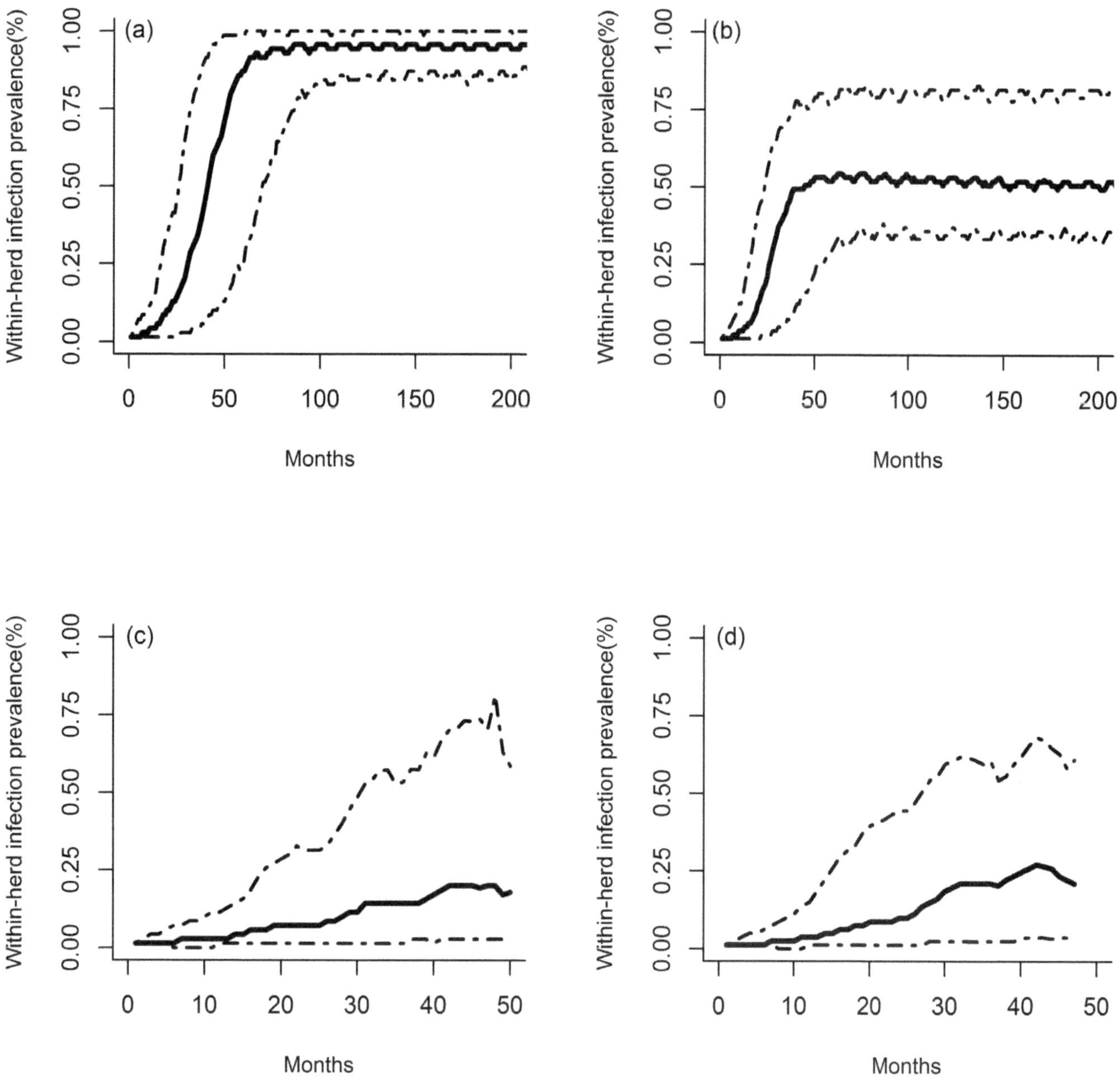


**Figure 6. Monthly evolution of the simulated proportion of infected animals (animals in E or I health state; solid line: median, dashed: 2.5% and 97.5% percentiles) in standard French beef (a and c, 70 animals) and dairy (b and d, 81 animals) herds without (a and b) or with (c and d) implementation of control program B (1,000 simulations per scenario).**

The observed percentage of animals with lesions was located between the 2.5% and 97.5% percentiles of the simulated proportion of animals with lesions in 92% (23 herds out of 25) and 83% (24 herds out of 29) of herds of datasets B and C, respectively. This percentage was 100% (29 herds out of 29) for dataset D.

## Effective reproductive ratio

Figures 8a and 8b show the evolution of R(t) values over time. In both types of herds, the shape of the curve was similar, with an increase in the values of R(t) that reached a plateau 72 months after disease introduction in dairy herds and at 96 months after disease introduction in beef herds. After 30 simulated years, which corresponds to a situation of enzootic circulation of *M. bovis* within the herd, the value of R(t) was slightly higher for beef herds than for dairy herds. The median value and the percentiles [2.5%–97.5%] of the predicted distribution of R(t) were 2.2 [1.8–2.8] for beef herds and 1.7 [1.5–2.2] for dairy herds (Figure 8a and Figure 8b). It was higher than 1 for each of the 1,000 simulations and higher than 2 for 11% of simulations in dairy herds and 79% in beef herds (Figure 8c and Figure 8d).

## Discussion

In this work, we have built a stochastic discrete-time model to simulate bTB dynamics and control in cattle herds. The model accounts for interactions among the three processes that influence the within-herd evolution of bTB: (i) the natural history of the disease (infectious process), (ii) husbandry practices (demographic

**Table 4.** Predicted median and percentiles [2.5%–97.5%] of the distribution of outbreak duration (in months) in standard French beef herds (70 animals) and dairy herds (81 animals) without or with control programs (1,000 simulations per scenario).

| | | Beef herds | Dairy herds |
|---|---|---|---|
| Time to natural extinction of infection | Without a simulated control program | 7 [3–30] | 8 [4–32] |
| | With a simulated control program | | |
| | No detection of infection | 6 [3–11] | 7 [4–16] |
| | Infection suspected but not confirmed[1] | 8 [3–25] | 8 [3–23] |
| Time to total slaughter of the herd after infection detection by routine skin testing | | 24 [9–50] | 23 [9–48] |
| Time to total slaughter of the herd after infection detection by slaughterhouse surveillance | | 24 [8–50] | 30 [9–47] |

[1]No isolation of *M. bovis*, total slaughter is thus not applied.

process) and (iii) control programs (detection and control process). The model has been designed to allow simulating control programs of arbitrary degree of complexity. Parameterization allows representing various herd types and farming practices (in particular the differences between dairy herd and beef herd management). Epidemiological parameters of the within-herd transmission of bTB (duration of the latent state, transmission parameters) were estimated using a first set and the model was validated using a second set of field data collected between 1980 and 2010. Herd management practices and bTB control programs have evolved during this period and the model could be successfully adapted by changing parameter values, suggesting a good level of genericity and an ability to simulate various herd management systems and bTB detection and control programs of arbitrary complexity. The simulated effective reproductive ratio R(t) may allow assessing the effectiveness (*i.e.* whether $R(t)<1$) of different bTB control strategies or the effect of changes in herd management practices on the bTB within-herd dynamic.

Internal validation showed a good reliability of the estimation procedure of model parameters by the ABC method. External validation showed that the model was able to reproduce field data regarding bTB collected between 1980 and 2010 in several French departments. This suggests that the transmission coefficients ($\beta_{inside}$ and $\beta_{outside}$) and the duration of the latent period ($1/\alpha$) have remained roughly constant between the 1980s and today, despite changes in parameters related to herd management practices such as the culling rate, the age at culling and the herd size.

Health states used in the proposed model (susceptible, latent and infectious) are consistent with the pathology of infection. However, published models often divide the latent state into two sub-states (latent, non-skin test-responsive and latent, skin test-responsive [13–15]) and use shorter time steps (day or week). Latent non-skin test-responsive animals spend an average of four weeks in this state before developing a positive skin test reaction [13,14,27]. Conlan et *al.* [28] estimated the median period of the latent non-skin test-responsive state at 28 days. As we used a monthly time step, it was thus not necessary to represent this health state. Infectious state $I$ was defined by the presence of detectable lesions at slaughter. The type (open or stabilised) and the location of the lesions were not taken into account. This may induce an underestimation of within-herd transmission: an animal with so-called "open tuberculosis lesions" may infect the rest of the herd in a short period of time. However, much is unclear about open tuberculosis lesions, including the probability of developing this type of lesion, its duration and its infectivity [15]. Therefore we did not incorporate the type of lesion in our model. This choice is in line with other epidemiological models of bTB [13–15].

Most of the models previously developed in the literature that simulated the within-herd spread of bTB used a single transmission parameter. In our model, we used two transmission parameters to distinguish transmission inside buildings and in pastures. Both transmission coefficients are related to herd practices, allowing to use the model for representing various farming systems by changing the grazing periods and the composition of batches. The within-herd transmission of bTB is considered higher in dairy herds than in beef herds [12]. This is attributed to a combination of several factors that contribute to the spread of infections in dairy herds such as the high contact rate, especially in the milking parlour [13], the high animal density and stress factors related to intense animal management [12,29–32]. The estimated value of $\beta_{inside}$ was greater than that of $\beta_{outside}$ in 96% of cases, suggesting that the definition of two transmission coefficients allowed capturing the difference between dairy herds and beef herds in terms of within-herd transmission of bTB. Indeed, although the 95% credible intervals of the two within-herd transmission coefficients of bTB ($\beta_{inside}$ and $\beta_{outside}$) overlapped, the median within-herd transmission coefficient inside the stable was 5 time as high as in the pasture: 0.43 month$^{-1}$ [0,16–0.84] and 0.08 month$^{-1}$ [0.01–0.32] for $\beta_{inside}$ and $\beta_{outside}$, respectively. These estimated medians were consistent with the estimates of within-herd transmission of bTB in Spain proposed by Alvarez et *al.* [12], with higher transmission coefficients in dairy herds (median 0.39 month$^{-1}$) than in beef herds (0.19 month$^{-1}$). The estimated values of the transmission coefficients are also in the same range as values obtained by other studies: 0.22 month$^{-1}$ in the study of Barlow et *al.* [13] and 0.18 months$^{-1}$ in the study of Perez et *al.* in Argentina [2].

The only transmission route taken into account in our work is the direct transmission between animals of the same batch. Indirect transmission, the potential role of wildlife, contacts with cattle from other herds at pastures and contacts between cattle of different batches are not taken into account. In particular, between-batch disease transmission was not taken into account because we assumed that the probability of contact between animals of different batches was low: different batches of the same herd are often placed by the breeder on different and distant pastures and even sometimes in separate buildings. It should be noted that the transmission between batches does nevertheless occur in the model: as bTB is a chronic, slowly evolving disease, between-batch transmission occurs when animals are moved from one batch to another. However, the closed herd assumption would

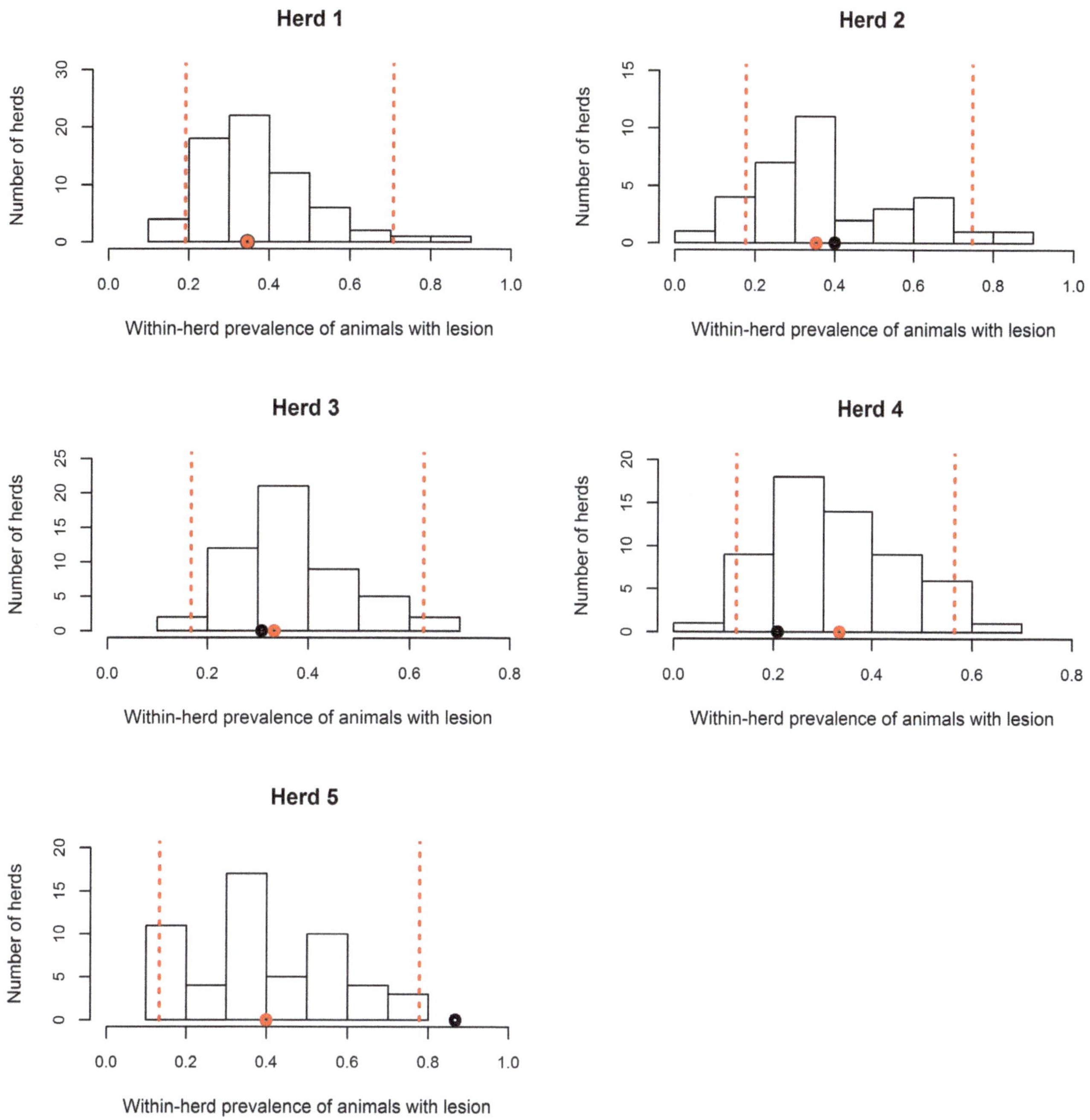


**Figure 7. Observed and simulated proportions of animals with detected lesions when total slaughter is performed, for the 5 herds of dataset A.** Black point: observed proportion of animals with bTB-like lesions in each herd; red point: simulated median proportion of animals with bTB-like lesions in each herd; dotted lines: 2.5% and 97.5% percentiles of the simulated distributions.

likely underestimate the persistence of *M. bovis* infection in herds, if wildlife or introduced animals can reintroduce infection [33].

The median duration of the estimated latency period (state *E*) was 3.5 months [2–8] months. This range of values is consistent with the chronic nature of bTB. However, our estimate of the latency duration was lower than the mean estimated duration of 24 months (95% confidence interval: 15–34 months) that was obtained by a simulation model of the within-herd transmission of bTB in Argentina [2]. This difference may be explained by differences among animals in the two countries, as the incubation period depends on the susceptibility of the host [34].

Model predictions for 2000 showed that the infected herds (dairy or beef) were in 15 to 20% of cases detected by slaughterhouse surveillance. This result was expected because the frequency of intradermal skin test applied in control program B was biennial. The percentage of spontaneous extinction of the disease (thanks to routine cull) was almost identical between dairy and beef herds.

When no control program was applied, simulated within-herd prevalence reached>50% in both types of farms. This simulated prevalence is high compared to the values observed in some parts of Africa and Asia (between 6% and 15%), where no control

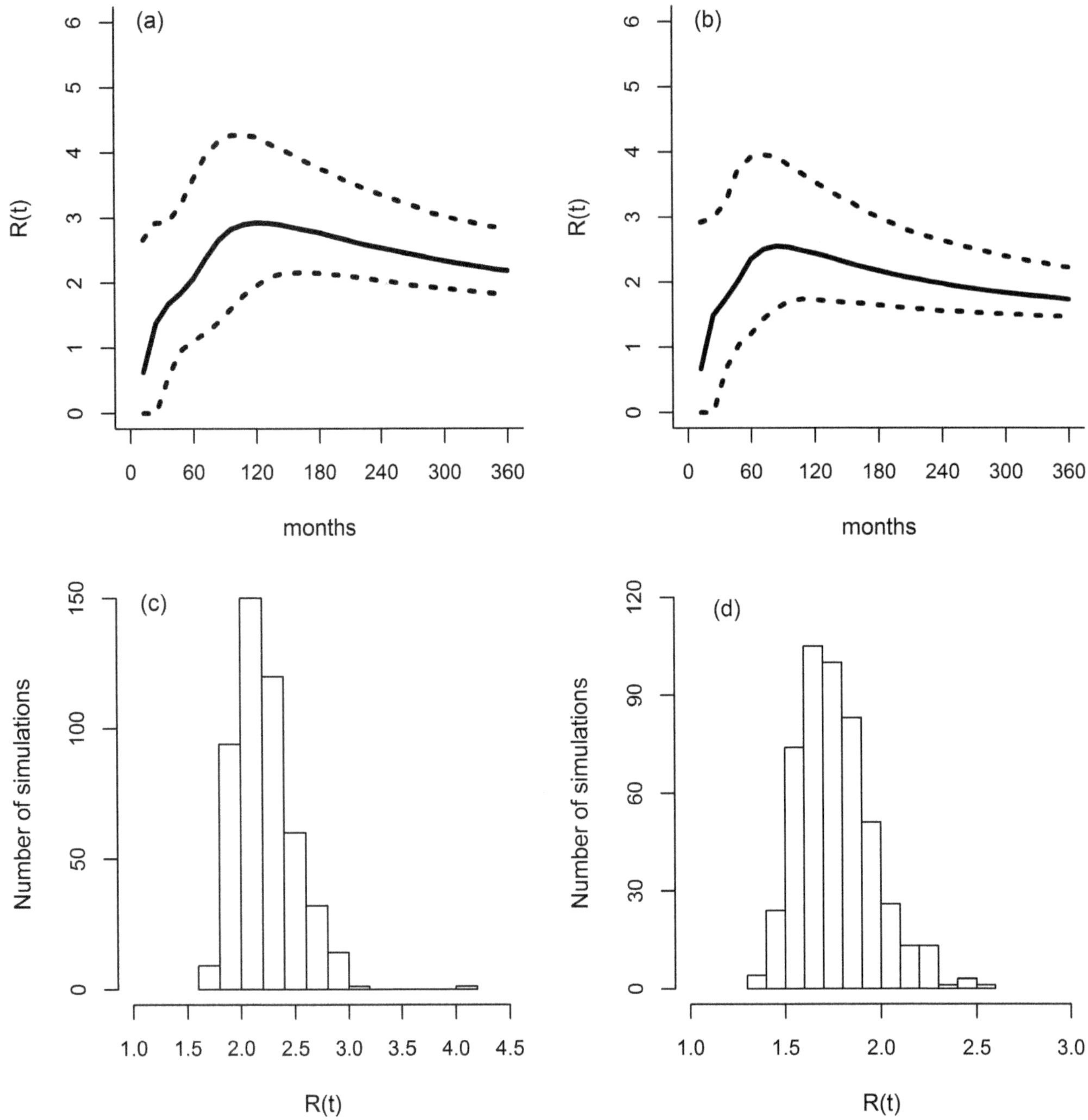


**Figure 8. Monthly evolution of the simulated effective reproductive ratio R(t) (a and b), and distribution of this R(t) (c and d) when the disease has become enzootic (after 30 years) in standard French beef (a and c, 70 animals) and dairy (b and d, 81 animals) herds.** Solid line: median; dashed: 2.5% and 97.5% percentiles.

program is applied [35,36]. In Africa and in Asia, most cattle are zebu (*B. indicus*) [37]. Ameni *et al.* [38] report that, under identical field husbandry conditions, a higher susceptibility to bTB was observed in Holstein cows than in zebus: (odds ratio: 2.32). In France dairy cattle are kept most of the time indoors, while beef cattle have stabling periods of approximately five months each year. Oppositely, in Africa and in Asia, cattle are most of the time kept outdoors, where contacts between animals are less intense than indoors, hence the lower observed within-herd prevalence in these regions.

The low predicted values of the effective reproductive ratio R(t) over time are consistent with field observations of a low bTB prevalence in French infected herds in the 2000s, and support the results of transmission trials indicating that cattle do not readily infect other cattle [39,40]. According to our estimation, when no control program is applied and once bTB has become enzootic within the herd, the median number of animals infected by an infectious animal is 2.2 in beef herds and 1.7 in dairy herds. This difference is due to a lower age at culling for dairy cows than for beef cows.

Several models have been proposed in the literature to simulate the dynamic of within-herd transmission of bTB: stochastic Reed-Frost model in discrete time [2,12], Reed-Frost model in continuous time [14], stochastic individual based model [15], compartmental stochastic model in continuous time [28] and compartmental stochastic model in discrete time [13,33,41]. We used a compartmental stochastic model operating in discrete time. Bovine TB health states are best represented using discrete health states (S-E-I), we thus chose to use a compartmental model. The scarcity and the discrete nature of the main simulated event which is the infection of animals (in a national survey in France between 2005 and 2007 in infected herds, the median of animals with lesion was 2), led us to use a stochastic approach rather than a deterministic approach. ABC methods are computationally cumbersome. Discrete time simulation was chosen as it appeared less computationally demanding than continuous time simulation. The duration of the time step (one month) was chosen because in literature the estimated duration of the latent state (in models with a single latent state [2] or with two states of latency [28]) was greater than one month. In addition, for the modeled screening, control and demographic processes, a one month time step also appeared satisfactory (frequency of tuberculin test, stabling period).

While this type of model is relatively easy to simulate, it is analytically difficult to manage when the number of states is high. In this case, numerical resolution by simulation is the only way to estimate the parameters [42]. The likelihood is difficult to calculate in our model, because the infection process is only partially observed (only a fraction of infectious animals is detected at the slaughterhouse, depending on the sensitivity of the visual inspection of carcasses). We thus chose to use the ABC method for the estimation of model parameters because this method does not require the calculation of the likelihood [43]. The choice of summary statistics and of the tolerance threshold are two sources of error in the ABC method [44]. To reduce errors linked to the choice of summary statistics, a preliminary step was performed in which several summary statistics were tested and compared using simulated data. Other methods have been proposed in the literature such as the algorithm developed by Barnes et *al.* [45] to choose a set of summary statistics from all possible summary statistics. This algorithm uses the mutual information as a selection tool [45]. Choosing a non-zero value for tolerance, $\varepsilon$, can bias the posterior estimation of the parameters [44]. To reduce errors due to non-zero values of $\varepsilon$, the local linear regression algorithm of the ABC method was used in our work for estimating the model parameters [19]. ABC-SMC (Sequential Monte Carlo) algorithm allows using specific methods to select the tolerance threshold [46,47]. But, according to Filippi et *al.* [48], more work needs to be done on the choice of tolerance threshold in the ABC-SMC algorithm. Besides, the use of the local linear regression algorithm is however not without problems, as the relationship between summary statistics and parameters is highly non-linear in most cases, which may bias the estimation of parameters [19].

The sensitivity analysis of a model is of great interest [42] because it helps to determine the most influential parameters. The sensitivity analysis performed here was used to independently assess the effect of fixed parameters values on the posterior parameter estimates. For practical reasons (of computation time required to perform this sensitivity analysis), the sensitivity analysis did not assess the effect of interactions among fixed parameters. Nevertheless, according to the literature, this method is the first step in exploring global associations (between the estimated and non-estimated parameters in the model), excluding interactions [49]. The results of the sensitivity analysis showed that fixed parameters have only limited effects on the posterior parameter estimates, except for the effect of the specificity of ESAT-6 $\gamma$IFN test on $\beta_{inside}$. The specificity of ESAT-6 $\gamma$IFN test is used in control program C (Figure 3), which includes a 2-steps confirmation of positive SITT using $\gamma$IFN test (steps 2 and 3 in Figure 3). When the specificity of ESAT-6 $\gamma$IFN is increased to 100% (see Table 2), the overall specificity of the 3rd step of control program C also reaches 100% (see Figure 3). Because the overall sensitivity of both $\gamma$IFN-based confirmation steps is low, the time required to detect bTB increases, which induces a decrease of $\beta_{inside}$ estimate.

We assumed that the sensitivity and the specificity of the screening tests were independent of cattle age. Proaño-Perez et *al.* [50] reported a negative correlation between the age of cattle and the response to avian tuberculin. This result suggests that young animals are more likely to show false positive reactions to SITT than adult animals. The impact on our results is however likely to be limited, as the sensitivity analysis showed that the specificity of screening test had a little effect on the estimated values of the epidemiological model parameters ($\alpha$, $\beta_{inside}$ and $\beta_{outside}$).

In conclusion, the model proposed here has been designed to be generic enough to allow simulating various herd management systems and bTB detection and control programs. Its parameters have been estimated using field data and it has been validated using an independent dataset. In the future, this model will be used to analyse the impacts of changes of control programs and of herd management practices on the dynamics of bTB in France, between the beginning of the control program (in 1965) and obtaining the bTB-free status (in 2000). Besides, it will be used to assess the effectiveness of bTB control programs, in order to identify alternative strategies to the total slaughter protocol currently applied in infected herds.

## Supporting Information

**Figure S1** Control program A applied in the Nord department between 1981 and 1983. Step 1: yearly bTB screening using SITT (the herd being considered disease-free). All animals are tested ($\mu$ i,k $Q(i,k) = 1$). Sensitivity (Se) and specificity (Sp): those of SITT. Transition to step 2 if the proportion of positive animals exceeds a predefined threshold ($n^{pos}/n^{test}$>P.ab), one month later ($\Delta t = 1$ months). Transition to step 3 if positive results are observed, the proportion being below the threshold. Step 2: total slaughter. All the animals ($\mu$ i,k $Q(i,k) = 1$)are slaughtered ($m = 1$). Step 3: selective slaughter. Positive animals of step 1 ($\mu$ i $Q(i,1) = 1$)are slaughtered ($m = 1$). Transition to step 1, 6 months later. See table 2 for the definition of the other parameters.

**Figure S2** Control program D applied in the Côte d'Or department between 2005 and 2009. Step 1: biennial bTB screening using SICCT (the herd being considered disease-free). All animals are tested ($\mu$ i,k $Q(i,k) = 1$). Sensitivity (Se) and specificity (Sp): those of SICCT. Immediate transition to step 2 if non-negative results are observed ($n^{pos}>0$). Step 2: interpretation of SICCT non-negative results. All the positive animals of step 1 ($\mu$ i $Q(i,1) = 1$) are concerned. Sensitivity: for an infected animal, probability that a non-negative positive result is not doubtful (pi: probability of a doubtful SICCT result for animals in health states E or I); specificity: for a susceptible animal, probability that a non-negative SICCT result is doubtful (ps: probability of a doubtful SICCT result for animals in health state S). Transition to step 3 if all the non-negative SICCT animals are doubtful, 2 months later; otherwise: transition to step 4, 3 months later. Step 3: confirmation of the positive results of step 1 using SICCT. All the positive animals of step 1 are tested ($\mu$ i $Q(i,1) = 1$), as well as 10% of the negative animals (i Q(i,0) = 0.1). Sensitivity (Se) and specificity (Sp): those of SICCT. Transition to step 5 if an animal is positive, three months later; otherwise: transition to step 1. Step 4: slaughter of the positive animals of step 2 and isolation of *M. bovis* from lesions. All the positive animals of step 2 (μ i Q(i,2) = 1) are slaughtered (m = 1) and bacterial culture is performed from observed lesions. Sensitivity (Se): sequential combination of a visual inspection at the slaughterhouse and of a bacterial culture. Transition to step 6 if an animal is positive, 1 month later (Δt = 1 month). Step 5: slaughter of the positive animals of step 3 and isolation of *M. bovis* from lesions. All the positive animals of step 3 (μ i Q(i,3) = 1) are slaughtered (m = 1) and bacterial culture is performed from observed lesions. Sensitivity (Se): sequential combination of a visual inspection at the slaughterhouse and of a bacterial culture. Transition to step 6 if an animal is positive, 1 month later (Δt = 1 month). Step 6: total slaughter. All the animals (μ i,k Q(i,k) = 1) are slaughtered (m = 1). Routine detection of lesions at slaughterhouse. Sensitivity ($Se_{ev}$): sequential combination of a visual inspection at the slaughterhouse and of a bacterial culture. If a positive animal is thus detected, regardless of the current stage of control program, transition to step 6. See table 2 for the definition of the other parameters.

**Appendix S1** Model equations.

**Appendix S2** Choice of summary statistics.

**Appendix S3** Effective reproductive rate R(t).

## Acknowledgments

The authors would like to thank the staff of the Departmental Director of Social Cohesion and Protection of Populations (DDCSPP) of Dordogne and the Departmental Director of Protection of Populations (DDPP) of Côte d'Or, for providing data. We would also like to thank Pr. Coureau from the Alfort National Veterinary School for giving us information on herd management practices in France.

## Author Contributions

Conceived and designed the experiments: JJB AC BD MEAB. Performed the experiments: AC BD MEAB. Analyzed the data: BD MEAB. Contributed reagents/materials/analysis tools: JJB BD. Contributed to the writing of the manuscript: AC BD MEAB.

## References

1. Radostits OM, Blood DC, Gay CC, Arundel JH, Ikede BO, et al. (1994) Veterinary Medicine: A Textbook of the Diseases of Cattle, Sheep, Pigs, Goats and Horses. London: Bailliere Tindall. 1763 p.
2. Perez AM, Ward MP, Charmandarian A, Ritacco V (2002) Simulation model of within-herd transmission of bovine tuberculosis in Argentine dairy herds. Prev Vet Med 54: 361–372.
3. Reynolds D (2006) A review of tuberculosis science and policy in Great Britain. Vet Microbiol 112: 119–126.
4. de la Rua-Domenech R (2006) Bovine Tuberculosis in the European Union and other countries: current status, control programmes and constrains to eradication. Gov Vet J 16: 19–45.
5. Wahlstrom H, Carpenter T, Giesecke J, Andersson M, Englund L, et al. (2000) Herd-based monitoring for tuberculosis in extensive swedish deer herds by culling and meat inspection rather than by intradermal tuberculin testing. Prev Vet Med 43: 103–116.
6. Cousins DV (2001) Mycobacterium bovis infection and control in domestic livestock. Rev Sci Tech 20: 71–85.
7. Bekara ME, Azizi L, Benet JJ, Durand B (2014) Spatial-temporal Variations of Bovine Tuberculosis Incidence in France between 1965 and 2000. Transbound Emerg Dis.
8. de la Rua-Domenech R, Goodchild AT, Vordermeier HM, Hewinson RG, Christiansen KH, et al. (2006) Ante mortem diagnosis of tuberculosis in cattle: a review of the tuberculin tests, gamma-interferon assay and other ancillary diagnostic techniques. Res Vet Sci 81: 190–210.
9. Corner LA (1994) Post mortem diagnosis of Mycobacterium bovis infection in cattle. Vet Microbiol 40: 53–63.
10. de Jong MCM (1995) Mathematical modelling in veterinary epidemiology: why model building is important. Prev Vet Med 25: 183–193.
11. Smith S, Anders B (1994) Tuberculosis mastoiditis caused by Mycobacterium bovis. Pediatr Infect Dis J 13: 538–539.
12. Alvarez J, Perez AM, Bezos J, Casal C, Romero B, et al. (2012) Eradication of bovine tuberculosis at a herd-level in Madrid, Spain: study of within-herd transmission dynamics over a 12 year period. BMC Vet Res 8: 100.
13. Barlow ND, Kean JM, Hickling G, Livingstone PG, Robson AB (1997) A simulation model for the spread of bovine tuberculosis within New Zealand cattle herds. Prev Vet Med 32: 57–75.
14. Griffin JM, Williams DH, Collins JD (1999) Tuberculosis Investigation Unit, University College Dublin. Selected papers: 55–58.
15. Fischer EA, van Roermund HJ, Hemerik L, van Asseldonk MA, de Jong MC (2005) Evaluation of surveillance strategies for bovine tuberculosis (Mycobacterium bovis) using an individual based epidemiological model. Prev Vet Med 67: 283–301.
16. Kao RR, Roberts MG, Ryan TJ (1997) A model of bovine tuberculosis control in domesticated cattle herds. Proc Biol Sci 264: 1069–1076.
17. Carnon F (1985) Contribution à l'étude épidemiologique de la tuberculose bovine dans le département du Nord. Maisons-Alfort: Ecole Nationale Vétérinaire d'Alfort. 210 p.
18. Toni T, Welch D, Strelkowa N, Ipsen A, Stumpf MP (2009) Approximate Bayesian computation scheme for parameter inference and model selection in dynamical systems. J R Soc Interface 6: 187–202.
19. Beaumont MA, Zhang W, Balding DJ (2002) Approximate Bayesian computation in population genetics. Genetics 162: 2025–2035.
20. Neill SD, Hanna J, Mackie DP, Bryson TG (1992) Isolation of Mycobacterium bovis from the respiratory tracts of skin test-negative cattle. Vet Rec 131: 45–47.
21. Aeschbacher S, Beaumont MA, Futschik A (2012) A novel approach for choosing summary statistics in approximate Bayesian computation. Genetics 192: 1027–1047.
22. Csillery K, Blum MG, Gaggiotti OE, Francois O (2010) Approximate Bayesian Computation (ABC) in practice. Trends Ecol Evol 25: 410–418.
23. Beaumont MA (2010) Approximate Bayesian computation in evolution and ecology. Annual Review of Ecology, Evolution, and Systematics 41: 379–406.
24. Csilléry K, François O, Blum MGB (2012) abc: an R package for approximate Bayesian computation (ABC). Methods in Ecology and Evolution 3: 475–479.
25. R Development Core Team (2013) R: A language and environment for statistical computing. Vienna, Austria: R Foundation for Statistical Computing.
26. Anderson RM, May RM (1991) Infectious diseases of humans: dynamics and control. Oxford and New York: Oxford University Press. 768 p.
27. Kleeberg HH (1960) The tuberculin test in cattle. J South African Vet Med Assoc 31: 213–225.
28. Conlan AJ, McKinley TJ, Karolemeas K, Pollock EB, Goodchild AV, et al. (2012) Estimating the hidden burden of bovine tuberculosis in Great Britain. PLoS Comput Biol 8: e1002730.
29. Karolemeas K, McKinley TJ, Clifton-Hadley RS, Goodchild AV, Mitchell A, et al. (2011) Recurrence of bovine tuberculosis breakdowns in Great Britain: risk factors and prediction. PrevVet Med 102: 22–29.
30. Porphyre T, Stevenson MA, McKenzie J (2008) Risk factors for bovine tuberculosis in New Zealand cattle farms and their relationship with possum control strategies. PrevVet Med 86: 93–106.
31. Ramirez-Villaescusa AM, Medley GF, Mason S, Green LE (2010) Risk factors for herd breakdown with bovine tuberculosis in 148 cattle herds in the south west of England. PrevVet Med 95: 224–230.
32. Vial F, Johnston WT, Donnelly CA (2011) Local cattle and badger populations affect the risk of confirmed tuberculosis in British cattle herds. PLoS One 6: e18058.
33. Smith RL, Schukken YH, Lu Z, Mitchell RM, Grohn YT (2013) Development of a model to simulate infection dynamics of Mycobacterium bovis in cattle herds in the United States. J Am Vet Med Assoc 243: 411–423.
34. Hagan WA, Bruner DW, Gillespie JH (1973) Hagan's Infectious diseases of domestic animals: with special reference to etiology, diagnosis, and biologic therapy. Californie: Comstock Pub. Associates. 1385 p.
35. Gumi B, Schelling E, Firdessa R, Aseffa A, Tschopp R, et al. (2011) Prevalence of bovine tuberculosis in pastoral cattle herds in the Oromia region, southern Ethiopia. Trop Anim Health Prod 43: 1081–1087.
36. Thakur A, Sharma M, Katoch VC, Dhar P, Katoch RC (2010) A study on the prevalence of Bovine Tuberculosis in farmed dairy cattle in Himachal Pradesh. Vet World 3: 408–413.
37. Ameni G, Aseffa A, Engers H, Young D, Hewinson G, et al. (2006) Cattle husbandry in Ethiopia is a predominant factor affecting the pathology of bovine tuberculosis and gamma interferon responses to mycobacterial antigens. Clin Vaccine Immunol 13: 1030–1036.

38. Ameni G, Aseffa A, Engers H, Young D, Gordon S, et al. (2007) High prevalence and increased severity of pathology of bovine tuberculosis in Holsteins compared to zebu breeds under field cattle husbandry in central Ethiopia. Clin Vaccine Immunol 14: 1356–1361.
39. Costello E, Doherty ML, Monaghan ML, Quigley FC, O'Reilly PF (1998) A study of cattle-to-cattle transmission of Mycobacterium bovis infection. Vet J 155: 245–250.
40. O'Reilly LM, Costello E (1988) Bovine tuberculosis with special reference to the epidemiological significance of pulmonary lesions. Irish Veterinary News: 11–21.
41. Brooks-Pollock E, Roberts GO, Keeling MJ (2014) A dynamic model of bovine tuberculosis spread and control in Great Britain. Nature 511: 228–231.
42. Pouillot R, Dufour B, Durand B (2004) A deterministic and stochastic simulation model for intra-herd paratuberculosis transmission. Vet Res 35: 53–68.
43. Tanaka MM, Francis AR, Luciani F, Sisson SA (2006) Using approximate Bayesian computation to estimate tuberculosis transmission parameters from genotype data. Genetics 173: 1511–1520.
44. Sunnaker M, Busetto AG, Numminen E, Corander J, Foll M, et al. (2013) Approximate Bayesian computation. PLoS Comput Biol 9: e1002803.
45. Barnes C, Filippi S, Stumpf MPH, Thorne T (2012) Considerate approaches to constructing summary statistics for ABC model selection. Statistics and Computing 22: 1181–1197.
46. Del Moral P, Doucet A, Jasra A (2012) An adaptive sequential Monte Carlo method for approximate Bayesian computation. Statistics and Computing 22: 1009–1020.
47. Drovandi CC, Pettitt AN (2011) Estimation of parameters for macroparasite population evolution using approximate bayesian computation. Biometrics 67: 225–233.
48. Filippi S, Barnes CP, Cornebise J, Stumpf MP (2013) On optimality of kernels for approximate Bayesian computation using sequential Monte Carlo. Stat Appl Genet Mol Biol 12: 87–107.
49. Vose D (2000) Risk Analysis, A quantitative guide. Chichester: Wiley and Sons.752 p.
50. Proano-Perez F, Benitez-Ortiz W, Celi-Erazo M, Ron-Garrido L, Benitez-Capistros R, et al. (2009) Comparative intradermal tuberculin test in dairy cattle in the north of Ecuador and risk factors associated with bovine tuberculosis. Am J Trop Med Hyg 81: 1103–1109.
51. Pollock JM, Girvin RM, Lightbody KA, Clements RA, Neill SD, et al. (2000) Assessment of defined antigens for the diagnosis of bovine tuberculosis in skin test-reactor cattle. Vet Rec 146: 659–665.
52. Courcoul A, Moyen JL, Brugere L, Faye S, Henault S, et al. (2014) Estimation of sensitivity and specificity of bacteriology, histopathology and PCR for the confirmatory diagnosis of bovine tuberculosis using latent class analysis. PLoS One 9: e90334.

# Increased Mortality in Groups of Cattle Administered the β-Adrenergic Agonists Ractopamine Hydrochloride and Zilpaterol Hydrochloride

**Guy H. Loneragan[1]*, Daniel U. Thomson[2], H. Morgan Scott[3]**

1 International Center for Food Industry Excellence, Department of Animal and Food Sciences, College of Agriculture and Natural Resources, Texas Tech University, Lubbock, Texas, United States of America, 2 Department of Clinical Sciences, College of Veterinary Medicine, Kansas State University, Manhattan, Kansas, United States of America, 3 Department of Diagnostic Medicine/Pathobiology, College of Veterinary Medicine, Kansas State University, Manhattan, Kansas, United States of America

## Abstract

The United States Food and Drug Administration (FDA) approved two β-adrenergic agonists (βAA) for in-feed administration to cattle fed in confinement for human consumption. Anecdotal reports have generated concern that administration of βAA might be associated with an increased incidence of cattle deaths. Our objectives, therefore, were to a) quantify the association between βAA administration and mortality in feedlot cattle, and b) explore those variables that may confound or modify this association. Three datasets were acquired for analysis: one included information from randomized and controlled clinical trials of the βAA ractopamine hydrochloride, while the other two were observational data on zilpaterol hydrochloride administration to large numbers of cattle housed, fed, and cared for using routine commercial production practices in the U.S. Various population and time at-risk models were developed to explore potential βAA relationships with mortality, as well as the extent of confounding and effect modification. Measures of effect were relatively consistent across datasets and models in that the cumulative risk and incidence rate of death was 75 to 90% greater in animals administered the βAA compared to contemporaneous controls. During the exposure period, 40 to 50% of deaths among groups administered the βAA were attributed to administration of the drug. None of the available covariates meaningfully confounded the relationship between βAA and increased mortality. Only month of slaughter, presumably a proxy for climate, consistently modified the effect in that the biological association was generally greatest during the warmer months of the year. While death is a rare event in feedlot cattle, the data reported herein provide compelling evidence that mortality is nevertheless increased in response to administration of FDA-approved βAA and represents a heretofore unquantified adverse drug event.

**Editor:** William Barendse, CSIRO, Australia

**Funding:** Initial funding for the evaluation of an association of ractopamine hydrochloride with death loss was provided by Elanco Animal Health, Greenfield, IN. In preparation for manuscript development, further analyses of ractopamine hydrochloride data and all analyses of zilpaterol hydrochloride data were performed solely in the service of the first author's employment at Texas Tech University. The funders had no role in study design, data collection and analysis, decision to publish, or preparation of the manuscript.

**Competing Interests:** I have read the journal's policy and have the following conflicts: GHL has accepted consulting fees for service on scientific advisory boards for Elanco Animal Health, and Intervet/Schering Plough (now known as Merck Animal Health) an dZoetis. In addition, GHL has received honoraria and paid travel expenses from Elanco Animal Health and Zoetis, and has served as a co-author with, and graduate advisor of, Elanco Animal Health personnel. DUT reports that Elanco Animal Health provided funds in support of the Beef Cattle Institute at Kansas State University of which DUT is the director. HMS reports no conflicts of interest related to either drug sponsor. The authors have no other competing interests such as employment, patents, products in development or marketed products.

* E-mail: Guy.Loneragan@TTU.edu

## Introduction

The United States Food and Drug Administration (FDA) approved two β-adrenergic agonists (βAA) for in-feed administration to cattle that are fed in confinement (i.e., typically feedlot operations) for human consumption [1,2]. When administered according to label directions, βAA result in well-characterized and predictable improvements in the rate and efficiency of weight gain, as well as increased leanness and yield of edible products derived from beef carcasses [3–7]. Ractopamine hydrochloride (RH) was the first βAA approved in cattle; further, RH may be used in a variety of dosages and has also been approved for administration to swine and turkeys. Ractopamine hydrochloride is included in cattle feed during the final 28 to 42 days of the fattening period. Zilpaterol hydrochloride (ZH), the second βAA approved by the FDA, may only be used at a single rate of inclusion in cattle feeds. It is included in cattle feed for 20 to 40 days prior to slaughter; however, in contrast to RH, a 3-day period during which ZH may not be administered must be observed prior to shipment to the abattoir (i.e., slaughter withholding).

In addition to their production uses in food-animal production, β-adrenergic agonists are routinely used in human clinical medicine for various conditions such as acute therapeutic intervention and maintenance care of asthma and chronic obstructive pulmonary disease (i.e., COPD). In human clinical medicine, these drugs are not innocuous in that side effects (or adverse unintended consequences) of approved medical uses of βAA have been observed and include an increased risk of asthma exacerbations and hospitalizations, arrhythmias, myocardial infarction, and death [8–12]. Unintended consequences have also been observed with βAA use in several animal species. For example, observations both in non-ruminant and ruminant species indicate that administration of βAA is associated, albeit somewhat inconsistently, with elevated heart rates, body temperature, physical activity, lameness or foot lesions, and aggression [13–18]. Given the widespread distribution of β-adrenoreceptors in the body, some of these unintended consequences, such as elevated heart rate, might (or ought to) be expected with βAA administration.

With some similarity to observations of increased risk of mortality associated with use of certain βAA in human clinical medicine [9,19], unpublished reports from some end-users of RH and ZH indicated the potential for an increase in mortality in cattle associated with FDA-approved administration of βAA. Mortality in feedlot cattle represents a clear and meaningful economic loss to producers. Death loss also raises broader societal concerns about the welfare of animals fed βAA in that progression from a healthy status to death may include in pain and suffering in affected animals. If a relationship between βAA administration and increased risk of mortality exists, it ought to stimulate discussion of the pros and cons of the use of drugs approved purely to improve the efficiencies of production yet offering no offsetting health benefits to the animals. As an example, some antibiotics are believed to improve production efficiency, at least in part, by controlling or preventing subclinical disease. To our knowledge, no beneficial 'side-effect' favoring improved health or welfare, or control or prevention of disease in response to FDA-approved βAA administration in cattle is believed to exist. Our objectives, therefore, were to: a) quantify the association between βAA administration and mortality in feedlot cattle, and b) explore those variables that may confound or modify this association.

## Methods

Three confidential datasets were received either through solicitation by the first author in the case of RH or following requests for analytical support from the owners of the data (i.e., feedlot operators) in the case of ZH. The authors adhere to all the PLOS ONE policies on sharing data and materials. Different models were constructed and where possible and appropriate, a number of covariates (as potential and plausible confounders and effect modifiers) were evaluated. Datasets analyzed included both experimental trials (i.e., clinical trials involving randomized treatment allocation) and observational studies (i.e., treatment allocation was determined by factors other than a random process).

### Description of Datasets

The first dataset (hereafter ***4-company RH dataset***) included information concerning administration of RH. A convenience sample of cattle-feeding companies was contacted to evaluate whether or not they had performed field-based experiments of RH administration and, if so, to ascertain their willingness to provide their data for further analysis. Inclusion criteria included: a) experimental observations of RH administered to cattle pens according to label directions so as to provide a target dose of 200 mg/animal/day for the 28 to 42 days immediately prior to shipment to the abattoir, b) inclusion of appropriate contemporaneous control pens, and c) randomization performed either at the group level, or at the individual level. Groups were then classified as either unexposed in that all animals within the group were fed the usual fattening diet, or else as exposed in that all animals within the group were fed the usual fattening diet with RH incorporated to provide the target dose.

This first aggregated dataset included information from four companies and collectively included 12 separate randomized experiments. All experiments used a randomized block design in that there were at least 2 groups per block, i.e., one administered RH and the other not. These data included information suitable for analysis on a total of 79,171 cattle. These animals were aggregated into 509 groups that averaged 155.5 (standard deviation [SD] = 70.9, minimum [MIN] = 42, maximum [MAX] = 381) animals per group. The number of studies and groups varied by company in that company A provided data on 1,510 animals that were enrolled in 1 study and housed in 24 groups, company B provided data on 5,696 animals across 3 studies and housed in 52 groups, company C provided data on 62,379 animals across 3 studies and housed in 329 groups, and company D provided data on 9,586 animals across 5 studies and housed in a total of 104 groups. The combined 6 studies conducted by companies B and C were conducted in 6 different feedlots whereas the 5 studies of company D were all conducted in the same feedlot. The vast majority of cattle were steers (i.e., castrated males; n = 72,868 in 441 groups) with the balance being females (i.e., heifers; n = 6,303 in 68 groups). Heifers were exclusively enrolled as the study population in 3 of the 5 experiments of company D.

The second dataset (***multi-feedlot ZH dataset***) concerned administration of ZH and included observational data from nine feedlots. Data on 722,704 animals were provided and the cattle were housed in 3,110 groups of an average size of 232.4 animals per group (SD = 91.3, MIN = 32, MAX = 943). Out of these animals, 79.3% (n = 573,076) were steers and 20.7% (n = 149,628) were heifers. The at-risk period of interest for exposed animals included both the period during which ZH was administered as well as the slaughter-withholding period. This slaughter-withholding period must be a minimum of 3 days but may be longer depending on various marketing strategies used by the feedlots. For the unexposed population in the multi-feedlot ZH dataset, the at-risk period was calculated by including the typical number of days ZH was administered to the exposed cohort and the typical slaughter-withholding period. Based on the observed and calculated at-risk periods for the exposed and unexposed cohorts, respectively, the mean at-risk periods consisted of the final 29.4 and 29.2 days prior to shipment, respectively. The data were unbalanced in that there were 2,775 groups comprised of 637,339 animals administered ZH (i.e., exposed cohort) and 335 groups comprised of 85,365 animals that served as comparative controls (i.e., unexposed cohort). The mean group sizes were 229.7 (SD = 90.7, MIN = 32, MAX = 943) and 254 (SD = 94.0, MIN = 55, MAX = 610), respectively. The number of animals per feedlot for which data were provided varied from 61,059 to 123,679. Heifers were represented in the data from 8 of the 9 feedlots.

The third dataset (***single-feedlot ZH dataset***) included observational data on 149,636 animals that were housed in 835 groups in a single feedlot. Of the population at risk, 88.7% (n = 132,725) were steers and 11.3% (n = 16,911) were heifers. The data were more balanced than the multi-feedlot ZH dataset in that 56.1% of the cattle (n = 83,865 in 470 groups) were administered ZH (i.e., exposed cohort) whereas 43.9% of the cattle (n = 65,711 in 365 groups) served as the contemporaneous control cohort. The mean group sizes were 178.4 (SD = 76.3, MIN = 30, MAX = 352) and 180.2 (SD = 63.9, MIN = 54, MAX = 382), respectively. The feedlot was managed in such a way that ZH was administered 21 days prior to shipment and a 3-day withdrawal period was observed. This 24-day period was considered the at-risk period.

Consequently, the final 24 days prior to shipment were considered the comparative at-risk period for those animals not administered ZH.

## Data Analyses

The primary outcome variable of interest across all datasets was the number of animals that died in each group during the at-risk period (consequently, the group may be considered the experimental unit of interest). This outcome variable, therefore, represents a count response and the approaches described herein to model count data within groups in which the outcome may be clustered have been described [20–23]. Two offset variables were used as denominators in the various statistical models. The first was the natural logarithm of the population within a group (i.e., the at-risk population) at the start of the exposure period; the use of this offset allows calculation of model-adjusted estimates of the proportion the population at risk that died within each cohort [20]. Where the at-risk period varied across groups, i.e., in the 4-company RH and multi-feedlot ZH datasets, the natural logarithm of time at-risk was used (in other words, the total number of cattle-days within a group during the period of interest); the use of this offset allows calculation of model-adjusted estimates of the incidence of death (i.e., deaths per unit time) within each cohort [20]. Because information on the day during the at-risk period that individual animals died was only available for Company C in the RH dataset, a uniform approach to estimating time at risk (expressed as cattle-days) was used for the 4-company RH and multi-feedlot ZH datasets. Time at risk was estimated as a function of the at-risk period (days) multiplied by the population at risk. To account for withdrawals due to death, half of the total possible days at-risk was subtracted from the group's time at risk for each animal that died within the group.

Consistent across all datasets, therefore, were deaths within a group, whether they were administered a βAA or not, and population at risk. In addition, for the 4-company RH and multi-feedlot ZH datasets, time at risk was calculated using a common approach. Generalized linear mixed models were constructed using commercially available statistical analysis software (SAS System for Windows release 9.3, SAS Institute, Cary, NC) and parameterized similarly across datasets to account for the hierarchical nature of the data. A Poisson distribution was used with a log-link function. In all models, a group-level term was forced into the model to account for potential over-dispersion of the data (i.e., extra-Poisson variation). In the model of the 4-company RH dataset, random intercept terms were included for company, study within company, and block within study. In the multi-feedlot ZH dataset, a random intercept term was included to account for potential clustering of the outcome within feedlots. To explore potential modification of βAA effect on death loss across companies in the 4-company RH dataset and across feedlots in the multi-feedlot ZH dataset, the highest-level random-intercept term was changed from a random effect to a fixed effect to explore the interaction with exposure. Because of model convergence issues due to sparsely populated cells in the former dataset, when evaluating the interaction of feedlot and RH administration, data from company A was dropped from the model (i.e., n = 1,510 animals in 24 groups where no deaths were reported in either cohort).

Covariates were variably recorded across the three datasets. For example, within the 4-company RH dataset, the number of deaths within a group prior to exposure was relatively consistently recorded whereas month of shipment was only recorded for Company C. Further, in the multi-feedlot ZH dataset, a variety of covariates were consistently recorded and included: sex of the animals within a group, percentage of a group that died prior to the at-risk period, percentage of a group that were treated prior to the at-risk period, percentage of cattle within a group that had a predominantly black hide, mean carcass weight of the surviving animals that were shipped to slaughter, and month in which the at-risk period ended (i.e., animals were shipped to an abattoir for slaughter). However, in the single-feedlot ZH dataset only sex of the animal and the month in which the animals were shipped to an abattoir for slaughter were available for analysis. For each dataset, therefore, covariates that could potentially confound or modify the association between βAA use and death loss were evaluated in univariate models to test their association with mortality. Model structures were similar to those described above. Covariates with $P$ values less than or equal to 0.20 were included as fixed effects in multivariable models that included main effects (i.e., potential confounders) and terms for the interaction of each of these main effects with exposure to βAA (i.e., effect-measure modifiers). While maintaining hierarchy within variables, terms were removed from the model in a backward, stepwise manner using $\alpha = 0.10$ level of significance for retention. For all models, measures of burden (proportion of population at risk and incidence of mortality per time at risk) and measures of effect (cumulative relative risk [RR] and incidence rate ratio [IRR], respectively) were computed from the model estimates and presented with their 95% confidence limits (CL) and $P$ values where appropriate.

In addition to the primary outcome of mortality, secondary outcomes were available for both the multi-feedlot ZH and single-feedlot ZH datasets. Secondary count-based outcomes included number of animals treated for illness during the at-risk period and the number of carcasses that were classified as *dark, firm and dry* (which is colloquially referred to as *dark cutter* in the beef cattle industry). In addition, the proportion of cattle that could not be shipped to slaughter because they were within a slaughter-withholding period at the time the rest of the group was shipped was available in the single-feedlot dataset. This so-called *medicine hold* results from administration of a therapeutic drug, such as an injectable antimicrobial drug to treat bacterial bronchopneumonia; most such FDA-approved antimicrobial drugs have slaughter-withholding periods and these must be observed prior to shipment to slaughter for human consumption.

To explore the extent of unexplained model variation attributable to the levels of company and study within company, a multilevel, hierarchical model was constructed for the 4-company RH dataset using commercially available software (MLwiN 2.26, Centre for Multilevel Modelling, University of Bristol, Bristol, UK). Four levels of organization were included in the model: company, study within company, block within study, and group within block; generalized linear mixed models within a Poisson distribution were constructed. The dependent and offset variables were those described above and RH administration was the independent variable of interest. After accounting for RH administration as a fixed effect in the model, unexplained variation was partitioned to the highest 3 level terms (i.e., level 1 variance was not calculated given the assumptions of the Poisson model) [22]. Model estimation was performed using reweighted iterative generalized least squares and $2^{nd}$-order penalized quasi-likelihood approximation, while allowing for over-dispersion of the data [24].

Multivariable semi-parametric survival analysis was performed on the data provided by Company C owing to the rich level of detail available (Stata version 12.1, Stata Corp., College Station, TX). The multivariable model accounted for fixed effects of study, month of shipment to slaughter, prior mortality events experienced within the cohort, and exposure to RH. In addition, the model accounted for the shared frailty experience of animals within each group. Backwards elimination was employed to yield the final model (accounting for potential confounding variables and effect modifiers) and the importance of the shared group frailty was determined using the parameter Theta.

## Results

### 4-company RH Dataset

Overall, 0.27% (n = 211) of the 79,171 cattle died during the at-risk period (Table 1). After accounting for various levels of clustering, the estimates of cumulative risk and the incidence rate of death were 0.26% (95% confidence limits [CL] = 0.16, 0.40) and 0.86 (95% CL = 0.58, 1.30) deaths per 10,000 cattle days, respectively. Model-adjusted estimates of risk and incidence of death for cattle administered RH were 0.34% (95% CL = 0.25, 0.45) and 1.12 (95% CL = 0.85, 1.45) deaths per 10,000 cattle days at risk, respectively. For cattle not administered RH, risk and incidence of death were 0.18% (95% CL = 0.13, 0.25) and 0.59 (95% CL = 0.43, 0.83) deaths per 10,000 cattle days, respectively.

After controlling for clustering within company, study, block and group, cattle administered RH were 91% more likely to die than control animals during the at-risk period (RR = 1.90 [95% CL = 1.38, 2.60]; $P<0.01$). A very similar measure of effect was observed for the incidence rate ratio (IRR = 1.90 [95% CL = 1.38, 2.61]; $P<0.01$). The only potential covariate collected across multiple studies was the number of animals that died in each group prior to the at-risk period. This variable neither modified the effect of exposure on mortality ($P=0.77$) nor did it confound the association between βAA and increased death loss ($P=0.63$). There was no evidence that the association between RH administration and increased death loss varied across the 3 companies that supplied data in which at least one death was observed ($P=0.66$; Figure 1). Moreover, in the multilevel hierarchical model that included all 4 companies, there was no unexplained variation attributed to unmeasured factors across companies (model variance = 0.00 [SE = 0.00]). At lower levels of organization, the unexplained model variation was 4-fold greater among blocks within studies (model variance = 0.169 [SE = 0.120]) than among studies within companies (model variance = 0.048 [SE = 0.072]).

Company C provided information on 167 deaths across 1,983,564 animal-days and within this company, the incidence of death was 83% greater among those animals administered RH compared to controls (IRR = 1.83; 95% CL = 1.32, 2.56; $P<0.01$). A graphical representation of the unadjusted Kaplan-Meier non-parametric survival experiences for each set of treatment cohorts is presented in Figure 2. The smoothed instantaneous force of mortality for each cohort set (i.e., the hazard function [h(t)]) is presented in Figure 3 and suggests that the constant hazard assumption was met for approximately the first 25–28 days of the at risk period. Moreover, the proportional hazards assumption was met throughout the entire at-risk period. In addition, the pen-level mortality rate prior to exposure was evaluated as a covariate but was neither associated with the incidence of death during exposure ($P=0.61$), nor did this variable confound the relation between exposure to RH and survivor function. Importantly, the multi-variable adjusted treatment hazard ratio (HR) of 2.00 (95% CL = 1.36, 2.96; $P<0.01$) was greater than that observed when the other significant covariates (i.e., fixed effects of feedlot and month of slaughter) and the shared frailties of group were ignored. Because the force of mortality (hazard) was relatively constant over time, the HR is comparable to IRR; the observed IRR was 1.83 and was included within the 95% confidence interval of the HR.

### Multi-feedlot ZH Dataset

Of the 722,704 cattle at risk, 0.51% (n = 3,657) of animals died during the exposure period (Table 1). After accounting for feedlot- and group-level clustering, the estimates of risk and incidence of death were 0.50% (95% CL = 0.44, 0.57) and 1.68 (95% CL = 1.51, 1.86) deaths per 10,000 animal-days, respectively. Model-adjusted estimates of the risk and incidence of death for cattle administered ZH were 0.53% (95% CL = 0.47, 0.59) and 1.77 (95% CL = 1.62, 1.92) deaths per 10,000 animal-days, respectively. Model-adjusted estimates among the unexposed cohort were 0.30 (95% CL = 0.25, 0.36) and 1.01 (95% CL = 0.85, 1.19) deaths per 10,000 animal-days, respectively. After adjusting for the various levels of clustering, administration of ZH was associated with a significant increase in the likelihood of death (RR = 1.76 [95% CL = 1.50, 2.05]; $P<0.01$). When time at risk during the exposure period was used as the offset variable instead of population at risk, the measure of effect was similar in that the IRR was 1.75 (95% CL = 1.50, 2.05; $P<0.01$).

Covariates included in the multivariable model as main effects (i.e., potential confounders) along with terms representing their interactions with exposure to ZH (i.e., effect modifiers) included (Table 2):

- Sex of the animals within a group,
- Percentage of a group that died prior to the at-risk period,
- Percentage of a group that were treated prior to the at-risk period,
- Percentage of cattle within a group that had a predominantly black hide,
- Mean carcass weight of the surviving animals that were shipped to slaughter, and
- Month in which the at-risk period ended (i.e., animals were shipped to an abattoir for slaughter).

Of the main effect terms, only percentage of a group that died prior to the at-risk period and mean carcass weight of animals shipped to slaughter were removed from the final model. The only effect modifier (i.e., interaction term) that was retained was month in which the animals were shipped to the abattoir ($P=0.07$). While controlling for the other covariates, RR estimates were 1.17 (95% CL = 0.64, 2.14; $P=0.60$; Figure 4); 1.56 (95% CL = 0.86, 2.82; $P=0.15$); 1.23 (95% CL = 0.87, 1.73; $P=0.24$); 2.69 (95% CL = 1.86, 3.90; $P<0.01$); 1.80 (95% CL = 1.28, 2.53; $P<0.01$); 1.86 (95% CL = 1.42, 2.43; $P<0.01$) and 1.71 (95% CL = 1.18, 2.49; $P=0.01$) for March, April, May, June, July, August, and September, respectively.

Averaged across the means of all other terms in the model, treatment remained significantly associated with mortality (RR = 1.66 [95% CL = 1.40, 1.96]; $P<0.01$). Thus, its effect did not appear to be confounded to any meaningful extent by other variables (those listed in bullet form above) in that the measure of effect from the multivariable model was similar to the reduced model in which only exposure was included as a fixed effect.

The same set of covariates listed above were associated with the incidence rate of death in univariate models; that is, the models in which when time at risk was included as the offset variable. These variables were subsequently included as covariates (main effects and effect modifiers) in the multivariable model of the incidence of death. While the same main effects were retained in the final model as those for the population at-risk model, no effect modifiers were retained in the final model (i.e., $P>0.14$ for all interaction terms). Controlling for the covariates that remained in the model (i.e., sex, month shipped to slaughter, percent of the group with a

**Table 1.** Summary statistics for death loss (counts, and crude and model-adjusted estimates) by dataset for all cattle, the exposed cohort (i.e., groups of cattle administered either ractopamine hydrochloride [RH] or zilpaterol hydrochloride [ZH]), and the unexposed cohort.

| | | Statistic | | | | | | |
|---|---|---|---|---|---|---|---|---|
| **Dataset** | **Cohort** | **Population at risk (n)** | **Deaths (n)** | **Crude estimate of cumulative risk (%)** | **Model-adjusted cumulative risk (%)** | **95% confidence limits** | **Model-adjusted incidence (deaths/10, 000 animal days)** | **95% confidence limits** |
| 4-company RH dataset | All cattle | 79,171 | 211 | 0.27 | 0.26 | 0.16, 0.40 | 0.86 | 0.58, 1.30 |
| | Exposed | 39,890 | 139 | 0.35 | 0.34 | 0.25, 0.41 | 1.12 | 0.85, 1.45 |
| | Unexposed | 39,281 | 72 | 0.18 | 0.18 | 0.13, 0.25 | 0.59 | 0.43, 0.83 |
| Multi-feedlot ZH dataset | All cattle | 722,704 | 3,657 | 0.51 | 0.50 | 0.44, 0.57 | 1.68 | 1.51, 1.86 |
| | Exposed | 637,339 | 3,405 | 0.53 | 0.53 | 0.44, 0.59 | 1.77 | 1.62, 1.92 |
| | Unexposed | 85,365 | 252 | 0.30 | 0.30 | 0.25, 0.36 | 1.01 | 0.85, 1.19 |
| Single-feedlot ZH dataset | All cattle | 149,636 | 571 | 0.38 | 0.38 | 0.35, 0.42 | * | * |
| | Exposed | 83,865 | 401 | 0.48 | 0.48 | 0.43, 0.43 | * | * |
| | Unexposed | 65,771 | 170 | 0.26 | 0.26 | 0.22, 0.31 | * | * |

* Not calculated as both cohorts had the same exposure period of 24 days.

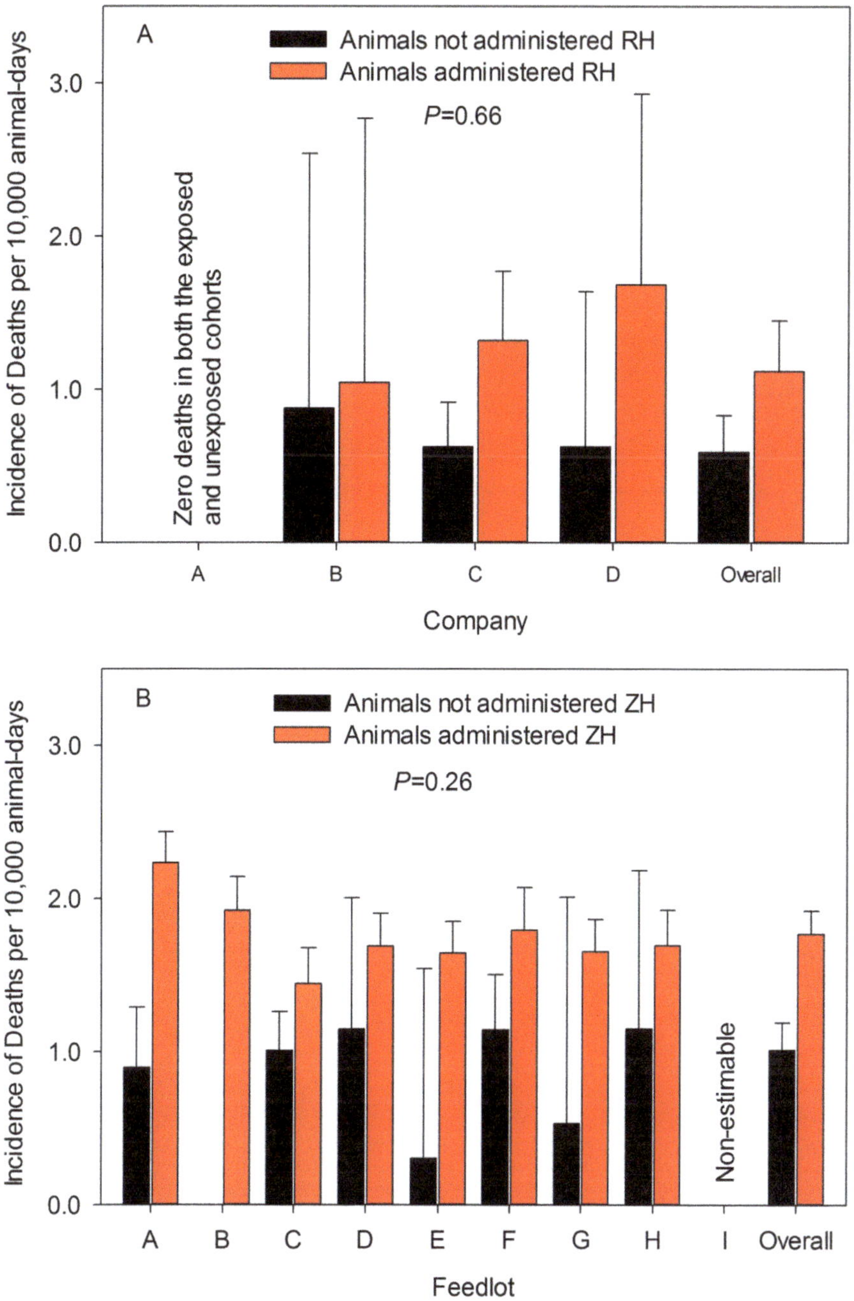


**Figure 1. Association between β-adrenergic agonist administration and mortality.** Model-adjusted estimates of incidence of death per 10,000 animal-days for cattle administered either ractopamine hydrochloride (RH – graph A) or zilpaterol hydrochloride (ZH – graph B) compared to a diet without a beta agonist. No deaths were reported for Company A (graph A) and rates for Feedlot I (graph B) were non-estimable. *P* values are those associated with interaction term for exposure by company (graph A) or feedlot (graph B). Bars represent upper 95% confidence limit.

black hide, and percent of pen treated prior to the at-risk period), the incidence of death was 80% greater in animals administered ZH than the comparative control cohort (IRR = 1.80 [95% CL = 1.55, 2.10]; $P<0.01$). The association between incidence rate of death and administration of βAA, therefore, did not appear to be confounded by the covariates available for analysis.

There was little evidence of feedlot-to-feedlot variation in the association between ZH and increased risk of death. In the population at-risk model, for example, the covariance parameter was much greater at the group-level (1.14 [SE = 0.03]) than that observed at the feedlot-level (0.04 [SE = 0.02]). In the alternative model that treated feedlot and its interaction with exposure as ZH fixed effects, there was no evidence that the association between administration of ZH and increased mortality was modified by which feedlot the animals were housed in either the population at risk ($P = 0.16$) or time at risk models ($P = 0.26$; Figure 1).

Two secondary outcomes were available for analysis (Table 3). Cattle administered ZH were 33% more likely (RR = 1.33 [95% CL = 1.18, 1.50]; $P<0.01$) to require treatment for illness during the at-risk period than animals not administered ZH. In addition, an association between ZH administration and an increased likelihood of the animal's beef being classified as *dark, firm and dry*

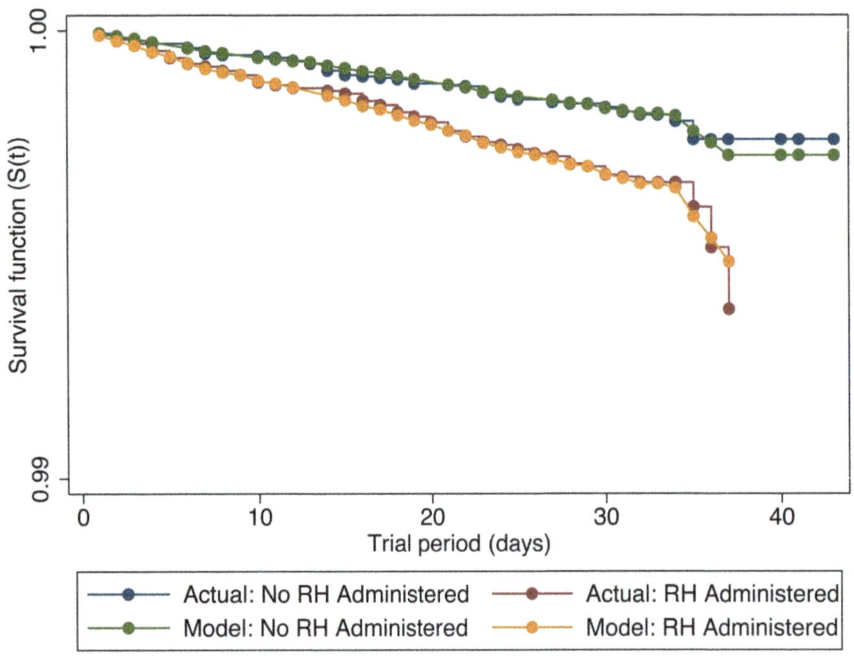


**Figure 2. Survival analysis for cattle administered a β-adrenergic agonist.** Kaplan-Meier non-parametric (actual) and Cox proportional hazards (predicted) survivor functions (S(t)) for cattle administered a diet containing ractopamine hydrochloride (RH) compared to a diet without RH in Company C.

was detected (Table 4); however, this latter association was modified by the sex of the animal ($P = 0.01$). The carcasses derived from steers administered ZH were 2.31 times more likely to be classified as *dark, firm and dry* compared to carcasses of steers not administered ZH (RR = 2.31 [95% CL = 1.77, 3.02]; $P<0.01$). Of the steer carcasses, 1.87 and 0.81%, of those administered ZH and the unexposed cohort, respectively, were classified as *dark, firm and dry*. No such association was observed with treatment among the carcasses derived from heifers ($P = 0.36$); of the heifer carcasses, 1.30 and 1.07%, of those administered ZH and the unexposed cohort, respectively, were classified as *dark, firm and dry*.

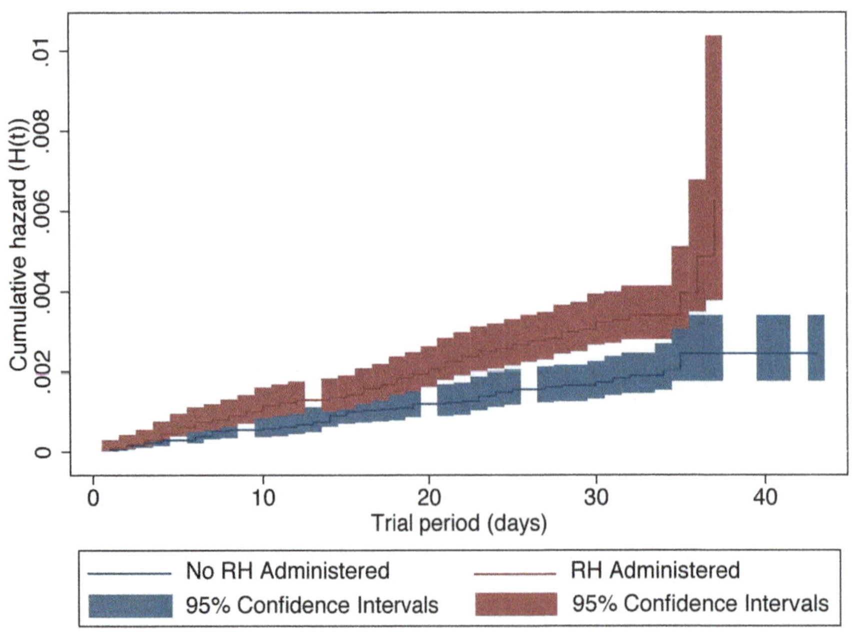


**Figure 3. Force of mortality among cattle administered a β-adrenergic agonist.** Empirical cumulative hazard function (H(t)) and 95% confidence intervals (during those time periods where mortalities occurred) for cattle administered a diet containing ractopamine hydrochloride (RH) compared to a diet without RH in Company C.

**Table 2.** Covariates evaluated for an association with mortality in groups of cattle either administered zilpaterol hydrochloride (ZH) or not in the multi-feedlot ZH dataset in univariate and multivariable models.

| | | Univariate model | | Multivariable model | | | |
|---|---|---|---|---|---|---|---|
| **Measure of mortality in groups of cattle** | **Covariate** | ***P* value** | **Included in multivariate model** | **Main effect *P* value*** | **Retained in final model** | ***P* value* of interaction term**** | **Retained in final model** |
| Cumulative risk (%) | Sex of the animals within a group | <0.01 | Yes | <0.01 | Yes | 0.46 | No |
| | Percentage of a group that died prior to the at-risk period | 0.13 | Yes | 0.31 | No | 0.39 | No |
| | Percentage of a group that were treated prior to the at-risk period | <0.01 | Yes | <0.01 | Yes | 0.41 | No |
| | Percentage of cattle within a group that had a predominantly black hide | <0.01 | Yes | <0.01 | Yes | 0.66 | No |
| | Mean carcass weight of the surviving animals that were shipped to slaughter | <0.01 | Yes | 0.77 | No | 0.73 | No |
| | Month in which the at-risk period ended | <0.01 | Yes | <0.01 | Yes | 0.07 | Yes |
| | Days at the feedlot prior to exposure | 0.23 | No | – | – | – | – |
| Incidence (deaths/10,000 animal-days) | Sex of the animals within a group | <0.01 | Yes | <0.01 | Yes | 0.42 | No |
| | Percentage of a group that died prior to the at-risk period | 0.05 | Yes | 0.70 | No | 0.49 | No |
| | Percentage of a group that were treated prior to the at-risk period | <0.01 | Yes | <0.01 | Yes | 0.50 | No |
| | Percentage of cattle within a group that had a predominantly black hide | <0.01 | Yes | <0.01 | Yes | 0.85 | No |
| | Mean carcass weight of the surviving animals that were shipped to slaughter | <0.01 | Yes | 0.50 | No | 0.95 | No |
| | Month in which the at-risk period ended | <0.01 | Yes | <0.01 | Yes | 0.14 | No |
| | Days at the feedlot prior to exposure | 0.91 | No | – | – | – | – |

Administration of ZH was associated with mortality ($P<0.01$) in all multivariable models.
**P* value indicated was that observed at the time of removal from the multivariable model if >0.10, or if retained, i.e., ≤0.10, its value in the final multivariable model.
** Interaction term of the covariate with administration of ZH, i.e., evaluation of potential modification of the association between death loss and administration.

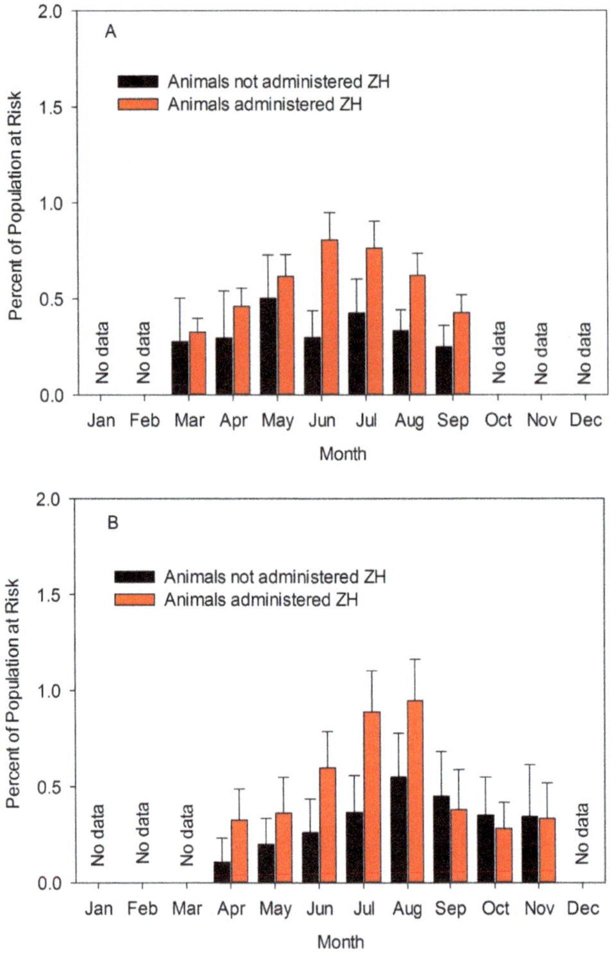


**Figure 4. Seasonal modification of the association between β-adrenergic agonist administration and mortality.** Model-adjusted estimates of the percentage of cattle that died among groups of animals administered zilpaterol hydrochloride (ZH) compared to a diet without ZH by month in which they were shipped to slaughter. Graph A represents 722,704 animals housed in 9 feedlots and graph B represents 149,636 animals housed in a single feedlot.

## Single-feedlot ZH Dataset

Of the 149,636 cattle at risk, 0.38% (n = 571) died during the 24-day exposure period and after accounting for potential over-dispersion in the data, the 95% CL of this estimate were 0.35% and 0.42%, respectively (Table 1). Model-adjusted estimates of the risk of death among those administered ZH and the comparative cohort were 0.48% (95% CL = 0.43, 0.53) and 0.26% (95% CL = 0.22, 0.31), respectively. After adjusting for clustering, administration of ZH was associated with an 85% increase in the risk of death (RR = 1.85 [95% CL = 1.51, 2.27]; $P<0.01$) during the at-risk period. Mortalities were classified as attributable to conditions either of the respiratory system or of the digestive system (this latter category also included all other attributable causes of death). After adjusting for clustering, administration of ZH was associated with an increase in the risk of death attributable to conditions of the respiratory (RR = 2.15 [95% CL = 1.60, 2.91]; $P<0.01$) and digestive (RR = 1.82 [95% CL = 1.31, 2.52]; $P<0.01$) systems.

Overall mortality was associated with sex and month of slaughter; in addition to their interactions with exposure, these

**Table 3.** Cumulative risk of treatment for disease among the exposed cohort (i.e., groups of cattle administered zilpaterol hydrochloride [ZH]) and the unexposed cohort.

| Dataset | Disease grouping | Cohort | Statistic | | | | |
|---|---|---|---|---|---|---|---|
| | | | Cumulative risk (%) | 95% confidence limits | Relative risk | 95% confidence limits | *P* value |
| Multi-feedlot ZH dataset | All diseases | Exposed | 1.35 | 1.13, 1.62 | 1.33 | 1.18, 1.50 | <0.01 |
| | All diseases | Unexposed | 1.01 | 0.83, 1.24 | – | – | – |
| Single-feedlot ZH dataset | All diseases | Exposed | 1.33 | 1.22, 1.46 | 1.23 | 1.07, 1.41 | <0.01 |
| | All diseases | Unexposed | 1.09 | 0.98, 1.21 | – | – | – |
| | Respiratory | Exposed | 0.86 | 0.78, 0.95 | 2.31 | 0.55, 0.79 | <0.01 |
| | Respiratory | Unexposed | 0.37 | 0.31, 0.44 | – | – | – |
| | Digestive and other | Exposed | 0.47 | 0.42, 0.54 | 0.66 | 0.55, 0.79 | <0.01 |
| | Digestive and other | Unexposed | 0.72 | 0.63, 0.81 | – | – | – |

main effects were included in a final multivariable model. The final model included sex, month and the interaction between ZH exposure and month. Averaged across other variables, animals exposed to ZH were 56% more likely to die than animals in the comparative cohort (RR = 1.56 [95% CL = 1.25, 1.95]; $P<0.01$). However, this effect was modified by the month in which animals were shipped to the abattoir ($P=0.01$). Model-adjusted RR estimates were 3.06 ($P=0.01$; Figure 4); 1.83 ($P=0.07$); 2.30 ($P<0.01$); 2.42 ($P<0.01$); 1.72 ($P=0.01$); 0.85 ($P=0.58$), 0.80 ($P=0.47$) and 0.97 ($P=0.94$) for April, May, June, July, August, September, October, and November, respectively.

A number of secondary outcomes were available for analysis in this third dataset (Table 3). Animals administered ZH were 23% more likely to require treatment for any condition during the exposure period (RR = 1.23 [95% CL = 1.07, 1.41]; $P<0.01$). The likelihood of treatment, however, varied by the body system to which the clinical signs were attributed. Animals were 34% less likely to require treatment for conditions attributed to the digestive system (RR = 0.66 [95% CL = 0.55, 0.79]; $P<0.01$) but 2.3 times more likely to require treatment for conditions attributed to the respiratory system (RR = 2.31 [95% CL = 1.89, 2.83]; $P<0.01$). In addition, animals administered ZH were 2.3 times more likely to require more than a single treatment regimen for respiratory disease than contemporaneous controls (RR = 2.34 [95% CL = 1.82, 3.04]; $P<0.01$). As a consequence of the increased burden of treatments in animals administered ZH, there was a 3-fold increase in the percentage of animals subject to a slaughter withholding period at the time the rest of their group was shipped to slaughter (RR = 3.00 [95% CL = 2.31, 3.90]; $P<0.01$). Those animals that could not be shipped to slaughter with their contemporaries were dispersed across many groups rather than clustered within a few; that is, 50.0% of groups administered ZH had one or more animals that could not be shipped with the rest of their group compared to 25.8% of groups not administered ZH.

Once at the abattoir, a greater percentage of carcasses from ZH administered cattle (1.59%) were classified as dark, firm and dry (Table 4), compared to those carcasses from animals not administered ZH (0.53%; RR = 3.02 [95% CL = 2.35, 3.89]; $P<0.01$). As opposed to the 9-feedlot dataset, sex was neither a modifier of this association ($P=0.21$) nor a significant covariate ($P=0.34$).

## Discussion

Death is a relatively rare event in feedlot cattle [25,26]. Even so, the data presented herein provide compelling evidence that administration of FDA-approved βAA to cattle increased both the cumulative incidence (risk) and incidence rate of death. The various measures of effect used to explore the relationship between βAA and mortality, i.e., RR, IRR and HR where applicable, were similar across the multiple datasets that included both randomized field trials and population-based observational data. The attributable fraction (AF), a measure of the proportion of deaths attributable to βAA administration among those cattle in the exposed cohort, varied little from dataset to dataset, remaining relatively constant at 40–50%.

Of considerable practical interest among those responsible for the care and well-being of cattle intended for slaughter for human consumption is identification of variables that could be exploited to help select or manage groups of cattle for which βAA drugs might be contraindicated. The most consistent modifier of the biological association between βAA administration and mortality was month of year (Figure 4). In the multi-feedlot ZH data set, month was statistically detected as an effect modifier in one model

**Table 4.** Prevalence of carcasses classified as *dry, dark, and firm* among the exposed cohort (i.e., groups of cattle administered zilpaterol hydrochloride [ZH]) and the unexposed cohort.

| | | Statistic | | | | | | | |
|---|---|---|---|---|---|---|---|---|---|
| Dataset | Cohort | Prevalence (%) | 95% confidence limits | *P* value associated with main effect comparison | *P* value associated with interaction with sex | Prevalence among steers (%) | 95% confidence limits | Prevalence among heifers (%) | 95% confidence limits |
| Multi-feedlot ZH dataset | Exposed | 1.75 | 1.11, 2.75 | <0.01 | 0.01 | 1.87 | 1.19, 2.95 | 1.30 | 0.82, 2.05 |
| | Unexposed | 0.86 | 0.53, 1.40 | - | - | 0.81 | 0.49, 1.33 | 1.07 | 0.60, 1.89 |
| Single-feedlot ZH dataset | Exposed | 1.59 | 1.43, 1.78 | <0.01 | 0.21 | 1.57 | 1.39, 1.78 | 1.70 | 1.30, 2.22 |
| | Unexposed | 0.53 | 0.42, 0.66 | - | - | 0.50 | 0.40, 0.64 | 0.93 | 0.44, 1.96 |

(i.e., population at-risk) but not in the other model (i.e., time at-risk). However, in the latter model, the observed $P$ value (0.14) provided some, albeit weak, evidence of variation in the association across months. In the single-feedlot ZH dataset, month was again detected as a significant modifier of the association between βAA administration and mortality ($P = 0.01$). Furthermore, the consistency of the temporal pattern of this association across the multi- and single-feedlot ZH datasets (Figure 4) provides compelling reasons to explore seasonal factors that might modify the effect of βAA administration on mortality. By this we mean that month is clearly a proxy for other variables with the most probable being thermal heat index. Given the variation in the measure of effect across months, at least for ZH, it seems plausible that greater heat indices may have contributed to a disproportionate increase in the risk of death among animals administered βAA when compared to those not administered this synthetic hormone. It was not possible, however, to test this association in our analysis; clearly this warrants additional prospective exploration. If month is indeed a proxy for a causal biological interaction between some measure of heat index and βAA administration, then mitigation strategies, as yet unknown, need to be developed for a substantial proportion of the year in that a significant relationship was detected in those cattle shipped for slaughter beginning in April (in the single-feedlot ZH dataset) and for those cattle shipped for slaughter through September (in the multi-feedlot ZH dataset).

Force of mortality, which is sometimes referred to as the hazard function, is an instantaneous measure of the risk of death in the next time period, conditioned on having survived to the present. This measure was greater for those animals administered RH than unexposed animals and was relatively constant for both groups across the majority of the exposure period for company C. The apparent late-stage increase (Figure 3) in the hazard function among the RH-treated groups of animals, when compared to the control groups of animals, did not lead to a violation of the proportional hazards assumption of the semi-parametric survival analysis. However, this increase appeared quite marked and bears further examination in any follow-up research. Unfortunately, days of exposure for each animal that died were only available for company C and included 3 large randomized trials of RH of 62,379 animals. Consequently, the force of mortality for those cattle administered ZH could not be evaluated.

A number of broad hypotheses might explain the observed association of βAA drug administration and increased mortality. The first potential hypothesis is that other unmeasured confounding variable(s) gave rise to a spurious relationship. For example, in the ZH datasets, the contemporaneous control cohorts consisted of those animals not administered a βAA. There are various reasons why these animals were not fed a βAA. For example, they may have been destined for a marketing program that precluded the use of a βAA or they were deemed not suitable for βAA administration (e.g., their body weight was sufficient enough that if fed a βAA hormone, their carcass weights would have been excessive). While it is possible that these unmeasured variables confounded the association of βAA use and increased death loss, it seems highly improbable given the strength of association and consistency of effect across feedlots, datasets, and the covariates evaluated in the models. This possibility seems even more implausible in the dataset involving RH. Each of the 12 studies included randomized allocation of animals or groups to the exposed or unexposed cohorts. One of the major benefits of randomization is the unbiased distribution of unmeasured confounders among the treatment groups. Furthermore, apart from company A in which no deaths in either cohort were reported, statistical variation among the measured associations of RH administration and mortality was not detected among companies B, C, and D. In addition, no statistical variation in the association between βAA and increased death loss was observed across feedlots in the multi-feedlot ZH dataset. This association, therefore, was relatively consistent and predictable from operation to operation despite unmeasured variation in farming practices and other attributes such as feedlot size, geographical location, animal husbandry, and cattle diets.

A second and related hypothesis is that the association was not necessarily due to the drug itself; rather, the association might have been a consequence of those management changes required to administer the βAA in the ration. Such collinearity of effects is virtually impossible to disentangle without purposively designing a study (e.g., a cross-over design) to deal with the phenomenon. For example, although not available in the data described herein, it is possible that the time at which feed was delivered changed for those animals administered a βAA. That is, most modern cattle feedlots have developed strategies to provide a consistent diet in a consistent and timely manner to the cattle. In this scheme, the amount of feed delivered is expected to be consumed within 24 hours. Because a minority of groups of cattle are fed βAA at any one time in an feedlot, one management strategy might be to feed the unexposed cattle first and subsequently feed the ration containing the βAA later in the day. If so, this strategy could have resulted in a relatively sudden change in the time at which the βAA-exposed cattle were fed (possibly up to a 2-hour delay for example). While cattle tend to adapt relatively quickly to changes in routine, such a change might have initially resulted in cattle that were hungry and thus over-consumed readily fermentable carbohydrates. However, the observed force of mortality was relatively constant over the exposure period (i.e., from day 0 up to 42). The feedlots that supplied data for the 2 observational datasets have had a number of years of experience in feeding the FDA-approved βAA and have had the opportunity to develop and adopt management strategies to minimize changes in routine.

A third hypothesis is that the βAA hormones themselves are causally associated with the increased mortality. Clearly, the measure of effects are relatively strong and consistent across the datasets; yet, given that the RH studies were not purposefully designed to investigate an association with mortality and the observational nature of the ZH data, it is difficult to definitively establish a causal relationship nor to identify the mechanistic explanations. However, evidence supporting adverse drug events can be drawn from the aggregated observations of βAA administration in human medicine. For example, authors of various studies that included randomized clinical trials and an FDA-performed meta-analysis of the available data, have concluded that long-acting $\beta_2$ adrenergic agonists used for asthma contribute to an increased risk of severe asthma events and death [8,9,11,19]. While this association might be somewhat ameliorated by the inclusion of a corticosteroid, the FDA determined this was a class effect and now requires a boxed warning to be included on the labels of all long-acting βAA intended to be administered routinely, e.g., daily for the control of asthma. Furthermore, in the datasets of ZH in which risk of treatment was available, exposed animals were more likely to require treatment during the at-risk period than unexposed animals.

Others have reported an increased risk of myocardial disease in certain patients administered βAA drugs [12]. If a similar association occurs in cattle, one might expect metabolic markers of such an event, elevated creatine phosphokinase (CPK) for example [27]. Indeed, ZH administration is associated with increased serum CPK [2]; unfortunately the isoform(s) were not reported so it is uncertain if the elevated serum CPK resulted from damage to striated muscle, myocardial tissue, or both.

After accounting for the effect of exposure, most of the unexplained variation in mortality occurred at the level of the group rather than at the organizational levels of the study, feedlot, or company depending on the dataset. Moreover, there was almost no (and at times zero) unexplained variation attributed to unmeasured company or feedlot level factors. In other words, the biological association did not vary sufficiently among companies or feedlots to be detected by the statistical methods described herein. Group-level factors, therefore, seem to be important determinants (or modifiers) of mortality and if discovered, may influence the design of management strategies to reduce mortality, particularly in pens administered βAA. Unfortunately, we could not identify any meaningful and consistent effect modifiers, other than month, in the data and further research is needed to identify factors that may modify the observed association between βAA and increased risk of death. Saliently, the statistical variation of groups is a function of both the individuals as well as the interaction among the individuals within the group. In addition to group-level management strategies, therefore, 2 additional lines of investigation ought to be pursued and include animal-level factors, such as genetic variation in response to βAA [28], and how individuals interact to influence the behavior and potentially the response to βAA of other animals.

The data reported have considerable strengths in that the number of observations permitted the detection of a change in a rare event (or events, if one considers the secondary outcomes). When seeking approval for products, it is extremely uncommon to have available such numbers to detect rare adverse drug events. Based on the results observed herein, the number needed to harm, a measure of extent of exposure required for a βAA-related death, was approximately 500 animals or 15,000 animal days. In most reports of well-controlled cattle experiments, the number of animals included is usually insufficient to speak to an association of βAA with mortality, given the rarity of death and estimates of the number needed to harm. As a result, most drug approvals require some post-approval monitoring, often termed the pharmaco-epidemiology or Type IV trials, and the self-reporting by patients of side-effects, either to the company, the FDA, or their doctor, is part and parcel of a holistic drug regulatory framework. Furthermore, our data are from commercial cattle fattening feedlots and as such, provide a degree of validity to the results. However, the number of companies and feedlots is relatively limited and consequently, selection bias is possible in that the decision to provide the data for inclusion in the analyses may have been influenced by a suspicion of increased mortality associated with βAA administration.

Despite the potential limitations of the data, we argue that given the magnitude of the data, and the strength and consistency of the various measures of effect, both RH and ZH are most likely causally associated with increased cumulative incidence, incidence rate, and hazard of death when they are administered in accordance with the FDA-approved label directions. The excess deaths attributed to βAA administration, and potentially the secondary outcomes of illness and occurrence of beef classified as *dark, firm and dry*, represent adverse drug events. If so, we believe a broad and inclusive dialogue that explores the balance between improved production efficiencies achieved through means such as βAA [29] and resultant adverse effects on the welfare of animals we raise for food is needed. This is particularly warranted for those drugs that are approved solely to improve the efficiencies of production yet offer no offsetting health benefits to the animals to which it is administered. For this dialogue to be sufficiently inclusive, it ought to include a broad collection of stakeholders such as animal scientists, cattle and beef producers, animal health specialists, welfarists, ethologists, and consumers.

## Author Contributions

Conceived and designed the experiments: GHL. Analyzed the data: GHL HMS. Wrote the paper: GHL DUT HMS.

## References


1. Health and Human Services, Food and Drug Administration website. Freedom of information summary. Original new animal drug application. NADA 141-221 ractopamine hydrochloride (OPTAFLEXX 45) type A medicated article for beef cattle. Available: http://www.fda.gov/downloads/AnimalVeterinary/Products/ApprovedAnimalDrugProducts/FOIADrugSummaries/ucm118030.pdf. Accessed 2013 July 10.
2. Health and Human Services, Food and Drug Administration website. Freedom of information summary. Original new animal drug application. NADA 141-258 ZILMAX (zilpaterol hydrochloride) type A medicated article for cattle fed in confinement for slaughter. Available: http://www.fda.gov/downloads/AnimalVeterinary/Products/ApprovedAnimalDrugProducts/FOIADrugSummaries/ucm051412.pdf. Accessed 2013 July 5.
3. Hilton GG, Montgomery JL, Krehbiel CR, Yates DA, Hutcheson JP, et al. (2009) Effects of feeding zilpaterol hydrochloride with and without monensin and tylosin on carcass cutability and meat palatability of beef steers. J Anim Sci 87: 1394–1406. 10.2527/jas.2008-1170.
4. Rathmann RJ, Bernhard BC, Swingle RS, Lawrence TE, Nichols WT, et al. (2012) Effects of zilpaterol hydrochloride and days on the finishing diet on feedlot performance, carcass characteristics, and tenderness in beef heifers. J Anim Sci 90: 3301–3311. 10.2527/jas.2011-4375.
5. Bohrer BM, Kyle JM, Boler DD, Rincker PJ, Ritter MJ, et al. (2013) Meta-analysis of the effects of ractopamine hydrochloride on carcass cutability and primal yields of finishing pigs. J Anim Sci 91: 1015–1023. 10.2527/jas.2012-5647.
6. Scramlin SM, Platter WJ, Gomez RA, Choat WT, McKeith FK, et al. (2010) Comparative effects of ractopamine hydrochloride and zilpaterol hydrochloride on growth performance, carcass traits, and longissimus tenderness of finishing steers. J Anim Sci 88: 1823–1829. 10.2527/jas.2009-2405.
7. Bryant TC, Engle TE, Galyean ML, Wagner JJ, Tatum JD, et al. (2010) Effects of ractopamine and trenbolone acetate implants with or without estradiol on growth performance, carcass characteristics, adipogenic enzyme activity, and blood metabolites in feedlot steers and heifers. J Anim Sci 88: 4102–4119. 10.2527/jas.2010-2901.
8. Nelson HS, Weiss ST, Bleecker ER, Yancey SW, Dorinsky PM, et al. (2006) The Salmeterol Multicenter Asthma Research Trial: a comparison of usual pharmacotherapy for asthma or usual pharmacotherapy plus salmeterol. Chest 129: 15–26. 10.1378/chest.129.1.15.
9. McMahon AW, Levenson MS, McEvoy BW, Mosholder AD, Murphy D (2011) Age and risks of FDA-approved long-acting beta(2)-adrenergic receptor agonists. Pediatrics 128: e1147–1154. 10.1542/peds.2010-1720.
10. Salpeter SR (2010) An update on the safety of long-acting beta-agonists in asthma patients using inhaled corticosteroids. Expert Opin Drug Saf 9: 407–419. 10.1517/14740330903535852.
11. Salpeter SR, Buckley NS, Ormiston TM, Salpeter EE (2006) Meta-analysis: effect of long-acting beta-agonists on severe asthma exacerbations and asthma-related deaths. Ann Intern Med 144: 904–912.
12. Salpeter SR, Ormiston TM, Salpeter EE (2004) Cardiovascular effects of beta-agonists in patients with asthma and COPD: a meta-analysis. Chest 125: 2309–2321.
13. Marchant-Forde JN, Lay DC Jr., Pajor EA, Richert BT, Schinckel AP (2003) The effects of ractopamine on the behavior and physiology of finishing pigs. J Anim Sci 81: 416–422.
14. Poletto R, Rostagno MH, Richert BT, Marchant-Forde JN (2009) Effects of a "step-up" ractopamine feeding program, sex, and social rank on growth performance, hoof lesions, and Enterobacteriaceae shedding in finishing pigs. J Anim Sci 87: 304–313. 10.2527/jas.2008-1188.
15. Poletto R, Meisel RL, Richert BT, Cheng HW, Marchant-Forde JN (2010) Behavior and peripheral amine concentrations in relation to ractopamine feeding, sex, and social rank of finishing pigs. J Anim Sci 88: 1184–1194. 10.2527/jas.2008-1576.
16. Wagner SH, Mostrom MS, Hammer CJ, Thorson JF, Smith DJ (2008) Adverse Effects of Zilpaterol Administration in Horses: Three Cases. J Equine Vet Sci 28: 238–243.
17. Avendano-Reyes L, Macias-Cruz U, Alvarez-Valenzuela FD, Aguila-Tepato E, Torrentera-Olivera NG, et al. (2011) Effects of zilpaterol hydrochloride on growth performance, carcass characteristics, and wholesale cut yield of hair-

breed ewe lambs consuming feedlot diets under moderate environmental conditions. J Anim Sci 89: 4188–4194. 10.2527/jas.2011–3904.
18. Baszczak JA, Grandin T, Gruber SL, Engle TE, Platter WJ, et al. (2006) Effects of ractopamine supplementation on behavior of British, Continental, and Brahman crossbred steers during routine handling. J Anim Sci 84: 3410–3414. 10.2527/jas.2006-167.
19. Beasley R (2006) A historical perspective of the New Zealand asthma mortality epidemics. J Allergy Clin Immunol 117: 225–228.
20. McDermott JJ, Schukken YH, Shoukri MM (1994) Study design and analytical methods for data collected from clusters of animals. Prev Vet Med 18: 175–191.
21. McDermott JJ, Schukken YH (1994) A review of methods used to adjust for cluster effects in explanatory epidemiological studies of animal populations. Prev Vet Med 18: 155–173.
22. Dohoo IR, Martin SW, Stryhn H (2009) Veterinary Epidemiological Research. Charlottetown, Prince Edward Island, Canada: VER Inc.
23. Brown H, Prescott R (2006) Applied mixed models in medicine; Senn S, Barnett V, editors. West Sussex, England: John Wiley & Sons Ltd.
24. University of Bristol, Centre for Multilevel Modeling website. A User's Guide to MLwiN Version 2.26. Available: http://www.bristol.ac.uk/cmm/software/mlwin/download/2-26/manual-web.pdf Available. Accessed 2012 November 2.
25. Loneragan GH, Dargatz DA, Morley PS, Smith MA (2001) Trends in mortality ratios among cattle in US feedlots. J Am Vet Med Assoc 219: 1122–1127.
26. Babcock AH, Cernicchiaro N, White BJ, Dubnicka SR, Thomson DU, et al. (2013) A multivariable assessment quantifying effects of cohort-level factors associated with combined mortality and culling risk in cohorts of U.S. commercial feedlot cattle. Prev Vet Med 108: 38–46. 10.1016/j.prevetmed.2012.07.008.
27. Guzy PM (1977) Creatine phosphokinase-MB (CPK-MB) and the diagnosis of myocardial infarction. West J Med 127: 455–460.
28. Kononoff PJ, Defoor PJ, Engler MJ, Swingle RS, James ST, et al. (2013) Impact of a leptin SNP and zilpaterol hydrochloride (ZH) on growth and carcass characteristics in finishing steers. J Anim Sci. 10.2527/jas.2012–5229.
29. Capper JL, Hayes DJ (2012) The environmental and economic impact of removing growth-enhancing technologies from U.S. beef production. J Anim Sci 90: 3527–3537. 10.2527/jas.2011–4870.

# 23

# Early-Life Hepatitis E Infection in Pigs: The Importance of Maternally-Derived Antibodies

**Mathieu Andraud[1,2]*, Maribel Casas[6,7], Nicole Pavio[3,4,5], Nicolas Rose[1,2]**

**1** Pig Epidemiology and Welfare Unit, Anses, Laboratoire de Ploufragan-Plouzané, Ploufragan, France, **2** Université Européenne de Bretagne, Rennes, France, **3** UMR 1161 Virology, Anses, Laboratoire de Santé Animale, Maisons-Alfort, France, **4** UMR 1161 Virology, INRA, Maisons-Alfort, France, **5** UMR 1161 Virology, Université Paris-Est, Ecole Nationale Vétérinaire d'Alfort, Maisons-Alfort, France, **6** Centre for Research in Environmental Epidemiology (CREAL), Barcelona, Spain, **7** Centre de Recerca en Sanitat Animal (CReSA), Barcelona, Spain

## Abstract

Passive immunity (PI), acquired through colostrum intake, is essential for piglet protection against pathogens. Maternally-derived antibodies (MDAs) can decrease the transmission of pathogens between individuals by reducing shedding from infected animals and/or susceptibility of naïve animals. Only a limited number of studies, however, have been carried out to quantify the level of protection conferred by PI in terms of transmission. In the present study, an original modeling framework was designed to estimate parameters governing the transmission of infectious agents in the presence and absence of PI. This epidemiological model accounts for the distribution of PI duration and two different forces of infection depending on the serological status of animals after colostrum intake. A Bayesian approach (Metropolis-Hastings algorithm) was used for parameter estimation. The impact of PI on hepatitis E virus transmission in piglets was investigated using longitudinal serological data from six pig farms. A strong impact of PI was highlighted, the efficiency of transmission being on average 13 times lower in piglets with maternally-derived antibodies than in fully susceptible animals (range: 5–21). Median infection-free survival ages, based on herd-specific estimates, ranged between 8.7 and 13.8 weeks in all but one herd. Indeed, this herd exhibited a different profile with a relatively low prevalence of infected pigs (50% at slaughter age) despite the similar proportions of passively immune individuals after colostrum intake. These results suggest that the age at HEV infection is not strictly dependent upon the proportion of piglets with PI but is also linked to farm-specific husbandry (mingling of piglets after weaning) and hygiene practices. The original methodology developed here, using population-based longitudinal serological data, was able to demonstrate the relative impact of MDAs on the transmission of infectious agents.

**Editor:** Dipshikha Chakravortty, Indian Institute of Science, India

**Funding:** This study was funded by the French Research Agency (ANR) as part of the research project called "HEVECODYN" (ANR-10-CESA-0010; http://www.agence-nationale-recherche.fr/en/anr-funded-project/?tx_lwmsuivibilan_pi2[CODE] = ANR-10-CESA-0010). The funders had no role in study design, data collection and analysis, decision to publish, or preparation of the manuscript.

**Competing Interests:** The authors have declared that no competing interests exist.

* Email: mathieu.andraud@anses.fr

## Introduction

Passive immunity (PI) is the primary protection against infections in early life for several species, including humans [1–4]. However, this protection is only partial, rarely totally preventing from infection, but slows down the transmission process among individuals and/or reduces clinical consequences whenever infections occur. Moreover, this protection is only temporary due to maternal antibodies waning and, in the absence of vaccination, individuals become fully susceptible to infection. Although vaccination could overcome this issue, several studies evidenced antagonistic effects between vaccine-induced immunity and maternally derived antibodies through inhibition of vaccine protection, which could potentially worsen the dynamics of infection [5–9]. Hence, timely vaccination in regards to passive immunity waning appears crucial to optimize vaccine efficacy.

In this context, we developed a methodological framework to analyze the main characteristics of agent-specific passive immunity, in terms of duration and protection, based on longitudinal data. Our model explicitly takes into account the distribution of passive immunity duration to define the transition rate between passively immune and susceptible states. The protective impact of passive immunity is assessed through the estimation of differential forces of infection (FOI), defined as the *per capita* rate of infection per time unit, in regards with the early life serological status (*i.e.* with or without maternally derived antibodies). A Bayesian approach, based on Monte-Carlo Markov Chains (MCMC), was used to estimate the protection conferred by maternally derived antibodies on the transmission process. Bayesian methodology is of particular interest for the estimations of parameters related to infectious disease dynamics [10–18], since such data are subject to uncertainties due, for example, to under-reporting of cases, partially observed processes, or time-aggregated data.

As passive immunity was shown to be partially protective in the context of Hepatitis E virus (HEV) infection in pigs [19], the developed methodology was applied to quantify the protection level conferred to new-born piglets by maternally derived antibodies (MDAs). Hepatitis E virus is a recognized zoonotic agent, which can cause enterically-transmitted hepatitis in humans, and for which domestic pigs are considered as the main reservoir [20]. Understanding the impact of passive immunity on infection dynamics is therefore of pivotal importance to identify strategies to decrease the prevalence of infected animals at slaughter time. To our knowledge, few studies have focused on quantifying the protection conferred by passive immunity (PI) in terms of pathogen transmission in pigs. In 1997, Bouma *et al.* calculated a reproduction number of 0.2 for pseudorabies virus in PI animals (*vs.* 6.3 in absence of MDAs) [5]. Another study on PCV-2 transmission demonstrated a two-fold reduction in pigs with MDAs and an estimated reproduction number of 1.5 [21]. Allerson *et al.* recently studied the transmission potential of swine influenza virus in the presence of homologous and heterologous MDAs (obtained from vaccinated dams) with regard to the challenge strain [22]. The results showed high protection with homologous MDAs but weaker protection in piglets born to sows vaccinated with a heterologous strain. The main risk regarding hepatitis E infection in humans in industrialized countries is the introduction of viremic pigs into the food chain, which is strongly related to the age at infection. Data from a cohort study were first used to characterize MDAs (IgG) kinetics in the early life of piglets (up to 14 weeks of age) and to estimate the distribution of the duration of passive immunity. A second dataset derived from a longitudinal study in 6 Spanish pig-herds [23] was then analyzed, accounting for previous estimates of MDAs waning time, to estimate forces of infection from herd-specific HEV prevalences.

## Materials and Methods

### Distribution of passive immunity duration

A non-linear mixed effect model was used to represent the decay of MDAs titer. The antibody titers $A$ in piglets which received maternally-derived antibodies (MDAs) were assumed to decrease exponentially with age according to the equation $dA/dt = -rA$. The antibody titers would thus depend on the initial level of antibodies delivered at birth $A_0$ and the antibody decay rate $r$. As the initial antibody titer would likely be related to the dam's serological status, dam antibody titers were consequently considered as a covariate related with the initial level of antibodies $A_0$ in offspring. Thus, the model describing serological titer of individual $i$ at observation time $t_{ij}$ (with a constant residual error model) is given by:

$$A_{ij} = A_0^{(i)} \exp(-r_i t_{ij}) + a\varepsilon_{ij},$$

where $A_0^{(i)}$ and $r_i$ are the individual parameters and $\boldsymbol{\varepsilon}_i$is a vector of standardized random variables. As described in [24], individual parameters were assumed to be log-normally distributed. The serological titer of the dam $A_{Dam}^{(i)}$one week after parturition was considered as covariate for the initial serological titer of each individual ($A_0^{(i)}$). The model for individual parameters is given by:

$$A_0^{(i)} = A_0^{pop} \left(A_{Dam}^{(i)}\right)^{\beta_{Dam}} e^{\eta_A^{(i)}}$$

and

$$r_i = r_{pop} e^{\eta_r^{(i)}},$$

where $i$ denotes the individual and $r_{pop}$ the median decay rate at the population level. $A_0^{pop}$ represents the typical initial serological titer in the population, the predicted initial serological titer for individual $i$ being given by $A_0^{pop}\left(A_{Dam}^{(i)}\right)^{\beta_{Dam}}$. $\eta_A^{(i)}$ and $\eta_r^{(i)}$ are vectors of random effects assumed as independent centered Gaussian vectors with variance $\Omega_A$ and $\Omega_r$, representing inter-individual variability. Parameters were estimated using MLE and the SAEM algorithm for the analysis of hierarchical nonlinear mixed-effects models [25].

Individual antibody kinetics were derived from the individual empirical parameter estimates. Subjects with antibody levels below the cut-off value of the ELISA test were considered seronegative (HEV ELISA 4.0v, MP Diagnostics,Illkirch,France, [26]). Different distributions were tested to model the time to immunity waning (Weibull, lognormal and gamma distributions). Distribution parameters were estimated using the maximum likelihood method and model selection was based on the AIC values. A nonparametric bootstrap procedure was used to determine the 95% confidence intervals for the parameter estimates and thus measure their accuracy. One thousand bootstrap replicates were generated by resampling individual profiles for each dataset. For each bootstrap replicate, selected model was refitted to get an estimate of the distribution parameters. The 95% confidence interval was constructed from the 2.5$^{th}$ and 97.5$^{th}$ percentiles of distribution parameters [27].

### Impact of passive immunity on transmission

PI is known to be a protective factor against infections in early life. However, this protection is rarely fully efficient. We developed a general model framework to represent the evolution of passive immunity with age of the animals. For this purpose, the transition from passively immune to susceptible states was considered as age-dependent according to the estimated distribution described in the previous subsection. The following procedure was adopted: Let $g(t)$ and $G(t)$ be the probability density and cumulative probability function of this distribution, respectively. Let $T_i$ be the interval duration between two observation times $[a_i, a_{i+1}]$.In the absence of infection, animals will then either be passively immune or susceptible and the transition between these two states is governed by:

$$M_{i+1} = M_0(1 - G(a_{i+1})) = M_i \frac{(1 - G(a_{i+1}))}{1 - G(a_i)} \quad (1)$$

and

$$S_{i+1} = S_i + (M_i - M_{i+1}) = S_i + \frac{M_i}{1 - G(a_i)}(G(a_{i+1}) - G(a_i)) \quad (2)$$

The second term in equation (2) corresponds to the proportion of individuals losing passive immunity between ages $a_i$ and $a_{i+1}$. $M_i$ (resp.$S_i$) represents the proportion of passively immune (resp. susceptible) animals of age $a_i$.

**Infectious process.** The present study was designed to quantify the protective impact MDAs from longitudinal serological data. For this purpose, and based on equations (1 - 2), the following epidemiological model was developed to account for passive immunity duration and differential forces of infection according to the serological status of individuals at birth.

Let $\lambda_M$ be the force of infection exerted on individuals with passive immunity and $\lambda$ the FOI in the absence of passive immunity (Full model). The probability that an animal with maternal antibodies will escape infection between $a_i$ and $a_{i+1}$ is therefore expressed by $\exp(-\lambda_M T_i)$ and the proportion of individuals remaining passively immune is given by:

$$M_{i+1} = M_i \frac{(1-G(a_{i+1}))}{1-G(a_i)} \exp(-\lambda_M T_i) \quad (3)$$

Individuals for which immunity waning occurs at time $\tau \in [a_i, a_{i+1}]$ are susceptible at age $a_{i+1}$ if they were not infected between $a_i$ and $\tau$ (*i.e.* when they were still passively immune), or between $\tau$ and $a_{i+1}$, when they were susceptible. Moreover, individuals that were susceptible at age $a_i$ remain so if they escaped infection during the time interval $[a_i, a_{i+1}]$ with a probability of $\exp(-\lambda T_i)$. The proportion of susceptible individuals is therefore represented by:

$$S_{i+1} = \frac{M_i}{1-G(a_i)} \int_0^{T_i} g(a_i+\tau) \exp(-\lambda_M \tau) \exp(-\lambda(T_i-\tau)) d\tau + S_i \exp(-\lambda T_i) \quad (4)$$

To clearly identify the impact of PI, we compared this model structure to a simplified version assuming that passive immunity played no role in the transmission process ($\lambda_M = \lambda$). In this sub-model, hereafter referred to as the single force of infection model (SFOI model), the proportion of individuals with passive immunity and susceptible individuals are governed by:

$$M_{i+1} = M_i \frac{(1-G(a_{i+1}))}{1-G(a_i)} \exp(-\lambda T_i) \quad (5)$$

$$S_{i+1} = \left( \frac{M_i}{1-G(a_i)} (G(a_{i+1}) - G(a_i)) + S_i \right) \exp(-\lambda T_i) \quad (6)$$

Whatever the assumption concerning the impact of passive immunity, the proportion of infected individuals $C_i$ is given by:

$$C_i = 1 - M_i - S_i. \quad (7)$$

Model (3-4-7), which considers two forces of infection, is hereafter termed the "full model", in contrast to model (5-6-7) which considers a single force of infection.

**Parameter inference.** Let $n$ be the number of sampled animals from a total number of $N$animals per herd. As described by Backer *et al.* [18], the number of infected individuals is not directly observed so the true number of infected individuals of age $a_i$ in herd $h$, and denoted by $J_i^{(h)}$ follows the distribution:

$$f\left(J_i^{(h)} | N, n, k_i^{(h)}\right) = \frac{\binom{J_i^{(h)}}{k_i^{(h)}} \binom{N-J_i^{(h)}}{n-k_i^{(h)}}}{\binom{N}{n}} \frac{n+1}{N+1},$$

with $k_i^{(h)}$ the observed prevalence at age $a_i$ in herd $h$.

Considering that the proportion of infected animals at each sampling time is given by equation (7), we can calculate the herd-specific contribution to the likelihood of the observations if there were $J^{(h)}$ infected animals and $N - J^{(h)}$ non infected pigs in herd $h$:

$$L\left(\mathbf{k}^{(\mathbf{h})}, n, N, \mathbf{a}, \mathbf{T} | \lambda_M^{(h)}, \lambda^{(h)}, \mathbf{J}^{(\mathbf{h})}\right) =$$

$$\Pi_i f\left(J_i^{(h)} \middle| N, n, k_i^{(h)}\right) \binom{N}{J_i^{(h)}} \left(C_i^{(h)}\right)^{J_i^{(h)}} \left(1 - C_i^{(h)}\right)^{N-J_i^{(h)}}$$

The global likelihood is thus expressed as the product of the herd-specific contributions:

$$L(\mathbf{k}, n, N, \mathbf{a}, \mathbf{T} | \boldsymbol{\lambda}_\mathbf{M}, \boldsymbol{\lambda}, \mathbf{J}) = \prod_h L\left(\mathbf{k}^{(\mathbf{h})}, n, N, \mathbf{a}, \mathbf{T} | \boldsymbol{\lambda}_\mathbf{M}, \boldsymbol{\lambda}, \mathbf{J}^{(\mathbf{h})}\right).$$

.

A Bayesian framework was used to infer the following parameters: the herd-specific forces of infection $\left(\lambda_M^{(h)}, \lambda^{(h)}\right)$, the actual number of infected individuals for each observation time $\left(\mathbf{J}^{(\mathbf{h})}\right)$ and the initial proportion of passively immune animals $\left(M_0^{(h)}\right)$. Flat uninformative priors were assumed for all parameters but $M_0^{(h)}$ for which a beta distribution was selected in line with the average percentage of seropositive sows in the sampled herds (59%, sd: 10; [23]). Parameters were updated using the Metropolis-Hastings algorithm with gamma proposal distributions for the forces of infection and normal proposal distributions for **J** for which the mean value corresponded to the last accepted value and using the variance as tuning parameter. Uniform proposal distributions, also centered on current value ($Y \sim U(M_{0;t} - \delta, M_{0;t} + \delta)$, Y denoting the proposal value and delta being the tuning parameter) were used to estimate $\mathbf{M}_0$. Three chains were run with random initial conditions, 50000 steps per chain, a burn-in of 1000 steps and a thinning parameter of 10. Convergence was assessed by visual inspection and diagnostic tests (autocorrelation plots, Geweke and Gelman-Rubin diagnostics).

## Application to Hepatitis E virus transmission in pigs

No data were specifically collected for the present analysis. Data were available from previous field studies which were described in references [23] and [28].

**Passive immunity duration.** Serological data, used to estimate the duration of PI, were derived from a longitudinal epidemiological study on HEV prevalence in France [28] by monitoring 120 piglets, from 3 consecutive batches in one herd known to be infected by HEV, from birth to 14 weeks of age. Ten sows were chosen at random from this herd and from each followed batch. One week post-farrowing, 4 piglets born to the selected sows were chosen at random and ear-tagged. At the same time, blood samples were collected from the sows to determine their serological status with regard to HEV infection. Blood samples were also taken from piglets at the ages of 1, 6, 10 and 14 weeks for serological analysis.

A commercial test validated for veterinary analyses (HEV ELISA 4.0v (MP Diagnostics, Illkirch, France)) was used to detect anti-HEV antibodies in pigs as described in [26]. This test is based on a double sandwich ELISA that allows the detection of all classes of immunoglobulins (IgG, IgA and IgM) regardless of the animal species. The HEV ELISA 4.0v utilizes a proprietary recombinant antigen, which is highly conserved between different HEV strains.

**Table 1.** Parameter estimates of HEV maternal antibodies kinetics.

| | Parameter estimate | SE | p-value |
|---|---|---|---|
| $A_0$ | 0.47 | 0.07 | |
| $r$ | 0.4 | 0.02 | |
| $\beta_{Dam}$ | 0.66 | 0.04 | $<10^{-3}$ |

$A_0$: initial antibodies concentration (ELISA test Optic density (OD)).
$r$: antibody decay rate ($week^{-1}$).
$\beta_{Sow}$: sow's serological titer covariate coefficient on piglets' initial antibody titer.

Analyses were performed according to the manufacturer's instructions except that 10 μl of sera were used. Samples were considered positive when the OD450 value of the sample was above the cut-off value (Co = mean of the negative control + 0.300).

**Impact of passive immunity on transmission.** The forces of infection in the presence and absence of PI were estimated from data provided by Casas *et al.* [23] because these data had shown evidence of early infections occurring in young piglets in the presence of residual maternal antibodies. Briefly, 6 farrow-to-finish Spanish swine herds of similar size (100–300 productive sows) and known to be infected by HEV were included in the study. In each farm, 20 piglets from 6 previously selected sows, were randomly selected and ear-tagged and serial blood samples were collected at 3, 7, 13, and 18 weeks. Blood was also collected at slaughter (25 weeks of age approximately) at bleeding. We assumed that this number of sampled animals corresponded to 10% of the total number of pigs in the batch (N = 200). In this study, the sera were tested for specific IgG and also IgM which allowed differentiation between passively derived antibodies and a post-infectious response following a recent infection (IgM) [29]. Based on individual serological results, the course of infection was determined from the cumulated number of IgM positive piglets among the 20 piglets examined at each time point.

## Results

### Antibody decay rate

The serological investigations in sows revealed a mean HEV seroprevalence of 77% at farrowing with variations (from 70 to 90%) between batches. Seventy percent of the selected piglets were positive for anti-HEV antibodies at one week of age. The MDA titers of piglets were correlated with the titer of their dam ($\beta_{Sow}$ = 0.66, SE = 0.04, p<0.001; Table 1). The estimated antibody decay rate was 0.4 per week (SE: 0.02, Table 1) which corresponded to a half life of 12.1 days. Individual fits of serological profiles are shown in Figure S1. The individual duration of PI, based on the individual parameter estimates, was determined by long term projection of the antibody kinetics (assuming a threshold value of 0.38 Optic Density (OD)). Different distributions were fitted to the individual durations of PI (Weibull, lognormal and gamma distributions). The mean durations of PI were similar for all distributions but the gamma distribution was selected, based on the Akaike Information Criterion (AIC). Thus, the duration of PI was fitted to a gamma distribution with a mean duration of 6.5 weeks (95% CI: [6.0; 7.2]; Figure 1). This distribution was then used to model the transition between the PI state and the fully susceptible state in the epidemiological model (Equations (3–7)).

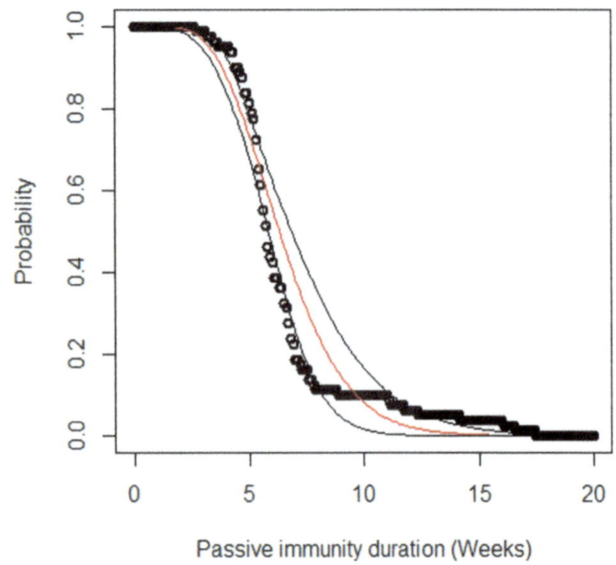


**Figure 1. Distribution of the duration of HEV passive immunity (Weeks).** We assumed gamma distributions for the duration of PI and used the maximum likelihood method to estimate the shape and scale parameters (red line). 95% confidence bands (black lines) were obtained by the bootstrapping method (see text for details).

### HEV Infectious process and passive immunity

Convergence of the MCMC Algorithm is shown in figure S2. A graphical representation of the posterior distribution of herd-specific prevalences, *i.e.* the proportion of infected individuals ($\mathbf{J}^{(\mathbf{h})}/N$), according to the model structure is provided in Figure 2. As previously highlighted by Casas *et al.* [23], all but one herd (herd 2) exhibited similar profiles with a huge increase in prevalence between 7 and 13 weeks of age, leading to asymptotic behavior close to 1. This increase in prevalence coincided with the estimated duration of PI, suggesting an impact on transmission dynamics. Visually, the estimated prevalence values based on the full model (blue boxplots) were better adapted to reproduce this behavior than the SFOI approach (black boxplots).

The force of infection estimates, obtained by using the SFOI model, ranged between 0.019 and 0.136 week$^{-1}$ (Table 2). They tended to overestimate the predicted prevalence of infected animals before 7 weeks of age and led to underestimations thereafter. The full model structure was able to disentangle the transmission process in the presence and absence of PI. Both visual inspection (Figure 2) and the DIC values (Table 2) led us to conclude that the model with two different forces of infection better captured the dynamics of infection observed in field conditions. With this full model, the forces of infection exerted on fully susceptible individuals were estimated on average 1.8 times higher than with the SFOI model (range: 1.3 – 2.0; Table 2). In contrast, transmission in piglets with PI was found, on average, 13 times lower than in fully susceptible animals, with a relatively large variability between herds (range: 5–21).

The impact of PI on time to infection was assessed by modeling the infection-free survival functions in piglets with MDAs (Figure 3; full line) and fully susceptible animals (Figure 3; dashed lines) for each herd. In the herds with high prevalence (all but herd 2), fully susceptible individuals escaped infection according to exponential distributions with median infection-free survival ages ranging between 2.7 and 7.6 weeks. In contrast, in passively immune animals, the combined gamma-distributed duration of PI

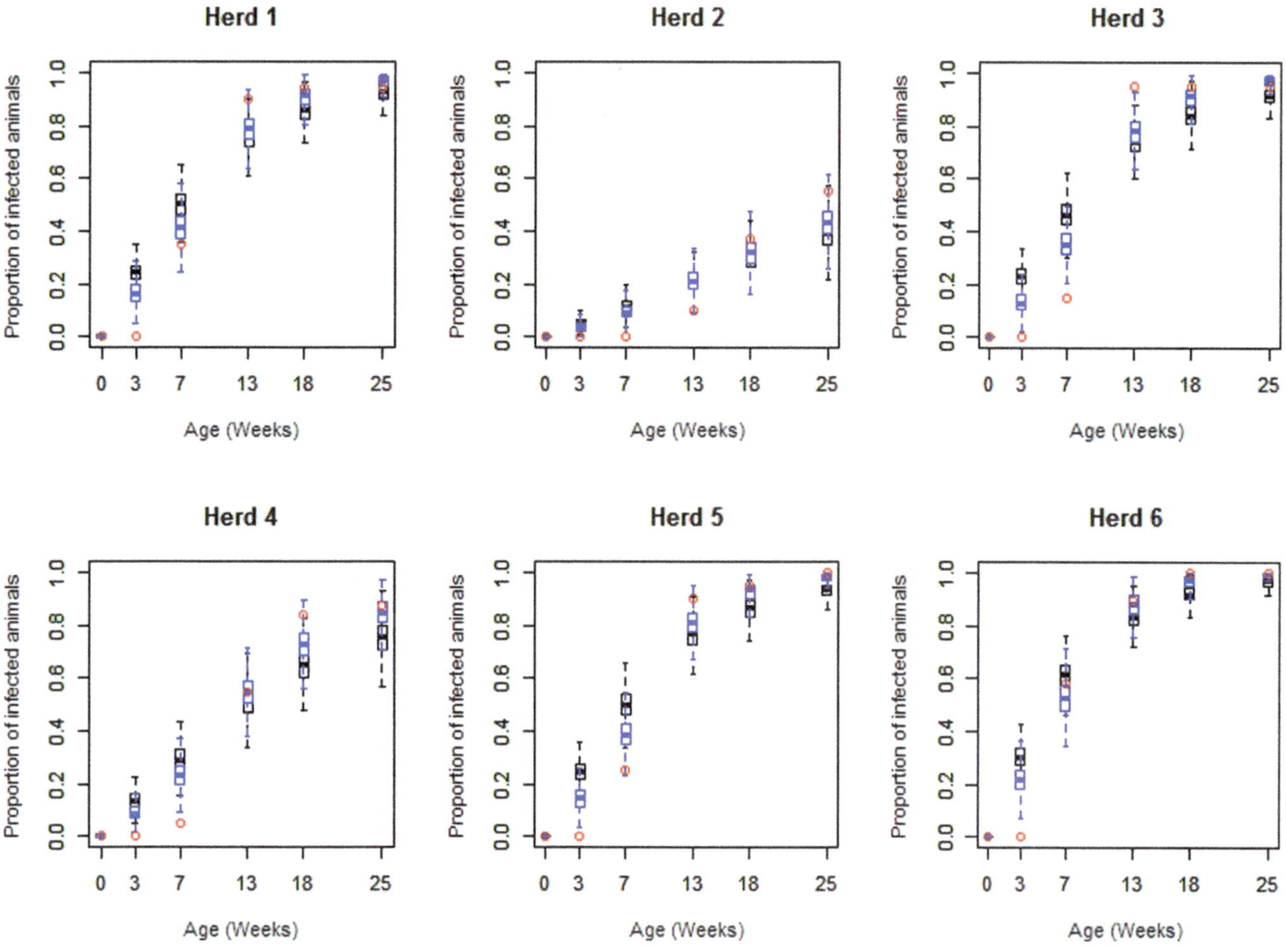


**Figure 2. Estimated age-specific HEV seroprevalences.** The distributions of estimated seroprevalences are shown according to model structure (SFOI model: black boxplots; Full model: blue boxplots). Observed prevalence data, derived from the longitudinal data in six pig herds in Spain [23], are represented by red dots.

and the differential forces of infection produced sigmoid survival curves with a slight probability of infection in the presence of MDAs (before 7 weeks of age), followed by an increased probability of infection thereafter. In consequence, the infection-free survival age in piglets which had received maternal antibodies was much higher than in fully susceptible animals, ranging between 8.7 and 13.8 weeks. In herd 2, the infection-free survival curves were similar in shape for both fully susceptible and passively immune animals (median: 27.5, 95% CI: [17.0; 53.4] and 32.0 weeks, 95% CI: [22.3; 51.5], respectively), but are not shown in Figure 3 since they extended beyond the normal slaughter age.

## Discussion

Piglets do not receive any maternal antibodies during gestation due to the epitheliochorial placentation in sows [30]. Neonate piglets are thus susceptible to the majority of infectious agents to which they are exposed in early life. It is therefore essential to provide new-born piglets with MDAs via colostrum to protect them against viral and bacterial infections before their own immune system develops [31]. However, the protection conferred by PI has been shown to be partial for several pathogens, reducing the clinical expression and/or within-host viral replication in infected piglets [21,32–35]. To our knowledge, the extent to which MDAs affect the transmission process has rarely been quantified to date [5,21,22]. We present an original approach to investigate and model the impact of PI on the transmission of an infectious agent in pigs based on batch-level prevalence data using two nested modeling structures: (i) differential forces of infection dependent on the serological status of piglets following colostrum intake (full model); and (ii) an equal force of infection exerted on fully susceptible and passively immune animals (SFOI model). This modeling approach, which combines a realistic distribution of the duration of PI and differential forces of infection, with and without maternal antibodies, provides a better understanding of early-life transmission of infectious agents. This approach was used to analyze serological data from a longitudinal study on hepatitis E prevalence in 6 pig herds.

The model only provides a simplified representation of the actual process of hepatitis E virus transmission with a main shortcut relying on the assumption of constant forces of infection. Indeed, the role of the environment on the course of infection was highlighted in a previous study, indicating that the force of infection represents a combination of transmission by direct contact between animals and ingestion of viral particles present in the environment [36]. Moreover, it is likely that the susceptibility to infectious agent in passively immune individuals varies as the level of maternally derived antibodies decreases with age. Thus,

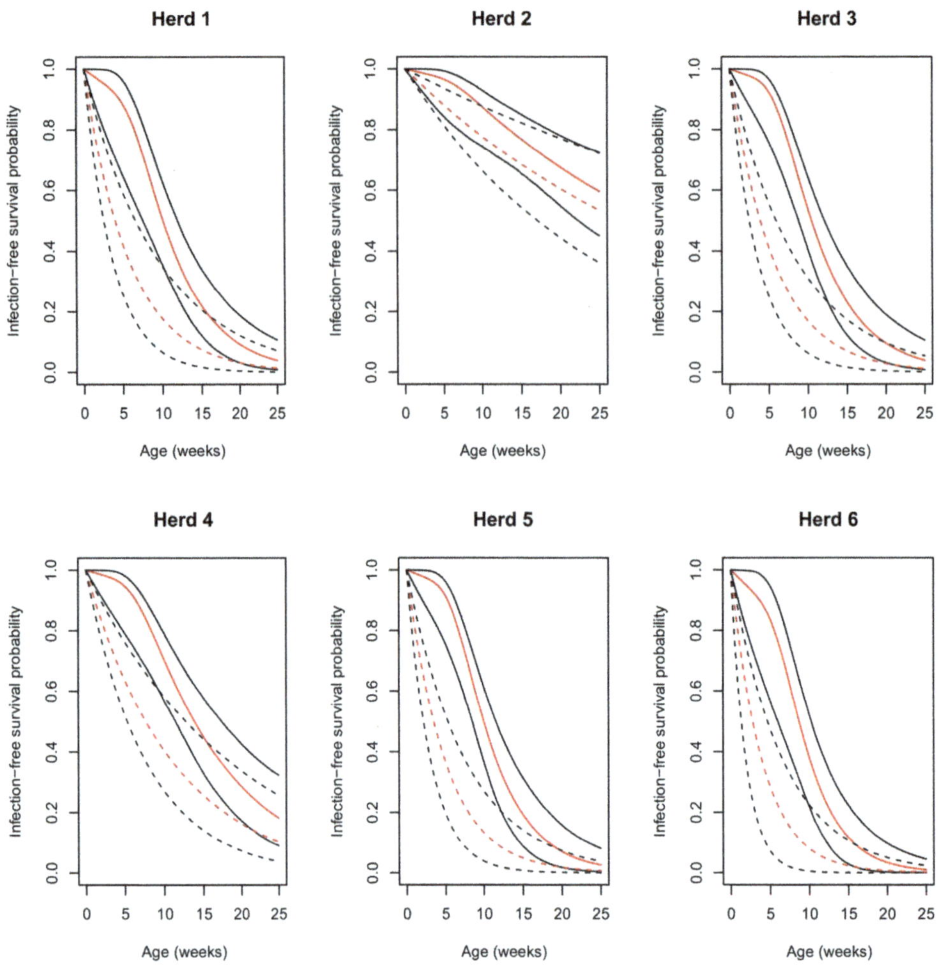


**Figure 3. HEV Infection-free survival function.** Median (red curves) and 95% credibility interval (black curves) of age specific probability to escape infection according to piglet's serological status after colostrum intake (dashed lines: fully susceptible animal; full lines: pig with HEV passive immunity).

**Table 2.** Estimates of forces of infection according to piglets' serological status.

| | Herd N° | λ[1] | $\lambda_M$[2] | DIC |
|---|---|---|---|---|
| **SFOI model**[3] | 1 | 0.105 (0.079;0.140) | – | 90 |
| | 2 | 0.019 (0.012;0.028) | – | 81 |
| | 3 | 0.099 (0.074;0.129) | – | 101 |
| | 4 | 0.055 (0.038;0.077) | – | 94 |
| | 5 | 0.106 (0.079;0.140) | – | 94 |
| | 6 | 0.136 (0.102;0.181) | – | 88 |
| **Full model**[4] | 1 | 0.181 (0.109;0.290) | 0.017 (9E-04;0.082) | 78 |
| | 2 | 0.025 (0.015;0.042) | 0.006 (5E-04;0.032) | 78 |
| | 3 | 0.190 (0.122;0.301) | 0.009 (7E-04;0.048) | 83 |
| | 4 | 0.092 (0.056;0.140) | 0.008 (7E-04;0.042) | 84 |
| | 5 | 0.215 (0.129;0.351) | 0.010 (5E-04;0.054) | 75 |
| | 6 | 0.274 (0.148;0.615) | 0.023 (0.001;0.112) | 75 |

[1]$\lambda$: Posterior median estimate with 95% credibility interval of the force of infection without PI (week$^{-1}$).
[2]$\lambda_M$: Posterior median estimate with 95% credibility interval of the force of infection in the presence of PI (week$^{-1}$).
[3]SFOI: Single force of infection model ($\lambda_M = \lambda$).
[4]Full model: differential forces of infection in the presence and absence of MDAs ($\lambda_M \neq \lambda$).

the estimated FOIs should be considered as the average infection pressure exerted on individuals to assess and quantify the protection conferred by PI. As all pigs within a specific herd were reared under the same conditions, the transmission process was summarized by forces of infection, by considering that a similar infection pressure was exerted on each individual independently of their serological status after colostrum intake. In consequence, the protection induced by PI is clearly related to a decrease in susceptibility.

The aim of the present study was to quantify the protective impact conferred by PI against Hepatitis E infection. Data from a longitudinal survey in France were used to characterize the kinetics of maternally-derived antibodies in 120 piglets. The serological titers of piglets at 1 week of age were highly correlated with those of the dam, thus supporting previous observations [37,38]. The duration of PI was modeled by a gamma distribution with a mean duration of 45.6 days [41.4; 50.4]. Although the duration was slightly shorter than in several epidemiological studies (8–9 weeks of age) [19,37,38], because of the gamma distribution, the probability of remaining passively immune fell below 0.1 by 60 days of age, as in these previous studies. In a recent study, the transmission parameters for hepatitis E virus in pigs were estimated from field data and the effectiveness of different hypothetical control strategies (early life vaccination and delayed vaccination around 10 weeks of age) was evaluated [18]. The author concluded that vaccination should lead to a 75% reduction of the main parameters governing the transmission process (host susceptibility, transmission rate or infectious period duration) to significantly decrease the expected prevalence at slaughter age. However, several studies showed interference between the maternally-derived antibodies and the immune response after vaccination [5,6,32,39,40]. Such interactions could have dramatic consequences on transmission. For example, the study by Bouma *et al.* revealed an increased reproduction number in piglets vaccinated against pseudorabies virus in the presence of PI as compared to vaccinated piglets without MDAs [5]. Further work should therefore be carried out to assess possible interactions between MDAs and future candidate vaccines before considering their usefulness in HEV control. However, a delayed vaccination strategy, as studied in [18], would certainly avoid such interactions as only 10% of piglets who received MDAs remained passively immune at 10 weeks of age.

Based on individual serological data for 120 pigs from 6 herds published by Casas *et al.* [23], we estimated the herd specific cumulative incidence by assuming that the first-positive IgM detection reflected a recent infection. Two nested model structures were tested assuming either an equal force of infection in the presence and absence of MDAs (SFOI) or two different forces of infection (full model). However, neither model fitted the null prevalence observed at 3 weeks of age for two main reasons: (i) the estimated observed initial proportion of passively immune animals was about 60% in all herds so that 40% of the animals were deemed fully susceptible at birth; (ii) even in the presence of PI, piglets are not fully protected against infection as shown by the non negligible FOIs estimated in the presence of maternal antibodies. Moreover, only 20 piglets in each herd were sampled to establish the herd-specific seroprevalence profiles, which might have led to considerable uncertainty, especially if prevalence was low. These considerations are supported by the good concordance between the estimates and the data from 7 weeks of age onwards. In herd 2, the prevalence increased slowly, continuously and exponentially. Indeed the cumulative prevalence observed at slaughter age was about 50%. As the level of PI was similar to that of the other herds, this difference reflects a low infection pressure which might be due to differences in husbandry practices and herd structure.

In models with a single force of infection, the inverse of the force of infection corresponds to the average age at infection [41]. However, this relationship cannot be applied to the full model due to the interplay between piglet serological status and the differential forces of infection. The impact of PI on time-to-infection was therefore assessed by modeling the infection-free survival functions, *i.e.* the probability of escaping infection according to age of the animal, in piglets with and without MDAs, for each herd. In a fully susceptible animal, transmission occurs much earlier than in pigs with MDAs. Several epidemiological studies have shown that peak HEV prevalence occurs after the nursery stage, at about 13 weeks of age, which corresponds to the estimated median age at infection in passively immune individuals [19,38,42]. However, the predicted average age-at-infection in herd 2 was considerably higher and beyond the usual slaughter age (25 weeks). As the level of PI in all herds was similar, this difference could presumably only be explained by differences in herd structures and husbandry practices which might help to reduce the infection pressure due to piglet mixing and the environmental viral load. These results suggest that the age at HEV infection does not strictly depend on the proportion of passive immune piglets but also relies on later events which condition the force of infection after PI waning [43].

The origins of most autochthonous HEV cases remain undetermined, although some cases have been related to the consumption of uncooked meat, liver or offal from domestic pigs or wild animals (*e.g.* wild boar or deer) [44–46]. In consequence, control measures in pig farms are aimed at preventing the presence of viremic animals at slaughter age. The modeling framework developed in this study was successfully used to evaluate the age at infection according to the serological status of piglets after colostrum intake in different pig farms. In herd 2, a large proportion of pigs remained susceptible before slaughter-age, mainly due to a lower force of infection after the wane of PI. These conditions could be associated with an increased risk as regards public health especially if later events, shortly before slaughter, enhance the infectious process and lead to massive infection of pigs. Further work needs to be carried out to fully understand which factors are involved in the resulting variations in forces of infection between herds.

We developed an original modeling framework to analyze the role of passive immunity in the transmission of infectious agents in animal populations based on batch-level incidence data. The protective characteristics of MDAs, acquired either through colostrum intake or transplacental passage, play an important part in several mammalian species [47–49]. In consequence, the methodology developed in this study could be extended to quantify the level of protection conferred by maternally-derived antibodies for different pathogens and species.

## Supporting Information

**Figure S1 Individual fits of decay of maternally derived antibodies (12 individual profiles are shown).**

**Figure S2 Convergence plot of Monte Carlo Makov Chains for Herd 1 parameters (lines correspond to smoothing averages for each chain).**

## Acknowledgments

We thank V. Dorenlor, F. Eono and E. Eveno for collecting field data to assess the antibodies kinetics in piglets' early life and M. Dumarest for laboratory analyses.

## Author Contributions

Conceived and designed the experiments: NR MA NP. Performed the experiments: MA NR NP. Analyzed the data: MA MC. Contributed reagents/materials/analysis tools: MC NP. Wrote the paper: MA NR NP MC.

## References


1. Brinkhof MWG, Mayorga O, Bock J, Heininger U, Herzog C (2013) Kinetics of maternally acquired anti-hepatitis A antibodies: Prediction of waning based on maternal or cord blood antibody levels. Vaccine 31: 1490–1495.
2. Elbers ARW, Stockhofe-Zurwieden N, van der Poel WHM (2014) Schmallenberg virus antibody persistence in adult cattle after natural infection and decay of maternal antibodies in calves. BMC Veterinary Research 10.
3. Pravieux JJ, Poulet H, Charreyre C, Juillard V (2007) Protection of Newborn Animals through Maternal Immunization. J Comp Pathol 137: S32–S34.
4. Swain P, Nayak SK (2009) Role of maternally derived immunity in fish. Fish and Shellfish Immunology 27: 89–99.
5. Bouma A, De Jong MCM, Kimman TG (1997) The influence of maternal immunity on the transmission of pseudorabies virus and on the effectiveness of vaccination. Vaccine 15: 287–294.
6. Kitikoon P, Nilubol D, Erickson BJ, Janke BH, Hoover TC, et al. (2006) The immune response and maternal antibody interference to a heterologous H1N1 swine influenza virus infection following vaccination. Vet Immunol Immunopathol 112: 117–128.
7. Pastoret PP (2007) Challenges and Issues of Early Life Vaccination in Animals and Humans. J Comp Pathol 137, Supplement 1: S2–S3.
8. Siegrist CA (2007) The Challenges of Vaccine Responses in Early Life: Selected Examples. J Comp Pathol 137, Supplement 1: S4–S9.
9. Siegrist CA, Barrios C, Martinez X, Brandt C, Berney M, et al. (1998) Influence of maternal antibodies on vaccine responses: inhibition of antibody but not T cell responses allows successful early prime-boost strategies in mice. Eur J Immunol 28: 4138–4148.
10. Britton T, Kypraios T, O'Neill PD (2011) Inference for Epidemics with Three Levels of Mixing: Methodology and Application to a Measles Outbreak. Scandinavian Journal of Statistics 38: 578–599.
11. Cauchemez S, Ferguson NM (2008) Likelihood-based estimation of continuous-time epidemic models from time-series data: application to measles transmission in London. Journal of the Royal Society, Interface/the Royal Society 5: 885–897.
12. Courcoul A, Vergu E, Denis JB, Beaudeau F (2010) Spread of Q fever within dairy cattle herds: key parameters inferred using a Bayesian approach. Proceedings Biological sciences/The Royal Society 277: 2857–2865.
13. Elderd BD, Dwyer G, Dukic V (2013) Population-level differences in disease transmission: A Bayesian analysis of multiple smallpox epidemics. Epidemics 5: 146–156.
14. Jewell CP, Kypraios T, Neal P, Roberts GO (2009) Bayesian analysis for emerging infectious diseases. 10.1214/09-ba417: 465–496.
15. McBryde ES, Gibson G, Pettitt AN, Zhang Y, Zhao B, et al. (2006) Bayesian modelling of an epidemic of severe acute respiratory syndrome. Bulletin of mathematical biology 68: 889–917.
16. O'Neill PD (2002) A tutorial introduction to Bayesian inference for stochastic epidemic models using Markov chain Monte Carlo methods. Mathematical Biosciences 180: 103–114.
17. Streftaris G, Gibson GJ (2004) Bayesian analysis of experimental epidemics of foot-and-mouth disease. Proceedings of the Royal Society B: Biological Sciences 271: 1111–1117.
18. Backer JA, Berto A, McCreary C, Martelli F, van der Poel WHM (2012) Transmission dynamics of hepatitis E virus in pigs: Estimation from field data and effect of vaccination. Epidemics 4: 86–92.
19. Kanai Y, Tsujikawa M, Yunoki M, Nishiyama S, Ikuta K, et al. (2010) Long-term shedding of hepatitis E virus in the feces of pigs infected naturally, born to sows with and without maternal antibodies. Journal of Medical Virology 82: 69–76.
20. Pavio N, Meng XJ, Renou C (2010) Zoonotic hepatitis E: Animal reservoirs and emerging risks. Veterinary Research 41.
21. Rose N, Blanchard P, Cariolet R, Grasland B, Amenna N, et al. (2007) Vaccination of porcine circovirus type 2 (PCV2)-infected sows against porcine Parvovirus (PPV) and Erysipelas: effect on post-weaning multisystemic wasting syndrome (PMWS) and on PCV2 genome load in the offspring. J Comp Pathol 136: 133–144.
22. Allerson M, Deen J, Detmer SE, Gramer MR, Joo HS, et al. (2013) The impact of maternally derived immunity on influenza A virus transmission in neonatal pig populations. Vaccine 31: 500–505.
23. Casas M, Cortes R, Pina S, Peralta B, Allepuz A, et al. (2011) Longitudinal study of hepatitis E virus infection in Spanish farrow-to-finish swine herds. Vet Microbiol 148: 27–34.
24. Snoeck E, Chanu P, Lavielle M, Jacqmin P, Jonsson EN, et al. (2010) A comprehensive hepatitis C viral kinetic model explaining cure. Clinical pharmacology and therapeutics 87: 706–713.
25. Lavielle M, Mentre F (2007) Estimation of population pharmacokinetic parameters of saquinavir in HIV patients with the MONOLIX software. J Pharmacokinet Pharmacodyn 34: 229–249.
26. Barnaud E, Rogee S, Garry P, Rose N, Pavio N (2012) Thermal inactivation of infectious hepatitis E virus in experimentally contaminated food. Appl Environ Microbiol 78: 5153–5159.
27. Efron B, Tibshirani J (1993) An introduction to the bootstrap. New York: Chapman & Hall, Inc.
28. Rose N, Lunazzi A, Dorenlor V, Merbah T, Eono F, et al. (2011) High prevalence of Hepatitis E virus in French domestic pigs. Comparative Immunology, Microbiology and Infectious Diseases 34: 419–427.
29. Peralta B, Casas M, de Deus N, Martín M, Ortuño A, et al. (2009) Anti-HEV antibodies in domestic animal species and rodents from Spain using a genotype 3-based ELISA. Vet Microbiol 137: 66–73.
30. Pluske JR, Le Dividich J (2003) Weaning the pig: concepts and consequences: Wageningen Academic Publishers.
31. Rooke JA, Bland IM (2002) The acquisition of passive immunity in the new-born piglet. Livest Prod Sci 78: 13–23.
32. Kitching RP, Salt JS (1995) The interference by maternally-derived antibody with active immunization of farm animals against foot-and-mouth disease. British Veterinary Journal 151: 379–389.
33. Loeffen WLA, Heinen PP, Bianchi ATJ, Hunneman WA, Verheijden JHM (2003) Effect of maternally derived antibodies on the clinical signs and immune response in pigs after primary and secondary infection with an influenza H1N1 virus. Vet Immunol Immunopathol 92: 23–35.
34. Vandeputte J, Too HL, Ng FK, Chen C, Chai KK, et al. (2001) Adsorption of colostral antibodies against classical swine fever, persistence of maternal antibodies, and effect on response to vaccination in baby pigs. Am J Vet Res 62: 1805–1811.
35. Ward LA, Rich ED, Besser TE (1996) Role of Maternally Derived Circulating Antibodies in Protection of Neonatal Swine against Porcine Group A Rotavirus. J Infect Dis 174: 276–282.
36. Andraud M, Dumarest M, Cariolet R, Aylaj B, Barnaud E, et al. (2013) Direct contact and environmental contaminations are responsible for HEV transmission in pigs. Vet Res 44: 102.
37. Meng XJ, Purcell RH, Halbur PG, Lehman JR, Webb DM, et al. (1997) A novel virus in swine is closely related to the human hepatitis E virus. Proc Natl Acad Sci U S A 94: 9860–9865.
38. de Deus N, Casas M, Peralta B, Nofrarias M, Pina S, et al. (2008) Hepatitis E virus infection dynamics and organic distribution in naturally infected pigs in a farrow-to-finish farm. Veterinary Microbiology 132: 19–28.
39. Francis MJ, Black L (1986) Response of young pigs to foot-and-mouth disease oil emulsion vaccination in the presence and absence of maternally derived neutralising antibodies. Res Vet Sci 41: 33–39.
40. Klinkenberg D, Moormann RJ, de Smit AJ, Bouma A, de Jong MC (2002) Influence of maternal antibodies on efficacy of a subunit vaccine: transmission of classical swine fever virus between pigs vaccinated at 2 weeks of age. Vaccine 20: 3005–3013.
41. Anderson RM, May RM (2004) Infectious diseases of humans: dynamics and control: Oxford University Press.
42. Fernandez-Barredo S, Galiana C, Garcia A, Vega S, Gomez MT, et al. (2006) Detection of hepatitis E virus shedding in feces of pigs at different stages of production using reverse transcription-polymerase chain reaction. Journal of Veterinary Diagnostic Investigation 18: 462–465.
43. Walachowski S, Dorenlor V, Lefevre J, Lunazzi A, Eono F, et al. (2013) Risk factors associated with the presence of hepatitis E virus in livers and seroprevalence in slaughter-age pigs: a retrospective study of 90 swine farms in France. Epidemiol Infect 10.1017/S0950268813003063: 1–11.
44. Colson P, Borentain P, Queyriaux B, Kaba M, Moal Vr, et al. (2010) Pig Liver Sausage as a Source of Hepatitis E Virus Transmission to Humans. J Infect Dis 202: 825–834.
45. Masuda J-I, Yano K, Tamada Y, Takii Y, Ito M, et al. (2005) Acute hepatitis E of a man who consumed wild boar meat prior to the onset of illness in Nagasaki, Japan. Hepatol Res 31: 178–183.
46. Tei S, Kitajima N, Takahashi K, Mishiro S (2003) Zoonotic transmission of hepatitis E virus from deer to human beings. Lancet 362: 371–373.
47. Hassig M, Stadler T, Lutz H (2007) Transition from maternal to endogenous antibodies in newborn calves. Veterinary Record 160: 234–235.
48. Tomlinson A, Chambers M, Delahay R (2012) Mycobacterium bovis infection in badger cubs: Re-assessing the evidence for maternally derived immunological protection from advanced disease. Vet Immunol Immunopathol 148: 326–330.
49. Van De Perre P (2003) Transfer of antibody via mother's milk. Vaccine 21: 3374–3376.

# Comparison of *spa* Types, SCC*mec* Types and Antimicrobial Resistance Profiles of MRSA Isolated from Turkeys at Farm, Slaughter and from Retail Meat Indicates Transmission along the Production Chain

**Birgit Vossenkuhl*, Jörgen Brandt¤a, Alexandra Fetsch, Annemarie Käsbohrer, Britta Kraushaar, Katja Alt¤b, Bernd-Alois Tenhagen**

Federal Institute for Risk Assessment, Berlin, Germany

## Abstract

The prevalence of MRSA in the turkey meat production chain in Germany was estimated within the national monitoring for zoonotic agents in 2010. In total 22/112 (19.6%) dust samples from turkey farms, 235/359 (65.5%) swabs from turkey carcasses after slaughter and 147/460 (32.0%) turkey meat samples at retail were tested positive for MRSA. The specific distributions of *spa* types, SCC*mec* types and antimicrobial resistance profiles of MRSA isolated from these three different origins were compared using chi square statistics and the proportional similarity index (Czekanowski index). No significant differences between *spa* types, SCC*mec* types and antimicrobial resistance profiles of MRSA from different steps of the German turkey meat production chain were observed using Chi-Square test statistics. The Czekanowski index which can obtain values between 0 (no similarity) and 1 (perfect agreement) was consistently high (0.79–0.86) for the distribution of *spa* types and SCC*mec* types between the different processing stages indicating high degrees of similarity. The comparison of antimicrobial resistance profiles between the different process steps revealed the lowest Czekanowski index values (0.42–0.56). However, the Czekanowski index values were substantially higher than the index when isolates from the turkey meat production chain were compared to isolates from wild boar meat (0.13–0.19), an example of a separated population of MRSA used as control group. This result indicates that the proposed statistical method is valid to detect existing differences in the distribution of the tested characteristics of MRSA. The degree of similarity in the distribution of *spa* types, SCC*mec* types and antimicrobial resistance profiles between MRSA isolates from different process stages of turkey meat production may reflect MRSA transmission along the chain.

**Editor:** Herminia de Lencastre, Rockefeller University, United States of America

**Funding:** This work was carried out within the Project MedVet-Staph funded by the German Bundesministerium für Bildung und Forschung, Grant Nr. 01KI1014C. The funders had no role in study design, data collection and analysis, decision to publish, or preparation of the manuscript.

**Competing Interests:** The authors have declared that no competing interests exist.

* E-mail: 43@bfr.bund.de

¤a Current address: Humboldt University, Unter den Linden 6, Berlin, Germany
¤b Current address: Robert Koch Institute, DGZ-Ring 1, Berlin, Germany

## Introduction

*Staphylococcus (S.) aureus* is a common cause of food poisoning due to the production of various enterotoxins. *S. aureus* is a frequent colonizer of the skin and mucous membranes and therefore, personnel and food-producing animals are the main sources of *S. aureus* in food [1]. The control of *S. aureus* is routinely considered in the food producing industry if standard food safety management systems are operated. In recent years, methicillin-resistant *Staphylococcus aureus* (MRSA), previously known as a multidrug resistant pathogen causing severe healthcare associated and community acquired infections, [2] has been observed worldwide in livestock husbandry as well as in food of different animal origins raising concerns about a possible farm to fork transmission.

First reported from pigs in the Netherlands [3] and France [4] a distinct MRSA lineage, Clonal Complex (CC) 398, has emerged in food producing animals in Europe especially in herds of pigs [5–8], veal calves [9] broiler flocks [10,11] and turkeys [12]. Therefore, the term "livestock-associated MRSA" (LA-MRSA) was introduced considering livestock to form a new and separate reservoir for MRSA [13]. In Asian countries, however, sequence type ST9, a separate genetic linage, is predominating among MRSA isolates from livestock animals [14,15]. Different DNA sequencing methods are used for typing MRSA strains. In order to define MRSA clones, Multilocus sequence typing (MLST), a method of classifying MRSA strains by the allelic profile of seven housekeeping genes, is used in conjunction with PCR analysis of the staphylococcal chromosomal cassette *mec* (SCC*mec*), a mobile genetic element that contains the *mec A* gene encoding for resistance to methicillin [16]. 11 different SCC*mec* types have been described, so far. The class of *mec* gene complex and the type of ccr gene complex carrying a set of recombinase genes responsible for integration and excision of the cassette characterize the different types of SCCmec elements [17]. Whereas SCCmec I-X harbor *mecA* SCCmec XI carries a divergent *mecA* homologue (*mecA*$_{LGA251}$) [18]. *Spa* typing differentiates MRSA strains by the number of tandem repeats and the sequence variation in region X of the protein A gene (spa) and can be used for reliable and discriminatory typing of MRSA [19]. As particular MLST have shown to be associated with specific repeats and repeat successions it is, with few exceptions, possible to infer an MLST type from the spa type. (http://www.spaserver.ridom.de). The frequent use of antimicrobials in animal production is suspected to facilitate the emergence and spread of MRSA due to antimicrobial selection pressure [20–22]. High stocking density in intensive food animal production holdings and intensive animal trading promote the rapid spread of MRSA between livestock populations [23,24]. LA-MRSA strains have also been detected in raw meat at retail including beef, veal, pork and poultry [25–33] indicating potential transmission along the chain due to cross contamination during slaughter and processing. However, the extent of this transmission is so far poorly understood.

In Germany, the national monitoring for zoonotic agents aims at characterizing the prevalence of potential zoonotic pathogens at different stages of various food chains. The monitoring is part of the official control of foodstuffs and fulfills the requirements of EU Directive 2003/99/EC [34]. In 2010, the turkey meat production chain was addressed in this monitoring scheme.

The objective of the present study was to use data from the national monitoring of zoonotic agents in the food chain to obtain a comprehensive insight into the presence and transmission of MRSA in the German turkey meat production chain. A new approach is proposed for analyzing a cross sectional MRSA data set from different stages of the food chain in order to draw conclusions on potential farm to fork transmission. For this purpose, the prevalence of MRSA and the distribution of *spa* types, SCC*mec* types and antimicrobial resistance profiles among MRSA isolated from different steps of the turkey meat production chain were compared. It is proposed that the degree of similarity in the distribution of *spa* types, SCC*mec* types and antimicrobial resistance profiles between the samples from the three process steps may be interpreted as reflecting MRSA transmission along the chain.

## Materials and Methods

### 1. Study Design

Sampling was conducted in 2010 by the competent authorities of the federal states according to a pre-defined protocol in the framework of the national monitoring for zoonotic agents. All participating competent authorities are listed in table S1. Dust samples from 112 German turkey flocks were collected in order to quantify the presence of MRSA in primary production and to assess the introduction of MRSA into the slaughterhouses. Samples at slaughterhouses (n = 359) were analyzed to estimate the transfer to carcasses during slaughter and to determine the transmission of MRSA from carcasses to fresh turkey meat during further processing. Finally, 460 turkey meat portions were sampled to evaluate the MRSA exposure of consumers via contaminated turkey meat.

Turkey pens were sampled by pooling 5 dust swab samples, collected from different sections representing an area of 500 cm$^2$, each. At the slaughterhouse, at least 30 g neck skin was sampled from turkey carcasses after slaughter and chilling, but prior to further processing. Samples of 25 g of fresh turkey meat (with or without skin) were collected at retail. In order to ensure a high level of representativity, the distribution of the samples in primary production and at slaughter across Germany was proportional to the number of turkey flocks and the slaughter capacity of the respective federal state. Meat samples at retail were distributed according to the human population size of the executive federal state. A more detailed description of the principles of the national monitoring for zoonotic agents has been published before [35].

### 2. MRSA Isolation

MRSA were isolated by the regional laboratories according to the recommended method of the National Reference Laboratory (NRL) for staphylococci including S. aureus at the Federal Institute for Risk Assessment (BfR). The dust samples were pooled per turkey house in 100 ml Mueller Hinton broth supplemented with 6.5% NaCl for pre-enrichment. Neck skin samples (at least 30 g), fresh meat (25 g) and meat preparations (25 g) were pre-enriched in 225 ml Mueller Hinton broth supplemented with 6.5% NaCl. After incubation for 16–20 h at 37°C, 1 ml pre-enrichment broth was transferred into 9 ml of tryptic soy broth supplemented with 50 mg/l aztreonam and 3.5 mg/l cefoxitin. After incubation of this selective-enrichment broth for a further 16–20 h at 37°C one loopful was plated onto sheep blood agar and chromogenic MRSA screening agar respectively, and incubated for 24–48 h at 37°C. Presumptive MRSA isolates were sent to the National Reference Laboratory (NRL) for staphylococci including *S. aureus* at the Federal Institute for Risk Assessment (BfR) for MRSA confirmation and characterization. The number of MRSA isolates included in further analyses is not exactly congruent to the amount of positive samples obtained within the national monitoring for zoonotic agents because first, the NRL did not always receive the corresponding isolate from the competent authorities of the federal states or second, isolates which did not exactly correspond to the monitoring sampling plan in terms of completeness of data reporting to the national level but were obtained from the correct matrix were excluded from prevalence estimations but included in further typing and strain comparisons.

Twenty one MRSA isolates from wild boar meat within the national monitoring for zoonotic agents of 2011 were used in the analyses as a control group (data not shown in detail). The control group was selected to ensure wide differences with the population under study concerning the distribution of MRSA strains in order to evaluate if the used analytical approach is appropriate to differentiate between the matrices.

### 3. Molecular Typing

Presumptive MRSA isolates were confirmed by an in-house multiplex PCR simultaneously targeting the 23S rDNA specific for *Staphylococcus* species [36], the nuclease gene *nuc* which is specific for *S. aureus*, and the resistance gene *mecA* [37]. Template DNA was extracted using the "RTP Bacteria DNA Mini Kit" (Invitek, Berlin, Germany). All MRSA isolates were further characterized using *spa* typing [38] and SCC*mec*-typing [39]. The method applied for typing of the *SCC*mec differentiates SCC*mec* types I to V and their subtypes. However, isolates of the CC398 characterized as type III by the method have been shown to rather be a variant of type V [40]. The software Ridom Staphytype (Ridom GmbH, Würzburg, Germany) was used to assign *spa* types. *Spa* types which have not been identified and assigned to a clonal complex (CC) by the NRL before were additionally subjected to multilocus sequence typing (MLST) [41].

### 4. Antimicrobial Susceptibility Testing

All isolates were tested for the susceptibility to antimicrobials using broth microdilution in accordance with Clinical and Laboratory Standards Institute guidelines [42]. Commercial microtitre plates were used (TREK Diagnostic Systems, Magellan Biosciences, West Sussex, England). Minimum inhibitory concentrations (MIC) were evaluated according to epidemiological cut-off values (ECOFFs) published for MRSA and *S. aureus* by the European committee for antimicrobial susceptibility testing (www.eucast.org). MIC values above the ECOFFs indicated microbiological resistance. MIC lower or equal to the ECOFFs characterised susceptible strains. *S. aureus* strain ATCC 25923 was used for quality assurance Resistance testing included gentamicin, kanamycin, streptomycin, chloramphenicol, ciprofloxacin, tetracycline, clindamycin, erythromycin, mupirocin, linezolid, vancomycin, quinupristin/dalfopristin, penicillin, fusidic acid, cefoxitin, trimethoprim, sulfamethoxazole, rifampicin and tiamulin.

## 5. Statistical Analysis

The chi square test of homogeneity was used to analyze differences in the distribution of *spa* types and antibiotic resistance profiles between MRSA strains from the turkey flocks, carcasses at slaughter and meat. Isolates were grouped according to their *spa* types and antibiotic resistance profiles to assure appropriate numbers of isolates in all categories. All *spa* types were aggregated in accordance to their frequency of occurrence. The phenotypic antimicrobial resistance profiles were grouped by hierarchical cluster analysis using Ward's minimum variance and squared Euclidean distance. The MIC values for each isolate were categorized into resistant or susceptible according to the ECOFFs to generate a binary data set. The final amount of clusters was determined using the Pseudo-F [43] and Pseudo-T [44] statistics. Both tests indicate possible breakpoints for splitting the data into the appropriate amount of clusters. The distribution of SCC*mec* types in the different matrices were compared using Fisher's exact test as 33.3% of the cells of the contingency table had an expected value below 5. P-values of $<0.05$ were considered statistically significant. Chi square test, Fisher's exact test and cluster analysis were calculated using the statistical software package SPSS 18.0 (SPSS Inc. Munich, Germany). Pseudo-F and Pseudo-T statistics were performed using SAS/STAT software 9.2 (SAS Institute Inc., Cary, NC, USA).

The degree of similarity between the frequency distributions of *spa* types, SCC*mec* types and resistance profiles of MRSA among the sample sets from the turkey primary production, carcasses at slaughterhouse and turkey meat at retail was estimated using the Czekanowski index or proportional similarity index (PSI) [45]. It is calculated by:

$$PS = 1 - 0{,}5\sum_i |p_i - q_i| = \sum_i \min(p_i, q_i)$$

where $p_i$ and $q_i$ represent the proportion of strains out of all strains among the data sets P and Q which agree in the realization i of the variable of interest. The values for PS range from 1 for identical frequency distributions of the variable of interest to zero for no similarities between the data sets. Since the size of the samples is rather small, a realization of the PSI index may deviate largely from its true value. Thus, the PSI was bootstrapped obtaining a probability density distribution from which we derived the 95% confidence interval for the PSI. The statistic open source software R (available at: http://www.R-project.org) was used to calculate the approximate confidence interval of the Czekanowski index using the non-parametric boostrap $BC_\alpha$ method utilizing 2000 iterations [46].

## Results

Twenty two (19.6%) of 112 dust samples from the turkey primary production, 235 (65.5%) of 359 turkey carcasses after slaughter and 147 (32.0%) of 460 turkey meat samples at retail were tested positive for MRSA [47]. A set of 32 isolates from dust samples, 248 isolates from turkey carcasses and 241 isolates from turkey meat was used for further laboratory analyses (Table 1).

A total of 16 different *spa* types were identified. The number of different *spa* types increased during processing from 5 different types in dust samples over 8 in carcasses to 15 different types in meat samples. The proportion of strains assigned to CC398 ranged between 85.9 and 90.6%. Among CC398, t011 (43.8–46.9%) and t034 (32.0–43.8%) were the predominating *spa*-types on every process step. *Spa* types t1430 (4.0–6.3%) and t002 (3.1–9.1%) were dominating within the group of non CC398 strains.

Most of the strains carried SCC*mec*-type V (58.1–71.9%) followed by type IVa (19–27.0%). Type III (0–1.2%) was identified sporadically (Table 1). However, there is evidence in former literature that CC398 strains which were identified as SCC*mec* type III by the typing scheme of Zhang et al. [39] are rather assigned to a separate variant of SCC*mec* type V [40,48,49]. In 5.7–17.6% of the strains the SCC*mec* type could not be identified by the method used.

Susceptibility to 19 different antimicrobial agents was determined (Figure 1). Throughout the turkey production chain, the vast majority of isolates was resistant to tetracycline (98.8%–100%). High resistance rates were obtained to clindamycin (79.4–93.8%), erythromycin (73.8–87.5%), trimethoprim (65.7–78.1%), quinupristin/dalfopristin (62.2–66.1%) and tiamulin (52.3–65.6%). Resistances to mupirocin, linezolid, sulfamethoxazole and rifampicin were observed sporadically in individual isolates from all steps of the process chain. All isolates were susceptible to vancomycin. Resistance to tiamulin (62.2 versus 8.2%), gentamicin (25.2 versus 6.6%) and trimethoprim (72.0 versus 36.1%) was considerably more frequent among CC398 than among non-CC398 strains. Resistance to ciprofloxacin was common among non-CC398 strains (98.4 versus 26.1% in CC398 strains) (Figure 1).

All 521 MRSA strains were included in further similarity estimations. In accordance to the frequency of their occurrence all *spa* types were aggregated in 4 different categories for further statistical analysis. The most prevalent *spa* types t011 and t034 built their own group whereas rare *spa* types of CC398 and all non CC398 strains were summarized in separate groups. The chi square distribution of the *spa* type groups did not significantly differ between primary production, carcasses at slaughter and meat at retail ($p = 0.06$). Likewise, no significant difference was identified in the distribution of SCC*mec* types between the origins using fisher's exact test ($p = 0.095$). A total of 101 different resistance profiles were identified among the MRSA isolates including resistance to 2 to 12 different antimicrobial substances. The hierarchical cluster algorithm of Wards minimum variance combined with squared Euclidean distance separated the antimicrobial resistance profiles into homogenous clusters. Identical resistance phenotypes did not appear in more than one cluster. Based on the Pseudo-F and Pseudo-T statistics the 3 cluster solution containing 33, 44 and 24 different phenotypic resistance profiles, respectively, was identified to best describe the binary data set. Detailed characteristics of the cluster composition, concerning antimicrobial resistance and the distribution of groups of *spa* types and SCC*mec* types, is summarized in table S2. The antimicrobial resistance clusters did not significantly differ in their chi square distribution between the MRSA samples from the three origins ($p = 0.295$).

**Table 1.** MRSA prevalence and distribution of *spa* types, SCC*mec* types and antimicrobial resistance clusters at different steps of the German turkey meat production chain in 2010.

| Process step | | Primary production | | Slaughter | | Meat | | total |
|---|---|---|---|---|---|---|---|---|
| Samples (n) | | 112 | | 359 | | 460 | | 931 |
| MRSA positive samples (n) | | 22 | | 235 | | 147 | | 404 |
| MRSA prevalence (%) | | 19,6 | | 65,5 | | 32 | | |
| No. of isolates included in further statistics[a] | | 32 | | 248 | | 241 | | 521 |
| Genetic Typing | | | | | | | | |
| CC398 | *spa* types | n | % | n | % | n | % | n |
| | t011 | 14 | 43.8 | 113 | 45.6 | 113 | 46.9 | 240 |
| | t034 | 14 | 43.8 | 105 | 42.3 | 77 | 32.0 | 196 |
| | t108 | - | | - | | 1 | 0.4 | 1 |
| | t571 | 1 | 3.1 | - | | 1 | 0.4 | 2 |
| | t899 | - | | 1 | 0.4 | 5 | 2.1 | 6 |
| | t1255 | - | | - | | 3 | 1.2 | 3 |
| | t1344 | - | | - | | 3 | 1.2 | 3 |
| | t1580 | - | | - | | 1 | 0.4 | 1 |
| | t2510 | - | | - | | 1 | 0.4 | 1 |
| | t2576 | - | | 1 | 0.4 | - | | 1 |
| | t2970 | - | | 3 | 1.2 | 1 | 0.4 | 4 |
| | t4652 | - | | 1 | 0.4 | 1 | 0.4 | 2 |
| | total | 29 | 90.6 | 224 | 90.3 | 207 | 85.9 | 460 |
| non CC398 | | | | | | | | |
| assigned MLST types | *spa* types | | | | | | | |
| ST5 | t002 | 1 | 3.1 | 14 | 5.6 | 22 | 9.1 | 37 |
| ST5 | t010 | - | | - | | 1 | 0.4 | 1 |
| ST45 | t015 | - | | - | | 1 | 0.4 | 1 |
| ST9 | t1430 | 2 | 6.3 | 10 | 4.0 | 10 | 4.1 | 22 |
| | total | 3 | 9.4 | 24 | 9.7 | 34 | 14.1 | 61 |
| total | | 32 | 100 | 248 | 100 | 241 | 100 | 521 |
| | SCC*mec*Types | | | | | | | |
| | n.t.[b] | 2 | 6.3 | 24 | 9.7 | 33 | 13.7 | 59 |
| | *mec* III | - | | 1 | 0.4 | 3 | 1.2 | 4 |
| | *mec* IVa | 7 | 21.9 | 47 | 19.0 | 65 | 27.0 | 119 |
| | *mec* V | 23 | 71.9 | 176 | 71.0 | 140 | 58.1 | 339 |
| total | | 32 | 100 | 248 | 100 | 241 | 100 | 521 |
| | Resistance profiles[c] | | | | | | | |

**Table 1.** Cont.

| Process step | | Primary production | | Slaughter | | Meat | | total |
|---|---|---|---|---|---|---|---|---|
| | Cluster A | 17 | 53.1 | 121 | 48.8 | 97 | 40.2 | 235 |
| | Cluster B | 10 | 31.3 | 82 | 33.1 | 88 | 36.5 | 180 |
| | Cluster C | 5 | 15.6 | 45 | 18.1 | 56 | 23.2 | 106 |
| total | | 32 | 100 | 248 | 100 | 241 | 100 | 521 |

[a]MRSA isolates which did not exactly correspond to the monitoring sampling plan in terms of completeness of data reporting to the national level were excluded from prevalence estimations but included in further typing and strain comparisons.
[b]Not typable.
[c]Resistance cluster were calculated using Ward's minimum variance with squared Euclidean distance.

The distribution of *spa* types, SCC*mec* types and antimicrobial resistance profiles within the sample collections from the three process steps and the control group were compared pair wise using the Czekanowski index (Table 2). High index values were obtained for the distribution of *spa* types (PSI 0.79–0.86) among MRSA from the turkey meat chain. The comparison of the distribution of antimicrobial resistance profiles resulted in the lowest index values (PSI 0.42–0.56). The distribution of *spa* types and antimicrobial resistance profiles showed remarkably higher similarity between the different production steps of the turkey meat chain as to samples from the control group (PSI 0.55–0.56 and 0.13–0.19 resp.). High similarity in the distributions of SCC*mec* types was calculated between all process steps of the turkey meat production chain (PSI 0.85–0.91). However, a strong association was also received with SCC*mec* types of the control group (PSI 0.83–0.85).

## Discussion

In the present study, a new approach is proposed for analyzing a cross sectional set of MRSA isolates originating from three consecutive stages of the turkey meat production chain in order to draw conclusions on a potential farm to fork transmission. In the course of the German national monitoring for zoonotic agents in 2010 MRSA was isolated at all stages of the turkey meat production chain with prevalences ranging from 19.6% to 65.5%. To our knowledge, this is the first representative national MRSA prevalence study in the turkey production chain. In a regional prevalence study among fattening turkeys in southern Germany in 2009, a considerably higher prevalence of 90% MRSA positive flocks was observed using the same sampling procedure [12]. The difference might be explained by the regional restriction of sampling and the small sample size in that study. The proportion of positive meat samples is in line with results from the Netherlands [25]. Outside of Europe, low MRSA contamination rates of 3.85% [32] and 1.7% [50] were reported among US turkey meat.

The high MRSA prevalence in turkey carcasses after slaughter in comparison to the flock prevalence is in contrast to the situation in pigs [12,51] and indicates that the turkey slaughter process may play an important role in the transmission of MRSA. Turkeys are slaughtered highly automated at a speed of line up to 3,600 turkey hens and up to 2,700 turkey toms per hour which leads to a permanent introduction of MRSA into the poultry processing plants [52]. During the process, MRSA on animal surfaces can get transmitted via direct contact or indirect via surface processing machinery, scalding water or the hands of staff. Scalding takes place at a constant water temperature between 50 and 65°C for 60 to 210 sec [52]. Although the surface of the carcasses is exposed to a heat treatment during scalding, the temperature and duration of the process might be insufficient to substantially reduce superficial MRSA counts. The selective growth of *S. aureus* after the elimination of less heat resistant microbial flora in the scalding water has been discussed [53]. As bacterial counts increase in the tanks throughout the slaughter day scalding can contribute to cross contamination [54]. After scalding, the birds go through the plucking machines consisting of revolving drums with rubber beaters or discs with plucking fingers. The birds are flailed and scraped for 30–90 sec while being sprayed with warm or cold water [52]. Plucking equipment is difficult to clean and a persisting microbiological flora can get established [55]. Cross contamination during slaughter and meat processing might lead to an extensive distribution of *spa* types between different animals and slaughter flocks. In addition, the increase in manual handling during processing facilitates the entry of human MRSA strains into

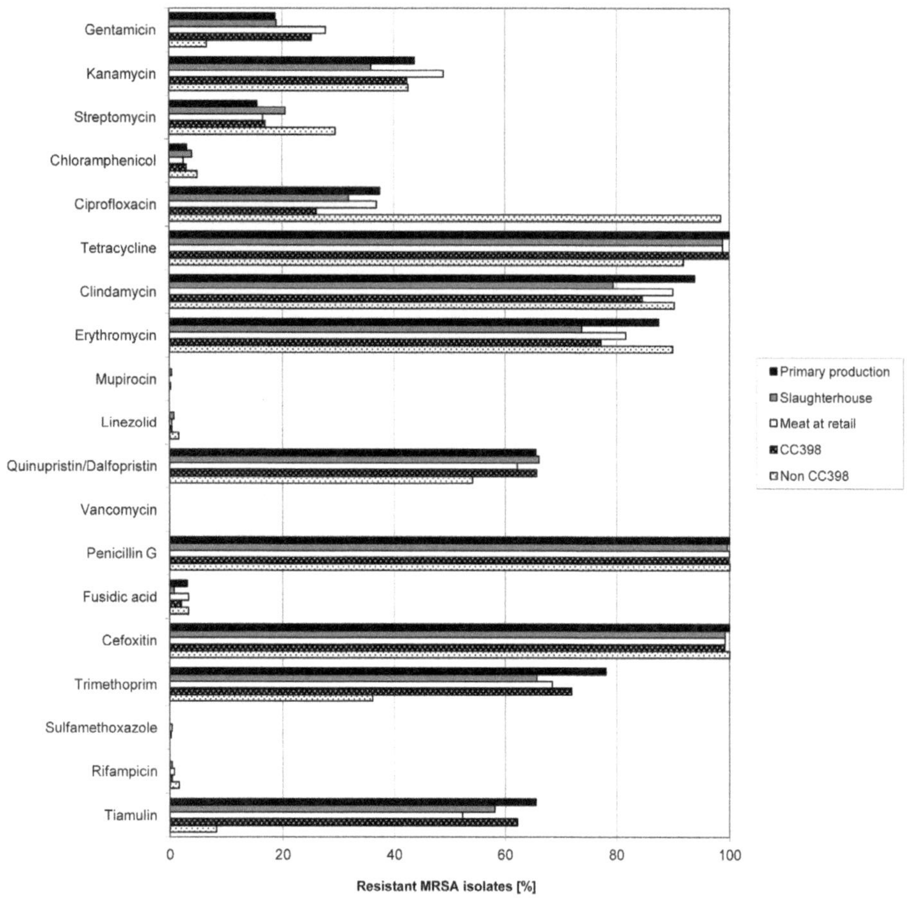


**Figure 1. Antimicrobial resistance of MRSA in the German turkey meat production chain.** Distribution of antimicrobial resistance of MRSA strains separated into CC398 and non CC 398 strains as well as different steps of the turkey meat production chain isolated from dust samples at turkey primary production (n = 32), carcasses at slaughter (n = 248) and meat at retail (n = 241). The MRSA strains were isolated in the course of the national monitoring for zoonotic agents in Germany in 2010.

the production units. This can explain the increase in the variability of *spa* types along the chain and is in line with the increase in the proportion of non CC398 strains in meat samples compared to dust or carcasses. *Spa* types t002 and t1430 were also present in primary production and therefore probably have been transmitted along the food chain. In contrast, *spa* types t010, t015 were first observed in meat samples.

The majority of MRSA from the German turkey production chain was assigned to the livestock associated CC398 with the predominant *spa* types t011 and t034. This is in line with results from other livestock like veal calves [9], dairy cattle [56,57] and pigs [6] as well as in food [25]. In the present study, 37 of the 521 MRSA strains (7.1%) were identified as t002. This *spa* type t002 is assigned to CC5. In Germany, CC5 is one of the epidemic MRSA strains among humans [58]. Finding t002 in turkey flocks and in turkey meat is in line with other studies from central Europe [12,25,59]. So far, it is not known, whether this strain originates from the "human" strain and is introduced into the food chain on different levels or whether it got established in the turkey population and is transmitted along the chain. Detailed molecular-epidemiological investigations are needed to compare strains both from human and farm to fork origin. In the present study, 4.2% of the MRSA isolates were characterized as *spa* type t1430, a MRSA strain which was also frequently isolated from chicken meat [25] and broilers at slaughter [60] in the Netherlands. However, it was has also been detected in turkey flocks at farm level [12]. The strain is assigned to ST9, a lineage genetically unrelated to ST398. ST9 is the predominating sequence type among MRSA from pigs in Asian countries [14,15,61–64]. Outside of Europe, MRSA contamination was reported among US turkey meat [32,50]. In both surveys, all isolates belonged to USA 300 (ST8), the most common community associated MRSA strain in the USA, suggesting human contamination during processing.

The frequent use of antimicrobials at farm is discussed as a risk factor for the wide dissemination of MRSA in livestock production chains [65]. In recent studies antimicrobials were identified to be used in more than 90% of the investigated turkey flocks and animals received on average 33 daily doses of antimicrobials during raising and fattening [66]. With a share of 21% β-lactams were most often used followed by polypeptides (15.2%), macrolides (13.4%), tetracyclines and aminoglycosides (12.4% both). Fluoroquinolones were used in 6.5% of the investigated flocks. The common application of antimicrobials via drinking water bears the risk of under dosing of individual animals and contamination of the barn environment with antimicrobials which also facilitates the selection of resistance [67].

**Table 2.** Similarity matrix of *spa* types, SCC*mec* types and resistance profiles of MRSA isolated from the German turkey meat production chain in the course of the national monitoring for zoonotic agents in 2010 (95% confidence intervals).

| | | Primary production | Slaughterhouse | Meat at retail | Control Group Wild boar meat |
|---|---|---|---|---|---|
| | | av. PSI[a] (CI 95%)[b] | av. PSI[a] (CI 95%)[b] | av. PSI[a] (CI 95%)[b] | av. PSI[a] (CI 95%)[b] |
| Primary production | *spa* types | 1 | | | |
| | SCC*mec* types | 1 | | | |
| | resistance profiles | 1 | | | |
| Slaughterhouse | *spa* types | 0.86 (0.72, 0.95) | 1 | | |
| | SCC*mec* types | 0.91 (0.79, 0.98) | 1 | | |
| | resistance profiles | 0.43 (0.30, 0.53) | 1 | | |
| Meat at retail | *spa* types | 0.79 (0.64, 0.90) | 0.86 (0.79, 0.92) | 1 | |
| | SCC*mec* types | 0.85 (0.70, 0.96) | 0.87 (0.79, 0.95) | 1 | |
| | resistance profiles | 0.42 (0.33, 0.51) | 0.56 (0.49, 0.62) | 1 | |
| Control Group | *spa* types | 0.55 (0.33, 0.71) | 0.56 (0.38, 0.74) | 0.56 (0.38, 0.74) | 1 |
| Wild boar meat | SCC*mec* types | 0.84 (0.62, 0.98) | 0.83 (0.64, 0.96) | 0.85 (0.70, 0.95) | 1 |
| | resistance profiles | 0.13 (0.03, 0.27) | 0.19 (0.06, 0.34) | 0.14 (0.04, 0.23) | 1 |

[a]PSI: Czekanowski index or proportional similarity index.
[b]CI 95%: 95% confidence interval.

Cluster analysis was used to better describe the multidimensional data set of antibiotic resistance profiles grouping all MRSA strains within 3 different clusters. As the ordinal MIC values generated by two-fold dilutions in substance concentration are difficult to describe by cluster analysis a binary interpretation of the data set was used. Ward's minimum variance with squared Euclidian distance was proven to be the best method to produce well separated cluster in binary antimicrobial resistance data sets [68,69]. No resistance phenotype simultaneously appeared in several clusters. The distribution of *spa* types, SCC*mec* types and the three clusters of antimicrobial resistance types did not significantly differ in the MRSA samples from the three origins. The chi square value was approaching significance with respect to the *spa*-types, which was presumably due to the slightly higher proportion of other CC398 and non CC398. However, considering all three features it cannot be rejected on the basis of the included data that the MRSA isolates from different steps of the turkey meat production chain originate from the same population of strains. This result might rather indicate farm to fork transmission of MRSA of the same pool of strains than development of separate MRSA populations at each step of the chain. The calculation of the Czekanowski index for *spa* type and SCC*mec* type data results in consistently high similarity values between the matrices whereas the comparison of antimicrobial resistance phenotypes observed medium index values. Higher values of similarity were obtained between the adjacent process steps primary production/slaughter and slaughter/meat than between samples from primary production and meat. This result was expected as an increase in the variability of the MRSA isolates might be conceivable at each process stage due to external introduction of new strains via human or environmental contamination or due to spontaneous mutations in the strains.

The lower values of similarity between the distribution of *spa* types and antimicrobial resistance profiles of samples from the turkey meat production chain and the control group indicate that that the proposed statistical method is valid to detect existing differences in the distribution of these characteristics of MRSA.

Concerning SCC*mec* types, high index values were also observed in comparison to the control group which might be explained by the insufficient discriminatory power of SCC*mec* typing. In addition, MRSA isolates with not typeable SCC*mec* cassettes were considered as equal that might lead to an overestimation of similarity.

It can be concluded that MRSA is present at every step of the turkey meat production chain in Germany. Using the Czekanowski index it is possible to quantify the similarity of the distribution of *spa* types, SCC*mec* types and antimicrobial resistance phenotypes between MRSA data sets from different stages of turkey meat production chain. Combined with chi square statistics, the high level of similarity suggests MRSA transmission along the chain.

## Supporting Information

**Table S1** List of competent authorities of the German federal states which were responsible for collecting the samples.

**Table S2** Distribution of resistance against 19 different antimicrobials, grouped *spa* types and SCC*mec* types within the binary phenotypic resistance clusters of 521 MRSA isolates sampled at different steps of the German turkey meat production chain in 2010.

## Author Contributions

Conceived and designed the experiments: AK BAT KA. Performed the experiments: BK. Analyzed the data: BV. Contributed reagents/materials/analysis tools: JB. Wrote the paper: BV. Designed the software used for bootstrap method: JB. Administration and organisation of the project: AF.

## References


1. Hennekinne JA, De Buyser ML, Dragacci S (2012) *Staphylococcus aureus* and its food poisoning toxins: Characterization and outbreak investigation. FEMS Microbiology Reviews 36: 815–836.
2. Köck R, Becker K, Cookson B, van Gemert-Pijnen JE, Harbarth S, et al. (2010) Methicillin-resistant *Staphylococcus aureus* (MRSA): burden of disease and control challenges in Europe. Euro surveillance: bulletin européen sur les maladies transmissibles = European communicable disease bulletin 15: 19688.
3. Voss A, Loeffen F, Bakker J, Klaassen C, Wulf M (2005) Methicillin-resistant *Staphylococcus aureus* in Pig Farming. Emerging Infectious Diseases 11: 1965–1966.
4. Armand-Lefevre L, Ruimy R, Andremont A (2005) Clonal comparison of Staphylococcus from healthy pig farmers, human controls, and pigs. Emerging Infectious Diseases 11: 711–714.
5. Crombé F, Willems G, Dispas M, Hallin M, Denis O, et al. (2012) Prevalence and antimicrobial susceptibility of methicillin-resistant *Staphylococcus aureus* among pigs in Belgium. Microbial Drug Resistance 18: 125–131.
6. EFSA (2009) Analysis of the baseline survey on the prevalence of methicillin-resistant *Staphylococcus aureus* (MRSA) in holdings with breeding pigs, in the EU, 2008, Part A: MRSA prevalence estimates; on request from the European Commission. The EFSA Jounal 11: 1376.
7. Khanna T, Friendship R, Dewey C, Weese JS (2008) Methicillin resistant *Staphylococcus aureus* colonization in pigs and pig farmers. Vet Microbiol 128: 298–303.
8. Smith TC, Male MJ, Harper AL, Kroeger JS, Tinkler GP, et al. (2009) Methicillin-resistant *Staphylococcus aureus* (MRSA) strain ST398 is present in midwestern U.S. swine and swine workers. PloS one 4.
9. Graveland H, Wagenaar JA, Heesterbeek H, Mevius D, van Duijkeren E, et al. (2010) Methicillin resistant *Staphylococcus aureus* ST398 in veal calf farming: human MRSA carriage related with animal antimicrobial usage and farm hygiene. PloS one 5.
10. Nemati M, Hermans K, Lipinska U, Denis O, Deplano A, et al. (2008) Antimicrobial resistance of old and recent *Staphylococcus aureus* isolates from poultry: First detection of livestock-associated methicillin-resistant strain ST398. Antimicrobial Agents and Chemotherapy 52: 3817–3819.
11. Persoons D, Van Hoorebeke S, Hermans K, Butaye P, De Kruif A, et al. (2009) Methicillin-resistant *Staphylococcus aureus* in poultry. Emerging Infectious Diseases 15: 452–453.
12. Richter A, Sting R, Popp C, Rau J, Tenhagen BA, et al. (2012) Prevalence of types of methicillin-resistant *Staphylococcus aureus* in turkey flocks and personnel attending the animals. Epidemiology and Infection 140: 2223–2232.
13. Reischl U, Frick J, Hoermansdorfer S, Melzl H, Bollwein M, et al. (2009) Single-nucleotide polymorphism in the SCC*mec*-orfX junction distinguishes between livestock-associated MRSA CC398 and human epidemic MRSA strains. Euro surveillance: bulletin european sur les maladies transmissibles = European communicable disease bulletin 14.
14. Anukool U, O'Neill CE, Butr-Indr B, Hawkey PM, Gaze WH, et al. (2011) Meticillin-resistant *Staphylococcus aureus* in pigs from Thailand. International Journal of Antimicrobial Agents 38: 86–87.
15. Wagenaar JA, Yue H, Pritchard J, Broekhuizen-Stins M, Huijsdens X, et al. (2009) Unexpected sequence types in livestock associated methicillin-resistant *Staphylococcus aureus* (MRSA): MRSA ST9 and a single locus variant of ST9 in pig farming in China. Veterinary Microbiology 139: 405–409.
16. Cookson BD, Robinson DA, Monk AB, Murchan S, Deplano A, et al. (2007) Evaluation of molecular typing methods in characterizing a European collection of epidemic methicillin-resistant *Staphylococcus aureus* strains: the HARMONY collection. J Clin Microbiol 45: 1830–1837.
17. (IWG-SCC) IWGotCoSCCE (2009) Classification of staphylococcal cassette chromosome mec (SCCmec): guidelines for reporting novel SCCmec elements. Antimicrobial Agents and Chemotherapy 53: 4961–4967.
18. Garcia-Alvarez L, Holden MT, Lindsay H, Webb CR, Brown DF, et al. (2011) Meticillin-resistant *Staphylococcus aureus* with a novel mecA homologue in human and bovine populations in the UK and Denmark: a descriptive study. Lancet Infect Dis 11: 595–603.
19. Frenay HM, Bunschoten AE, Schouls LM, van Leeuwen WJ, Vandenbroucke-Grauls CM, et al. (1996) Molecular typing of methicillin-resistant *Staphylococcus aureus* on the basis of protein A gene polymorphism. Eur J Clin Microbiol Infect Dis 15: 60–64.
20. de Neeling AJ, van den Broek MJ, Spalburg EC, van Santen-Verheuvel MG, Dam-Deisz WD, et al. (2007) High prevalence of methicillin resistant *Staphylococcus aureus* in pigs. Vet Microbiol 120: 366–372.
21. Pires SM, Evers EG, van PW, Ayers T, Scallan E, et al. (2009) Attributing the human disease burden of foodborne infections to specific sources. Foodborne PathogDis 6: 417–424.
22. WHO (2001) World Health Organisation Global Principles for the Containment of Antimicrobial Resistance in Animals intended for Food. http://wwwwhoint/drugresistance/WHO_Global_Strategyhtm/en/.
23. Alt K, Fetsch A, Schroeter A, Guerra B, Hammerl J, et al. (2011) Factors associated with the occurrence of MRSA CC398 in herds of fattening pigs in Germany. BMC Veterinary Research 7: 69.
24. Broens EM, Graat EAM, Van der Wolf PJ, van de Giessen AW, de Jong MCM (2011) Prevalence and risk factor analysis of livestock associated MRSA-positive pig herds in The Netherlands. Preventive Veterinary Medicine 102: 41–49.
25. de Boer E, Zwartkruis-Nahuis JTM, Wit B, Huijsdens XW, de Neeling AJ, et al. (2009) Prevalence of methicillin-resistant *Staphylococcus aureus* in meat. International Journal of Food Microbiology 134: 52–56.
26. Hanson BM, Dressler AE, Harper AL, Scheibel RP, Wardyn SE, et al. (2011) Prevalence of *Staphylococcus aureus* and methicillin-resistant *Staphylococcus aureus* (MRSA) on retail meat in Iowa. Journal of Infection and Public Health 4: 169–174.
27. Lim SK, Nam HM, Park HJ, Lee HS, Choi MJ, et al. (2010) Prevalence and characterization of methicillin-resistant *Staphylococcus aureus* in raw meat in Korea. Journal of Microbiology and Biotechnology 20: 775–778.
28. Lozano C, López M, Gómez-Sanz E, Ruiz-Larrea F, Torres C, et al. (2009) Detection of methicillin-resistant *Staphylococcus aureus* ST398 in food samples of animal origin in Spain. Journal of Antimicrobial Chemotherapy 64: 1325–1326.
29. O'Donoghue M, Chan M, Ho J, Moodley A, Boost M. Prevalence of Mehicillin-Resistant Staphylococcus aureus in Meat from Hong Kong Shops and Markets. 27.
30. Pu S, Han F, Ge B (2009) Isolation and characterization of methicillin-resistant *Staphylococcus aureus* strains from louisiana retail meats. Applied and Environmental Microbiology 75: 265–267.
31. Van Loo IHM, Diederen BMW, Savelkoul PHM, Woudenberg JHC, Roosendaal R, et al. (2007) Methicillin-resistant *Staphylococcus aureus* in meat products, the Netherlands. Emerging Infectious Diseases 13: 1753–1755.
32. Waters AE, Contente-Cuomo T, Buchhagen J, Liu CM, Watson L, et al. (2011) Multidrug-resistant *staphylococcus aureus* in US meat and poultry. Clinical Infectious Diseases 52: 1227–1230.
33. Weese JS, Reid-Smith R, Rousseau J, Avery B (2010) Methicillin-resistant *Staphylococcus aureus* (MRSA) contamination of retail pork. Canadian Veterinary Journal-Revue Veterinaire Canadienne 51: 749–752.
34. EC (2003) Directive 2003/99/EC of the European Parliament and of the Council of 17 November 2003 on the monitoring of zoonoses and zoonotic agents. Official Journal of the European Union, L325/31. http://eur-lex.europa.eu/LexUriServ/LexUriServ.do?uri = OJ:L:2003:325:0031:0040:EN:PDF.
35. Käsbohrer A, Wegeler C, Tenhagen BA (2009) EU-weite und nationale Monitoringprogramme zu Zoonoseerregern in Deutschland. Journal für Verbraucherschutz und Lebensmittelsicherheit. EU-weite und nationale Monitoringprogramme zu Zoonoseerregern in Deutschland. 41–45.
36. Straub JA, Hertel C, Hammes WP (1999) A 23S rDNA-targeted polymerase chain reaction-based system for detection of *Staphylococcus aureus* in meat starter cultures and dairy products. Journal of Food Protection 62: 1150–1156.
37. Poulsen AB, Skov R, Pallesen LV (2003) Detection of methicillin resistance in coagulase-negative staphylococci and in staphylococci directly from simulated blood cultures using the EVIGENE MRSA Detection Kit. Journal of Antimicrobial Chemotherapy 51: 419–421.
38. Shopsin B, Gomez M, Montgomery SO, Smith DH, Waddington M, et al. (1999) Evaluation of protein A gene polymorphic region DNA sequencing for typing of *Staphylococcus aureus* strains. Journal of Clinical Microbiology 37: 3556–3563.
39. Zhang K, McClure JA, Elsayed S, Louie T, Conly JM (2005) Novel multiplex PCR assay for characterization and concomitant subtyping of staphylococcal cassette chromosome *mec* types I to V in methicillin-resistant *Staphylococcus aureus*. Journal of Clinical Microbiology 43: 5026–5033.
40. Jansen MD, Box ATA, Fluit AC (2009) SCCmec typing in methicillin-resistant *Staphylococcus aureus* strains of animal origin. Emerging Infectious Diseases 15: 136.
41. Enright MC, Day NPJ, Davies CE, Peacock SJ, Spratt BG (2000) Multilocus sequence typing for characterization of methicillin-resistant and methicillin-susceptible clones of *Staphylococcus aureus*. Journal of Clinical Microbiology 38: 1008–1015.
42. CLSI (2006) Performance Standards for Antimicrobial Disk Susceptibility Tests; Approved Standard.
43. Calinski T, Harabasz J (1974) A dendrite method for cluster analysis. Communications in Statistics: Taylor & Francis. 1–27.
44. Duda RO, Hart PE (1973) Pattern Classification and Scene Analysis. New York: Wiley.
45. Rosef O, Kapperud G, Lauwers S, Gondrosen B (1985) Serotyping of *Campylobacter jejuni*, *Campylobacter coli*, and *Campylobacter laridis* from domestic and wild animals. Applied and Environmental Microbiology 49: 1507–1510.
46. Efron B, Tibshirani R (1986) Bootstrap methods for standard errors, confidence invervals and other measures of statistical accuracy. Statistical Science 1: 54–75.
47. Bundesamt für Verbraucherschutz und Lebensmittelsicherheit (2012) Berichte zur Lebensmittelsicherheit 2010. BVL Reporte http://www.bvl.bund.de/SharedDocs/Downloads/01_Lebensmittel/04_Zoonosen_Monitoring/Zoonosen_Monitoring_Bericht_2010.pdf?__blob = publicationFile&v = 6.
48. Kreausukon K, Fetsch A, Kraushaar B, Alt K, Müller K, et al. (2012) Prevalence, antimicrobial resistance, and molecular characterization of methicillin-resistant *Staphylococcus aureus* from bulk tank milk of dairy herds. Journal of Dairy Science 95: 4382–4388.
49. Argudin MA, Rodicio MR, Guerra B (2010) The emerging methicillin-resistant *Staphylococcus aureus* ST398 clone can easily be typed using the Cfr9I SmaI-neoschizomer. Lett Appl Microbiol 50: 127–130.

50. Bhargava K, Wang X, Donabedian S, Zervos M, da Rocha L, et al. (2011) Methicillin-resistant *staphylococcus aureus* in retail meat, Detroit, Michigan, USA. Emerging Infectious Diseases 17: 1135–1137.
51. Lassok B, Tenhagen BA (2013) From pig to pork: Methicillin-resistant *staphylococcus aureus* in the pork production chain. Journal of Food Protection 76: 1095–1108.
52. Löhren U (2012) Overview on current practices of poultry slaughtering and poultry meat inspection Supporting Publications 2012: EN-298. http://wwwefsaeuropaeu/en/supporting/doc/298epdf.
53. Hentschel S, Kusch D, Sinell HJ (1979) *Staphylococcus aureus* in Poultry-biochemical characteristics, antibiotic resistance and phage pattern. Zentralblatt für Bakteriologie, Parasitenkunde, Infektionskrankheiten und Hygiene. Erste Abteilung Originale. Reihe B: Hygiene, Betriebshygiene, präventive Medizin 07, 168 (5–6): 546–561.
54. Groβklaus D, Lessing G (1972) Hygieneprobleme beim Schlachtgeflügel. Fleischwirtschaft 52: 1011–1013.
55. Berrang ME, Buhr RJ, Cason JA, Dickens JA (2001) Broiler carcass contamination with Campylobacter from feces during defeathering. Journal of Food Protection 64: 2063–2066.
56. Spohr M, Rau J, Friedrich A, Klittich G, Fetsch A, et al. (2011) Methicillin-resistant *Staphylococcus aureus* (MRSA) in three dairy herds in southwest Germany. Zoonoses and Public Health 58: 252–261.
57. Vanderhaeghen W, Hermans K, Haesebrouck F, Butaye P (2010) Methicillin-resistant *Staphylococcus aureus* (MRSA) in food production animals. Epidemiol Infect 138: 606–625.
58. Robert Koch-Institut (2011) Auftreten und Verbreitung von MRSA in Deutschland 2010. Epidemiologisches Bulletin 26: 233–244.
59. Feβler AT, Kadlec K, Hassel M, Hauschild T, Eidam C, et al. (2011) Characterization of methicillin-resistant *Staphylococcus aureus* isolates from food and food products of poultry origin in Germany. Applied and Environmental Microbiology 77: 7151–7157.
60. Mulders MN, Haenen APJ, Geenen PL, Vesseur PC, Poldervaart ES, et al. (2010) Prevalence of livestock-associated MRSA in broiler flocks and risk factors for slaughterhouse personnel in the Netherlands. Epidemiology and Infection 138: 743–755.
61. Cui S, Li J, Hu C, Jin S, Li F, et al. (2009) Isolation and characterization of methicillin-resistant *Staphylococcus aureus* from swine and workers in China. Journal of Antimicrobial Chemotherapy 64: 680–683.
62. Larsen J, Imanishi M, Hinjoy S, Tharavichitkul P, Duangsong K, et al. (2012) Methicillin-resistant *Staphylococcus aureus* ST9 in pigs in Thailand. PloS one 7.
63. Neela V, Zafrul AM, Mariana NS, Van Belkum A, Liew YK, et al. (2009) Prevalence of ST9 methicillin-resistant *Staphylococcus aureus* among pigs and pig handlers in Malaysia. Journal of Clinical Microbiology 47: 4138–4140.
64. Tsai HY, Liao CH, Cheng A, Liu CY, Huang YT, et al. (2012) Isolation of meticillin-resistant *Staphylococcus aureus* sequence type 9 in pigs in Taiwan. International Journal of Antimicrobial Agents 39: 449–451.
65. Schwarz S, Chaslus-Dancla E (2001) Use of antimicrobials in veterinary medicine and mechanisms of resistance. Veterinary Research 32: 201–225.
66. State Office for Consumer Protection and Food Safety in Lower Saxony G (2011) Bericht über den Antibiotikaeinsatz in der landwitschaftlichen Nutztier-haltung in Niedersachsen November 2011. http://wwwmlniedersachsende/portal/livephp?navigation_id = 27751&article_id = 102202&_psmand = 7.
67. Richter A, Hafez HM, Böttner A, Gangl A, Hartmann K, et al. (2009) Applications of antibiotics in poultry. Tierarztliche Praxis Ausgabe G: Grosstiere – Nutztiere Verabreichung von Antibiotika in Geflügelbeständen. 321–329.
68. Berge ACB, Atwill ER, Sischo WM (2003) Assessing antibiotic resistance in fecal *Escherichia coli* in young calves using cluster analysis techniques. Preventive Veterinary Medicine 61: 91–102.
69. Milligan GW (1981) A monte carlo study of thirty internal criterion measures for cluster analysis. Psychometrika 46: 187–199.

# PERMISSIONS

All chapters in this book were first published in PLOS ONE, by The Public Library of Science; hereby published with permission under the Creative Commons Attribution License or equivalent. Every chapter published in this book has been scrutinized by our experts. Their significance has been extensively debated. The topics covered herein carry significant findings which will fuel the growth of the discipline. They may even be implemented as practical applications or may be referred to as a beginning point for another development.

The contributors of this book come from diverse backgrounds, making this book a truly international effort. This book will bring forth new frontiers with its revolutionizing research information and detailed analysis of the nascent developments around the world.

We would like to thank all the contributing authors for lending their expertise to make the book truly unique. They have played a crucial role in the development of this book. Without their invaluable contributions this book wouldn't have been possible. They have made vital efforts to compile up to date information on the varied aspects of this subject to make this book a valuable addition to the collection of many professionals and students.

This book was conceptualized with the vision of imparting up-to-date information and advanced data in this field. To ensure the same, a matchless editorial board was set up. Every individual on the board went through rigorous rounds of assessment to prove their worth. After which they invested a large part of their time researching and compiling the most relevant data for our readers.

The editorial board has been involved in producing this book since its inception. They have spent rigorous hours researching and exploring the diverse topics which have resulted in the successful publishing of this book. They have passed on their knowledge of decades through this book. To expedite this challenging task, the publisher supported the team at every step. A small team of assistant editors was also appointed to further simplify the editing procedure and attain best results for the readers.

Apart from the editorial board, the designing team has also invested a significant amount of their time in understanding the subject and creating the most relevant covers. They scrutinized every image to scout for the most suitable representation of the subject and create an appropriate cover for the book.

The publishing team has been an ardent support to the editorial, designing and production team. Their endless efforts to recruit the best for this project, has resulted in the accomplishment of this book. They are a veteran in the field of academics and their pool of knowledge is as vast as their experience in printing. Their expertise and guidance has proved useful at every step. Their uncompromising quality standards have made this book an exceptional effort. Their encouragement from time to time has been an inspiration for everyone.

The publisher and the editorial board hope that this book will prove to be a valuable piece of knowledge for researchers, students, practitioners and scholars across the globe.

# LIST OF CONTRIBUTORS

**Sonja Hartnack and Fraser Iain Lewis**
Section of Epidemiology, Vetsuisse Faculty, Zurich, Switzerland

**Christina Nathues**
Veterinary Public Health Institute, Vetsuisse Faculty, Liebefeld, Switzerland

**Heiko Nathues**
Department of Production and Population Health, Royal Veterinary College London, London, United Kingdom
Department of Clinical Veterinary Medicine, Vetsuisse Faculty, Zurich, Switzerland

**Elisabeth Grosse Beilage**
Field Station for Epidemiology, University of Veterinary Medicine Hannover, Foundation, Bakum, Germany

**Marian E. H. Bos and Dick J. J. Heederik**
Division of Environmental Epidemiology, Institute for Risk Assessment Sciences, Utrecht University, Utrecht, The Netherlands

**Femke J. Taverne and Ingeborg M. van Geijlswijk**
Pharmacy, Faculty of Veterinary Medicine, Utrecht University, Utrecht, The Netherlands

**Johan W. Mouton**
Department of Medical Microbiology, Radboud University Nijmegen Medical Centre, Nijmegen, The Netherlands

**Dik J. Mevius**
Department of Infectious Diseases and Immunology, Faculty of Veterinary Medicine, Utrecht University, Utrecht, The Netherlands
Central Veterinary Institute of Wageningen UR, Lelystad, The Netherlands

**Macarena P. Quintana-Hayashi and Siddhartha Thakur**
Department of Population Health and Pathobiology, College of Veterinary Medicine, North Carolina State University, Raleigh, North Carolina, United States of America

**Joost Hordijk, Jaap A. Wagenaar and Dik Mevius**
Department of Infectious Diseases and Immunology, Faculty of Veterinary Medicine, Utrecht University, Utrecht, The Netherlands
Central Veterinary Institute, part of Wageningen UR, Lelystad, The Netherlands

**Arie Kant, Alieda van Essen-Zandbergen, Cindy Dierikx and Kees Veldman**
Central Veterinary Institute, part of Wageningen UR, Lelystad, The Netherlands

**Ben Wit**
Netherlands Food and Consumer Product Safety Authority, Utrecht, The Netherlands

**Martin Reist and Sara Schärrer**
Veterinary Public Health Institute, Vetsuisse Faculty University of Bern, Bern, Switzerland

**Nadine Geser, Herbert Hächler and Roger Stephan**
Institute for Food Safety and Hygiene, Vetsuisse Faculty University of Zurich, Zurich, Switzerland

**Henrike C. Jäger, Trevelyan J. McKinley, James L. N. Wood, Gareth P. Pearce and Alexander W. Tucker**
Department of Veterinary Medicine, University of Cambridge, Cambridge, United Kingdom,

**Susanna Williamson**
Veterinary Laboratories Agency (AHVLA), Bury St Edmunds, Suffolk, United Kingdom

**Benjamin Strugnell and Stanley Done**
Veterinary Laboratories Agency (AHVLA), West House, Thirsk, North Yorkshire, United Kingdom

**Henrike Habernoll**
BQP Ltd, Stradbroke Business Centre, Stradbroke, Suffolk, United Kingdom

**Andreas Palzer**
Ludwig-Maximilians-Universität München, Clinic for Swine, Sonnenstrasse, Oberschleissheim Germany

**Carol H. Sandt**
Bureau of Laboratories, Pennsylvania Department of Health, Exton, Pennsylvania, United States of America

**Paula J. Fedorka-Cray**
United States Department of Agriculture, Agricultural Research Service, Athens, Georgia, United States of America

**Stephen Ostroff and Nkuchia M. M'ikanatha**
Division of Infectious Disease Epidemiology, Pennsylvania Department of Health, Harrisburg, Pennsylvania, United States of America

**Deepanker Tewari**
Pennsylvania Veterinary Laboratory, Pennsylvania Department of Agriculture, Harrisburg, Pennsylvania, United States of America

**Kevin Joyce**
National Antimicrobial Resistance Monitoring System, Centers for Disease Control and Prevention, Atlanta, Georgia, United States of America

**Alexander D. C. Berriman, Helen E. Clough and Robert M. Christley**
Institute of Infection and Global Health, Department of Epidemiology and Population Health, University of Liverpool, Neston, Cheshire, United Kingdom

**Damian Clancy**
Department of Mathematical Sciences, University of Liverpool, Liverpool, United Kingdom

**Derek Armstrong**
BPEX, Stoneleigh Park, Warwickshire, United Kingdom

**Kaarina Matilainen, Esa A. Mäntysaari, Martin H. Lidauer and Ismo Strandén**
Biotechnology and Food Research, MTT Agrifood Research Finland, Jokioinen, Finland

**3Robin Thompson**
Biomathematics and Bioinformatics Department, Rothamsted Research, Harpenden, United Kingdom

**Vera H. Hausner**
Department of Arctic and Marine Biology, University of Tromsø, Tromsø, Norway

**Per Fauchald**
Department of Arctic Ecology, Norwegian Institute for Nature Research, FRAM Centre, Tromsø, Norway

**Johnny-Leo Jernsletten**
Centre for Sámi Studies, Teorifagbygget, Tromsø, Norway

**Alisa L. Gallant**
United States Geological Survey, Earth Resources Observation and Science Center, Sioux Falls, South Dakota, United States of America

**Ned H. Euliss Jr.**
United States Geological Survey, Northern Prairie Wildlife Research Center, Jamestown, North Dakota, United States of America

**Zac Browning**
Browning's Honey Company, Inc., Jamestown, North Dakota, United States of America

**Maud Marsot and Benoit Durand**
Université Paris-Est, French Agency for Food, Environmental and Occupational Health and Safety (Anses), Laboratoire de Santé Animale, Epidemiology Unit, Maisons- Alfort, France

**Séverine Rautureau**
Animal Health Unit, French General Directorate for Food (DGAL), Ministry of Agriculture, Food, Fisheries and Rural Affairs, Paris, France

**Barbara Dufour**
EPIMAI, Alfort National Veterinary School (ENVA), USC Anses, Maisons-Alfort, France

**Jillian P. Fry**
Department of Environmental Health Sciences, Johns Hopkins Bloomberg School of Public Health, Baltimore, Maryland, United States of America
Center for a Livable Future, Johns Hopkins Bloomberg School of Public Health, Baltimore, Maryland, United States of America

**Linnea I. Laestadius and Clare Grechis**
Center for a Livable Future, Johns Hopkins Bloomberg School of Public Health, Baltimore, Maryland, United States of America
Department of Health Policy & Management, Johns Hopkins Bloomberg School of Public Health, Baltimore, Maryland, United States of America

**Keeve E. Nachman and Roni A. Neff**
Department of Environmental Health Sciences, Johns Hopkins Bloomberg School of Public Health, Baltimore, Maryland, United States of America
Center for a Livable Future, Johns Hopkins Bloomberg School of Public Health, Baltimore, Maryland, United States of America
Department of Health Policy & Management, Johns Hopkins Bloomberg School of Public Health, Baltimore, Maryland, United States of America

**Richard Kasch, Florian Schmidt, Harry R. Merk and Ralph Kayser**
Clinic and Outpatient Clinic for Orthopedics and Orthopedic Surgery, University Medicine Greifswald, Greifswald, Germany

**Birger Mensel**
Department of Diagnostic Radiology and Neuroradiology, University Medicine Greifswald, Greifswald, Germany

**Wolf Drescher**
Department of Orthopedic and Trauma Surgery, RWTH Aachen University, Aachen, Germany

**Ralf Pfuhl**
Leibnitz Institute for Farm Animal Biology, Dummerstorf, Germany

**Sebastian Ruetten**
Department of Spine Surgery and Pain Therapy, Center for Orthopedics and Traumatology, St. Anna-Hospital Herne, Herne, Germany

**Germinal J. Canto Alarcon, Yezenia Rubio Venegas, Andrea M. Olvera Ramírez, Feliciano Milian Suazo and Susana Sosa Gallegos**
Facultad de Ciencias Naturales, Universidad Autónoma de Querétaro, Querétaro, Qro., México

**Luis Bojorquez Narvaez**
Productora Nacional de Biológicos Veterinarios (PRONABIVE), Distrito Federal, México

**Oscar E. Pizano Martínez, Leticia García Casanova and Alejandro Nava Vargas**
Centro Nacional de Investigación Disciplinaria en Fisiología y Mejoramiento Animal, INIFAP, Ajuchitlán, Qro., México

**Katerina Karolemeas, James L. N. Wood, Trevelyan J. McKinley and Andrew J. K. Conlan**
Disease Dynamics Unit, Department of Veterinary Medicine, University of Cambridge, Cambridge, United Kingdom

**Ricardo de la Rua-Domenech**
Animal Health and Veterinary Laboratories Agency and Bovine Tuberculosis Programme of the Department for Environment, Food and Rural Affairs, London, United Kingdom

**Roderick Cooper**
Animal Health and Veterinary Laboratories Agency, West Midlands Regional Office, Stafford, United Kingdom

**Anthony V. Goodchild, Richard S. Clifton-Hadley, Andrew P. Mitchell and R. Glyn Hewinson**
Animal Health and Veterinary Laboratories Agency, Weybridge, United Kingdom

**Christl A. Donnelly**
MRC Centre for Outbreak Analysis and Modelling, Department of Infectious Disease Epidemiology, Imperial College London, London, United Kingdom

**Jillian P. Fry**
Center for a Livable Future, Johns Hopkins University, Baltimore, Maryland, United States of America
Department of Environmental Health Sciences, Johns Hopkins Bloomberg School of Public Health, Baltimore, Maryland, United States of America

**Linnea I. Laestadius**
Center for a Livable Future, Johns Hopkins University, Baltimore, Maryland, United States of America
Joseph J. Zilber School of Public Health, University of Wisconsin-Milwaukee, Milwaukee, Wisconsin, United States of America

**Clare Grechis**
Center for a Livable Future, Johns Hopkins University, Baltimore, Maryland, United States of America
Department of Health Policy and Management, Johns Hopkins Bloomberg School of Public Health, Baltimore, Maryland, United States of America

**Keeve E. Nachman and Roni A. Neff**
Center for a Livable Future, Johns Hopkins University, Baltimore, Maryland, United States of America
Department of Environmental Health Sciences, Johns Hopkins Bloomberg School of Public Health, Baltimore, Maryland, United States of America
Department of Health Policy and Management, Johns Hopkins Bloomberg School of Public Health, Baltimore, Maryland, United States of America

**Sara Schärrer and Patrick Presi**
Veterinary Public Health Institute/University of Berne, Berne, Switzerland

**Jan Hattendorf and Jakob Zinsstag**
Swiss Tropical and Public Health Institute/University of Basel, Basel, Switzerland

**Nakul Chitnis**
Swiss Tropical and Public Health Institute/University of Basel, Basel, Switzerland
Fogarty International Center, National Institutes of Health, Bethesda, Maryland, United States of America

**Martin Reist**
Federal Food Safety and Veterinary Office, Bern, Switzerland

**Francisco Sanchez-Bayo**
Faculty of Agriculture and Environment, The University of Sydney, Eveleigh, New South Wales, Australia

**Koichi Goka**
National Institute for Environmental Sciences, Tsukuba Ibaraki, Japan

**Debasis Upadhyay**
Department of Botany, Budge Budge College, Kolkata, West Bengal, India
Centre of Advanced Studies, Department of Botany, University of Calcutta, West Bengal, India

**Swapan Bhattacharya**
Department of Microbiology, Moulana Azad College, Kolkata, West Bengal, India

**David K. Ferguson**
Department of Paleontology, University of Vienna, Vienna, Austria

**Subir Bera**
Centre of Advanced Studies, Department of Botany, University of Calcutta, West Bengal, India

**Mohammed El Amine Bekara, Aurélie Courcoul and Benoit Durand**
University Paris Est, Anses, Laboratory of Animal Health, Epidemiology Unit, Maisons-Alfort, France

**Jean-Jacques Bénet**
University Paris Est, National Veterinary School of Alfort (ENVA), EpiMAI Unit, Maisons-Alfort, France

**H. Loneragan**
International Center for Food Industry Excellence, Department of Animal and Food Sciences, College of Agriculture and Natural Resources, Texas Tech University, Lubbock, Texas, United States of America

**Daniel U. Thomson**
Department of Clinical Sciences, College of Veterinary Medicine, Kansas State University, Manhattan, Kansas, United States of America

**H. Morgan Scott**
Department of Diagnostic Medicine/Pathobiology, College of Veterinary Medicine, Kansas State University, Manhattan, Kansas, United States of America

**Mathieu Andraud and Nicolas Rose**
Pig Epidemiology and Welfare Unit, Anses, Laboratoire de Ploufragan-Plouzané, Ploufragan, France
Université Européenne de Bretagne, Rennes, France

**Nicole Pavio**
UMR 1161 Virology, Anses, Laboratoire de Santé Animale, Maisons-Alfort, France
UMR 1161 Virology, INRA, Maisons-Alfort, France
UMR 1161 Virology, Université Paris-Est, Ecole Nationale Vétérinaire d9Alfort, Maisons-Alfort, France

**Maribel Casas**
Centre for Research in Environmental Epidemiology (CREAL), Barcelona, Spain
Centre de Recerca en Sanitat Animal (CReSA), Barcelona, Spain

**Birgit Vossenkuhl, Jörgen Brandt, Alexandra Fetsch, Annemarie Käsbohrer, Britta Kraushaar,**
Katja Alt and Bernd-Alois Tenhagen
Federal Institute for Risk Assessment, Berlin, Germany

# Index